T0291588

Essential Numerical Methods in Electromagnetics

Essential Numerical Methods in Electromagnetics

A derivative of Handbook of Numerical Analysis,
Special Volume: Numerical Methods
in Electromagnetics, Vol XIII

General Editor:

P.G. Ciarlet

Laboratoire Jacques-Louis Lions, Université Pierre et Marie Curie,
4 Place Jussieu, 75005 PARIS, France

and

Department of Mathematics, City University of Hong Kong,
Tat Chee Avenue, KOWLOON, Hong Kong

Guest Editors:

W.H.A. Schilders

Philips Research Laboratories, IC Design, Prof. Holstlaan 4,
5656 AA, Eindhoven, The Netherlands

E.J.W. ter Maten

Philips Research Laboratories,
Electronic Design & Tools/Analogue Simulation,
Prof. Holstlaan 4, 5656 AA, Eindhoven, The Netherlands

Amsterdam • Boston • Heidelberg • London • New York
Oxford • Paris • San Diego • San Francisco • Singapore
Sydney • Tokyo

ELSEVIER

Elsevier
30 Corporate Drive, Suite 400, Burlington, MA 01803, USA
Linacre House, Jordan Hill, Oxford OX2 8DP, UK
Radarweg 29, PO Box 211, 1000 AE Amsterdam, The Netherlands

British Library Cataloguing in Publication Data
A catalogue record for this book is available from the British Library

Library of Congress Cataloging-in-Publication Data
A catalog record for this book is available from the Library of Congress

ISBN: 978-0-444-53756-0

For information on all Elsevier publications
visit our web site at *books.elsevier.com*

Typeset by: diacriTech, India

Printed and bound in USA
11 12 13 10 9 8 7 6 5 4 3 2 1

Working together to grow
libraries in developing countries

www.elsevier.com | www.bookaid.org | www.sabre.org

ELSEVIER   BOOK AID
            International   Sabre Foundation

General Preface

In the early eighties, when Jacques-Louis Lions and I considered the idea of a *Handbook of Numerical Analysis*, we carefully laid out specific objectives, outlined in the following excerpts from the "General Preface" which has appeared at the beginning of each of the volumes published so far:

> During the past decades, giant needs for ever more sophisticated mathematical models and increasingly complex and extensive computer simulations have arisen. In this fashion, two indissociable activities, *mathematical modeling* and *computer simulation*, have gained a major status in all aspects of science, technology, and industry.
>
> In order that these two sciences be established on the safest possible grounds, mathematical rigor is indispensable. For this reason, two companion sciences, *Numerical Analysis* and *Scientific Software*, have emerged as essential steps for validating the mathematical models and the computer simulations that are based on them.
>
> *Numerical Analysis* is here understood as the part of *Mathematics* that describes and analyzes all the numerical schemes that are used on computers; its objective consists in obtaining a clear, precise, and faithful, representation of all the "information" contained in a mathematical model; as such, it is the natural extension of more classical tools, such as analytic solutions, special transforms, functional analysis, as well as stability and asymptotic analysis.
>
> The various volumes comprising the *Handbook of Numerical Analysis* will thoroughly cover all the major aspects of Numerical Analysis, by presenting accessible and in-depth surveys, which include the most recent trends.
>
> More precisely, the Handbook will cover the *basic methods of Numerical Analysis*, gathered under the following general headings:
>
> – Solution of Equations in $\mathbb{R}^n$,
> – Finite Difference Methods,
> – Finite Element Methods,
> – Techniques of Scientific Computing.

It will also cover the *numerical solution of actual problems of contemporary interest in Applied Mathematics*, gathered under the following general headings:
  – Numerical Methods for Fluids,
  – Numerical Methods for Solids.

In retrospect, it can be safely asserted that Volumes I to IX, which were edited by both of us, fulfilled most of these objectives, thanks to the eminence of the authors and the quality of their contributions.

After Jacques-Louis Lions' tragic loss in 2001, it became clear that Volume IX would be the last one of the type published so far, i.e., edited by both of us and devoted to some of the general headings defined above. It was then decided, in consultation with the publisher, that each future volume will instead be devoted to a single *"specific application"* and called for this reason a *"Special Volume." "Specific applications"* will include Mathematical Finance, Meteorology, Celestial Mechanics, Computational Chemistry, Living Systems, Electromagnetism, Computational Mathematics etc. It is worth noting that the inclusion of such "specific applications" in the *Handbook of Numerical Analysis* was part of our initial project.

To ensure the continuity of this enterprise, I will continue to act as Editor of each Special Volume, whose conception will be jointly coordinated and supervised by a Guest Editor.

P.G. CIARLET
July 2002

Preface

The electronics industry has shown extremely rapid advances over the past 50 years, and it is largely responsible for the economic growth in that period. It all started with the invention of the bipolar transistor based on silicon at the end of the 1940s, and since then the industry has caused another evolution for mankind. It is hard to imagine a world without all the achievements of the electronics industry.

In order to be able to continue these rapid developments, it is absolutely necessary to perform virtual experiments rather than physical experiments. Simulations are indispensable in the electronics industry nowadays. Current electronic circuits are extremely complex, and its production requires hundreds of steps that altogether take several months of fabrication time. The adagio is "first time right," and this has its repercussions for the way designers work in the electronics industry. Nowadays, they make extensive use of software tools embedded in virtual design environments. The so-called "virtual fab" has made an entry, and it is foreseen that its importance will only grow in the future.

Numerical methods are a key ingredient of a simulation environment, whence it is not surprising that the electronics industry has become one of the most fertile working environments for numerical mathematicians. Since the 1970s, there is a strong demand for efficient and robust software tools for electronic circuit simulation. Initially, this development started with the analysis of large networks of resistors, capacitors, and inductors, but soon other components such as bipolar transistors and diodes were added. Specialists made models for these components, but the problems associated with the extreme nonlinearities introduced by these models had to be tackled by numerical analysts. It was one of the first serious problems that were encountered in the field, and it initiated research into damped Newton methods for extremely nonlinear problems. In the past 30 years, electronic circuit simulation has become a very mature subject, with many beautiful results (both from the engineering and the mathematical point of view), and it still is a very active area of mathematical research. Nowadays, hot topics are the research into differential algebraic equations and the efficient calculation of (quasi-)periodic steady states.

Although circuit simulation was one of the first topics to be addressed by numerical mathematicians in the electronics industry, the simulation of semiconductor devices quickly followed at the end of the 1970s. Transistors rapidly became more complex, and a multitude of different devices was discovered. Transistors of the MOS-type (metal-oxide-semiconductor) became much more popular, and are now mainly responsible for the rapid developments in the industry. In order to be able to simulate the behavior of these devices, research into semiconductor device simulation was carried out. Soon it became clear that this was a very demanding problem from the numerical point of view, and it took many years and many conferences before some light was seen at the end of the tunnel. Applied mathematicians analyzed the famous drift-diffusion problem, and numerical mathematicians developed algorithms for its discretization and solution. During the 1990s, extended models were introduced for the modeling of semiconductor devices (hydrodynamic models, quantum effects), and nowadays this development is still continuing.

Parallel to these developments in the area of electronic circuits and devices, the more classical electromagnetics problems were also addressed. Design of magnets for loudspeakers and magnet design for MRI (magnetic resonance imaging) were important tasks, for which we can also observe a tendency towards heavy usage of simulation tools and methods. The field also generated many interesting mathematical and numerical results, whereas the role of the numerical mathematician was again indispensable in this area.

Whether it is by coincidence or not, the fields of circuit/device simulation and the more classical electromagnetics simulation, have come very close to each other in recent years. Traditionally, researchers working in the two areas did not communicate much, and separate conferences were organized with a minimum of cross-fertilization. However, owing to the increased operating frequencies of devices and the shrinking dimensions of electronics circuits, electromagnetic effects have started to play an important role. These effects influence the behavior of electronic circuits, and it is foreseen that these effects may be dramatic in the future if they are not understood well and precautions are taken. Hence, recent years show an increased interest in combined simulations of circuit behavior with electromagnetics that, in turn, has led to new problems for numerical mathematicians. One of these new topics is model order reduction, which is the art of reducing large discrete systems to a much smaller model that nevertheless exhibits behavior similar to the large system. Model order reduction is a topic at many workshops and conferences nowadays, with a multitude of applications also outside the electronics industry.

From the foregoing, it is clear that the electronics industry has always been, and still is, a very fruitful area for numerical mathematics. On the one hand, numerical mathematicians have played an important role in enabling the set-up of virtual design environments. On the other hand, many new methods have been developed as a result of the work in this specialist area. Often, the methods developed to solve the electronics problems can also be applied in other application areas. Therefore, the reason for this special volume is twofold. The first aim is to give insight in the way numerical methods are being used to solve the wide variety of problems in the electronics industry. The second aim is to give researchers from other fields of application the opportunity to benefit from the results, which have been obtained in the electronics industry.

This special volume of the Handbook of Numerical Analysis gives a broad overview of the use of numerical methods in the electronics industry. Since it is not assumed that all readers are familiar with the concepts being used in the field, Chapter 1 gives a detailed overview of models being used. The starting point is the set of Maxwell equations, and from this all models can be derived. The chapter serves as the basis for the other chapters, so that readers can always go back to Chapter 1 for a physical explanation, or a derivation of the models.

The remaining chapters discuss the use of numerical methods for different applications within the electronics industry. In the following we give a short summary of the remaining chapters.

Chapter 2 also discusses methods for discretising the Maxwell equations, using the finite difference time domain method that is extremely popular nowadays. The authors of this chapter have widespread experience in applying the method to practical problems, and the chapter discusses a multitude of related topics.

In Chapter 3, the first step towards coupled circuit/device simulations with electromagnetic effects is made by considering the problem of analyzing the electromagnetic behavior of printed circuit boards. The chapter discusses in detail the efficient evaluation of the interaction integrals, and shows the use of some numerical techniques that are not very well known.

Chapter 4 is of a more theoretical character, which does not mean that its contents are less important. The model order reduction methods discussed in Chapter 4 are equally important, since they provide a sound basis for enabling the coupled simulations required in present-day design environments. Strangely enough, it turns out that the techniques used in the area of model order reduction are intimately related to the solution methods for linear systems. In this respect, the last chapter is closely related, though very different in character.

We hope that this volume will inspire readers, and that the presentation given in the various chapters is of interest to a large community of researchers and engineers. It is also hoped that the volume reflects the importance of numerical mathematics in the electronics industry. In our experience, we could attach tags to almost all electronic products with the statement: "Mathematics inside." Let this be an inspiration for young people to not only benefit from the developments of the electronics industry, but also contribute physically to the developments in the future by becoming an enthusiastic numerical mathematician!

Eindhoven, June 2004

Wil Schilders
Jan ter Maten

List of Contributors

Zhaojun Bai, *University of California, Department of Computer Science, One Shields Avenue, 3023 Engineering II, Davis, CA, USA; e-mail: bai@cs.ucdavis.edu* (Ch. 4).

Patrick M. Dewilde, *TU Delft, Fac. Eletrotechniek Vakgroep CAS (Room L2.500), Mekelweg 4, 2628 CD, Delft, The Netherlands; e-mail: p.dewile@dimes.tudelft.nl* (Ch. 4).

Roland W. Freund, *University of California, Davis, Department of Mathematics, One Shields Avenue, Davis, CA 95616, USA; e-mail: freund@math.ucdavis.edu* (Ch. 4).

Stephen D. Gedney, *University of Kentucky, College of Engineering, Department of Electrical and Computer Engineering, 687C F. Paul Anderson Tower Lexington, KY 40506-0046, USA; e-mail: gedney@engr.uky.edu* (Ch. 2).

Susan C. Hagness, *University of Wisconsin-Madison, College of Engineering, Department of Electrical and Computer Engineering, 3419 Engineering Hall, 1415 Engineering Drive, Madison, WI 53706, USA; e-mail: hagness@engr. wisc.edu* (Ch. 2).

Wim Magnus, *IMEC, Silicon Process and Device Technology Division (SPDT), Quantum Device Modeling Group (QDM), Kapeldreef 75, Flanders, B-3001 Leuven, Belgium; e-mail: wim.magnus@imec.be* (Ch. 1).

W.H.A. Schilders, *Philips Research Laboratories, IC Design, Prof. Holstlaan 4, 5656 AA, Eindhoven, The Netherlands; e-mail: wil.schilders@philips.com* (Ch. 3).

Wim Schoenmaker, *MAGWEL N.V., Kapeldreef 75, B-3001 Leuven, Belgium* (Ch. 1).

Allen Taflove, *Northwestern University, Computational Electromagnetics Lab (NUEML), Department of Electrical and Computer Engineering, 2145 Sheridan Road, Evanston, IL 60208-3118, USA; e-mail: taflove@ece.northwestern.edu* (Ch. 2).

A.J.H. Wachters, *Philips Research Laboratories, Prof. Holstlaan 4, 5656 AA, Eindhoven, The Netherlands; e-mail: wachters@natlab.research.philips.com* (Ch. 3).

Table of Contents

Introduction to Electromagnetism

Wim Magnus

IMEC, Silicon Process and Device Technology Division (SPDT), Quantum Device Modeling Group (QDM), Kapeldreef 75, Flanders, B-3001 Leuven, Belgium
E-mail address: wim.magnus@imec.be

Wim Schoenmaker

MAGWEL N.V., Kapeldreef 75, B-3001 Leuven, Belgium

List of symbols

$\mathbf{A}$	vector potential
$\mathbf{A}_{EX}$	external vector potential
$\mathbf{A}_{IN}$	induced vector potential
$\mathbf{B}, \mathbf{B}_{IN}$	magnetic induction
C	capacitance
c	speed of light, concentration
$\mathbf{dr}$	line element
$\mathbf{dS}$	surface element
ds	elementary distance in Riemannian geometry
$d\tau$	volume element
D_n	electron diffusion coefficient
D_p	hole diffusion coefficient
$\mathbf{D}$	electric displacement vector
E	energy
E_F	Fermi energy
$E_{\alpha\mathbf{k}}(W)$	electron energy

Essential Numerical Methods in Electromagnetics
Special Volume (W.H.A. Schilders and E.J.W. ter Maten, Guest Editors) of
HANDBOOK OF NUMERICAL ANALYSIS, VOL. XIII
P.G. Ciarlet (Editor)

e	elementary charge
$\mathbf{E}$	electric field
$\mathbf{E}_C$	conservative electric field
$\mathbf{E}_{EX}$	external electric field
$\mathbf{E}_{IN}$	induced electric field
$\mathbf{E}_{NC}$	nonconservative electric field
$\mathbf{e}_z$	unit vector along z-axis
$\mathbf{e}_\phi$	azimuthal unit vector
$F_{\mu\nu}$	electromagnetic field tensor
$f, f_{\mathrm{n}}, f_{\mathrm{p}}$	(Boltzmann) distribution function
G	conductance, generation rate
G_Q	quantized conductance
$g_{\mu\nu}$	metric tensor
H	Hamiltonian
$H_{\mathbf{p}'\mathbf{p}}$	Hamiltonian scattering matrix element
h	Planck's constant
$\hbar$	reduced Planck constant ($h/2\pi$)
I	electric current
i	imaginary unit
J_G	gate leakage current
$\mathbf{J}, \mathbf{J}_{\mathrm{n}}, \mathbf{J}_{\mathrm{p}}$	electric current density
$\mathbf{H}$	magnetic field intensity
k	wavenumber
k_B	Boltzmann's constant
$\mathbf{k}$	electron wave vector
L	inductance, Lagrangian, length
L_x, L_y	length
$\mathbf{L}$	total angular momentum
l	subband index, angular momentum quantum number, length
m	angular momentum quantum number
m, m_n	charge carrier effective mass
m_0	free electron mass
$m_{\mathrm{n}}, m_{g\alpha x}, m_{g\alpha y},$ $m_{g\alpha z}, m_{1,\mathrm{ox},\alpha},$ $m_{2,\mathrm{ox},\alpha}, m_{3,\mathrm{ox},\alpha}, \ldots,$ $m_{\alpha x}, m_{\alpha y}, m_{\alpha z}$	electron effective mass
m_{p}	hole effective mass
$\mathbf{M}$	magnetization vector
$\mathbf{m}$	magnetic moment
N	number of particles, coordinates, or modes
N_A	acceptor doping density
n	electron concentration
$\mathbf{n}$	unit vector
$p, p_i, \mathbf{p}, \mathbf{p}_i, P,$ $P_i, \mathbf{P}, \mathbf{P}_i, \ldots$	generalized momenta

p	hole concentration
$\mathbf{P}$	total momentum, electric polarization vector
$\mathbf{p}$	momentum, electric dipole moment
$q, q_i, \mathbf{q}, \mathbf{q}_i, Q,$ $Q_i, \mathbf{Q}, \mathbf{Q}_i, \dots$	generalized coordinates
Q	electric charge
q_n	carrier charge
Q_A	electric charge residing in active area
R	resistance, recombination rate
R_H	Hall resistance
R_K	von Klitzing resistance
R_L	lead resistance
R_Q	quantized resistance
$R^{\mu}_{\rho\lambda\sigma}$	Riemann tensor
$\mathbb{R}$	set of real numbers
(r, θ, ϕ)	spherical coordinates
$\mathbf{r}, \mathbf{r}_n$	position vector
S	action, entropy
$S(\mathbf{p}, \mathbf{p}')$	transition rate
$\mathbf{S}$	Poynting vector
$\mathbf{S}_n, \mathbf{S}_p$	energy flux vector
t	time
T	lattice temperature
T_n	electron temperature
T_p	hole temperature
$\mathbf{T}, \mathbf{T}_{\alpha\beta}$	EM energy momentum tensor
U_E	electric energy
U_M	magnetic energy
U_{EM}	EM energy
$U(y), U(z)$	potential energy
u_{EM}	EM energy density
V	scalar electric potential
V_H	Hall voltage
V_G	gate voltage
$\mathbf{v}_n$	carrier velocity
$\mathbf{v}_n, \mathbf{v}_p$	drift velocity
$\mathbf{v}$	drift velocity, velocity field
$W, W_{\alpha l}(W)$	subband energy
w, w_n, w_p	carrier energy density
(x, y, z)	Cartesian coordinates
Y	admittance
Z	impedance
α	summation index, valley index, variational parameters
β	summation index, $1/k_B T$

$\partial\Omega$	boundary surface of Ω
$\partial\Omega_\infty$	boundary surface of Ω_∞
ε_0	electric permittivity of vacuum
ε	electric permittivity
$\varepsilon_r, \varepsilon_S$	relative electric permittivity
Γ	closed curve inside a circuit
$\Gamma_{\alpha l}$	resonance width
$\Gamma^\alpha_{\mu\nu}$	affine connection
κ	wavenumber
κ_n, κ_p	thermal conductivity
Λ_{EM}	EM angular momentum density
μ	magnetic permeability, chemical potential
μ_0	magnetic permeability of vacuum
μ, μ_n, μ_p	carrier mobility
μ_r	relative magnetic permeability
Ω	connected subset of $\mathbb{R}^3$, volume, circuit region
Ω_∞	all space
ω	angular frequency
π_{EM}	EM momentum density
ρ	electric charge density
(ρ, ϕ, z)	cylindrical coordinates
σ	electrical conductivity, spin index
$\tau, \tau_0, \tau_e, \tau_p,$	
τ_{en}, τ_{ep}	relaxation time
$\tau_{\alpha l}$	resonance lifetime
χ	gauge function
χ_e	electric susceptibility
$\chi_k(y)$	wave function
χ_m	magnetic susceptibility
Φ_D	electric flux (displacement)
Φ_E	electric flux (electric field)
Φ_{ex}	external magnetic flux
Φ, Φ_M	magnetic flux
$\psi(\mathbf{r}), \psi_\alpha(\mathbf{r}, z),$	
$\psi_{\alpha\mathbf{k}}(\mathbf{r}, z),$	
$\phi_\alpha(W, z), \psi(x, y)$	wave function
∇	gradient
$\nabla\cdot$	divergence
$\nabla\times$	curl
∇^2	vectorial Laplace operator
∇^2	Laplace operator
$\mathcal{L}$	Lagrange density, inductance per unit length
V_ε	electromotive force

1. Preface

Electromagnetism, formulated in terms of the Maxwell equations, and quantum mechanics, formulated in terms of the Schrödinger equation, constitute the physical laws by which the bulk of natural experiences are described. Apart from the gravitational forces, nuclear forces, and weak decay processes, the description of the physical facts starts with these underlying microscopic theories. However, knowledge of these basic laws is only the beginning of the process to apply these laws in realistic circumstances and to determine their quantitative consequences. With the advent of powerful computer resources, it has become feasible to extract information from these basic laws with unprecedented accuracy. In particular, the complexity of realistic systems manifests itself in the nontrivial boundary conditions, such that without computers, reliable calculation are beyond reach.

The ambition of physicists, chemists, and engineers, to provide tools for performing calculations, does not only boost progress in technology but also has a strong impact on the formulation of the equations that represent the physics knowledge and hence provides a deeper understanding of the underlying physics laws. As such, computational physics has become a cornerstone of theoretical physics and we may say that without a computational recipe, a physics law is void or at least incomplete. Contrary to what is sometimes claimed, that after having found the unifying theory for gravitation and quantum theory, there is nothing left to investigate, we believe that physics has just started to flourish and there are wide fields of research waiting for exploration.

This volume is dedicated to the study of electrodynamic problems. The Maxwell equations appear in the form

$$\Delta(\text{field}) = \text{source}, \tag{1.1}$$

where Δ describes the near-by field variable correlation of the field that is induced by a source or field disturbance. Near-by correlations can be mathematically expressed by differential operators that probe changes going from one location the a neighboring one. It should be emphasized that "near-by" refers to space and time.

One could "easily" solve these equations by construction the inverse of the differential operator. Such an inverse is usually known as a Green function.

There are two main reasons that prevent a straightforward solution of the Maxwell equations. First of all, realistic structure boundaries may be very irregular, and therefore the corresponding boundary conditions cannot be implemented analytically. Secondly, the sources themselves may depend on the values of the fields and will turn the problem in a highly nonlinear one, as may be seen from Eq. (1.1) that should be read as

$$\Delta(\text{field}) = \text{source(field)}. \tag{1.2}$$

The bulk of this volume is dedicated to find solutions to equations of this kind. In particular, Chapters I and II are dealing with above type of equations. A considerable amount of work deals with obtaining the details of the right-hand side of Eq. (1.2), namely how the source terms, being charges and currents depend in detail on the values of the field variables.

Whereas, the microscopic equation describe the physical processes in great detail, i.e., at every space–time point field and source variables are declared, it may be profitable to collect a whole bunch of these variables into a single basket and to declare for each basket a few representative variables as the appropriate values for the fields and the sources. This kind of reduction of parameters is the underlying strategy of circuit modeling. Here, the Maxwell equations are replaced by Kirchhoff's network equations.

The "basket" containing a large collection of fundamental degrees of freedom of field and source variables, should not be filled at random. Physical intuition suggests that we put together in one basket degrees of freedom that are "alike." Field and source variables at near-by points are candidates for being grabbed together, since physical continuity implies that a all elements in the basket will have similar values.[1]

The baskets are not only useful for simplifying the continuous equations. They are vital to the discretization schemes. Since any computer has only a finite memory storage, the continuous or infinite collection of degrees of freedom must be mapped onto a finite subset. This may be accomplished by appropriately positioning and sizing of all the baskets. This procedure is named "grid generation" and the construction of a good grid is often of great importance to obtain accurate solutions.

After having mapped the continuous problem onto a finite grid one may establish a set of algebraic equations connecting the grid variables (basket representatives) and explicitly reflecting the nonlinearity of the original differential equations. The solution of large systems of nonlinear algebraic equations is based on Newton's iterative method. To find the solution of the set of nonlinear equations $\mathbf{F}(\mathbf{x}) = \mathbf{0}$, an initial guess is made: $\mathbf{x} = \mathbf{x}_{init} = \mathbf{x}_0$. Next the guess is (hopefully) improved by looking at the equation:

$$\mathbf{F}(\mathbf{x}_0 + \Delta\mathbf{x}) \simeq \mathbf{F}(\mathbf{x}_0) + \mathbf{A} \cdot \Delta\mathbf{x}, \tag{1.3}$$

where the matrix $\mathbf{A}$ is

$$\mathbf{A}_{ij} = \left(\frac{\partial F_i(\mathbf{x})}{\partial x_j}\right)_{\mathbf{x}_0}. \tag{1.4}$$

In particular, by assuming that the correction brings us close to the solution, i.e., $\mathbf{x}_1 = \mathbf{x}_0 + \Delta\mathbf{x} \simeq \mathbf{x}^*$, where $\mathbf{F}(\mathbf{x}^*) = 0$, we obtain that

$$0 = \mathbf{F}(\mathbf{x}_0) + \mathbf{A} \cdot \Delta\mathbf{x} \quad \text{or}$$
$$\Delta\mathbf{x} = -\mathbf{A}^{-1} \cdot \mathbf{F}(\mathbf{x}_0). \tag{1.5}$$

Next we repeat this procedure, until convergence is reached. A series of vectors, $\mathbf{x}_{init} = \mathbf{x}_0, \mathbf{x}_1, \mathbf{x}_2, \dots, \mathbf{x}_{n-1}, \mathbf{x}_n = \mathbf{x}_{final}$, is generated, such that $|\mathbf{F}(\mathbf{x}_{final})| < \varepsilon$, where ε is some prescribed error criterion. In each iteration a large linear matrix problem of the type $\mathbf{A}|\mathbf{x}\rangle = |\mathbf{b}\rangle$ needs to be solved.

---

[1]It should be emphasized that such a picture works at the classical level. Quantum physics implies that near-by field point may take any value and the continuity of fields is not required.

2. The microscopic Maxwell equations

2.1. The microscopic Maxwell equations in integral and differential form

In general, any electromagnetic field can be described and characterized on a microscopic scale by two vector fields $\mathbf{E}(\mathbf{r}, t)$ and $\mathbf{B}(\mathbf{r}, t)$ specifying respectively the electric field and the magnetic induction in an arbitrary space point $\mathbf{r}$ at an arbitrary time t. All dynamical features of these vector fields are contained in the well-known Maxwell equations (MAXWELL [1954a], MAXWELL [1954b], JACKSON [1975] FEYNMAN, LEIGHTON, and SANDS [1964a])

$$\nabla \cdot \mathbf{E} = \frac{\rho}{\varepsilon_0}, \tag{2.1}$$

$$\nabla \cdot \mathbf{B} = 0, \tag{2.2}$$

$$\nabla \times \mathbf{E} = -\frac{\partial \mathbf{B}}{\partial t}, \tag{2.3}$$

$$\nabla \times \mathbf{B} = \mu_0 \mathbf{J} + \varepsilon_0 \mu_0 \frac{\partial \mathbf{E}}{\partial t}. \tag{2.4}$$

They describe the spatial and temporal behavior of the electromagnetic field vectors and relate them to the sources of electric charge and current that may be present in the region of interest. Within the framework of a microscopic description, the electric charge density ρ and the electric current density $\mathbf{J}$ are considered spatially localized distributions residing in vacuum. As such they represent both mobile charges giving rise to macroscopic currents in solid-state devices, chemical solutions, plasmas, etc., and bound charges that are confined to the region of an atomic nucleus. In turn, the Maxwell equations in the above presented form explicitly refer to the values taken by $\mathbf{E}$ and $\mathbf{B}$ in vacuum and, accordingly, the electric permittivity ε_0 and the magnetic permeability μ_0 appearing in Eqs. (2.1) and (2.4) correspond to vacuum.

From the mathematical point of view, the solution of the differential equations (2.1)–(2.4) together with appropriate boundary conditions in space and time, should in principle unequivocally determine the fields $\mathbf{E}(\mathbf{r}, t)$ and $\mathbf{B}(\mathbf{r}, t)$. In practice however, analytical solutions may be achieved only in a limited number of cases and, due to the structural and geometrical complexity of modern electronic devices, one has to adopt advanced numerical simulation techniques to obtain reliable predictions of electromagnetic field profiles. In this light, the aim is to solve Maxwell's equations on a discrete set of mesh points using suitable discretization techniques which are often taking advantage of integral form of Maxwell's equations. The latter may be derived by a straightforward application of Gauss' and Stokes' theorems. In particular, one may integrate Eqs. (2.1) and (2.1) over a simply connected region $\Omega \in \mathbb{R}^3$ bounded by a closed surface $\partial\Omega$ to obtain

$$\int_{\partial\Omega} \mathbf{E}(\mathbf{r}, t) \cdot d\mathbf{S} = \frac{1}{\varepsilon_0} Q(t), \tag{2.5}$$

$$\int_{\partial\Omega} \mathbf{B}(\mathbf{r}, t) \cdot d\mathbf{S} = 0, \tag{2.6}$$

where $Q(t)$ denotes the instantaneous charge residing in the volume Ω, i.e.,

$$Q(t) = \int_{\Omega} \rho(\mathbf{r}, t)\, d\tau. \tag{2.7}$$

Eq. (2.5) is nothing but Gauss' law stating that the total outward flux of the electric field threading the surface $\partial\Omega$ equals the total charge contained in the volume Ω up to a factor ε_0 whereas Eq. (2.6) reflects the absence of magnetic monopoles.

Similarly, introducing an arbitrary, open and simply connected surface Σ bounded by a simple, closed curve Γ, one may extract the induction law of Faraday and Ampère's law by integrating respectively Eqs. (2.3) and (2.4) over Σ:

$$\oint_{\Gamma} \mathbf{E}(\mathbf{r}, t) \cdot d\mathbf{r} = -\frac{d\Phi_M(t)}{dt}, \tag{2.8}$$

$$\oint_{\Gamma} \mathbf{B}(\mathbf{r}, t) \cdot d\mathbf{r} = \mu_0\left(I(t) + \varepsilon_0 \frac{d\Phi_E(t)}{dt}\right). \tag{2.9}$$

The variables $\Phi_E(t)$ and $\Phi_M(t)$ are representing the time-dependent electric and magnetic fluxes piercing the surface Σ and are defined as:

$$\Phi_E(t) = \int_{\Sigma} \mathbf{E}(\mathbf{r}, t) \cdot d\mathbf{S}, \tag{2.10}$$

$$\Phi_M(t) = \int_{\Sigma} \mathbf{B}(\mathbf{r}, t) \cdot d\mathbf{S}, \tag{2.11}$$

while the circulation of the electric field around Γ is the instantaneous electromotive force $V_\varepsilon(t)$ along Γ is:

$$V_\varepsilon(t) = \oint_{\Gamma} \mathbf{E}(\mathbf{r}, t) \cdot d\mathbf{r}. \tag{2.12}$$

The right-hand side of Eq. (2.9) consists of the total current flowing through the surface Σ

$$I(t) = \int_{\Sigma} \mathbf{J}(\mathbf{r}, t) \cdot d\mathbf{S} \tag{2.13}$$

and the so-called displacement current which is proportional to the time derivative of the electric flux. The sign of the above line integrals depends on the orientation of the closed loop Γ, the positive traversal sense of which is uniquely defined by the orientation of the surface Σ imposed by the vectorial surface element $d\mathbf{S}$. Apart from this restriction it should be noted that the surface Σ can be chosen freely so as to extract meaningful physical information from the corresponding Maxwell equation. In particular, though being commonly labeled by the symbol Σ, the surfaces appearing in Faraday's and Ampère's laws (Eqs. (2.8)–(2.9)) will generally be chosen in a different way as can be illustrated by the example of a simple electric circuit. In the case of Faraday's law, one usually wants $\Phi_M(t)$ to be the magnetic flux threading the circuit and therefore Σ would be chosen to "span" the circuit while Γ would be located in the interior of the circuit area. On the other hand, in order to exploit Ampère's law, the surface Σ should be pierced by the current density in the circuit in order to make $I(t)$ the current flowing through the circuit.

2.2. Conservation laws

Although a complete description of the electromagnetic field requires the full solution of the Maxwell equations in their differential form, one may extract a number of conservation laws may by simple algebraic manipulation. The differential form of the conservation laws takes the generic form

$$\mathbf{\nabla} \cdot \mathbf{F} + \frac{\partial \mathbf{G}}{\partial t} = \mathbf{K}, \tag{2.14}$$

where $\mathbf{F}$ is the generalized flow tensor associated with the field $\mathbf{G}$ and $\mathbf{K}$ is related to any possible external sources or sinks.

2.2.1. Conservation of charge – the continuity equation
Taking the divergence of Eq. (2.4) and the time derivative of Eq. (2.1) and combining the resulting equations, one easily obtains the charge-current continuity equation expressing the conservation of electric charge:

$$\mathbf{\nabla} \cdot \mathbf{J} + \frac{\partial \rho}{\partial t} = 0. \tag{2.15}$$

Integration over a closed volume Ω yields

$$\int_{\partial \Omega} \mathbf{J} \cdot \mathbf{dS} = -\frac{\partial}{\partial t} \int_{\Omega} \rho \, d\tau, \tag{2.16}$$

which states that the total current flowing through the bounding surface $\partial \Omega$ equals the time rate of change of all electric charge residing within Ω.

2.2.2. Conservation of energy – Poynting's theorem
The electromagnetic energy flow generated by a time dependent electromagnetic field is most adequately represented by the well-known Poynting vector given by

$$\mathbf{S} = \frac{1}{\mu_0} \mathbf{E} \times \mathbf{B}. \tag{2.17}$$

Calculating the divergence of $\mathbf{S}$ and using the Maxwell equations, one may relate the Poynting vector to the electromagnetic energy density u_{EM} through the energy conservation law

$$\mathbf{\nabla} \cdot \mathbf{S} + \frac{\partial u_{EM}}{\partial t} = -\mathbf{J} \cdot \mathbf{E}, \tag{2.18}$$

which is also known as the Poynting theorem. The energy density u_{EM} is given by

$$u_{EM} = \frac{1}{2} \left(\varepsilon_0 E^2 + \frac{B^2}{\mu_0} \right). \tag{2.19}$$

The energy conservation expressed in Eq. (2.18) refers to the total energy of the electromagnetic field and all charged particles contributing to the charge and current distributions. In particular, denoting the mechanical energy of the charged particles residing in the volume Ω by E_{MECH} one may derive for both classical and quantum mechanical

systems that the work done per unit time by the electromagnetic field on the charged
volume is given by

$$\frac{dE_{\text{MECH}}}{dt} = \int_{\Omega} \mathbf{J} \cdot \mathbf{E} \, d\tau. \tag{2.20}$$

Introducing the total electromagnetic energy associated with the volume Ω as $E_{\text{EM}} = \int_{\Omega} u_{\text{EM}} \, d\tau$ one may integrate Poynting's theorem to arrive at

$$\frac{d}{dt}(E_{\text{MECH}} + E_{\text{EM}}) = - \int_{\partial\Omega} \mathbf{S} \cdot d\mathbf{S}. \tag{2.21}$$

It should be emphasized that the above result also covers most of the common situations
where the energy of the charged particles is relaxed to the environment through dissipative processes. The latter may be accounted for by invoking appropriate constitutive
equations expressing the charge and current densities as linear or nonlinear responses
to the externally applied electromagnetic fields and other driving force fields. As an
example, we mention Ohm's law, proposing a linear relation between the macroscopic
electric current density and the externally applied electric field in a nonideal conductor:

$$\mathbf{J}_M = \sigma \mathbf{E}_{\text{EXT}}. \tag{2.22}$$

Here, the conductivity σ is assumed to give an adequate characterization of all microscopic elastic and inelastic scattering processes that are responsible for the macroscop-
ically observable electric resistance. The derivation of constitutive equations will be
discussed in greater detail in Section 4.

2.3. Conservation of linear momentum – the electromagnetic field tensor

In an analogous way, an appropriate linear momentum density π_{EM} may be assigned
to the electromagnetic field, which differs from the Poynting vector merely by a factor
$\varepsilon_0 \mu_0 = 1/c^2$:

$$\pi_{\text{EM}} = \varepsilon_0 \mathbf{E} \times \mathbf{B} = \frac{1}{c^2} \mathbf{S}. \tag{2.23}$$

The time evolution of π_{EM} is not only connected to the rate of change of the mechanical
momentum density giving rise to the familiar Lorentz force term, but also involves the
divergence of a second rank tensor $\mathbf{T}$ which is usually called the Maxwell stress tensor
(JACKSON [1975], LANDAU and LIFSHITZ [1962]). The latter is defined most easily by
its Cartesian components

$$\mathbf{T}_{\alpha\beta} = \varepsilon_0 \left(\frac{1}{2} |\mathbf{E}|^2 \delta_{\alpha\beta} - E_\alpha E_\beta \right) + \frac{1}{\mu_0} \left(\frac{1}{2} |\mathbf{B}|^2 \delta_{\alpha\beta} - B_\alpha B_\beta \right) \tag{2.24}$$

with $\alpha, \beta = x, y, z$.
   A straightforward calculation yields:

$$\frac{\partial \pi_{\text{EM}}}{\partial t} = -\rho \mathbf{E} - \mathbf{J} \times \mathbf{B} - \nabla \cdot \mathbf{T}. \tag{2.25}$$

2.3.1. Angular momentum conservation

The angular momentum density of the electromagnetic field and its corresponding flux may be defined respectively by the relations

$$\Lambda_{EM} = \mathbf{r} \times \boldsymbol{\pi}_{EM}, \qquad \boldsymbol{\Gamma} = \mathbf{r} \times \mathbf{T}. \tag{2.26}$$

The conservation law that governs the angular momentum, reads

$$\frac{\partial \Lambda_{EM}}{\partial t} = -\mathbf{r} \times (\rho \mathbf{E} + \mathbf{J} \times \mathbf{B}) - \boldsymbol{\nabla} \cdot \boldsymbol{\Gamma}. \tag{2.27}$$

3. Potentials and fields, the Lagrangian

Not only the Maxwell equations themselves but also all related conservation laws have been expressed with the help of two key observables describing the microscopic electromagnetic field, namely **E** and **B**. Strictly speaking, all relevant physics involving electromagnetic phenomena can be described correctly and completely in terms of the variables **E** and **B** solely, and from this point of view there is absolutely no need of defining auxiliary potentials akin to **E** and **B**. Nevertheless, it proves quite beneficial to introduce the scalar potential $V(\mathbf{r}, t)$ and the vector potential $\mathbf{A}(\mathbf{r}, t)$ as alternative electrodynamical degrees of freedom.

3.1. The scalar and vector potential

From the Maxwell equation $\boldsymbol{\nabla} \cdot \mathbf{B} = 0$ and Helmholtz' theorem it follows that, within a simply connected region Ω, there exists a regular vector field **A** – called vector potential – such that

$$\mathbf{B} = \boldsymbol{\nabla} \times \mathbf{A}, \tag{3.1}$$

which allows us to rewrite Faraday's law (2.8) as

$$\boldsymbol{\nabla} \times \left(\mathbf{E} + \frac{\partial \mathbf{A}}{\partial t} \right) = 0. \tag{3.2}$$

The scalar potential V emerges from the latter equation and Helmholtz' theorem stating that, in a simply connected region Ω there must exist a regular scalar function V such that

$$\mathbf{E} = -\boldsymbol{\nabla} V - \frac{\partial \mathbf{A}}{\partial t}. \tag{3.3}$$

Although V and **A** do not add new physics, there are at least three good reasons to introduce them anyway. First, it turns out that (JACKSON [1975], FEYNMAN, LEIGHTON, and SANDS [1964a]) the two potentials greatly facilitate the mathematical treatment of classical electrodynamics in many respects. For instance, the choice of an appropriate gauge [2] allows one to convert the Maxwell equations into convenient wave equations for V and **A** for which analytical solutions can be derived occasionally. Moreover, the scalar

---

[2] Gauge transformations will extensively be treated in Section 7.

potential V provides an natural link to the concept of macroscopic potential differences that are playing a crucial role in conventional simulations of electric circuits.

Next, most quantum mechanical treatments directly invoke the "potential" picture to deal with the interaction between a charged particle and an electromagnetic field. In particular, adopting the path integral approach, one accounts for the presence of electric and magnetic fields by correcting the action functional S related to the propagation from $(\mathbf{r}_0, t_0)$ to $(\mathbf{r}_1, t_1)$ along a world line, according to

$$S[V, \mathbf{A}] = S[0, 0] + q \left(\int_{\mathbf{r}_1}^{\mathbf{r}_2} \mathbf{A} \cdot d\mathbf{r} - \int_{t_0}^{t_1} dt \, V(\mathbf{r}, t) \right), \tag{3.4}$$

while the field-dependent Hamiltonian term appearing in the nonrelativistic, one-particle Schrödinger equation $i\hbar(\partial \psi / \partial t) = H\psi$, takes the form

$$H = \frac{1}{2m}(\mathbf{p} - q\mathbf{A})^2 + qV \tag{3.5}$$

with $\mathbf{p} = -i\hbar \nabla$. Furthermore, the canonical quantization of the electromagnetic radiation field leads to photon modes corresponding to the quantized transverse modes of the vector potential.

Finally, the third motivation for adopting scalar and vector potentials lies in the perspective of developing new numerical simulation techniques. For example, it was observed recently (SCHOENMAKER, MAGNUS, and MEURIS [2002]) that the magnetic field generated by a steady current distribution may alternatively be extracted from the fourth Maxwell equation (Ampère's law),

$$\nabla \times \nabla \times \mathbf{A} = \mu_0 \mathbf{J} \tag{3.6}$$

by assigning discretized vector potential variables to the *links* connecting adjacent nodes. This will be discussed in Section 8.

3.2. Gauge invariance

In contrast to the electric field and the magnetic induction, neither the scalar nor the vector potential are uniquely defined. Indeed, performing a so-called gauge transformation

$$\mathbf{A}'(\mathbf{r}, t) = \mathbf{A}(\mathbf{r}, t) + \nabla \chi(\mathbf{r}, t),$$
$$V'(\mathbf{r}, t) = V(\mathbf{r}, t) - \frac{\partial \chi(\mathbf{r}, t)}{\partial t}, \tag{3.7}$$

where the gauge field $\chi(\mathbf{r}, t)$ is an arbitrary regular, real scalar field, one clearly observes that the potentials are modified while the electromagnetic fields $\mathbf{E}(\mathbf{r}, t)$ and $\mathbf{B}(\mathbf{r}, t)$ remain unchanged. Similarly, any quantum mechanical wave function $\psi(\mathbf{r}, t)$ transforms according to

$$\psi'(\mathbf{r}, t) = \psi(\mathbf{r}, t) \exp\big(iq\chi(\mathbf{r}, t)\big),$$
$$\psi'^*(\mathbf{r}, t) = \psi^*(\mathbf{r}, t) \exp\big(-iq\chi(\mathbf{r}, t)\big),$$

whereas the quantum mechanical probability density $|\psi(\mathbf{r}, t)|^2$ and other observable quantities are invariant under a gauge transformation, as required.

3.3. Lagrangian for an electromagnetic field interacting with charges and currents

While the Maxwell equations are the starting point in the so-called *inductive approach*, one may alternatively adopt the *deductive approach* and try to "derive" the Maxwell equations from a proper variational principle. As a matter of fact it is possible indeed to postulate a Lagrangian density $\mathcal{L}(\mathbf{r}, t)$ and an action functional $S[\mathcal{L}, t_0, t_1] = \int_{t_0}^{t_1} \mathcal{L}(\mathbf{r}, t) \, d\tau$ such that the Maxwell equations emerge as the Euler–Lagrange equations that make the action

$$\delta S = 0 \tag{3.8}$$

stationary. While such a "derivation" is of utmost importance for the purpose of basic understanding from the theoretical point of view, the Lagrangian and Hamiltonian formulation of electromagnetism may look redundant when it comes to numerical computations. However, we have quoted the Lagrangian density of the electromagnetic field not only for the sake of completeness but also to illustrate the numerical potential of the underlying variational principle.

The Lagrangian density for the interacting electromagnetic field is conventionally postulated as a quadratic functional of the scalar and vector potential and their derivatives:

$$\mathcal{L} = \frac{1}{2}\varepsilon_0 \left| \nabla V + \frac{\partial \mathbf{A}}{\partial t} \right|^2 - \frac{1}{2\mu_0} |\nabla \times \mathbf{A}|^2 + \mathbf{J} \cdot \mathbf{A} - \rho V, \tag{3.9}$$

where the field variables V and $\mathbf{A}$ are linearly coupled to the charge and current distribution ρ and $\mathbf{J}$.

It is now straightforward to obtain the Maxwell equations as the Euler–Lagrange equations corresponding to Eq. (3.9) provided that the set of field variables is chosen to be either V or A_α. The first possibility gives rise to

$$\sum_{\beta=x,y,z} \frac{\partial}{\partial x_\beta} \left[\frac{\partial \mathcal{L}}{\partial \left(\frac{\partial V}{\partial x_\beta} \right)} i \right] + \frac{\partial}{\partial t} \left[\frac{\partial \mathcal{L}}{\partial \left(\frac{\partial V}{\partial t} \right)} \right] = \frac{\partial \mathcal{L}}{\partial V}. \tag{3.10}$$

Inserting all nonzero derivatives, we arrive at

$$\varepsilon_0 \sum_\beta \frac{\partial}{\partial x_\beta} \left(\frac{\partial V}{\partial x_\beta} + \frac{\partial A_\beta}{\partial t} \right) = -\rho, \tag{3.11}$$

which clearly reduces to the first Maxwell equation

$$\varepsilon_0 \nabla \cdot \mathbf{E} = \rho \quad \text{(Gauss' law).} \tag{3.12}$$

Similarly, the three Euler–Lagrange equations

$$\sum_{\beta=x,y,z} \frac{\partial}{\partial x_\beta} \left[\frac{\partial \mathcal{L}}{\partial \left(\frac{\partial A_\alpha}{\partial x_\beta} \right)} \right] + \frac{\partial}{\partial t} \left[\frac{\partial \mathcal{L}}{\partial \left(\frac{\partial A_\alpha}{\partial t} \right)} \right] = \frac{\partial \mathcal{L}}{\partial A_\alpha}, \quad \alpha = x, y, z \tag{3.13}$$

lead to the fourth Maxwell equation

$$\frac{1}{\mu_0} \left(\nabla \times \mathbf{B} - \varepsilon_0 \frac{\partial \mathbf{E}}{\partial t} \right) = \mathbf{J} \quad \text{(Ampère–Faraday's law).} \tag{3.14}$$

It should be noted that, within the deductive approach, the electric and magnetic field vectors are *defined* by the equations

$$\mathbf{E} = -\nabla V - \frac{\partial \mathbf{A}}{\partial t}, \qquad \mathbf{B} = \nabla \times \mathbf{A}, \tag{3.15}$$

whereas the latter are directly resulting from the Maxwell equations in the inductive approach. Mutatis mutandis, the two remaining Maxwell equations $\nabla \cdot \mathbf{B} = 0$ and $\nabla \times \mathbf{E} = -\partial \mathbf{B}/\partial t$ are a direct consequence of the operation of the vector identities (A.34) and (A.35) on Eqs. (3.15). It should also be noted that the Lagrangian density may be written as

$$\mathcal{L} = \frac{1}{2}\varepsilon_0 \mathbf{E}^2 - \frac{1}{2\mu_0}\mathbf{B}^2. \tag{3.16}$$

So far, we have considered the Maxwell equations from the perspective that the charge and the current densities are given and the fields should be determined. However, as was already mentioned in the introduction, the charge and current densities may also be influenced by the fields. In order to illustrate the opposite cause–effect relation, we consider the Lagrangian of N charged particles moving in an electromagnetic field. The Lagrangian is

$$L = \sum_{n=1}^{N} \frac{1}{2}m_n v_n^2 + \frac{1}{2}\int d\tau \left(\varepsilon_0 \mathbf{E}^2 - \frac{1}{\mu_0}\mathbf{B}^2\right) - \int d\tau\, \rho V + \int d\tau\, \mathbf{J} \cdot \mathbf{A}, \tag{3.17}$$

where we defined the charge and current densities as

$$\rho(\mathbf{r}, t) = \sum_{n=1}^{N} q_n \delta(\mathbf{r} - \mathbf{r}_n),$$

$$\mathbf{J}(\mathbf{r}, t) = \sum_{n=1}^{N} q_n \mathbf{v}_n \delta(\mathbf{r} - \mathbf{r}_n) \tag{3.18}$$

and the particles' velocities as $\mathbf{v}_n = d\mathbf{r}_n/dt$. Applying the Euler–Lagrange equations:

$$\frac{d}{dt}\left(\frac{\partial L}{\partial \mathbf{v}_n}\right) - \frac{\partial L}{\partial \mathbf{r}_n} = 0, \tag{3.19}$$

gives

$$m_n \frac{d^2 \mathbf{r}_n}{dt^2} = q_n \mathbf{E}(\mathbf{r}_n, t) + q_n \mathbf{v}_n \times \mathbf{B}(\mathbf{r}_n, t). \tag{3.20}$$

The last term is recognized as the Lorentz force.

3.4. Variational calculus

Although the numerical implementation of the variational principle leading to the Maxwell equations is not a common practice in numerical analysis, it may nevertheless turn out to be a useful approximation technique for particular classes of problems.

The exact solution of the Euler–Lagrange equations determines an extremum of the action functional which becomes stationary with respect to *any arbitrary* variations of the field functions that meet the boundary conditions invoked. On the other hand, being inspired by physical intuition or analogy with similar problems, one may be able to propose a class of trial functions satisfying the boundary conditions and exhibiting the expected physical behavior. If these trial functions can be characterized by one or more adjustable parameters $\alpha_1, \ldots, \alpha_n$, then one may calculate the values of $\alpha_1, \ldots, \alpha_n$ for which the action integral becomes stationary. Although the corresponding numerical value of the action will generally differ from the true extremum that is attained by the exact solution, the resulting trial function may surprisingly lead to rather accurate estimates of the physical quantities of interest. A nice example of this phenomenon is given in FEYNMAN, LEIGHTON, and SANDS [1964a] (Part II, Chapter 19) where a variational calculation of the capacitance of a cylindrical coaxial cable is presented and compared with the exact formula for various values of the inner and outer radii of the cable.

As an illustration, we have worked out the case of a long coaxial cable with a square cross section, for which the inductance is estimated within the framework of variational calculus.

Consider an infinitely long coaxial cable centered at the z-axis, consisting of a conducting core, a magnetic insulator, and a conducting coating layer. Both the core and the coating layer have a square cross section of sizes a and b, respectively. The core carries a current I in the z-direction which is flowing back to the current source through the coating layer, thereby closing the circuit as depicted in Fig. 3.1. Neglecting skin effects we assume that the current density is strictly localized at the outer surface of the core and the inner surface of the coating layer, respectively. Moreover, the translational symmetry in the z-direction reduces the solution of Maxwell's equations essentially to a two-dimensional problem whereas the square symmetry of the cable allows us to divide an arbitrary cable cross-section into four identical triangles and to work out the solution for just one triangular area. In particular, we will focus on the region Δ (see Fig. 3.2) bounded by

$$x \geqslant 0; \quad -x \leqslant y \leqslant x. \tag{3.21}$$

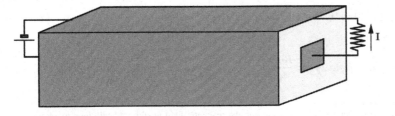

FIG. 3.1. Infinitely long coaxial cable carrying a stationary surface current.

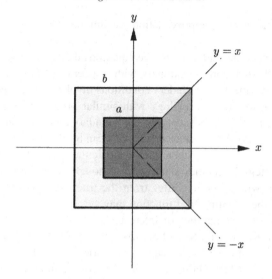

FIG. 3.2. Cross section of the coaxial cable.

Within this region, the current density takes the form

$$\mathbf{J}(x, y) = J_z(x)\mathbf{e}_z,$$

$$J_z(x) = \frac{I}{4}\left[\frac{1}{a}\delta\left(x - \frac{a}{2}\right) - \frac{1}{b}\delta\left(x - \frac{b}{2}\right)\right],$$

(3.22)

where the factor 4 indicates that the region Δ accounts for only a quarter of the total current flowing through the cable's cross-section. The particular shape of the current density reflects the presence of perfect shielding requiring that the magnetic field be vanishing for $x < a/2$ and $x > b/2$ whereas B_y should abruptly jump[3] to a nonzero value at $x = a/2 + \varepsilon$ and $x = b/2 - \varepsilon$ where $\varepsilon \to 0^+$. The nonzero limiting values of B_y are used to fix appropriate boundary for B_y simply by integrating the z-component of the Maxwell equation $\nabla \times \mathbf{B} = \mathbf{0}$ over the intervals $[a/2 - \varepsilon, a/2 + \varepsilon]$ and $[b/2 - \varepsilon, b/2 + \varepsilon]$, respectively. For instance, from

$$\int_{a/2-\varepsilon}^{a/2+\varepsilon} dx \left[\frac{\partial B_y(x, y)}{\partial x} - \frac{\partial B_x(x, y)}{\partial y}\right] = \frac{\mu_0 I}{4a}$$

(3.23)

and

$$B_y(x, y) = 0 \quad \text{for } x < a/2,$$

(3.24)

it follows that

$$\lim_{x \to 1/2a^+} B_y(x, y) = \frac{\mu_0 I}{4a}$$

(3.25)

---

[3]If the current density were smeared out, the magnetic field would gradually tend to zero inside the core and the coating layer.

and similarly

$$\lim_{x \to 1/2b^-} B_y(x, y) = \frac{\mu_0 I}{4b}. \tag{3.26}$$

Finally, the boundary conditions reflecting the connection of adjacent triangular areas are directly dictated by symmetry considerations requiring that the magnetic field vector be orthogonal to the segments $y^2 = x^2$:

$$B_y(x, \pm x) = \mp B_x(x, \pm x) \quad \text{for } \frac{a}{2} < x < \frac{b}{2}. \tag{3.27}$$

Next, we propose a set of trial functions for B_x and B_y that meet the above boundary conditions as well as the symmetry requirement that B_x change sign at $y = 0$:

$$B_x(x, y) = \frac{-\mu_0 I}{8} y \left[\frac{1}{x^2} + \alpha \frac{x^2 - y^2}{a^4} \right], \tag{3.28}$$

$$B_y(x, y) = \frac{\mu_0 I}{8x}, \tag{3.29}$$

if (x, y) lies inside the trapezoid $a/2 < x < b/2$, $|y| \leqslant x$ and $B_x = B_y = 0$ elsewhere. The parameter α is a variational parameter that will be chosen such that the action functional attains a minimum with respect to the class of trial functions generated by Eqs. (3.28) and (3.29). Since no dynamics is involved in the present problem, the time integral occurring in the action integral becomes irrelevant and the least action principle amounts to the minimization of the magnetic energy stored in the insulator.

We may calculate the inductance L of an electric circuit by equating $1/2LI^2$ to the magnetic energy stored in the circuit:

$$\begin{aligned}
\frac{1}{2} L I^2 = U_M &= \frac{1}{2\mu_0} \int_\Omega d\tau \, |\mathbf{B}|^2 \\
&= \frac{4l}{2\mu_0} \int_{a/2}^{b/2} dx \int_{-x}^{x} dy \left[B_x^2(x, y) + B_y^2(x, y) \right],
\end{aligned} \tag{3.30}$$

where Ω refers to the volume of the insulator and l is the length of the cable and the prefactor 4 accounts for the identical contributions from the four identical trapezoidal areas. From Eq. (3.30) we obtain the following expression for $\mathcal{L}$, the inductance per unit length:

$$\mathcal{L} \equiv \frac{L}{l} = \frac{4}{\mu_0 I^2} \int_{a/2}^{b/2} dx \int_{-x}^{x} dy \left[B_x^2(x, y) + B_y^2(x, y) \right]. \tag{3.31}$$

Since the trial functions defined in Eqs. (3.28) and (3.29) are chosen to meet the boundary conditions, the variational problem is reduced to the minimization of U_M, or equivalently, $\mathcal{L}$ with respect to α. The calculation of $\mathcal{L}(\alpha)$ is elementary and here we only quote the final result:

$$\frac{\mathcal{L}(\alpha)}{\mu_0} = \frac{1}{6} \log u + \frac{(u^4 - 1)}{215040} \left[112\alpha + (u^4 + 1)\alpha^2 \right] \tag{3.32}$$

W. *Magnus* and W. *Schoenmaker*

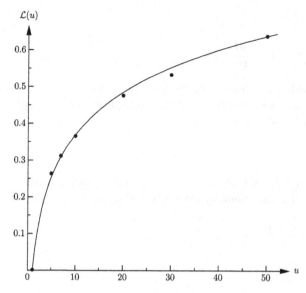

FIG. 3.3. Inductance per unit length: variational estimate (full line) versus numerical evaluation (•).

with $u = b/a$. Clearly, the required minimum corresponding to $\partial \mathcal{L}(\alpha)/\partial \alpha = 0$, is obtained for

$$\alpha = -\frac{56}{1 + u^4}. \tag{3.33}$$

Finally, inserting the above result into Eq. (3.31), we obtain the inductance per length as follows:

$$\frac{\mathcal{L}}{\mu_0} = \frac{1}{6} \log u - \frac{7}{480} \frac{(u^4 - 1)}{(u^4 + 1)}. \tag{3.34}$$

The variational result is plotted against the "exact" numerical evaluation of the inductance in Fig. 3.3. Being a variational estimate, Eq. (3.34) provides a rigorous upper bound for the true inductance.

4. The macroscopic Maxwell equations

4.1. Constitutive equations

The Maxwell equations contain source terms being the charge densities and the currents. In this section we will present the physics behind these terms and derive their precise form. We will see that the charge and current formulas depend very much on the medium in which these charges and currents are present. For solid media we can distinguish between insulators, semiconductors, and conductors. The corresponding expressions differ considerably for the different materials. Furthermore in the gas phase or the liquid phase again other expressions will be found. In the latter case we enter the realm of plasma physics and magnetohydrodynamics. These topics are beyond the present scope.

Before starting to derive the constitutive equations we need to address another machinery, namely statistical physics. From a philosophical point of view, statistical physics is a remarkable part of natural science. It does not contribute to a deeper understanding of the fundamental forces of nature, yet it introduces a fundamental constant of nature, the Boltzmann constant $k_B = 1.3805 \times 10^{23}$ J/K. Furthermore, there has been a discussion over several generations of physicists, debating the reality of irreversibility. The dispute in a nutshell is whether the idea of entropy increase is a sensible one, considering the fact that the microscopic dynamics is time-reversal invariant. As has been demonstrated in MAGNUS and SCHOENMAKER [1993] the time reversal invariance is broken in the limit of infinitely many degrees of freedom. In practice, "infinity" is already reached for 30 degrees of freedom in the study of MAGNUS and SCHOENMAKER [1993]. Therefore, we believe that the dispute is settled and statistical physics is "solid as a rock."

4.2. Boltzmann transport equation

In this section we will consider the assumptions that lead to the Boltzmann transport equation. This equation serves as the starting point for deriving the formules for the constitutive equation for the currents in metals, semiconductors, and insulators.

When describing the temporal evolution of many particles, one it not interested in the detailed trajectory of each individual particle in space and time. First of all, the particles are identical and therefore their trajectories are interchangeable. Secondly, the individual trajectories exhibit stochastic motion on a short time scale that is irrelevant on a larger time scale. In a similar way, the detailed knowledge at a short length scale is also not of interest for understanding the behavior at larger length scales. Thus we must obtain a procedure for eliminating the short-distance fluctuations from the description of the many particle system. In fact, to arrive at a manageable set of equations such a procedure should also reduce the number of variables for which the evolution equations need to be formulated.

There are a number of schemes that allow for such a reduction. All methods apply some kind of coarse graining, i.e., a number of microscopic variables are bundled and are represented by a single effective variable. In this section, we discuss the method that is due to Boltzmann and that leads to the Boltzmann transport equation.

Consider N particles with generalized coordinates $\mathbf{q}_i$, $i = 1, \ldots, N$, and generalized momenta $\mathbf{p}_i$, $i = 1, \ldots, N$. Each particle can be viewed as a point of the so-called μ-space, a six-dimensional space, spanned by the coordinates $\mathbf{q}, \mathbf{p}$. In this light, the N particles will trace out N curves in phase space as time evolves. Let us now subdivide the phase space into cells of size $\Delta\Omega = \Delta q^3 \Delta p^3$. Each cell can be labeled by a pair of coordinates $\mathbf{Q}_i$ and momenta $\mathbf{P}_i$. The number of particles that is found in the cell Ω_i is given by $f(\mathbf{P}_i, \mathbf{Q}_i, t)$. We can illustrate the role of the cell size setting $\Delta\Omega$. The function $f(\mathbf{P}_i, \mathbf{Q}_i, t)$ is given by

$$f(\mathbf{P}_i, \mathbf{Q}_i, t) = \sum_{i=1}^{N} \int_{\Delta\Omega} \mathrm{d}^3 p\, \mathrm{d}^3 q\, \delta(\mathbf{p} - \mathbf{p}_i(t)) \delta(\mathbf{q} - \mathbf{q}_i(t)). \qquad (4.1)$$

We can illustrate the role of the coarse-graining scaling parameter $\Delta\Omega$. If we take the size of the cell arbitrary small then we will occasionally find a particle in the cell. Such a choice of $\Delta\Omega$ corresponds to a fully microscopic description of the mechanical system and we will not achieve a reduction in degrees of freedom.

On the other hand, if we choose $\Delta\Omega$ arbitrary large, then all degrees of freedom are represented by one (static) point f, and we have lost all knowledge of the system. Therefore $\Delta\Omega$ must be chosen such that it acts as the "communicator" between the microscopic and macroscopic worlds. This connection can be obtained by setting the size of the cell large enough such that each cell contains a number of particles. Within each cell the particles are considered to be in a state of thermal equilibrium. Thus for each cell a temperature T_i and a chemical potential μ_i can be given. The (local) thermal equilibrium is realized if there occurs a thermalization, i.e., within the cell collisions should occur within a time interval Δt. Therefore, the cell should be chosen such that its size exceeds at least a few mean-free path lengths.

On the macroscopic scale, the cell labels $\mathbf{P}_i$ and $\mathbf{Q}_i$ are smooth variables. The cell size is the denoted by the differential $\mathrm{d}\Omega = \mathrm{d}^3 p\, \mathrm{d}^3 q$. Then we may denote the distribution functions as $f(\mathbf{P}, \mathbf{Q}, t) \equiv f(\mathbf{p}, \mathbf{q}, t)$. From the distribution function $f(\mathbf{p}, \mathbf{q}, t)$, the particle density function can be obtained from

$$\int \mathrm{d}^3 p\, f(\mathbf{p}, \mathbf{q}, t) = \rho(\mathbf{q}, t). \qquad (4.2)$$

As time progresses from t to $t + \delta t$, all particles in a cell at $\mathbf{p}, \mathbf{q}$ will be found in a cell at $\mathbf{p}', \mathbf{q}'$, provided that no collisions occurred. Hence

$$f(\mathbf{p}, \mathbf{q}, t)\,\mathrm{d}^3 p\, \mathrm{d}^3 q = f(\mathbf{p} + \mathbf{F}\delta t, \mathbf{q} + \mathbf{v}\delta t, t + \delta t)\,\mathrm{d}^3 p'\, \mathrm{d}^3 q'. \qquad (4.3)$$

According to Liouville's theorem (FOWLER [1936], HUANG [1963]), the two volume elements $\mathrm{d}^3 p\, \mathrm{d}^3 q$ and $\mathrm{d}^3 p'\, \mathrm{d}^3 q'$ are equal, which may appear evident if there are no external forces. If there are forces that do not explicitly depend on time, any cubic element deforms into a parallelepiped but with the same volume as the original cube. Taking also into account the effect of collisions that may kick particles in or out of the cube in the time interval δt, we arrive at the following equation for the distribution function

$$\left(\frac{\partial}{\partial t} + \frac{\mathbf{p}}{m} \cdot \nabla_{\mathbf{q}} + \mathbf{F} \cdot \nabla_{\mathbf{p}} \right) f(\mathbf{p}, \mathbf{q}, t) = \left(\frac{\partial f}{\partial t} \right)_{\mathrm{c}}, \qquad (4.4)$$

where the "collision term" $(\partial f / \partial t)_{\mathrm{c}}$ *defines* the effects of scattering. A quantitative estimate of this term is provided by studying the physical mechanisms that contribute to this term. As carriers traverse, their motion is frequently disturbed by scattering due to collisions with impurity atoms, phonons, crystal defects, other carriers, or even with foreign particles (cosmic rays). The frequency at which such events occur can be estimated by assuming that these events take place in an uncorrelated way; in other words two such events are statistically independent. Each physical mechanism is described by an interaction Hamiltonian or potential function, $U_S(\mathbf{r})$ that describes the details of the scattering process. The matrix element that describes the transition from a carrier in a state with momentum $|\mathbf{p}\rangle$ to a state with momentum $|\mathbf{p}'\rangle$ is

$$H_{\mathbf{p}'\mathbf{p}} = \frac{1}{\Omega} \int d\tau\, e^{-\frac{i}{\hbar}\mathbf{p}'\cdot\mathbf{r}} U_S(\mathbf{r}) e^{\frac{i}{\hbar}\mathbf{p}\cdot\mathbf{r}}, \tag{4.5}$$

where Ω is a box that is used to count the number of momentum states. This box is of the size Δq^3 as defined above.

The evaluation of the transition amplitude relies on Fermi's Golden Rule. The transition rate then becomes

$$S(\mathbf{p}', \mathbf{p}) = \frac{2\pi}{\hbar} |H_{\mathbf{p}'\mathbf{p}}|^2 \delta\big(E(\mathbf{p}') - E(\mathbf{p}) - \Delta E\big), \tag{4.6}$$

where ΔE is the change in energy related to the transition. If $\Delta E = 0$, the collision is *elastic* The collision term is the result of the balance between kick-in and kick-out of the transitions that take place per unit time:

$$\left(\frac{\partial f}{\partial t}\right)_c = \sum_{\mathbf{p}'} \big(S(\mathbf{p}', \mathbf{p}) f(\mathbf{q}, \mathbf{p}', t) - S(\mathbf{p}, \mathbf{p}') f(\mathbf{q}, \mathbf{p}, t)\big). \tag{4.7}$$

Once more it should be emphasized that although this balance picture is heuristic, looks reasonable and leads to a description of irreversibility it does not explain the latter. The collision term can be further fine-tuned to mimic the consequences of Pauli's exclusion principle by suppression of multiple occupation of states:

$$\left(\frac{\partial f}{\partial t}\right)_c = \sum_{\mathbf{p}'} \big[S(\mathbf{p}', \mathbf{p}) f(\mathbf{q}, \mathbf{p}', t)\big(1 - f(\mathbf{q}, \mathbf{p}, t)\big)$$
$$- S(\mathbf{p}', \mathbf{p}) f(\mathbf{q}, \mathbf{p}, t)\big(1 - f(\mathbf{q}, \mathbf{p}', t)\big)\big]. \tag{4.8}$$

4.3. Currents in metals

In many materials, the conduction current that flows due to the presence of an electric field, $\mathbf{E}$, is proportional to $\mathbf{E}$, so that

$$\mathbf{J} = \sigma \mathbf{E}, \tag{4.9}$$

where the electrical conductivity σ is a material parameter. In metallic materials, Ohm's law, Eq. (4.9) is accurate. However, a fast generalization should be allowed for anisotropic conducting media. Moreover, the conductivity may depend on the frequency mode such that we arrive at

$$\mathbf{J}_i(\omega) = \sigma_{ij}(\omega) \mathbf{E}_j(\omega) \tag{4.10}$$

and σ is a second-rank tensor. The derivation of Ohm's law from the Boltzmann transport equation was initiated by Drude. In Drude's model (DRUDE [1900a], DRUDE [1900b]), the electrons move as independent particles in the metallic region suffering from scattering during their travel from the cathode to the anode. The distribution function is assumed to be of the following form:

$$f(\mathbf{q}, \mathbf{p}, t) = f_0(\mathbf{q}, \mathbf{p}, t) + f_A(\mathbf{q}, \mathbf{p}, t), \tag{4.11}$$

where f_0 is the equilibrium distribution function, being symmetric in the momentum variable $\mathbf{p}$, and f_A is a perturbation due to an external field that is antisymmetric in the momentum variable. The collision term in Drude's model is crudely approximated by the following assumptions:

- only kick-out,
- all $S(\mathbf{p}, \mathbf{p}')$ are equal,
- no Pauli exclusion principle,
- no carrier heating, i.e., low-field transitions.

The last assumption implies that only the antisymmetric part participates in the collision term (LUNDSTROM [1999]). Defining a characteristic time $\tau_\mathbf{p}$, the momentum-relaxation time, we find that

$$\left(\frac{\partial f}{\partial t}\right)_c = -\frac{f_A}{\tau_\mathbf{p}} \quad \text{and} \quad \frac{1}{\tau_\mathbf{p}} = \sum_{\mathbf{p}'} S(\mathbf{p}, \mathbf{p}'). \tag{4.12}$$

Furthermore, assuming a constant electric field $\mathbf{E}$ and a spatially uniform charge electron distribution, the Boltzmann transport equation becomes

$$-q\mathbf{E} \cdot \nabla(f_0 + f_A) = -\frac{f_A}{\tau_\mathbf{p}}. \tag{4.13}$$

Finally, if we assume that $f \simeq f_0 \propto \exp(-p^2/2mk_B T)$ then

$$f_A = q\tau_\mathbf{p}\mathbf{E} \cdot \nabla_\mathbf{p} f_0 = \frac{q\tau_\mathbf{p}}{k_B T}\mathbf{E} \cdot \mathbf{v} f_0. \tag{4.14}$$

Another way of looking at this result is to consider $f = f_0 + f_A$ as a Taylor series for f_0:

$$f(\mathbf{p}) = f_0(\mathbf{p}) + (q\tau_\mathbf{p}\mathbf{E}) \cdot \nabla_\mathbf{p} f_0(\mathbf{p}) + \cdots = f_0(\mathbf{p} + q\tau_\mathbf{p}\mathbf{E}). \tag{4.15}$$

This is a *displaced* Maxwellian distribution function in the direction opposite to the applied field $\mathbf{E}$. The current density is $\mathbf{J} = qn\mathbf{v}$ follows from the averaged velocity

$$\mathbf{J} = qn\frac{\int d^3 p\,(\mathbf{p}/m) f(\mathbf{p})}{\int d^3 p\, f(\mathbf{p})} = \frac{q^2\tau_\mathbf{p}}{m}n\mathbf{E}. \tag{4.16}$$

The electron mobility, μ_n, is defined as the proportionality constant in the constitutive relation $\mathbf{J} = q\mu_n n\mathbf{E}$, such that

$$\mu_n = \frac{q\tau_\mathbf{p}}{m}. \tag{4.17}$$

So we have been able to "deduce" Ohm's law from the Boltzmann transport equation.

It is a remarkable fact that Drude's model is quite accurate, given the fact that no reference was made to Pauli's exclusion principle and the electron waves do not scatter while traveling in a perfect crystal lattice. Indeed, it was recognized by Sommerfeld that ignoring these effects will give rise to errors in the calculation of the order of 10^2, but both these errors cancel. Whereas Drude's model explains the existence of resistance, more advanced models are needed to accommodate for the nonlinear current-voltage characteristics, the frequency dependence, and the anisotropy of the conductance for some materials. A "modern" approach to derive conductance properties was initiated by KUBO [1957]. His theory naturally leads to the inclusion of anisotropy, nonlinearity, and frequency dependence. Kubo's approach also serves as the starting point to calculate transport properties in the quantum theory of many particles at finite temperature (MAHAN [1981]). These approaches start from the quantum-Liouville equation and the Gibb's theory of assembles on phase space. The latter has a more transparent generalization to the many-particle Hilbert space of quantum states.

Instead of reproducing here text book presentations of these various domains of physics, we intend to give the reader some sense of alertness, that the validity of some relations is limited. In order to push back the restrictions, one needs to re-examine the causes of the limitations. Improved models can be *guessed* by widening the defining expression as in the foregoing case where the scalar σ was substituted by the conductivity tensor $\boldsymbol{\sigma}$. The consequences of these guesses can be tested in simulation experiments. Therefore, simulation plays an important role to obtain improved models.

In the process of purchasing model improvements a few guidelines will be of help. First of all, the resulting theory should respect some fundamental physical principles. The *causality* principle is an important example. It states that there is a retarded temporal relation between cause and effect. The causality principle is a key ingredient to derive the Kramers–Kronig relations, that put severe limitations on the real and imaginary parts of the material parameters. Yet these relationships are not sufficient to determine the models completely, but one needs to include additional physical models.

4.4. Charges in metals

Metallic materials are characterized as having an appreciable conductivity. Any excess free charge distribution in the metal will decay exponentially to zero in a small time. Combining Gauss' law with the current continuity equation

$$\nabla \cdot (\varepsilon \mathbf{E}) = \rho, \qquad \nabla \cdot (\sigma \mathbf{E}) = \frac{\partial \rho}{\partial t} \tag{4.18}$$

and considering ε and σ constant, we find

$$\frac{\partial \rho}{\partial t} = -\frac{\sigma}{\varepsilon}\rho, \qquad \rho = \rho_0 \exp\left(-\frac{\sigma}{\varepsilon}t\right). \tag{4.19}$$

In metallic materials, the decay time $\tau = \varepsilon/\sigma$ is of the order of 10^{-18} s, such that $\rho = 0$ at any instant.

For conducting materials one usually assumes $\nabla \cdot \mathbf{D} = 0$ and for constant ε and ρ, the electric field $\mathbf{E}$ and current density $\mathbf{J}$ are constant (COLLIN [1960]). A subtlety arises when ε and ρ are varying in space. Considering the steady-state version of above set of equations, we obtain

$$\nabla \cdot (\varepsilon \mathbf{E}) = \rho, \qquad \nabla \cdot (\sigma \mathbf{E}) = 0. \tag{4.20}$$

The field **E** should simultaneously obey two equations. Posed as a boundary-value problem for the scalar potential, V, we may determine V from the second equation and determine ρ as a "postprocessing" result originating from the first equation.

4.5. Semiconductors

Intrinsic semiconductors are insulators at zero temperature. This is because the band structure of semiconductors consists of bands that are either filled or empty. At zero temperature, the chemical potential falls between the highest filled band which is called the valence band and the lowest empty band which is named the conduction band. The separation of the valance and conduction band is sufficiently small such that at some temperature, there is an appreciable amount of electrons that have an energy above the conduction band onset. As a consequence these electron are mobile and will contribute to the current if a voltage drop is put over the semiconducting material. The holes in the valance band act as positive charges with positive effective mass and therefore they also contribute to the net current. Intrinsic semiconductors are rather poor conductors but their resistance is very sensitive to the temperature ($\sim \exp(-A/T)$). By adding dopants to the intrinsic semiconductor, the chemical potential of the electrons and holes may by shifted up or down with respect to the band edges. Before going into further descriptions of dopant distributions, we would like to emphasize the following fact: *Each thermodynamic system in thermal equilibrium has constant intensive conjugated variables.* In particular, the temperature, T, conjugated to the internal energy of the system and the chemical potential, μ, conjugated to the number of particles in the systems are constant for a system in equilibrium. Therefore, if the dopant distribution varies in the device and the distance between the chemical potential and the band edges is modulated, then for the device being in equilibrium, the band edges must vary in accordance with the dopant variations, as illustrated in Fig. 4.1.

4.6. Currents in semiconductors

Whereas in metals the high conductivity prevents local charge accumulation at an detectable time scale, the situation in semiconductors is quite different. In uniformly doped semiconductors, the decay of an excess charge spot occurs by a diffusion process, that takes place on much longer time scale. In nonuniformly doped semiconductors,

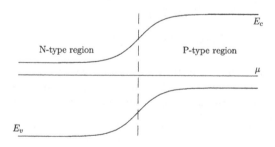

FIG. 4.1. Band edge modulation by doping.

there are depletion layers, or accumulation layers of charges that permanently exists even in thermal equilibrium.

The charge and current densities in semiconductors follow also from the general Boltzmann transport theory, but this theory needs to be complemented with specific details such as the band gap, the dopant distribution, and the properties related to the interfaces to other materials.

Starting from the Boltzmann transport equation, the *moment expansion* considers variables that are averaged quantities as far as the momentum dependence is concerned. The generic expression for the moment expansion is

$$\frac{1}{\Omega}\sum_{\mathbf{p}}Q(\mathbf{p})\left(\frac{\partial}{\partial t}+\frac{\mathbf{p}}{m}\cdot\nabla_{\mathbf{q}}+\mathbf{F}\cdot\nabla_{\mathbf{p}}\right)f(\mathbf{p},\mathbf{q},t)=\frac{1}{\Omega}\sum_{\mathbf{p}}Q(\mathbf{p})\left(\frac{\partial f}{\partial t}\right)_{c},\quad(4.21)$$

where $Q(\mathbf{p})$ is an polynomial in the components of $\mathbf{p}$ and the normalization $1/\Omega$ allows for a smooth transition to integrate over all momentum states in the Brillouin zone

$$\frac{1}{\Omega}\sum_{\mathbf{p}}\rightarrow\frac{1}{4\pi^3}\int_{BZ}d^3k.\qquad(4.22)$$

The zeroth order expansion gives (LUNDSTROM [1999])

$$\frac{\partial n}{\partial t}-\frac{1}{q}\nabla\cdot\mathbf{J}_n=-U,$$
$$\frac{\partial p}{\partial t}+\frac{1}{q}\nabla\cdot\mathbf{J}_p=U\qquad(4.23)$$

and where the various variables are:

electrons | *holes*

$$n(\mathbf{r},t)=\frac{1}{\Omega}\sum_{\mathbf{p}}f_n(\mathbf{p},\mathbf{r},t),\qquad p(\mathbf{r},t)=\frac{1}{\Omega}\sum_{\mathbf{p}}f_p(\mathbf{p},\mathbf{r},t),$$

$$\mathbf{J}_n(\mathbf{r},t)=-qn(\mathbf{r},t)\mathbf{v}_n(\mathbf{r},t),\qquad \mathbf{J}_p(\mathbf{r},t)=qp(\mathbf{r},t)\mathbf{v}_p(\mathbf{r},t),\qquad(4.24)$$

$$\mathbf{v}_n(\mathbf{r},t)=\frac{1}{\Omega}\sum_{\mathbf{p}}\frac{\mathbf{p}}{m}f_n(\mathbf{p},\mathbf{r},t),\qquad \mathbf{v}_p(\mathbf{r},t)=\frac{1}{\Omega}\sum_{\mathbf{p}}\frac{\mathbf{p}}{m}f_p(\mathbf{p},\mathbf{r},t)$$

and

$$U=\frac{1}{\Omega}\sum_{\mathbf{p}}\left(\frac{\partial f}{\partial t}\right)_{c}=R-G.\qquad(4.25)$$

The particle velocities give an expression for the current densities but by choosing $Q(\mathbf{p})=\mathbf{p}$, we obtain the first moment of the expansion that can be further approximated to give alternative expressions for the current densities. Defining the momentum relaxation time τ_p as a characteristic time for the momentum to reach thermal equilibrium from a nonequilibrium state and the electron and hole temperature tensors (FORGHIERI,

GUERRRI, CIAMPOLINI, GNUDI, and RUDAN [1988])

$$\frac{1}{2} n k_B T_{n,ij}(\mathbf{r}, t) = \frac{1}{\Omega} \sum_{\mathbf{p}} \frac{1}{2m} (p_i - m v_{n,i})(p_j - m v_{n,j}) f_n(\mathbf{p}, \mathbf{r}, t)$$

$$= \frac{1}{2} n k_B T_n(\mathbf{r}, t) \delta_{ij},$$

$$\frac{1}{2} p k_B T_{p,ij}(\mathbf{r}, t) = \frac{1}{\Omega} \sum_{\mathbf{p}} \frac{1}{2m} (p_i - m v_{p,i})(p_j - m v_{p,j}) f_p(\mathbf{p}, \mathbf{r}, t)$$

$$= \frac{1}{2} p k_B T_p(\mathbf{r}, t) \delta_{ij},$$

(4.26)

where the last equality follows from assuming an isotropic behavior, then one arrives at the following constitutive equation for the currents in semiconducting materials

$$\mathbf{J}_n + n \tau_{pn} \frac{d}{dt} \left(\frac{\mathbf{J}_n}{n} \right) = q \mu_n n \left(\mathbf{E} + \frac{k_B}{q} \nabla T_n \right) + q D_n \nabla n,$$

$$\mathbf{J}_p + p \tau_{pp} \frac{d}{dt} \left(\frac{\mathbf{J}_p}{p} \right) = q \mu_p p \left(\mathbf{E} - \frac{k_B}{q} \nabla T_p \right) - q D_p \nabla p.$$

(4.27)

The momentum relaxation times, the electron and hole mobilities, and the electron and hole diffusivities are related through the Einstein relations

$$D = \frac{k_B T}{q} \mu = \frac{k_B T}{m} \tau.$$

(4.28)

The second terms on the left-hand sides of Eq. (4.27) are the *convective currents*. The procedure of taking moments of the Boltzmann transport equation always involves a truncation, i.e., the nth order equation in the expansion demands information of the $(n + 1)$th order moment to be supplied. For the second-order moment, one thus needs to provide information on the third moment

$$\frac{1}{\Omega} \sum_{\mathbf{p}} p_i p_j p_k f(\mathbf{p}, \mathbf{r}, t).$$

(4.29)

In the above scheme the second-order expansion leads to the *hydrodynamic model* (FORGHIERI, GUERRRI, CIAMPOLINI, GNUDI, and RUDAN [1988]). In this model the carrier temperatures are determined self-consistently with the carrier densities. The closure of the system of equations is achieved by assuming a model for the term (4.29) that only contains lower order variables. The thermal flux Q, being the energy that gets transported through thermal conductance can be expressed as

$$\mathbf{Q} = \frac{1}{\Omega} \sum_{\mathbf{p}} \frac{1}{2m} |\mathbf{p} - m\mathbf{v}|^2 \left(\frac{\mathbf{p}}{m} - \mathbf{v} \right) = -\kappa \nabla T,$$

(4.30)

where $\kappa = \kappa_n, \kappa_p$ are the thermal conductivities.

Besides the momentum flux, a balance equation is obtained for the energy flux:

$$\frac{\partial(nw_n)}{\partial t} + \nabla \cdot \mathbf{S}_n = \mathbf{E} \cdot \mathbf{J}_n + n\left(\frac{\partial w_n}{\partial t}\right)_c,$$

$$\frac{\partial(pw_p)}{\partial t} + \nabla \cdot \mathbf{S}_p = \mathbf{E} \cdot \mathbf{J}_p + p\left(\frac{\partial w_p}{\partial t}\right)_c. \tag{4.31}$$

The energy flux is denoted as $\mathbf{S}$ and w is the energy density. In the isotropic approximation, the latter reads

$$w_n = \frac{3}{2}k_B T_n + \frac{1}{2}m_n v_n^2, \qquad w_p = \frac{3}{2}k_B T_p + \frac{1}{2}m_p v_p^2. \tag{4.32}$$

The energy flux can be further specified as

$$\mathbf{S}_n = \kappa_n \nabla T_n - (w_n + k_B T_n)\frac{\mathbf{J}_n}{q},$$

$$\mathbf{S}_p = \kappa_p \nabla T_p + (w_p + k_B T_p)\frac{\mathbf{J}_p}{q}. \tag{4.33}$$

Just as for the momentum, one usually assumes a characteristic time, τ_e, for a non-equilibrium energy distribution to relax to equilibrium. Then the collision term in the energy balance equation becomes

$$n\left(\frac{\partial w_n}{\partial t}\right)_c = -n\frac{w_n - w^*}{\tau_{en}} - Uw_n,$$

$$p\left(\frac{\partial w_p}{\partial t}\right)_c = -p\frac{w_p - w^*}{\tau_{ep}} - Uw_p \tag{4.34}$$

and w^* is the carrier mean energy at the lattice temperature. In order to complete the hydrodynamic model the thermal conductivities are given by the Wiedemann–Franz law for thermal conductivity

$$\kappa = \left(\frac{k_B}{q}\right)^2 T\sigma(T)\Delta(T). \tag{4.35}$$

Herein is $\Delta(T)$ a value obtained from evaluating the steady-state Boltzmann transport equation for uniform electric fields and $\sigma(T) = q\mu c$ the electrical conductivity ($c = n, p$). If a power-law dependence for the energy relaxation times can be assumed, i.e.,

$$\tau_e = \tau_0 \left(\frac{w}{k_B T^*}\right)^\nu, \tag{4.36}$$

then $\Delta(T) = 5/2 + \nu$. Occasionally, ν is considered to be a constant ($\nu = 0.5$). However, this results into too restrictive an expression for the $\tau_e(w)$. Therefore $\Delta(T)$ is often tuned towards Monte-Carlo data.

Comparing the present elaboration on deriving constitutive equations from the Boltzmann transport equation with the derivation of the currents in metals we note that we did not refer to a displaced Maxwellian distribution. Such a derivation is also possible for semiconductor currents. The method was used by STRATTON [1962]. A difference pops up in the diffusion term of the carrier current. For the above results we obtained

$$\mathbf{J}(\text{diffusive part}) \propto \mu \nabla T. \tag{4.37}$$

In Stratton's model one obtains

$$\mathbf{J}(\text{diffusive part}) \propto \nabla (\mu T), \tag{4.38}$$

the difference being a term

$$\xi = \frac{\partial \log \mu(T)}{\partial \log(T)}. \tag{4.39}$$

Stratton's model is usually referred to as the *energy transport* model.

For the semiconductor environment, the Scharfetter–Gummel scheme provides a means to discretize the current equations on a grid (SCHARFETTER and GUMMEL [1969]). In the case that no carrier heating effects are considered (T is constant) the diffusion equations are

$$\mathbf{J} = q \mu c \mathbf{E} \pm k T \mu \nabla c, \tag{4.40}$$

where the plus (minus) sign refers to negatively (positively) charged particles and c denotes the corresponding carrier density. It is assumed that both the current $\mathbf{J}$ and the electric field $\mathbf{E}$ are constant along a link and that the potential V varies linearly along the link. Adopting a local coordinate axis u with $u = 0$ corresponding to node i, and $u = h_{ij}$ corresponding to node j, we may integrate Eq. (4.40) along the link ij to obtain

$$J_{ij} = q \mu_{ij} c \left(\frac{V_i - V_j}{h_{ij}} \right) \pm k T \mu_{ij} \frac{dc}{du}, \tag{4.41}$$

which is a first-order differential equation in c. The latter is solved using the aforementioned boundary conditions and gives rise to a nonlinear carrier profile. The current J_{ij} can then be rewritten as

$$\frac{J_{ij}}{\mu_{ij}} = -\frac{\alpha}{h_{ij}} B \left(\frac{-\beta_{ij}}{\alpha} \right) c_i + \frac{\alpha}{h_{ij}} B \left(\frac{\beta_{ij}}{\alpha} \right) c_j, \tag{4.42}$$

using the Bernoulli function

$$B(x) = \frac{x}{e^x - 1}. \tag{4.43}$$

Furthermore, we used $\alpha = \pm k T$ and $\beta_{ij} = q(V_i - V_j)$.

Before turning to the consideration of insulating materials, we briefly discuss the influence of strong magnetic fields on the currents. These fields will bend the trajectories due to the Lorentz force. In the derivation of the macroscopic current densities from the Boltzmann transport equation, we should include this force. The result is that in the constitutive current expression we must make the replacement: $\mathbf{E} \to \mathbf{E} + q \mathbf{v} \times \mathbf{B}$. Since $\mathbf{J} = q c \mathbf{v}$, we arrive at the following *implicit* relation for $\mathbf{J}$:

$$\mathbf{J} = \sigma \mathbf{E} + \mu \mathbf{J} \times \mathbf{B}, \tag{4.44}$$

where $\sigma = q\mu c$ is the conductivity and μ is the mobility. This relation can be made *explicit* by solving the following set of linear equations:

$$\begin{bmatrix} 1 & -\mu B_z & \mu B_y \\ \mu B_z & 1 & -\mu B_x \\ -\mu B_y & \mu B_x & 1 \end{bmatrix} \cdot \begin{bmatrix} J_x \\ J_y \\ J_z \end{bmatrix} = \begin{bmatrix} \sigma E_x \\ \sigma E_y \\ \sigma E_x \end{bmatrix} \tag{4.45}$$

of which the solution is:

$$\mathbf{J} = \left[\sigma\mathbf{E} + \mu\sigma\mathbf{E} \times \mathbf{B} + \mu^2\sigma(\mathbf{E} \cdot \mathbf{B})\mathbf{B} \right] / (1 + \mu^2 B^2). \tag{4.46}$$

Above considerations are required for the description of Hall sensors. Here we will not further elaborate on this extension, nor will we consider the consequences of anisotropic conductivity properties.

4.7. Insulators

So far, we have been rather sloppy in classifying materials as being an insulator, semiconductor, or metal. We have referred to the reader's qualitative awareness of the conduction quality of a material under consideration. For the time being we will sustain in this practice and define insulators as having a negligible conductivity. Therefore, in an insulating material there are no conduction currents. The constitutive equation for $\mathbf{J}$ becomes trivial.

$$\mathbf{J} = \mathbf{0}. \tag{4.47}$$

Recently, there is an increased interest in currents in insulating materials. The gate dielectric material SiO_2 that has been used in mainstream CMOS technology has a band gap of 3.9 eV and therefore acts as a perfect insulator for normal voltage operation conditions around 3 V and using 60 Å thick oxides. However, the continuous down scaling of the transistor architecture requires that the oxides thicknesses are also reduced. With the current device generation (100 nm gate length), the oxide thickness should be less than 20 Å. For these thin layers, direct tunneling through the layer barrier becomes a dominating current leakage in integrated CMOS devices.

4.7.1. Subband states and resonances

A planar p-type silicon metal-insulator-semiconductor (MIS) capacitor consisting of a gate electrode, a gate stack, and a silicon substrate is considered. The gate stack has a thickness T_{ox} ranging from 15 to 40 Å and contains N_{ox} layers of insulating material such as SiO_2, Si_3Ni_5, etc. When a positive gate voltage V_G is applied to the gate electrode, the electrons residing in the electron inversion layer formed near the Si/insulator interface, are coupled to both the gate and the gate stack through nonvanishing tunneling amplitudes. As a result, measurable tunneling currents are observed that involve a net migration of electrons from the leaky inversion layer to the gate electrode.

In this section, we have summarized the approach followed in MAGNUS and SCHOEN-MAKER [2000a] and MAGNUS and SCHOENMAKER [2002] to calculate these tunneling currents.

W. Magnus and W. Schoenmaker

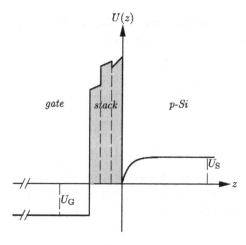

FIG. 4.2. Conduction band profile of a MIS capacitor. (Figure reproduced by permission of the Americal Institute of Physics and Springer Verlag.)

The z-axis is chosen to be perpendicular to the SiO_2-interface that is taken to be the (x, y)-plane. The gate, gate stack, and semiconductor region are defined by $-\infty \leqslant z < t_{ox}$, $-t_{ox} \leqslant z < 0$, and $0 \leqslant z \leqslant +\infty$, respectively, as depicted in Fig. 4.2.

All electron energies including the chemical potential, are measured with respect to the edge of the conduction band at the Si/insulator interface. The potential energy takes a uniform value in the gate region whereas it approaches the limit U_S in the bulk substrate.

The whole MIS capacitor can be treated as a single quantum mechanical entity for which the Schrödinger equation needs to be solved. Adopting the effective mass approximation for the electrons in the different valleys, and the Hartree approximation to describe the electron–electron interaction in the inversion layer, the three-dimensional time-independent Schrödinger equation for the semiconductor region takes the form

$$-\frac{\hbar^2}{2}\left(\frac{1}{m_{\alpha x}}\frac{\partial^2}{\partial x^2} + \frac{1}{m_{\alpha y}}\frac{\partial^2}{\partial y^2} + \frac{1}{m_{\alpha z}}\frac{\partial^2}{\partial z^2}\right)\psi_\alpha(\mathbf{r}, z) + \left[U(z) - E\right]\psi_\alpha(\mathbf{r}, z) = 0,$$
$$(4.48)$$

where $\mathbf{r} = (x, y)$, α is a valley index and $m_{\alpha x}$, $m_{\alpha y}$, and $m_{\alpha z}$ denote the components of the effective mass tensor along the principle directions of the silicon valleys. The same equation applies to the other regions upon insertion of appropriate effective masses. Assuming translational invariance in the lateral directions, one may write each one-electron wave function as a plane wave modulated by a one-dimensional envelope wave function $\phi_\alpha(W, z)$ and the corresponding one-electron eigenenergy $E_{\alpha k}(W)$ as follows:

$$\psi_{\alpha k}(W, \mathbf{r}, z) = \frac{1}{\sqrt{L_x L_y}}e^{i\mathbf{k}\cdot\mathbf{r}}\phi_\alpha(W, z),$$

$$E_{\alpha k}(W) = \frac{\hbar^2}{2}\left(\frac{k_x^2}{m_{\alpha x}} + \frac{k_y^2}{m_{\alpha y}}\right) + W,$$
$$(4.49)$$

where $\mathbf{k} = (k_x, k_y)$ and $\phi_\alpha(W, z)$ is an eigenfunction of the one-dimensional Schrödinger equation

$$-\frac{\hbar^2}{2m_{\alpha z}} \frac{\mathrm{d}^2 \phi_\alpha(W, z)}{\mathrm{d}z^2} + \left[U(z) - W\right]\phi_\alpha(W, z) = 0 \tag{4.50}$$

corresponding to the energy eigenvalue W.

Since the size of the whole system is assumed to be large in all directions, the energy spectrum will be dense and in particular the eigenvalues W can take all real values exceeding U_G. Moreover, the complete set of wave functions solving Eq. (4.50) constitutes an orthogonal, continuous basis for which a proper delta-normalization is invoked:

$$\langle \phi_\alpha(W') | \phi_\alpha(W) \rangle \equiv \int_\infty^\infty \mathrm{d}z \, \phi_\alpha^*(W, z)\phi_\alpha(W, z) = \delta(W' - W). \tag{4.51}$$

Although the insulating layers are relatively thin, the energy barriers separating the inversion layer from the gate electrode are generally high enough to prevent a flood of electrons leaking away into the gate. In other words, in most cases of interest the potential well, hosting the majority of inversion layer electrons, will be coupled only weakly to the gate region. It follows from ordinary quantum mechanics (FLUEGGE [1974]) that the relative probability of finding an electron in the inversion layer well should exhibit sharply peaked maxima for a discrete set of W-values. The latter are the resonant energies corresponding to a set of virtually bound states, also called quasi-bound states, that may be regarded as the subband states of the coupled system. This becomes intuitively clear when the thickness of the barrier region is arbitrarily increased so that the coupling between the gate electrode and the semiconductor region vanishes. In this limiting case, the resonant energies will coincide with the true subband energies of the isolated potential well while the resonant wave functions drop to zero at the interface plane $z = 0$. Similarly, the spectral widths of the resonant wave functions tend to zero and the resonance peaks turn into genuine delta functions of W.

The above picture provides a way to investigate the subband structure of an inversion layer. By applying a transfer matrix approach to a piecewise constant potential profile and tracing the maxima of the squared wave function amplitudes as a function of W the continuous wave functions can be calculated. Once the sequence of resonant subband energies $\{W_{\alpha l} \mid l = 1, 2, \ldots\}$ and the corresponding wave functions are found, one may analytically determine the spectral widths that are directly related to the second derivative of the wave functions, with respect to W, evaluated at the resonant energies.

Within the Hartree approximation, the potential energy profile $U(z)$ needs to be determined by solving self-consistently the above mentioned Schrödinger equation (4.50) and the one-dimensional Poisson equation

$$\frac{\mathrm{d}^2 U(z)}{\mathrm{d}z^2} = -\frac{e^2}{\varepsilon_S}\left[n(z) - p(z) + N_A(z)\right], \tag{4.52}$$

where $n(z)$, $p(z)$, $N_A(z)$, and ε_S denote, respectively, the electron, hole, and acceptor concentrations and the permittivity in the silicon part of the structure. In the present work we have not treated the occurrence of free charges in the gate and the gate stack. On the other hand, charges trapped by interface states are incorporated through a surface charge density D_{it}.

The potential energy is modeled by a piecewise constant profile defined on a one-dimensional mesh reflecting the gate stack layers and a user-defined number of substrate layers. In this light the self-consistent link between $n(z)$ and $U(z)$ is not provided for each point in the inversion layer but rather for their averages over the subsequent cells of the mesh. This approach is adequate whenever the number of cells is sufficiently large and it has been successfully employed in the past (JOOSTEN, NOTEBORN, and LENSTRA [1990], NOTEBORN, JOOSTEN, LENSTRA, and KASKI [1990]). In the following however, we focus on the procedure to extract the resonant energies and spectral widths.

The solutions to the Schrödinger equation for the layered structure can now compactly be written as linear combinations of u_1 and u_2, being generic basis functions in each cell.

In order to trace the resonance peaks and spectral widths, a numerically stable probability function scanning the presence of an electron in the inversion layer as a function of W, needs to be determined. Rewriting the gate and substrate wave functions as

$$\phi_\alpha(W, z) = \begin{cases} C_{g,\alpha} \sin(k_{g,\alpha}(z + t_{ox}) + \theta_\alpha) & \text{for } z < -t_{ox}, \\ C_{s,\alpha} \exp(-k_{s,\alpha}(z - a)) & \text{for } z > a, \end{cases} \quad (4.53)$$

one obtains the relative probability of an electron for being in the inversion layer:

$$P_\alpha(W) \equiv \left| \frac{C_{s,\alpha}(W)}{C_{g,\alpha}(W)} \right|^2. \quad (4.54)$$

Emerging as resonance energies in the continuous energy spectrum, the subband energies $W_{\alpha l}$ correspond to distinct and sharply peaked maxima of the $P_\alpha(W)$, or well defined minima of $P_\alpha^{-1}(W)$, even for oxide thicknesses as low as 10 Å. As a consequence, expanding $P_\alpha^{-1}(W)$ in a Taylor series around $W = W_{\alpha l}$, we may replace $P_\alpha(W)$ by a sum of Lorentz-shaped functions:

$$P_\alpha(W) \rightarrow \sum_l P_\alpha(W_{\alpha l}) \frac{\Gamma_{\alpha l}^2}{(W - W_{\alpha l})^2 + \Gamma_{\alpha l}^2}, \quad (4.55)$$

where the resonance widths $\Gamma_{\alpha l}^2$ are related to the second derivative of $P_\alpha^{-1}(W)$ through

$$\Gamma_{\alpha l}^2 = 2 P_\alpha^{-1}(W_{\alpha l}) \left[\frac{\partial^2 P_\alpha^{-1}}{\partial W^2}(W_{\alpha l}) \right]^{-1} \quad (4.56)$$

and can be directly extracted from the transmission matrices and their derivatives, evaluated at $W = W_{\alpha l}$.

4.7.2. Tunneling gate currents

The subband structure of a p-type inversion layer channel may be seen to emerge from an enumerable set of sharp resonances appearing in the continuous energy spectrum of the composed system consisting of the gate contact, the gate stack (insulating layers), the inversion layer and the substrate contact. In particular, the discreteness of the subband states is intimately connected with the presence of energy barriers in the gate stack that restrict the coupling between the channel and the gate regions and therefore the amplitude for electrons tunneling through the barriers (see Fig. 4.2). Clearly, the smallness of the above mentioned coupling is reflected in the size of the resonance width – or equivalently, the resonance lifetime $\tau_{\alpha l} = \hbar/2\Gamma_{\alpha l}$ – as compared to the resonance energy.

It is tempting to identify the gate leakage current as a moving ensemble of electrons originating from decaying subband states. However, before such a link can be established, a conceptual problem should be resolved. Although intuition obviously suggests that an electron residing in a particular subband $\rangle \alpha l$ should contribute an amount $-e/\tau_{\alpha l}$ to the gate current, this is apparently contradicted by the observation that *the current density corresponding to each individual subband wave function identically vanishes*. The latter is due to the nature of the resonant states. Contrary to the case of the doubly degenerate running wave states having energies above the bottom of the conduction band in the substrate, the inversion layer resonant states are nondegenerate and virtually bound, and the wave functions are rapidly decaying into the substrate area. As a consequence, all wave functions are real (up to an irrelevant phase factor) and the diagonal matrix elements of the current density operator vanishes. The vanishing of the current for the envelope wave functions was also noted in SUNE, OLIVIO, and RICCO [1991], MAGNUS and SCHOENMAKER [1999]. Therefore, we need to establish a sound physical model (workaround) resolving the current paradox and connecting the resonance lifetimes to the gate current. Since we do not adopt a plane-wave hypothesis for the inversion layer electrons in the perpendicular direction, our resolution of the paradox differs from the one that is proposed in SUNE, OLIVIO, and RICCO [1991].

The paradox can be resolved by noting that the resonant states, though diagonalizing the electron Hamiltonian in the presence of the gate bias, are constituting a *nonequilibrium* state of the whole system which is not necessarily described by a Gibbs-like statistical operator, even not when the steady state is reached. There are at least two alternatives to solve the problem in practice.

The most rigorous approach aims at solving the full time dependent problem starting from a MIS capacitor that is in thermal equilibrium ($V_G = 0$) until some initial time $t = 0$. Before $t = 0$, the potential profile is essentially determined by the gate stack barriers and, due to the absence of an appreciable inversion layer potential well, all eigen solutions of the time independent Schrödinger equation are linear combinations of transmitted and reflected waves. In other words, almost all states are carrying current, although the thermal average is of course zero (equilibrium). However, it should be possible to calculate the time evolution of the creation and annihilation operators related to the unperturbed states. The perturbed resonant states, defining the subband structure for $V_G > 0$, would serve as a set of intermediate states participating in all transitions between the unperturbed states caused by the applied gate voltage. Although such an approach is conceptually straightforward, it is probably rather cumbersome to be carried out in practice.

One may consider a strategy that is borrowed from the theory of nuclear decay (MERZBACHER [1970], LANDAU and LIFSHITZ [1958]). The resulting model leads to

a concise calculation scheme for the gate current. Under the assumption that the resonance widths of the virtual bound states are much smaller than their energies, the corresponding real wave functions can be extended to the complex plane if the resonance
energies and the corresponding resonance widths are combined to form complex energy
eigenvalues of the Schrödinger equation (MAGNUS and SCHOENMAKER [2000a]). Such
an extension enables us to mimic both the supply (creation) and the decay (disinte-
gration) of particles in a resonant bound state by studying the wave functions in those
regions of space where the real, i.e., noncomplex, wave functions would be standing
waves either asymptotically or exactly.

Within the scope of the this work, scattering by phonons, or any other material dependent interactions is neglected. Moreover, electron–electron interaction is treated in the
Hartree approximation that, in practice, amounts to a self-consistent solution of the one-
particle Schrödinger equation and Poisson's equation. Therefore, bearing in mind that
normal transport through the gate stack is limited by tunneling events, the time-reversal
symmetry breaking between decaying and loading states can be inserted through the
boundary conditions for the statistical operator corresponding to the noninteracting
Liouville equation. Consequently, the gate current density is given by

$$J_G = -\frac{e}{\pi \hbar^2 \beta} \sum_{\alpha l} \frac{\sqrt{m_{\alpha x} m_{\alpha y}}}{\tau_{\alpha l}} \log \frac{1 + \exp(\beta(E_F - W_{\alpha l} - eV_G))}{1 + \exp(\beta(E_F - W_{\alpha l}))}. \tag{4.57}$$

It is clear from Eq. (4.57) that the resonance lifetimes are the key quantities building
up the new formula for the gate leakage current. These variables apparently replace the
familiar transmission coefficients that would emerge from traveling states contributing
to the current in accumulation mode. This feature reflects the scope of nuclear decay
theory which is a fair attempt to resolve the leakage current paradox. Although the
latter theory produces a dynamical evolution of the one-particle wave functions, one
can eventually insert a time independent, yet nonequilibrium, statistical operator to calculate the averages. It would be desirable to verify the success of this procedure on
the grounds of sound time-dependent nonequilibrium theory. The same recommenda-
tion can be made regarding a more systematic investigation of the agreement between
the results of the present calculation and the simulations based on Bardeen's approach
(BARDEEN [1961]).

The above considerations have been used to evaluate the gate current numerically (MAGNUS and SCHOENMAKER [2000a], MAGNUS and SCHOENMAKER [2000b]).
In Fig. 4.3 the simulation results are compared with a gate current characteristic that
was obtained from measurements on a large MIS transistor with a NO insulator and
grounded source and drain contacts. The latter serve as huge electron reservoirs capable
of replacing the channel electrons (inversion) that participate in the gate tunneling current, such that the assumption on instantaneous injection or absorption compensating
for migrating electrons is justified.

The following parameters are used: $T = 300$ K, $T_{ox} = 25$ Å, $m_{g\alpha x} = m_{g\alpha y} = m_{g\alpha z} = 0.32m_0$, $N_{ox} = 3$, $m_{1,ox,\alpha} = m_{2,ox,\alpha} = m_{3,ox,\alpha} = 0.42m_0$. The barrier height and the
dielectric constant of the NO layer are taken to be 3.15 and 3.9 eV, respectively, while
the acceptor concentration N_A is 4×10^{17} cm^{-3}.Fig. 4.4 shows typical current-voltage

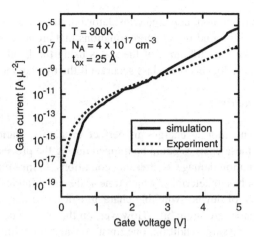

FIG. 4.3. Gate tunneling current vs. gate voltage for a NO layer with thickness of 25 Å. The doping is 4×10^{17} cm^{-3} and $T = 300$ K. (Figure reproduced by permission of The American Institute of Physics and Springer Verlag.)

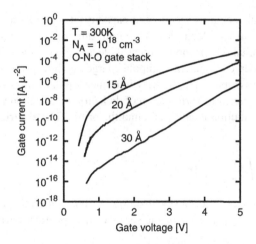

FIG. 4.4. Gate tunneling current vs. gate voltage for a NO layer with thickness of 15, 20, and 30 Å, the substrate doping being 10^{18} cm^{-3}. All other parameters are the same as in Fig. 4.3. (Figure reproduced by permission of The American Institute of Physics and Springer Verlag.)

characteristics for oxide thicknesses of 15, 20, and 30 Å and $N_A = 10^{18}$ cm^{-3}. The simulation results show a good agreement with the experimental data in the range 1–4 V. It should be noted that the results are based on a set of "default" material parameters (BRAR, WILK, and SEABAUGH [1996], DEPAS, VANMEIRHAEGHE, LAFLERE, and CARDON [1994]). In particular for the effective electron mass in SiO$_2$, we used the results from Brar et al. The latter ones were obtained by measurements on *accumulation* layers. We suspect that the overestimation of the gate leakage currents for higher voltages is partly caused by the depletion layer in the poly-crystalline gate region ("poly-depletion") such that a shift in the gate potential at the gate/insulator interface occurs. Another origin of the discrepancy may be found in the approximations that are used in the method. The evaluation of the resonance lifetimes of the states using the Breit–Wigner expansion (BREIT and WIGNER [1936]) becomes less accurate if the overlap increases.

4.8. Charges in insulators

Although there are no mobile charges in perfect insulators, static charges may be present. Physically, these charges could be trapped during the processing of the insulator, or caused by radiation damage or stressing conditions. In the simulation of charges in insulators one first has to determine which time scale one is interested in. On the time scale of the operation of device switching characteristics, one may safely assume that the charges in insulators are immobile. However, on the time scale of the device lifetime or accelerated stressing condition, one must consider tunneling currents and trap generations that definitely can be traced to mobile charges.

4.9. Dielectric media

A dielectric material increases the storage capacity of a condenser or a capacitor by neutralizing charges at the electrodes that would otherwise contribute to the external field. Faraday identified this phenomenon as dielectric polarization. The polarization is caused by a microscopic alignment of dipole charges with respect to the external field. Looking at the macroscopic scale, we may introduce a polarization vector field, $\mathbf{P}$.

In order to give an accurate formulation of dielectric polarization we first consider an arbitrary charge distribution localized around the origin. The electric potential in some point $\mathbf{r}$, is

$$V(\mathbf{r}) = \frac{1}{4\pi\varepsilon_0} \int \frac{\rho(\mathbf{r}')}{|\mathbf{r} - \mathbf{r}'|} \, d\tau'. \tag{4.58}$$

Now let $\mathbf{r}$ be a point outside the localization region of the charge distribution, i.e., $|\mathbf{r}| > |\mathbf{r}'|$. From the completeness of the series of the spherical harmonics, $Y_{lm}(\theta, \phi)$, one obtains

$$\frac{1}{|\mathbf{r} - \mathbf{r}'|} = 4\pi \sum_{l=0}^{\infty} \sum_{m=-l}^{l} \frac{1}{2l+1} \frac{|\mathbf{r}'|^l}{|\mathbf{r}|^{l+1}} Y_{lm}^*(\theta', \phi') Y_{lm}(\theta, \phi), \tag{4.59}$$

where

$$Y_{lm}(\theta, \phi) = \sqrt{\frac{2l+1}{4\pi} \frac{(l-m)!}{(l+m)!}} P_l^m(\cos\theta) e^{im\phi} \tag{4.60}$$

and

$$P_l^m(x) = (-1)^m (1-x^2)^{m/2} \frac{d^m}{dx^m} P_l(x) \tag{4.61}$$

are the associated Legendre polynomials. Using above expansion, the potential of the charge distribution can be written as:

$$V(\mathbf{r}) = \frac{1}{4\pi\varepsilon_0} \sum_{l=0}^{\infty} \sum_{m=-l}^{l} \frac{4\pi}{2l+1} q_{lm} \frac{Y_{lm}(\theta,\phi)}{r^{l+1}} \tag{4.62}$$

and

$$q_{lm} = \int Y_{lm}^*(\theta',\phi')(r')^l \rho(\mathbf{r}') \, d\tau' \tag{4.63}$$

are the *multipole moments* of the charge distribution. The zeroth-order expansion coefficient

$$q_{00} = \frac{1}{4\pi} \int \rho(\mathbf{r}) \, d\tau = \frac{Q}{4\pi} \tag{4.64}$$

corresponds to total charge of the localized charge distribution. The total charge can be referred to as the electric *monopole* moment. The electric dipole moment

$$\mathbf{p} = \int \mathbf{r}\rho(\mathbf{r}) \, d\tau \tag{4.65}$$

and the first order expansion coefficients are related according to

$$\begin{aligned}
q_{1,1} &= -\sqrt{\frac{3}{8\pi}} \, (p_x - ip_y), \\
q_{1,-1} &= \sqrt{\frac{3}{8\pi}} \, (p_x + ip_y), \\
q_{1,0} &= \sqrt{\frac{3}{4\pi}} \, p_z.
\end{aligned} \tag{4.66}$$

The higher-order moments depend on the precise choice of the origin inside the charge distribution and therefore their usage is mainly restricted to cases where a preferred choice of the origin is dictated by the physical systems.[4] The potential of the charge distribution, ignoring second and higher order terms is

$$V(\mathbf{r}) = \frac{1}{4\pi\varepsilon_0}\left(\frac{q}{r} + \frac{\mathbf{p}\cdot\mathbf{r}}{r^3}\right) \tag{4.67}$$

and the electric field of a dipole $\mathbf{p}$ located at the origin is

$$\mathbf{E}(\mathbf{r}) = \frac{3\hat{\mathbf{n}}(\mathbf{p}\cdot\hat{\mathbf{n}}) - \mathbf{p}}{4\pi\varepsilon_0 r^3} \tag{4.68}$$

and $\hat{\mathbf{n}} = \mathbf{r}/|\mathbf{r}|$. This formula is correct provided that $\mathbf{r} \neq \mathbf{0}$. An idealized dipole sheet at $x = 0$ is described by a charge distribution

$$\rho(\mathbf{r}) = \frac{\sigma}{4\pi\varepsilon_0}\delta'(x), \tag{4.69}$$

---

[4]For example, the center of a nucleus provides a preferred choice of the origin. The quadrupole moment of a nucleus is an important quantity in describing the nuclear structure.

where δ' is the derivative of the delta function. The corresponding electric field is

$$\mathbf{E}(\mathbf{r}) = -\frac{\sigma}{4\pi\varepsilon_0}\delta(x). \tag{4.70}$$

We will now consider the polarization of dielectric media and derive the macroscopic version of Gauss' law. If an electric field is applied to a medium consisting of a large number of atoms and molecules, the molecular charge distribution will be distorted. In the medium an electric polarization is produced. The latter can be quantitatively described as a macroscopic variable or cell variable such as $\mathbf{P} = \Delta\mathbf{p}/\Delta V$, i.e., as the dipole moment per unit volume. On a macroscopic scale, we may consider the polarization as a vector field, i.e., $\mathbf{P}(\mathbf{r})$. The potential $V(\mathbf{r})$ can be constructed by linear superposition of the contributions from each volume element $\Delta\Omega$ located at $\mathbf{r}'$. Each volume element gives a contribution originating from the net charge and a contributions arising from the dipole moment.

$$\Delta V(\mathbf{r}) = \frac{1}{4\pi\varepsilon_0}\left(\frac{\rho(\mathbf{r}')}{|\mathbf{r}-\mathbf{r}'|}\Delta\Omega + \frac{\mathbf{P}(\mathbf{r}')\cdot(\mathbf{r}-\mathbf{r}')}{|\mathbf{r}-\mathbf{r}'|^3}\right). \tag{4.71}$$

Adding all contributions and using the fact that

$$\nabla'\left(\frac{1}{|\mathbf{r}-\mathbf{r}'|}\right) = \frac{\mathbf{r}-\mathbf{r}'}{|\mathbf{r}-\mathbf{r}'|^3}, \tag{4.72}$$

we obtain

$$V(\mathbf{r}) = \frac{1}{4\pi\varepsilon_0}\int d\tau' \frac{1}{|\mathbf{r}-\mathbf{r}'|}\left(\rho(\mathbf{r}') - \nabla'\cdot\mathbf{P}(\mathbf{r}')\right). \tag{4.73}$$

This corresponds to the potential of a charge distribution $\rho - \nabla\cdot\mathbf{P}$. Since the microscopic equation $\nabla\times\mathbf{E} = 0$ does apply also on the macroscopic scale, we conclude that $\mathbf{E}$ is still derivable from a potential field, $\mathbf{E} = -\nabla V$, and

$$\nabla\cdot\mathbf{E} = \frac{1}{\varepsilon_0}(\rho - \nabla\cdot\mathbf{P}). \tag{4.74}$$

This result can be easily confirmed by using

$$\nabla^2\left(\frac{1}{|\mathbf{r}-\mathbf{r}'|}\right) = -4\pi\delta(\mathbf{r}-\mathbf{r}'). \tag{4.75}$$

The electric displacement, $\mathbf{D}$, is defined as

$$\mathbf{D} = \varepsilon_0\mathbf{E} + \mathbf{P} \tag{4.76}$$

and the first Maxwell equation becomes

$$\nabla\cdot\mathbf{D} = \rho. \tag{4.77}$$

If the response of the medium to the electric field is linear and isotropic then the coefficient of proportionality is the electric susceptibility, χ_e and the polarization reads

$$\mathbf{P} = \varepsilon_0\chi_e\mathbf{E}. \tag{4.78}$$

and consequently,

$$\mathbf{D} = \varepsilon_0(1 + \chi_e)\mathbf{E} = \varepsilon_0\varepsilon_r\mathbf{E}. \tag{4.79}$$

This is a *constitutive* relation connecting $\mathbf{D}$ and $\mathbf{E}$, necessary to solve the field equations. Here we have limited ourselves to consider an elementary connection. However, in general the connection can be nonlinear and anisotropic, such that $\mathbf{P} = \mathbf{P}(\mathbf{E})$ will involve a nontrivial expression.

It is instructive to apply above terminology to a parallel-plate capacitor. The storage capacity C of two electrodes with charges $\pm Q$ in vacuum is $C = Q/V$, where V is the voltage drop. Filling the volume between the plates with a dielectric material results into a voltage drop

$$V = \frac{Q/\varepsilon_r}{C}. \tag{4.80}$$

This equation may be interpreted as stating that of the total charge Q, the *free* charge Q/ε_r contributes to the voltage drop, whereas the *bound* charge $(1 - 1/\varepsilon_r)Q$, is neutralized by the polarization of the dielectric material. The electric susceptibility, χ_e emerges as the ratio of the bound charge and the free charge:

$$\chi_e = \frac{(1 - 1/\varepsilon_r)Q}{Q/\varepsilon_r} = \varepsilon_r - 1. \tag{4.81}$$

The displacement and the polarization both have the dimension [charge/area]. These variables correspond to electric flux densities. Given an infinitesimal area element $\mathbf{dS}$ on an electrode, the normal component of $\mathbf{D}$ corresponds to the charge $dQ = \mathbf{D} \cdot \mathbf{dS}$ on the area element and the normal component of $\mathbf{P}$ represents the bound charge $(1 - 1/\varepsilon_r)dQ$ on the area element. Finally, the normal component of $\varepsilon_0\mathbf{E}$ corresponds to the free charge dQ/ε_r residing on the area element. The question arises how the displacement $\mathbf{D}$, the polarization $\mathbf{P}$, and $\varepsilon_0\mathbf{E}$ can be associated to flux densities while there is no flow. In fact, the terminology is justified by analogy or mathematical equivalence with real flows. Consider for instance a stationary flow of water in $\mathbb{R}^3$. There exists a one-parameter family of maps $\phi_t : \mathbb{R}^3 \to \mathbb{R}^3$ that takes the molecule located at the position $\mathbf{r}_0$ at t_0 to the position $\mathbf{r}_1$ at t_1. Associated to the flow there exists a flux field

$$\mathbf{J}(\mathbf{r}) = \frac{d\mathbf{r}}{dt}. \tag{4.82}$$

The velocity field describes the streamlines of the flow. For an incompressible stationary flow we have that for any volume Ω

$$\oint_{\partial\Omega} \mathbf{J} \cdot \mathbf{dS} = 0 \quad \text{or} \quad \nabla \cdot \mathbf{J} = 0. \tag{4.83}$$

The number of water molecules that enter a volume exactly balances the number of water molecules that leave the volume. Now suppose that it is possible that water molecules are created or annihilated, e.g., by a chemical reaction $2H_2O \leftrightarrow O_2 + 2H_2$ in some volume. This process corresponds to a source/sink, Σ in the balance equation

$$\nabla \cdot \mathbf{J}(\mathbf{r}) = \Sigma(\mathbf{r}). \tag{4.84}$$

Comparing this equation with the first Maxwell equation, we observe the mathematical equivalence. The charge density ρ acts as a source/sink for the flux field $\mathbf{D}$.

4.10. Magnetic media

A stationary current density, $\mathbf{J}(\mathbf{r})$, generates a magnetic induction given by

$$\mathbf{B}(\mathbf{r}) = \frac{\mu_0}{4\pi} \int d\tau' \, \mathbf{J}(\mathbf{r}') \times \frac{\mathbf{r} - \mathbf{r}'}{|\mathbf{r} - \mathbf{r}'|^3}. \tag{4.85}$$

This result is essentially the finding of Biot, Savart, and Ampère. With the help of Eq. (4.72) we may write (4.85) as

$$\mathbf{B}(\mathbf{r}) = \frac{\mu_0}{4\pi} \nabla \times \int d\tau' \, \frac{\mathbf{J}(\mathbf{r}')}{|\mathbf{r} - \mathbf{r}'|}. \tag{4.86}$$

An immediate consequence is $\nabla \cdot \mathbf{B} = 0$. Using the identity $\nabla \times \nabla \times \mathbf{A} = \nabla(\nabla \cdot \mathbf{A}) - \nabla^2 \mathbf{A} = 0$, and the fact that $\mathbf{J} = 0$, as well as Eq. (4.75) one obtains that

$$\nabla \times \mathbf{B} = \mu_0 \mathbf{J}. \tag{4.87}$$

Helmholtz' theorem implies that there will be a vector field $\mathbf{A}$ such that $\mathbf{B} = \nabla \times \mathbf{A}$ and a comparison with Eq. (4.86) shows that

$$\mathbf{A}(\mathbf{r}, t) = \frac{\mu_0}{4\pi} \int d\tau' \, \frac{\mathbf{J}(\mathbf{r}')}{|\mathbf{r} - \mathbf{r}'|} + \nabla \chi(\mathbf{r}, t), \tag{4.88}$$

where χ is an arbitrary scalar function. The arbitrariness in the solution (4.88) for $\mathbf{A}$ illustrates the freedom to perform gauge transformations. This freedom however is lifted by fixing a gauge condition, i.e., by inserting an additional constraint that the component of $\mathbf{A}$ should obey, such that not all components are independent anymore. A particular choice is the Coulomb gauge, $\nabla \times \mathbf{A} = 0$. In that case, χ is a solution of Laplace's equation $\nabla^2 \chi = 0$. Provided that there are no sources at infinity and space is unbounded, the unique solution for χ is a constant, such that

$$\mathbf{A}(\mathbf{r}, t) = \frac{\mu_0}{4\pi} \int d\tau' \, \frac{\mathbf{J}(\mathbf{r}')}{|\mathbf{r} - \mathbf{r}'|}. \tag{4.89}$$

We will now consider a localized current distribution around some origin, $\mathbf{0}$. Then we may expand Eq. (4.89) for $|\mathbf{r}| > |\mathbf{r}'|$ using

$$\frac{1}{|\mathbf{r} - \mathbf{r}'|} = \frac{1}{|\mathbf{r}|} + \frac{\mathbf{r} \cdot \mathbf{r}'}{|\mathbf{r}|^3} + \cdots \tag{4.90}$$

as

$$\mathbf{A}(\mathbf{r}) = \frac{\mu_0}{4\pi r} \int d\tau' \, \mathbf{J}(\mathbf{r}') + \frac{\mu_0}{4\pi r^3} \int d\tau' \, (\mathbf{r} \cdot \mathbf{r}') \mathbf{J}(\mathbf{r}'). \tag{4.91}$$

The first integral is zero, i.e., $\int d\tau \, \mathbf{J}(\mathbf{r}) = 0$, whereas the second integral gives

$$\mathbf{A}(\mathbf{r}) = \frac{\mu_0}{4\pi} \frac{\mathbf{m} \times \mathbf{r}}{r^3}, \quad \mathbf{m} = \frac{1}{2} \int d\tau \, \mathbf{r} \times \mathbf{J}(\mathbf{r}). \tag{4.92}$$

The variable $\mathbf{m}$ is the *magnetic moment* of the current distribution. Following a similar reasoning as was done for the dielectric media, we consider the macroscopic effects of magnetic materials. Since $\nabla \cdot \mathbf{B} = 0$ at the microscopic scale, this equation also is valid at macroscopic scale. Therefore, Helmholtz' theorem is still applicable. By dividing space into volume elements ΔV, we can assign to each volume element a magnetic moment

$$\Delta \mathbf{m} = \mathbf{M}(\mathbf{r})\Delta V, \tag{4.93}$$

where $\mathbf{M}$ is the magnetization or magnetic moment density. For a substance consisting of k different atoms or molecules with partial densities ρ_i $(i = 1, \ldots, k)$ and with magnetic moment $\mathbf{m}_i$ for the ith atom or molecule, the magnetization is

$$\mathbf{M}(\mathbf{r}) = \sum_{i=1}^{k} \rho_i(\mathbf{r})\mathbf{m}_i. \tag{4.94}$$

The free-charge current density and the magnetization of the volume element ΔV at location $\mathbf{r}'$, give rise to a contribution to the vector potential at location $\mathbf{r}$ being

$$\Delta \mathbf{A}(\mathbf{r}) = \frac{\mu_0}{4\pi} \frac{\mathbf{J}(\mathbf{r}')}{|\mathbf{r} - \mathbf{r}'|} \Delta V + \frac{\mu_0}{4\pi} \frac{\mathbf{M}(\mathbf{r}') \times (\mathbf{r} - \mathbf{r}')}{|\mathbf{r} - \mathbf{r}'|^3} \Delta V. \tag{4.95}$$

Adding all contributions

$$\mathbf{A}(\mathbf{r}) = \frac{\mu_0}{4\pi} \int d\tau' \frac{\mathbf{J}(\mathbf{r}') + \nabla \times' \mathbf{M}(\mathbf{r}')}{|\mathbf{r} - \mathbf{r}'|}. \tag{4.96}$$

This corresponds to the vector potential of a current distribution $\mathbf{J} + \nabla \times \mathbf{M}$ and therefore

$$\nabla \times \mathbf{B} = \mu_0(\mathbf{J} + \nabla \times \mathbf{M}). \tag{4.97}$$

The magnetic *field* is defined as

$$\mathbf{H} = \frac{1}{\mu_0}\mathbf{B} - \mathbf{M}. \tag{4.98}$$

Then the stationary macroscopic equations become

$$\nabla \times \mathbf{H} = \mathbf{J}, \qquad \nabla \cdot \mathbf{B} = 0. \tag{4.99}$$

If we follow a strict analogy with the discussion on electrical polarization we should adopt a linear relation between the magnetization $\mathbf{M}$ and the induction $\mathbf{B}$ in order to obtain a constitutive relation between $\mathbf{H}$ and $\mathbf{B}$. However, historically it has become customary to define the *magnetic susceptibility* χ_m as the ratio of the magnetization and the magnetic field

$$\mathbf{M} = \chi_m \mathbf{H}. \tag{4.100}$$

Then we obtain

$$\mathbf{B} = \mu_0(\mathbf{H} + \mathbf{M}) = \mu_0(1 + \chi_m)\mathbf{H} = \mu_0\mu_r\mathbf{H} = \mu\mathbf{H}. \tag{4.101}$$

In here, μ is the *permeability* and μ_r is the *relative* permeability.

Just as is the case for electrical polarization, the constitutive relation, $\mathbf{B} = \mathbf{B(H)}$, can be anisotropic and nonlinear. In fact, the $\mathbf{B(H)}$ relation may be multiple-valued depending on the history of the preparation of the material or the history of the applied magnetic fields (hysteresis).

In deriving the macroscopic field equations, we have so far been concerned with stationary phenomena. Both the charge distributions and the current distributions were assumed to be time-independent. The resulting equations are

$$\nabla \times \mathbf{E} = 0, \tag{4.102}$$

$$\nabla \cdot \mathbf{B} = 0, \tag{4.103}$$

$$\nabla \cdot \mathbf{D} = \rho, \tag{4.104}$$

$$\nabla \times \mathbf{H} = \mathbf{J}. \tag{4.105}$$

Faraday's law that was obtained from experimental observation, relates the circulation of the electric field to the time variation of the magnetic flux

$$\oint \mathbf{E} \cdot \mathbf{dr} = -\frac{d}{dt} \int \mathbf{B} \cdot \mathbf{dS}, \tag{4.106}$$

or

$$\nabla \times \mathbf{E} + \frac{\partial \mathbf{B}}{\partial t} = 0. \tag{4.107}$$

Magnetic monopoles have never been observed nor mimiced by time-varying fields. Therefore, the equation $\nabla \cdot \mathbf{B} = 0$ holds in all circumstances. Maxwell observed that the simplest generalization of Eqs. (4.104) and (4.105) that apply to time-dependent situations and that are consistent with charge conservation, are obtained by substituting $\mathbf{J}$ in Eq. (4.105) by $\mathbf{J} + \partial \mathbf{D}/\partial t$, since using the charge conservation and Gauss' law gives

$$\nabla \cdot \left(\mathbf{J} + \frac{\partial \mathbf{D}}{\partial t}\right) = 0, \tag{4.108}$$

such that the left- and right-hand side of

$$\nabla \times \mathbf{H} = \mathbf{J} + \frac{\partial \mathbf{D}}{\partial t} \tag{4.109}$$

are both divergenceless. Eqs. (4.103), (4.107), (4.104), and (4.109) are referred to as the (macroscopic) *Maxwell equations*. From a theoretical point of view, the Maxwell equations (4.103) and (4.107) found their proper meaning within the geometrical interpretation of electrodynamics, where they are identified as the Bianci identities for the curvature (see Section 8).

5. Wave guides and transmission lines

An important application of the Maxwell theory concerns the engineering of physical devices that are capable of transporting electromagnetic energy. This transport takes place in a wave-like manner. The static limit does not take into account the wave behavior of the Maxwell equations. The easiest way to implement this feature is by confining the field in two dimensions, allowing it to move freely along the third dimension (i.e., longitudinal sections are much larger than transversal directions). In this way, guided waves are recovered. A particular case of this model is the transmission line.

The wave guide consists of boundary surfaces that are good conductors. In practical realizations these surfaces are metallic materials such that the ohmic losses will be low. In the description of wave guides one usually assumes that the surfaces are perfectly conducting in a first approximation and that for large but finite conductivity, the ohmic losses can be calculated by perturbative methods. Besides the (idealized) boundary surfaces, the wave guide consists of a dielectric medium with no internal charges ($\rho = 0$), no internal currents ($\mathbf{J} = \mathbf{0}$). Furthermore, for an idealized description it is assumed that the conductivity of the dielectric medium vanishes ($\sigma = 0$). Finally, a wave guide is translational invariant in one direction. It has become customary, to choose the z-axis parallel to this direction.

In order to solve the Maxwell equations for wave guides, one considers harmonic fields (modes). The generic solution may be obtained as a superposition of different modes. The physical fields $\mathbf{E}(\mathbf{r}, t)$ and $\mathbf{H}(\mathbf{r}, t)$ are obtained from

$$\mathbf{E}(\mathbf{r}, t) = \Re\left(\mathbf{E}(\mathbf{r})e^{i\omega t}\right), \qquad \mathbf{H}(\mathbf{r}, t) = \Re\left(\mathbf{H}(\mathbf{r})e^{i\omega t}\right), \tag{5.1}$$

where $\mathbf{E}(\mathbf{r})$ and $\mathbf{H}(\mathbf{r})$ are complex phasors. The Maxwell equations governing these phasors are

$$\begin{aligned} \mathbf{\nabla} \cdot \mathbf{E} &= 0, & \mathbf{\nabla} \cdot \mathbf{H} &= 0, \\ \mathbf{\nabla} \times \mathbf{E} &= -i\omega\mu\mathbf{H}, & \mathbf{\nabla} \times \mathbf{H} &= i\omega\varepsilon\mathbf{E}. \end{aligned} \tag{5.2}$$

Defining $\omega\mu = k\zeta$ and $\omega\varepsilon = k/\zeta$ then $k = \omega\sqrt{\mu\varepsilon}$ and $\zeta = \sqrt{\mu/\varepsilon}$. From Eqs. (5.2) it follows that the phasors satisfy the following equation:

$$(\nabla^2 + k^2)\left\{\begin{matrix} \mathbf{E} \\ \mathbf{H} \end{matrix}\right\} = 0. \tag{5.3}$$

The translational invariance implies that if $\mathbf{E}(\mathbf{r}), \mathbf{H}(\mathbf{r})$ is a solution of Eq. (5.3), then $\mathbf{E}(\mathbf{r} + \mathbf{a}), \mathbf{H}(\mathbf{r} + \mathbf{a})$ with $\mathbf{a} = a\mathbf{e}_z$, is also a solution of Eq. (5.3). We may therefore introduce a shift operator, $\hat{S}(a)$ such that

$$\hat{S}(a)\left\{\begin{matrix} \mathbf{E}(\mathbf{r}) \\ \mathbf{H}(\mathbf{r}) \end{matrix}\right\} = \left\{\begin{matrix} \mathbf{E}(\mathbf{r} + \mathbf{a}) \\ \mathbf{H}(\mathbf{r} + \mathbf{a}) \end{matrix}\right\}. \tag{5.4}$$

Performing a Taylor series expansion gives

$$\mathbf{E}(\mathbf{r} + \mathbf{a}) = \sum_{n=0}^{\infty} \frac{a^n}{n!}\frac{\partial^n}{\partial z^n}\mathbf{E}(\mathbf{r}) = \exp\left(a\frac{\partial}{\partial z}\right)\mathbf{E}(\mathbf{r}) \tag{5.5}$$

and therefore $\hat{S}(a) = \exp\left(a\frac{\partial}{\partial z}\right) = \exp(ia\hat{k})$ with $\hat{k} = -i\frac{\partial}{\partial z}$. The Helmholtz operator $\hat{H} = \nabla^2 + k^2$ commutes with $\hat{k}$, i.e., $[\hat{H}, \hat{k}] = 0$. As a consequence we can write the solutions of Eq. (5.3) in such a way that they are simultaneously eigenfunctions of $\hat{H}$ and $\hat{k}$. The eigenfunctions of $\hat{k}$ are easily found to be

$$f(z) = e^{i\kappa z}, \tag{5.6}$$

since

$$-i\frac{d}{dz}f(z) = \kappa f(z). \tag{5.7}$$

Thus from the translational invariance we may conclude that it suffices to consider solutions for **E** and **H** of the form $\mathbf{E}(x, y)e^{i\kappa z}$ and $\mathbf{H}(x, y)e^{i\kappa z}$. Defining explicitly the transversal and the longitudinal components of the fields

$$\mathbf{E}(x, y) = \mathbf{E}_T(x, y) + \mathbf{E}_L(x, y), \qquad \mathbf{E}_L(x, y) = E_z(x, y)\mathbf{e}_z,$$
$$\mathbf{H}(x, y) = \mathbf{H}_T(x, y) + \mathbf{H}_L(x, y), \qquad \mathbf{H}_L(x, y) = H_z(x, y)\mathbf{e}_z, \tag{5.8}$$

and

$$\nabla^2 = \nabla_T^2 + \frac{\partial}{\partial z^2} = \nabla_T^2 - \kappa^2, \tag{5.9}$$

where the subscript T stands for a transverse field in the x–y-plane, while the subscript L denotes the longitudinal fields along the z-axis, we obtain

$$\left(\nabla_T^2 + k^2 - \kappa^2\right)\begin{Bmatrix} \mathbf{E}_T(x, y) \\ \mathbf{H}_T(x, y) \end{Bmatrix} = 0,$$
$$\left(\nabla_T^2 + k^2 - \kappa^2\right)\begin{Bmatrix} E_z(x, y) \\ H_z(x, y) \end{Bmatrix} = 0. \tag{5.10}$$

The transverse equations correspond to an eigenvalue problem with fields vanishing at the boundaries in the transverse directions. The characteristic equations that need to be solved are the Helmholtz equations resulting into eigenvalue problems, where the eigenvalues are $p^2 = k^2 - \kappa^2$. The boundary conditions for the fields on the boundary surfaces are

$$\mathbf{n} \times \mathbf{E} = 0, \qquad \mathbf{n} \cdot \mathbf{H} = 0. \tag{5.11}$$

For the transverse components, going back to the full Maxwell equations, we get from Eq. (5.2)

$$\nabla_T E_z - \frac{\partial}{\partial z}\mathbf{E}_T = -i\omega\mu\mathbf{e}_z \times \mathbf{H}_T \tag{5.12}$$

and

$$\nabla_T H_z - \frac{\partial}{\partial z}\mathbf{H}_T = i\omega\varepsilon\mathbf{e}_z \times \mathbf{E}_T. \tag{5.13}$$

Combining (5.12) and (5.13) gives

$$p^2\mathbf{E}_T = i\omega\mu\mathbf{e}_z \times \nabla_T H_z + i\kappa\nabla_T E_z,$$
$$p^2\mathbf{H}_T = -i\omega\varepsilon\mathbf{e}_z \times \nabla_T E_z + i\kappa\nabla_T H_z.$$

We may define the transversal fields as

$$\mathbf{E}_T \propto V(z)e_t^{(1)}, \qquad \mathbf{H}_T \propto I(z)e_t^{(2)}, \tag{5.14}$$

where $e_t^{(1)}$ and $e_t^{(2)}$ are transversal vectors independent of z.

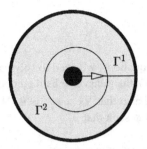

FIG. 5.1. Contours for evaluating voltage drops and currents of a two-conductor system in a TEM mode.

5.1. TEM modes

Inspired by waves in free space, we might look for modes that have a transverse behavior for both electric as magnetic field component, i.e., $E_z = H_z = 0$. These solutions are the transverse electromagnetic or TEM modes.

$$\left[\nabla_T^2 + p^2\right]\mathbf{E}_T = \mathbf{0}, \qquad \left[\nabla_T^2 + p^2\right]\mathbf{H}_T = \mathbf{0}. \tag{5.15}$$

For the TEM mode, the Maxwell equations result into $\kappa = k$. As a consequence Eqs. (5.15) are void. However, one also obtains from the Maxwell equations (5.12) and (5.13) that

$$\nabla \times \mathbf{E}_T = 0, \qquad \nabla \cdot \mathbf{E}_T = 0, \qquad \mathbf{H}_T = \frac{1}{\zeta}\mathbf{e}_z \times \mathbf{E}_T. \tag{5.16}$$

Therefore the TEM modes are as in an infinite medium. Since $E_z = 0$, the surfaces are equipotential boundaries and therefore at least two surfaces are needed to carry the wave. Since in any plane with constant z, we have a static potential, we can consider an arbitrary path going from one conductor to another. The voltage drop will be

$$V(z) = \int_{\Gamma^1} \mathbf{E}_T \cdot \mathbf{dr}. \tag{5.17}$$

The current in one conductor can be evaluated by taking a closed contour around the conductor and evaluate the field circulation. This is illustrated in Fig. 5.1.

$$I(z) = \oint_{\Gamma^2} \mathbf{H}_T \cdot \mathbf{dr}. \tag{5.18}$$

5.2. TM modes

When we look at solutions for which the longitudinal magnetic field vanishes ($H_z = 0$ everywhere), the magnetic field is always in the transverse direction. These solutions are the transverse magnetic or TM modes.

$$\left[\nabla_T^2 + p^2\right]E_z = \mathbf{0}, \tag{5.19}$$

$$p^2 \mathbf{E}_\mathrm{T} = i\kappa \boldsymbol{\nabla}_\mathrm{T} E_z, \tag{5.20}$$

$$p^2 \mathbf{H}_\mathrm{T} = -i\omega\varepsilon \mathbf{e}_z \times \boldsymbol{\nabla}_\mathrm{T} E_z. \tag{5.21}$$

To find the solution of these equations, we need to solve a Helmholtz equation for H_z, and from Eqs. (5.20) and (5.21), the transverse field components are derived. Eq. (5.20) implies that $\boldsymbol{\nabla}_\mathrm{T} \times \mathbf{E}_\mathrm{T} = 0$ and also that $\boldsymbol{\nabla}_\mathrm{T} \times e_T^{(1)} = \mathbf{0}$. Therefore, we may introduce a (complex) transverse potential ϕ such that

$$e_\mathrm{t}^{(1)} = -\boldsymbol{\nabla}_\mathrm{T}\phi. \tag{5.22}$$

This potential is proportional to E_z, i.e.,

$$E_z = -\frac{p^2}{i\kappa} V(z)\phi. \tag{5.23}$$

Substitution of the (5.14) and (5.14) into (5.13) gives that $e_\mathrm{t}^{(2)} = \mathbf{e}_z \times e_\mathrm{t}^{(1)}$ and $V(z) = -(\kappa/\omega\varepsilon)I(z)$.

5.3. TE modes

Similarly, when we look at solutions for which the longitudinal electric field vanishes ($E_z = 0$ everywhere), the electric field is always in the transverse direction. These solutions are the transverse electric or TE modes.

$$\left[\boldsymbol{\nabla}_\mathrm{T}^2 + p^2\right]B_z = 0, \tag{5.24}$$

$$p^2 \mathbf{E}_\mathrm{T} = i\omega\mu \mathbf{e}_z \times \boldsymbol{\nabla}_\mathrm{T} H_z, \tag{5.25}$$

$$p^2 \mathbf{H}_\mathrm{T} = i\kappa \boldsymbol{\nabla}_\mathrm{T} H_z. \tag{5.26}$$

To find the solution of these equations, we need to solve a Helmholtz equation for B_z, and from Eqs. (5.25) and (5.26), the transverse field components are derived. Since in this case $\boldsymbol{\nabla}_\mathrm{T} \times \mathbf{H}_\mathrm{T} = \mathbf{0}$ there exists a scalar potential ψ such that

$$e_\mathrm{t}^{(2)} = -\boldsymbol{\nabla}_\mathrm{T}\psi. \tag{5.27}$$

Following a similar reasoning as above we obtain that

$$H_z = \frac{p^2}{ik\zeta} V(z)\psi. \tag{5.28}$$

Furthermore, we find that $e_\mathrm{t}^{(1)} = -\mathbf{e}_z \times e_\mathrm{t}^{(2)}$ and $V(z) = -(\omega\mu/\kappa)I(z)$.

5.4. Transmission line theory – S parameters

The structure of the transverse components of the electric and magnetic fields gives rise to an equivalent-circuit description. In order to show this, we will study the TM mode, but the TE description follows the same reasoning. By assuming the generic transmission-line solutions

$$V(z) = V_+ e^{-i\kappa z} + V_- e^{i\kappa z}, \tag{5.29}$$

$$I(z) = \frac{1}{Z_c} \left(V_+ e^{-i\kappa z} - V_- e^{i\kappa z} \right), \tag{5.30}$$

where Z_c is the characteristic impedance of the transmission line or the "telegraph" equations

$$\frac{dV(z)}{dz} = -ZI(z), \tag{5.31}$$

$$\frac{dI(z)}{dz} = -YV(z). \tag{5.32}$$

In these equations, the series impedance is denoted by Z and Y is the shunt admittance of the equivalent transmission line model. Each propagating mode corresponds to an eigenvalue p and we find that

$$Z = \frac{p^2 - k^2}{i\omega\varepsilon}, \qquad Y = i\omega\varepsilon. \tag{5.33}$$

From these expressions, the resulting equivalent circuit can be constructed.

6. From macroscopic field theory to electric circuits

6.1. Kirchhoff's laws

Electronic circuits consist of electronic components or devices integrated in a network. The number of components may range form a few to several billion. In the latter case the network is usually subdivided in functional blocks and each block has a unique functional description. The hierarchal approach is vital to the progress of electronic design and reuse of functional blocks (sometimes referred to as intellectual property) determines the time-to-market of new electronic products. Besides the commercial value of the hierarchical approach, there is also a scientific benefit. It is not possible to design advanced electronic circuits by solving the Maxwell equations using the boundary conditions that are imposed by the circuit. The complexity of the problem simply does not allow such an approach taking into account the available compute power and the constraints that are imposed on the design time. Moreover, a full solution of the Maxwell equations is often not very instructive in obtaining insight into the operation of the circuit. In order to understand the operation or input/output response of a circuit, it is beneficial to describe the circuit in effective variables. These coarse-grained variables (in the introduction we referred to these variables as "baskets") should be detailed enough such that a physical meaning can be given to them, whereas on the other hand they should be sufficiently "coarse" so as to mask details that are not relevant for understanding the circuit properties. The delicate balancing between these two requirements has resulted into "electronic circuit theory." The latter is based on the physical laws that are expressed by Maxwell's equations, and the laws of energy and charge conservation. The purpose of this section is to analyze how the circuit equations may be extracted from these microscopic physical laws. It should be emphasized that the extraction is not a rigorous derivation in the mathematical sense but relies on the validity of a number of approximations and assumptions reflecting the ideal behavior of electric circuits. These assumptions should be critically revised if one wishes to apply the circuit equations in areas that are outside the original scope of circuit theory. A simple example is a capacitor consisting of two large, conducting parallel plates separated by a relatively thin insulating layer: its capacity may be a suitable, characteristic variable for describing its impact in a circuit at low and moderately high frequencies. However, at extremely large frequencies the same device may act as a wave guide or an antenna, partly radiating the stored electromagnetic energy.

Being aware of such pitfalls, we continue our search for effective formulations of the circuit equations. In fact, the underlying prescriptions are given by the following (plausible) statements:

- A circuit can be represented by a topological network that consists of branches and nodes.
- **Kirchhoff's voltage law** (KVL) – The algebraic sum of all voltages along any arbitrary loop of the network equals zero at every instant of time.
- **Kirchhoff's current law** (KCL) – The algebraic sum of all currents entering or leaving any particular network node equals zero at every instant of time.

In order to make sense out of these statements we first need to have a clear understanding of the various words that were encountered; in particular, we must explain what is meant by a node, a branch, a voltage, and a current. For that purpose we consider the most elementary circuit: a battery and a resistor that connects the poles of the battery. The circuit is depicted in Fig. 6.1. We have explicitly taken into account the finite resistance of the leads. In fact, a more realistic drawing is presented in Fig. 6.2, where we account for the fact that the leads have a finite volume. In particular, we have divided the full circuit volume into four different regions: (1) the battery region Ω_B, (2) the left lead region Ω_{1L}, (3) the right lead region Ω_{2L}, and (4) the resistor region Ω_A.

We will now consider the power supplied by the battery to the circuit volume. The work done by the electromagnetic field on all charges in the circuit volume per unit time

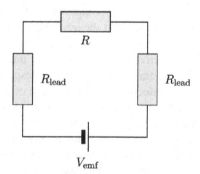

FIG. 6.1. Closed electric circuit containing a resistor connected to a DC power supply through two resistive leads.

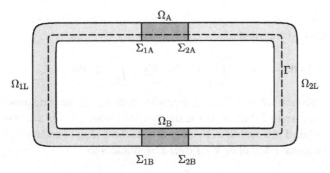

FIG. 6.2. The electric circuit of FIG. 6.1, taking into account the spatial extension of the leads. Γ is a circuit loop, i.e., an internal, closed loop encircling the "hole" of the circuit. (Figure reproduced by permission of the American Physical Society and Springer Verlag.)

is given by

$$\frac{dE_{\mathrm{MECH}}}{dt} = \int_{\Omega} \mathbf{J} \cdot \mathbf{E}\, d\tau. \tag{6.1}$$

This corresponds to the dissipated power in steady-state conditions for which $(\partial \rho / \partial t) = 0$. As a consequence, $\nabla \cdot \mathbf{J} = 0$ and therefore we may apply the $\mathbf{J} \cdot \mathbf{E}$ theorem (see Appendix Appendix A.1). We obtain:

$$\int_{\Omega} \mathbf{J} \cdot \mathbf{E}\, d\tau = \left(\oint_{\Sigma} \mathbf{J} \cdot d\mathbf{S} \right) \left(\oint_{\Gamma} \mathbf{E} \cdot d\mathbf{r} \right), \tag{6.2}$$

where Σ is an arbitrary cross section of the circuit and Γ is a circuit loop, i.e., an arbitrary closed path inside the circuit region. We identify the first integral of the right-hand side of Eq. (6.2) as the *current* in the circuit. The second integral of the right-hand side of Eq. (6.2) is identified as the *electromotive force* (EMF) or the *voltage* that is supplied by the battery, V_{ε}. The latter is nothing but the work done per unit charge by the electric field when the charge has made one full revolution around the circuit. Note the integral $\oint_{\Gamma} \mathbf{E} \cdot d\mathbf{r}$ is *nonzero*, although $\nabla \times \mathbf{E} = \mathbf{0}$. This is possible because the circuit is not a simply connected region in $\mathbb{R}^3$. More precisely, the topology of the circuit is that of a manifold of genus one, say a torus or a toroidal region with one "hole." We may now consider the left-hand side of Eq. (6.2) and consider the contributions to Eq. (6.2). For region (2) we obtain:

$$\int_{\Omega_{1L}} \mathbf{J} \cdot \mathbf{E}\, d\tau = -\int_{\Omega_{1L}} (\nabla V) \cdot \mathbf{J}\, d\tau = -\int_{\Omega_{1L}} \nabla \cdot (V\mathbf{J})\, d\tau + \int_{\Omega_{1L}} V \nabla \cdot \mathbf{J}\, d\tau. \tag{6.3}$$

The first equality is valid since $\mathbf{E} = -\nabla V$, in a simply-connected region such as Ω_{1L}, Ω_{2L}, Ω_A, or Ω_B. The last integral is equal to zero, since $\nabla \cdot \mathbf{J} = 0$ and the one-but-last integral is

$$-\int_{\Omega_{1L}} \nabla \cdot (V\mathbf{J})\, d\tau = -\oint_{\partial \Omega_{1L}} V\mathbf{J} \cdot d\mathbf{S}. \tag{6.4}$$

If we now *assume* that the potential is constant on a cross-section of the circuit, then this integral has two contributions:

$$-\oint_{\partial\Omega_{1L}} V\mathbf{J}\cdot d\mathbf{S} = -V_{\Sigma_{1B}}\int_{\Sigma_{1B}} \mathbf{J}\cdot d\mathbf{S} - V_{\Sigma_{1A}}\int_{\Sigma_{1A}} \mathbf{J}\cdot d\mathbf{S}. \tag{6.5}$$

Using Gauss' theorem we may identify the two remaining surface integrals can be identified as the total current I. Indeed, in the steady state regime ($\partial\rho/\partial t = 0$) the divergence of $\mathbf{J}$ vanishes while $\mathbf{J}$ is assumed to be tangential to the circuit boundary $\partial\Omega$. Therefore, the vanishing volume integral of $\nabla\cdot\mathbf{J}$ over Ω_{1L} reduces to

$$0 = \int_{\partial\Omega_{1L}} \mathbf{J}\cdot d\mathbf{S} = \int_{\Sigma_{1A}} \mathbf{J}\cdot d\mathbf{S} - \int_{\Sigma_{1B}} \mathbf{J}\cdot d\mathbf{S}, \tag{6.6}$$

which justifies the identification

$$\int_{\Sigma_{1A}} \mathbf{J}\cdot d\mathbf{S} = \int_{\Sigma_{1B}} \mathbf{J}\cdot d\mathbf{S} \equiv I \tag{6.7}$$

whence

$$-\oint_{\partial\Omega_{1L}} V\mathbf{J}\cdot d\mathbf{S} = I(V_{\Sigma_{1A}} - V_{\Sigma_{1B}}). \tag{6.8}$$

The regions (3) and (4) can be evaluated in a similar manner. As a consequence we obtain:

$$I(V_{\Sigma_{2B}} - V_{\Sigma_{2A}}) + I(V_{\Sigma_{2A}} - V_{\Sigma_{1A}}) + I(V_{\Sigma_{1A}} - V_{\Sigma_{1B}}) + \int_{\Omega_B} \mathbf{J}\cdot\mathbf{E}\,d\tau = IV_\varepsilon. \tag{6.9}$$

The final integral that applies to the battery region, is also equal to zero. This is because the electric field consists of two components: a conservative component and a nonconservative component, i.e., $\mathbf{E} = \mathbf{E}_C + \mathbf{E}_{NC}$. The purpose of the ideal[5] battery is to cancel the conservative field, such that after a full revolution around the circuit a net energy supply is obtained from the electric field. Then we finally arrive at the following result:

$$V_\varepsilon = V_{\Sigma_{2B}} - V_{\Sigma_{1B}}. \tag{6.10}$$

Eq. (6.10) is not a trivial result: having been derived from energy considerations, it relates the EMF of the battery, arising from a nonconservative field, to the potential difference at its terminals, i.e., a quantity characterizing a conservative field. Physically, it reflects the concept that an ideal battery is capable of maintaining a constant potential difference at its terminals even if a current is flowing through the circuit. This example illustrates how Kirchhoff's laws can be extracted from the underlying physical laws. It should be emphasized that we achieved more than what is provided by Kirchhoff's laws. Often Kirchhoff's voltage law is presented as a *trivial* identity, i.e., by putting N nodes on a closed path, as we have done by selecting a series of cross sections, it is

---

[5]The internal resistance of a real battery is neglected here.

always true that

$$(V_1 - V_2) + (V_2 - V_3) + \cdots + (V_{N-1} - V_N) + (V_N - V_1) = 0. \tag{6.11}$$

Physics enters this identity (turning it into a useful equation) by relating the potential differences to their physical origin. In the example above, the potential difference, $V_{\Sigma_{2B}} - V_{\Sigma_{1B}}$, is the result of a power supply.

By the in-depth discussion of the simple circuit, we have implicitly provided a detailed understanding of what is understood to be a voltage, a current, a node, and a branch in a Kirchhoff network. The nodes are geometrically idealized regions of the circuit to which network branches can be attached. The nodes can be electrically described by a single voltage value. A branch is also a geometrical idealization. Knowledge of the current *density* inside the branch is not required. All that counts is the total current in the branch. We also have seen that at some stages only progress could be made by making simplifying assumptions and finally that all variables are time independent. The last condition is a severe limitation. In the next section we will discuss the consequences of eliminating this restriction. We can insert more physics in the network description. So far, we have not exploited Ohm's law, $\mathbf{J} = \sigma\mathbf{E}$. For a resistor with length L, cross sectional area A, and constant resistivity σ, we find that

$$\int_{\Omega_A} \mathbf{J} \cdot \mathbf{E}\, d\tau = \sigma \int (\nabla V)^2\, d\tau = \sigma L \cdot A \left(\frac{V_{\Sigma_{2A}} - V_{\Sigma_{1A}}}{L} \right)^2 = I(V_{\Sigma_{2A}} - V_{\Sigma_{1A}}). \tag{6.12}$$

As a result, we can "define" the resistance as the ratio of the potential difference and the current:

$$V_{\Sigma_{2A}} - V_{\Sigma_{1A}} = RI, \qquad R = \frac{L}{\sigma A}. \tag{6.13}$$

6.2. Circuit rules

In the foregoing section we have considered DC steady-state currents, for which $\nabla \cdot \mathbf{J} = 0$ and $\partial\mathbf{B}/\partial t = \mathbf{0}$, such that the $\mathbf{J} \cdot \mathbf{E}$ theorem could be applied. In general, these conditions are not valid and the justification of using the Kirchhoff's laws becomes more difficult. Nevertheless, the guiding principles remain unaltered, i.e., the conservation of charge and energy will help us in formulating the circuit equations. On the other hand, as was already mentioned in the previous section, the idealization of a real circuit involves a number of approximations and assumptions that are summarized below in a – non-exhaustive – list of circuit rules:

(1) An electric circuit, or more generally, a circuit network, is a manifold of genus $N \geqslant 1$, i.e., a multiply connected region with N holes. The branches of this manifold consist of distinct circuit segments or devices, mainly active and passive components, interconnecting conductors, and seats of EMF.

(2) The active components typically include devices that are actively processing signals, such as transistors, vacuum tubes, operational amplifiers, A/D converters.

(3) Passive components refer to ohmic resistors, capacitors and inductors or coils, diodes, tunneling junctions, Coulomb blockade islands, etc. They are representing energy dissipation, induction effects, quantum mechanical tunneling processes, and many other phenomena.

(4) The seats of EMF include both DC and AC power supplies, i.e., chemical batteries, EMFs induced by externally applied magnetic fields, all different kinds of current and voltages sources and generators, etc. The electromagnetic power supplied by the EMF sources is dissipated entirely in the circuit. No energy is released to the environment of the circuit through radiation or any other mechanism.

(5) In compliance with the previous rule, all circuit devices are assumed to behave in an ideal manner. First, all conductors are taken to be perfect conductors. Considering perfect conduction as the infinite conductivity limit of realistic conduction ($\mathbf{J} = \sigma\mathbf{E}$), it is clear that no electric fields can survive inside a perfect conductor which therefore can be considered an equipotential volume. Clearly, from $\nabla \cdot \mathbf{E} = \rho/\varepsilon$ it follows that the charge density also vanishes inside the conductor. Furthermore, a perfect conductor is perfectly shielded from any magnetic field. Strictly speaking, this is not a direct consequence of Maxwell's third equation, since $\nabla \times \mathbf{E} = \mathbf{0}$ would only imply $\partial\mathbf{B}/\partial t = \mathbf{0}$ but the effect of static magnetic fields on the circuit behavior will not be considered here. It should also be noted that a perfect conductor is not the same as a superconductor. Although for both devices the penetration of magnetic fields is restricted to a very narrow boundary layer, called penetration depth, only the superconductor hosts a number of "normal" electrons (subjected to dissipative transport) and will even switch entirely to the normal state when the supercurrent attains its critical value. Furthermore, a supercurrent can be seen as a coherent, collective motion of so-called Cooper pairs of electrons, i.e., *bosons* while perfect conduction is carried by unpaired electrons or holes, i.e., *fermions*. Next, all energy dissipation exclusively takes place inside the circuit resistors. This implies that all capacitors and inductors are assumed to be made of perfect conductors. Inside the windings of an inductor and the plates of a capacitor, no electric or magnetic fields are present. The latter exist only in the cores of the inductors[6] while the corresponding vector potential and induced electric field are localized in the inductor. Similarly, the electric charge on the plates of a capacitor are residing in a surface layer and the corresponding, conservative electric field is strictly localized between the plates while all stray fields are ignored. Finally, the ideal behavior of the seats of EMF is reflected in the absence of internal resistances and the strict localization of the nonconservative electric fields that are causing the EMFs.

(6) The current density vector $\mathbf{J}$ defines a positive orientation of the circuit loop Γ. It corresponds the motion of a positive charge moving from the anode to the cathode outside the EMF seat and from cathode to anode inside the EMF seat.

---

[6]Topologically, the cores are not part of the circuit region Ω.

6.3. Inclusion of time dependence

The previous set of rules will guide us towards the derivation of the final circuit equations. However, before turning to the latter, it is worth to have a second look at Eq. (6.11). In the continuum, this identity can be given in the following way:

$$\oint_\Gamma d\mathbf{r} \cdot \nabla V(\mathbf{r}, t) = 0, \tag{6.14}$$

where Γ is an arbitrary closed loop. Note that above equation includes time-dependent fields $V(\mathbf{r}, t)$. In order to validate the first Kirchhoff law (KVL), we insert into Eq. (6.14) the potential that corresponds to

$$\nabla V = -\mathbf{E} - \frac{\partial \mathbf{A}}{\partial t}. \tag{6.15}$$

Of course, if we were to plug this expression into Eq. (6.14), we would just arrive at Faraday's law. The transition to the circuit equations is realized by cutting the loop into discrete segments (rule 1) and assigning to each segment appropriate lumped variables. To illustrate this approach we revisit the circuit of Fig. 6.1, where we have now folded the resistor of the left lead into a helix and, according to the circuit rules, its resistance is taken to be zero whereas the top resistor is replaced by a capacitor. The resulting, idealized circuit depicted in Fig. 6.3 has four segments. The battery region, that now may produce a time-dependent EMF, and the right-lead region can be handled as was done in the foregoing section. According to the circuit rules, it is assumed that all resistance is concentrated in the resistor located between node 3 and node 4, while both the inductor and the capacitor are made of perfect conductors and no leakage current is flowing between the capacitor plates. Starting from the identities

$$V_1 - V_2 + V_2 - V_3 + V_3 - V_4 + V_4 - V_1 = 0, \tag{6.16}$$

$$\oint_\Gamma \mathbf{E} \cdot d\mathbf{r} + \frac{\partial}{\partial t} \oint_\Gamma \mathbf{A} \cdot d\mathbf{r} = 0, \tag{6.17}$$

we decompose the electric field into a conservative, an external, and induced component:

$$\mathbf{E} = \mathbf{E}_C + \mathbf{E}_{EX} + \mathbf{E}_{IN}, \tag{6.18}$$

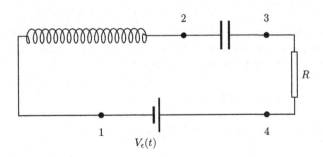

FIG. 6.3. The electric circuit of Fig. 6.2 with a helix-shaped "resistor."

where

$$\mathbf{A} = \mathbf{A}_{EX} + \mathbf{A}_{IN}, \qquad \mathbf{E}_C = -\nabla V,$$

$$\mathbf{E}_{EX} = -\frac{\partial}{\partial t}\mathbf{A}_{EX}, \qquad \mathbf{E}_{IN} = -\frac{\partial}{\partial t}\mathbf{A}_{IN}. \qquad (6.19)$$

Since the battery and the inductor are perfect conductors, the total electric field in these devices is identically zero:

$$\int_1^4 \mathbf{dr} \cdot \mathbf{E} = 0, \qquad (6.20)$$

$$\int_2^1 \mathbf{dr} \cdot \mathbf{E} = 0. \qquad (6.21)$$

Following the circuit rules, we assume that the induced electric field and the external field are only present in the inductor region and the battery region, respectively. Then Eq. (6.21) can be evaluated as

$$\int_2^1 \mathbf{dr} \cdot \mathbf{E} = \int_2^1 \mathbf{dr} \cdot (\mathbf{E}_C + \mathbf{E}_{IN}) = V_2 - V_1 + \int_2^1 \mathbf{dr} \cdot \mathbf{E}_{IN} = 0 \qquad (6.22)$$

and therefore

$$V_1 - V_2 = \int_2^1 \mathbf{dr} \cdot \mathbf{E}_{IN}. \qquad (6.23)$$

For the battery region we obtain:

$$\int_1^4 \mathbf{dr} \cdot \mathbf{E} = \int_1^4 \mathbf{dr} \cdot (\mathbf{E}_C + \mathbf{E}_{EX}) = V_1 - V_4 + \int_1^4 \mathbf{dr} \cdot \mathbf{E}_{EX} = 0 \qquad (6.24)$$

and therefore

$$V_4 - V_1 = \int_1^4 \mathbf{dr} \cdot \mathbf{E}_{EX} = \oint \mathbf{dr} \cdot \mathbf{E}_{EX} = V_\varepsilon. \qquad (6.25)$$

Inside the capacitor, the induced and external fields are zero, and therefore we obtain

$$\int_3^2 \mathbf{dr} \cdot \mathbf{E} = \int_3^2 \mathbf{dr} \cdot \mathbf{E}_C = V_3 - V_2. \qquad (6.26)$$

On the other hand, the potential difference between the capacitor is assumed to be proportional to charge Q stored on one of the plates, i.e., $Q = CV$, where C is the *capacitance*. The resistor is treated in an analogous manner:

$$V_4 - V_3 = \int_4^3 \mathbf{dr} \cdot \mathbf{E}_C = IR. \qquad (6.27)$$

Insertion of all these results into Eq. (6.16) gives:

$$\int_1^2 \mathbf{E}_{IN} \cdot \mathbf{dr} = -V_\varepsilon + IR + \frac{Q}{C}, \qquad (6.28)$$

where we anticipated that the electric field between the capacitor plates is given by $Q/(Cd)$ and d is the thickness of the dielectric. The integral at the left-hand side of Eq. (6.28) can be obtained by using Faraday's law once again:

$$\int_2^1 \mathbf{E}_{\text{IN}} \cdot d\mathbf{r} \simeq \oint_\Gamma \mathbf{E}_{\text{IN}} \cdot d\mathbf{r} = -\frac{\partial}{\partial t} \oint_\Gamma \mathbf{A}_{\text{IN}} \cdot d\mathbf{r} = -\frac{\partial}{\partial t} \int_{S(\Gamma)} \mathbf{B}_{\text{IN}} \cdot d\mathbf{S}, \qquad (6.29)$$

where $S(\Gamma)$ is the area enclosed by the loop Γ. Since the magnetic field $\mathbf{B}$ is only appreciably different from zero inside the core of the inductor, the integral may be identified as the magnetic self-flux Φ_{M} of the inductor. This flux is proportional to the circuit current I that also flows through the windings of the coil. Hence, $\Phi_{\text{M}} = LI$, where L is the *inductance* of the inductor and therefore Eq. (6.23) becomes:

$$V_1 - V_2 = -L\frac{dI}{dt}. \qquad (6.30)$$

We are now in the position to write down the circuit equation for the simple circuit of Fig. 6.3. Starting from the identity of Eq. (6.16), we find

$$-L\frac{dI}{dt} + V_\varepsilon - IR - \frac{Q}{C} = 0. \qquad (6.31)$$

So far, we have not considered energy conservation for the time-dependent circuit equations. However, this conservation law is important for determining explicit expressions for the inductances and capacitances. Integrating the electromagnetic energy density u_{EM} over an arbitrarily large volume Ω_∞ with a boundary surface $\partial\Omega_\infty$, we obtain the total energy content of the electromagnetic field:

$$U_{\text{EM}} = \frac{1}{2}\int_{\Omega_\infty} d\tau \left(\varepsilon E^2 + \frac{B^2}{\mu}\right) = \frac{1}{2}\int_{\Omega_\infty} d\tau \, (\mathbf{E}\cdot\mathbf{D} + \mathbf{B}\cdot\mathbf{H}). \qquad (6.32)$$

Replacing $\mathbf{E}$ and $\mathbf{B}$ by $-\nabla V - \partial\mathbf{A}/\partial t$ and $\nabla \times \mathbf{A}$, respectively, we may rewrite Eq. (6.32) as

$$U_{\text{EM}} = \frac{1}{2}\int_{\Omega_\infty} d\tau \left[-\left(\nabla V + \frac{\partial\mathbf{A}}{\partial t}\right)\cdot\mathbf{D} + \mathbf{H}\cdot\nabla\times\mathbf{A}\right]. \qquad (6.33)$$

Next, exploiting the vector identity (A.40), we applying Gauss' theorem to the volume Ω_∞ thereby neglecting all fields at the outer surface $\partial\Omega_\infty$, i.e.,

$$\int_{\partial\Omega_\infty} d\mathbf{S}\cdot(V\mathbf{D}) = 0, \qquad (6.34)$$

we obtain:

$$-\int_{\Omega_\infty} d\tau\,\nabla V\cdot\mathbf{D} = \int_{\Omega_\infty} d\tau\, V\nabla\cdot\mathbf{D} = \int_{\Omega_\infty} d\tau\,\rho V, \qquad (6.35)$$

where the last equality follows from the first Maxwell equation $\nabla\cdot\mathbf{D} = \rho$.

Similarly, using the identity (A.39) and inserting the fourth Maxwell equation, we may convert the volume integral of $\mathbf{H}\cdot\nabla\times\mathbf{A}$ appearing in Eq. (6.33):

$$\int_{\Omega_\infty} d\tau\,\mathbf{H}\cdot\nabla\times\mathbf{A} = \int_{\Omega_\infty} d\tau\,\mathbf{A}\cdot\left(\mathbf{J} + \frac{\partial\mathbf{D}}{\partial t}\right). \qquad (6.36)$$

Putting everything together, we may express the total electromagnetic energy as follows:

$$U_{EM} = \frac{1}{2} \int_{\Omega_\infty} d\tau \left[\rho V - \frac{\partial \mathbf{A}}{\partial t} \cdot \mathbf{D} + \mathbf{A} \cdot \left(\mathbf{J} + \frac{\partial \mathbf{D}}{\partial t} \right) \right] \tag{6.37}$$

$$= \frac{1}{2} \int_{\Omega} d\tau \left[\rho V - \frac{\partial \mathbf{A}}{\partial t} \cdot \mathbf{D} + \mathbf{A} \cdot \left(\mathbf{J} + \frac{\partial \mathbf{D}}{\partial t} \right) \right], \tag{6.38}$$

where the last integral is restricted to the circuit region Ω in view of the circuit rules stating that all electromagnetic fields are vanishing outside the circuit region. It is easy to identify in Eq. (6.38) the "electric" and "magnetic" contributions respectively referring to E^2 and B^2 in Eq. (6.32):

$$U_{EM} = U_E + U_M, \tag{6.39}$$

$$U_E = \frac{1}{2} \int_{\Omega} d\tau \left(\rho V - \frac{\partial \mathbf{A}}{\partial t} \cdot \mathbf{D} \right), \tag{6.40}$$

$$U_M = \frac{1}{2} \int_{\Omega} d\tau \, \mathbf{A} \cdot \left(\mathbf{J} + \frac{\partial \mathbf{D}}{\partial t} \right). \tag{6.41}$$

Neglecting the magnetic field inside the ideal circuit conductors according to the circuit rules, we take $\nabla \times \mathbf{A}$ to be zero inside the circuit. Moreover, bearing in mind that the identity

$$\nabla \cdot \left(\mathbf{J} + \frac{\partial \mathbf{D}}{\partial t} \right) = 0 \tag{6.42}$$

is generally valid, we may now apply the $\mathbf{J} \cdot \mathbf{E}$ theorem to the combination $\mathbf{A} \cdot (\mathbf{J} + \partial \mathbf{D}/\partial t)$:

$$U_M = \frac{1}{2} \left(\oint_\Gamma d\mathbf{r} \cdot \mathbf{A} \right) \left(\int_\Sigma d\mathbf{S} \cdot \left(\mathbf{J} + \frac{\partial \mathbf{D}}{\partial t} \right) \right). \tag{6.43}$$

The loop integral clearly reduces to the total magnetic flux, which consists of the self-flux Φ_M and the external flux Φ_{ex}. Furthermore, due to Eq. (6.42), the surface integral of Eq. (6.41) can be calculated for any cross-section Σ that does not contain accumulated charge. Taking Σ in a perfectly conducting lead, we have $\mathbf{D} = \mathbf{0}$ and the integral reduces to the total current $I = \int_\Sigma d\mathbf{S} \cdot \mathbf{J}$. On the other hand, if we were choosing Σ to cross the capacitor dielectric, the current density would vanish and the integral would be equal to $d\Phi_D(t)/dt$ where

$$\Phi_D(t) = \int_\Sigma d\mathbf{S} \cdot \mathbf{D}(\mathbf{r}, t) \tag{6.44}$$

is the flux of the displacement vector. Since both choices of Σ should give rise to identical results, we conclude that

$$I(t) = \frac{d\Phi_D(t)}{dt} \tag{6.45}$$

which confirms the observation that the circuit of Fig. 6.3 where the capacitor is in series with the other components, can only carry charging and discharging currents. In any case, we are left with

$$U_M = \tfrac{1}{2}(\Phi_M + \Phi_{ex})I \qquad\qquad (6.46)$$

or, reusing the "definition" of inductance, i.e., $\Phi_M = LI$,

$$U_M = \tfrac{1}{2}LI^2 + \tfrac{1}{2}\Phi_{ex}I, \qquad\qquad (6.47)$$

where $(1/2)LI^2$ is the familiar expression for the magnetic energy stored in the core of the inductor.

The electric energy may be rewritten in terms of capacitances in an analogous manner. The contribution of $\partial A/\partial t \cdot D$ in Eq. (6.39) vanishes because $\partial A/\partial t$, representing the nonconservative electric field, is nonzero only inside the inductor and the generator regions, where the total electric field reduces to zero. On the other hand, for perfectly conducting leads that are also equipotential domains, the first term gives:

$$U_E = \frac{1}{2}\sum_n Q_n V_n, \qquad\qquad (6.48)$$

where V_n generally denotes the potential of the nth (ideal) conductor, containing a charge Q_n. Being expressed in terms of bare potentials, the result of Eq. (6.48) seems to be gauge dependent at a first glimpse. It should be noted however, that Eq. (6.48) has been derived within the circuit approximation, which implies that the charged conductors are not arbitrarily distributed in space, but are all part of the – localized – circuit. In particular, the charges Q_n are assumed to be stored on the plates of the capacitors of the circuit, and as such the entire set $\{Q_n\}$ can be divided into pairs of opposite charges $\{(Q_j, -Q_j)\}$. Hence, Eq. (6.48) should be read

$$U_E = \frac{1}{2}\sum_j Q_j(V_{1j} - V_{2j}) = \frac{1}{2}\sum_j C_j(V_{1j} - V_{2j})^2, \qquad\qquad (6.49)$$

where $V_{1j} - V_{2j}$ is the gauge-invariant potential difference between the plates of the jth capacitor.

The second Kirchhoff law (KCL), follows from charge conservation. The branches in the network can not store charge, unless capacitors are included. The integrated charge is denoted by Q_n and

$$\frac{dQ_j}{dt} = -\int J \cdot dS = \sum_k I_{jk}, \qquad\qquad (6.50)$$

where the surface integral is over a surface around charge-storage domain and I_{jk} is the current flowing from the charge-storage region j into the jth circuit branch. As in the steady-state case, the Kirchhoff laws, in particular the expressions for the various voltage differences could only be obtained if some simplifying assumptions are made. For the inductor it was assumed that the induced magnetic field is only different from zero inside the core. For the capacitor, in a similar way it was assumed that the energy of storing the charge is localized completely between the plates. These assumptions need to be carefully checked before applying the network equations. As an illustration of this remark we emphasize that we ignored the volume integrals that are not parts of the circuits. In particular, the integral of the electric energy outside the circuit is the kinetic part of the radiation energy:

$$U_E^{\text{rad}} = -\frac{1}{2} \int_{\Omega_\infty \backslash \Omega} d\tau \, \frac{\partial \mathbf{A}}{\partial t} \cdot \mathbf{D} = \frac{1}{2} \varepsilon \int_{\Omega_\infty \backslash \Omega} d\tau \, \frac{\partial \mathbf{A}}{\partial t} \cdot \frac{\partial \mathbf{A}}{\partial t},$$ (6.51)

and the potential energy of the radiation field:

$$U_M^{\text{rad}} = -\frac{1}{2\mu} \int_{\Omega_\infty \backslash \Omega} d\tau \, (\nabla \times \mathbf{A}) \cdot (\nabla \times \mathbf{A})$$ (6.52)

are not considered at the level of circuit modeling.

7. Gauge conditions

The Maxwell theory of electrodynamics describes the interaction between radiation and charged particles. The electromagnetic fields are described by six quantities, the vector components of $\mathbf{E}$ and $\mathbf{B}$. The sources of the radiation fields are represented by the charge density ρ and the current density $\mathbf{J}$. If the sources are prescribed functions $\rho(\mathbf{r}, t)$ and $\mathbf{J}(\mathbf{r}, t)$, then the evolution of $\mathbf{E}(\mathbf{r}, t)$ and $\mathbf{B}(\mathbf{r}, t)$ is completely determined. The fields $\mathbf{E}$ and $\mathbf{B}$ may be obtained from a scalar potential V and a vector potential $\mathbf{A}$ such that

$$\mathbf{E} = -\nabla V - \frac{\partial \mathbf{A}}{\partial t}, \qquad \mathbf{B} = \nabla \times \mathbf{A}.$$ (7.1)

As was mentioned already in Section 3, the potentials $(V, \mathbf{A})$ are not unique. The choice

$$V \to V' = V - \frac{\partial \Lambda}{\partial t}, \qquad \mathbf{A} \to \mathbf{A}' = \mathbf{A} + \nabla \Lambda$$ (7.2)

gives rise two the same fields $\mathbf{E}$ and $\mathbf{B}$. A change in potential according to Eq. (7.2) is a gauge transformation. The Lagrangian density

$$\mathcal{L} = \frac{1}{2} \varepsilon_0 \left(\nabla V + \frac{\partial \mathbf{A}}{\partial t} \right)^2 - \frac{1}{2\mu_0} (\nabla \times \mathbf{A})^2 + \mathbf{J} \cdot \mathbf{A} - \rho V$$ (7.3)

gives rise to an action integral

$$S = \int dt \int d^3r \, \mathcal{L}(\mathbf{r}, t)$$ (7.4)

that is gauge invariant under the transformation (7.2). The gauge invariance of the Maxwell equations has been found a posteriori. It was the outcome of a consistent theory for numerous experimental facts. In modern physics invariance principles play a key role in order to classify experimental results. One often postulates some symmetry or some gauge invariance and evaluates the consequences such that one can decide whether the supposed symmetry is capable of correctly ordering the experimental data.

The equations of motion that follow from the variation of the action S are

$$-\varepsilon_0\left(\nabla^2 V + \nabla\cdot\frac{\partial\mathbf{A}}{\partial t}\right) = \rho, \tag{7.5}$$

$$\frac{1}{\mu_0}\nabla\times\nabla\times\mathbf{A} = \mathbf{J} - \varepsilon_0\frac{\partial}{\partial t}\left(\nabla V + \frac{\partial\mathbf{A}}{\partial t}\right). \tag{7.6}$$

These equations may be written as

$$M * \begin{bmatrix} V \\ \mathbf{A} \end{bmatrix} = \begin{bmatrix} \rho \\ \mathbf{J} \end{bmatrix}, \tag{7.7}$$

where the matrix operator M is defined as

$$M = \begin{bmatrix} -\varepsilon_0\nabla^2 & -\varepsilon_0\nabla\cdot\frac{\partial}{\partial t} \\ \varepsilon_0\nabla\cdot\frac{\partial}{\partial t} & \varepsilon_0\frac{\partial^2}{\partial t^2} + \frac{1}{\mu_0}\nabla\times\nabla\times \end{bmatrix}. \tag{7.8}$$

This operator is *singular*, i.e., there exist nonzero fields $(X, \mathbf{Y})$ such that

$$M * \begin{bmatrix} X \\ \mathbf{Y} \end{bmatrix} = \begin{bmatrix} 0 \\ \mathbf{0} \end{bmatrix}. \tag{7.9}$$

An example is the pair $(X, \mathbf{Y}) = (-\partial\Lambda/\partial t, \nabla\Lambda)$, where $\Lambda(\mathbf{r}, t)$ is an arbitrary scalar field.

The matrix M corresponds to the second variation of the action integral and therefore $\mathcal{L}$ corresponds to a singular Lagrangian density. The singularity of M implies that there does not exist an unique inverse matrix M^{-1} and therefore, Eq. (7.7) cannot be solved for the fields $(V, \mathbf{A})$ for given sources $(\rho, \mathbf{J})$. The singularity of the Lagrangian density also implies that not all the fields $(V, \mathbf{A})$ are independent. In particular, the canonical momentum conjugated to the generalized coordinate $V(\mathbf{r}, t)$ vanishes

$$\frac{\partial\mathcal{L}}{\partial\left(\frac{\partial V}{\partial t}\right)} = 0.$$

In fact, Gauss' law can be seen as a constraint for the field degrees of freedom and we are forced to restrict the set of field configurations by a gauge condition.

A gauge condition breaks the gauge invariance but it should not effect the theory such that the physical outcome is sensitive to it. In different words: the gauge condition should not influence the results of the calculation of the fields $\mathbf{E}$ and $\mathbf{B}$ and, furthermore, it must not make any field configurations of $\mathbf{E}$ and $\mathbf{B}$ "unreachable." Finally, the gauge condition should result into a nonsingular Lagrangian density such that the potentials can be uniquely determined from the source distributions. We will now discuss a selection of gauge conditions that can be found in the physics literature.

7.1. The Coulomb gauge

The Coulomb gauge is a constraint on the components of the vector potential such

$$C[\mathbf{A}] \equiv \nabla\cdot\mathbf{A} = 0. \tag{7.10}$$

The constraint can be included in the action, S, by adding a term to the Lagrangian that explicitly breaks the gauge invariance of the action. The new action becomes "gauge-conditioned." We set:

$$S \rightarrow S_{\text{g.c.}} = S_0 + \frac{\lambda}{\mu_0} \int dt\, d\tau\, C^2[\mathbf{A}], \tag{7.11}$$

where $S_{\text{g.c.}}$ it the gauge-conditioned action, S_0 is the gauge-invariant action, and λ is a dimensionless parameter. Then the equations for the potentials are

$$-\varepsilon_0 \left(\nabla^2 V + \mathbf{\nabla} \cdot \frac{\partial \mathbf{A}}{\partial t} \right) = \rho, \tag{7.12}$$

$$\frac{1}{\mu_0} \mathbf{\nabla} \times \mathbf{\nabla} \times \mathbf{A} - 2\frac{\lambda}{\mu_0} \mathbf{\nabla}(\mathbf{\nabla} \cdot \mathbf{A}) = \mathbf{J} - \varepsilon_0 \frac{\partial}{\partial t}\left(\mathbf{\nabla} V + \frac{\partial \mathbf{A}}{\partial t} \right). \tag{7.13}$$

The parameter λ, can be chosen freely. Exploiting the constraint in Eqs. (7.10) and (7.12), we obtain

$$-\varepsilon_0 \nabla^2 V = \rho, \tag{7.14}$$

$$\left(\varepsilon_0 \frac{\partial^2}{\partial t} - \frac{1}{\mu_0} \nabla^2 \right) \mathbf{A} = \mathbf{J} - \varepsilon_0 \frac{\partial}{\partial t}(\mathbf{\nabla} V), \tag{7.15}$$

$$\mathbf{\nabla} \cdot \mathbf{A} = 0. \tag{7.16}$$

Eq. (7.14) justifies the name of this gauge: the scalar potential is the instantaneous Coulomb potential of the charge distribution.

Eqs. (7.14) and (7.15) can be formally solved by Green functions. In general, a Green function corresponding to a differential operator Δ is the solution of the following equation:

$$\Delta * G(\mathbf{r}, \mathbf{r}') = \delta(\mathbf{r} - \mathbf{r}'). \tag{7.17}$$

We have already seen that the Coulomb problem can be solved by the Green function $G(\mathbf{r}, \mathbf{r}') = -(1/4\pi)\delta(\mathbf{r} - \mathbf{r}')$. It should be emphasized that the Green function is not only determined by the structure of the differential operator but also by the boundary conditions. The wave equation (7.15) can also be formally solved by a Green function obeying

$$\left(\frac{1}{c^2} \frac{\partial^2}{\partial t^2} - \nabla^2 \right) G(\mathbf{r}, t, \mathbf{r}', t') = \delta(\mathbf{r} - \mathbf{r}')\delta(t - t'), \tag{7.18}$$

such that

$$\mathbf{A}(\mathbf{r}, t) = \int_{-\infty}^{\infty} dt' \int d\tau'\, G(\mathbf{r}, t, \mathbf{r}', t')\left(\mathbf{J}(\mathbf{r}', t') - \varepsilon \frac{\partial}{\partial t} \mathbf{\nabla} V \right). \tag{7.19}$$

In free space the Green function is easily found by carrying out a Fourier expansion

$$G(\mathbf{r}, t, \mathbf{r}', t') = \frac{1}{(2\pi)^4} \int_{-\infty}^{\infty} d\omega \int d^3\mathbf{k}\, G(\omega, \mathbf{k}) \exp[\mathrm{i}(\omega(t - t') - \mathbf{k} \cdot (\mathbf{r} - \mathbf{r}'))]. \tag{7.20}$$

Defining $k^2 = (\omega/c)^2 - |\mathbf{k}|^2$, the Green function is $G(\omega, \mathbf{k}) = k^{-2}$. In order to respect physical causality the $(\omega, \mathbf{k})$ – integration should be done in such a way that the *retarded* Green function is obtained. This can be done by adding an infinitesimal positive shift to the poles of the Green function or propagator in the momentum representation, i.e., $G(\omega, \mathbf{k}) = 1/(k^2 - i\varepsilon)$. The ω-integral then generates a step function in the difference of the time arguments

$$\frac{1}{2\pi} \int_{-\infty}^{\infty} d\omega \, \frac{e^{i\omega(t-t')}}{\omega - \omega_0 - i\varepsilon} = i\theta(t - t')e^{i\omega_0(t-t')}. \tag{7.21}$$

7.2. The Lorenz gauge

The next most commonly used gauge condition is the Lorenz gauge. In this gauge the scalar potential and vector potential are treated on an equal footing. The condition reads

$$C[\mathbf{A}, V] \equiv \mathbf{\nabla} \cdot \mathbf{A} + \frac{1}{c^2} \frac{\partial V}{\partial t} = 0, \tag{7.22}$$

where $c^{-1} = \sqrt{\mu_0 \varepsilon_0}$ is the (vacuum) speed of light. The generic equations of motion (7.5) and (7.6) then lead to

$$\left(\frac{1}{c^2} \frac{\partial^2}{\partial t^2} - \nabla^2 \right) V = \frac{\rho}{\varepsilon_0}, \tag{7.23}$$

$$\left(\frac{1}{c^2} \frac{\partial^2}{\partial t^2} - \nabla^2 \right) \mathbf{A} = \mu_0 \mathbf{J}. \tag{7.24}$$

The Lorenz gauge is very suitable for performing calculations in the radiation regime. First of all, the similar treatment of all potentials simplifies the calculations and next, the traveling time intervals of the waves are not obscured by the "instantaneous" adaption of the fields to the sources as is done in the Coulomb gauge. This point is not manifest for free-field radiation, since for sourceless field solutions the absence of charges leads to $\mathbf{\nabla} \cdot \mathbf{E} = 0$ which is solved by $V(\mathbf{r}, t) = 0$. Therefore the Coulomb gauge is suitable to handle plane electromagnetic waves. These waves have two transverse polarization modes. In the case of extended charge distributions, Gauss' law gets modified and as a consequence the scalar potential cannot be taken identically equal to zero anymore. In the Lorenz gauge, there are four fields participating in the free-field solution. Definitely two of these fields are fictitious and, as such, they are called "ghost" fields. The longitudinal polarization of an electromagnetic wave corresponds to a ghost field. Care must be taken that these unphysical fields do not have an impact on the calculation of the physical quantities $\mathbf{E}$ and $\mathbf{B}$.

7.3. The Landau gauge

Various derivations of the integer quantum Hall effect (IQHE) are based on the Landau gauge. The IQHE that was discovered by VON KLITZING, DORDA, and PEPPER [1980] may generally occur in two-dimensional conductors with a finite width, such as the conduction channel in the inversion layer of a metal-oxide-semiconductor field-effect transistor (MOSFET) or the potential well of a semiconductor heterojunction.

Consider a two-dimensional electron gas (2DEG) confined to a ribbon $0 \leqslant x \leqslant L$, $|y| \leqslant W/2$, $z = 0$ carrying an electron current I in the x-direction. When a homogeneous magnetic field $\mathbf{B}$ is applied perpendicularly to the strip, the electrons are deflected by the Lorentz force $-e\mathbf{v} \times \mathbf{B}$ and start piling up at one side of the strip leaving a positive charge at the other side. As a result, a transverse Hall voltage V_H arises and prevents any further lateral transfer of deflected electrons. This phenomenon is of course nothing but the classical Hall effect for which the Hall field is probed by the Hall resistance being defined as the ratio of the Hall voltage and the longitudinal current I:

$$R_H = \frac{V_H}{I}. \tag{7.25}$$

However, if the ribbon is cooled down to cryogenic temperatures and the density of the 2DEG is systematically increased by changing the gate voltage, one may observe subsequent plateaus in the Hall resistance, corresponding to a series of quantized values

$$R_H = \frac{h}{2e^2\nu} = \frac{R_K}{\nu}, \tag{7.26}$$

where $R_K = h/2e^2 = 25812.8 \ \Omega$ is the von Klitzing resistance and ν is a positive integer.

Moreover, each time the Hall resistance attains a plateau, the longitudinal resistance of the ribbon drops to zero, which is a clear indication of ballistic, scattering free transport. For extensive discussions on the theory of the quantum Hall effect, we refer to BUTCHER, MARCH, and TOSI [1993], DATTA [1995], DITTRICH, HAENGGI, INGOLD, KRAMER, SCHOEN, and ZWERGER [1997], EZAWA [2000], and all references therein. Here we would merely like to sketch how the choice of a particular gauge may facilitate the description of electron transport in terms of spatially separated, current carrying states (edge states).

The one-electron Hamiltonian reads

$$H = \frac{(\mathbf{p} + e\mathbf{A})^2}{2m} + U(y), \tag{7.27}$$

where $\mathbf{A}$ is the vector potential incorporating the external magnetic field and $U(y)$ describes the confining potential in the lateral direction. In view of the longitudinal, macroscopic current, it is quite natural to inquire whether the eigensolutions of $H\psi(x, y, z) = E\psi(x, y, z)$ are modulated by plane waves propagating along the x-direction, i.e.,

$$\psi(x, y) = \frac{1}{\sqrt{L}}e^{ikx}\chi_k(y), \tag{7.28}$$

where the wave number k would be an integer multiple of $2\pi/L$ to comply with periodic boundary conditions. Clearly, the establishment of full translational invariance for the Hamiltonian proposed in Eq. (7.27) is a prerequisite and so we need to construct a suitable gauge such that the nonzero components of $\mathbf{A}$ do not depend on x. The simplest gauge meeting this requirement is the Landau gauge, which presently takes the form

$$\mathbf{A} = (-By, 0, 0), \tag{7.29}$$

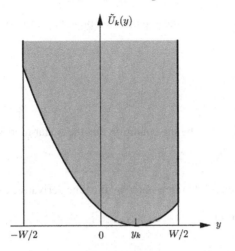

FIG. 7.1. Effective confinement potential in a Hall bar (shaded area). The bare confinement is invoked by a "hard wall" restricting the lateral motion to the interval $|y| < W/2$.

thereby giving rise to the correct magnetic field $\nabla \times \mathbf{A} = B\mathbf{e}_z$. Combining Eqs. (7.27), (7.28), and (7.29), we obtain an effective Schrödinger equation for the "transverse" wave functions $\chi_k(y)$:

$$-\frac{\hbar^2}{2m}\frac{d^2\chi_k(y)}{dy^2} + \left[\tilde{U}_k(y) - E\right]\chi_k(y) = 0 \tag{7.30}$$

with

$$\tilde{U}_k(y) = U(y) + \frac{1}{2}m\omega_c^2(y - y_k)^2. \tag{7.31}$$

$\tilde{U}_k(y)$ acts as an effective confinement potential, centered around its minimum at $y = y_k$ (see Fig. 7.1) where

$$y_k = \frac{\hbar k}{eB} \tag{7.32}$$

and $\omega_c = eB/m$ is the cyclotron frequency. For strong magnetic fields, the eigenfunctions of Eq. (7.30) corresponding to a given wave number k are strongly peaked around $y = y_k$ where the probability of finding an electron outside the effective potential well falls off very rapidly. In particular, when $|k|$ increases, y_k will become of the same order of magnitude as the ribbon half-width or get even larger, so that the corresponding eigenstates – the so-called "edge states" – are strongly localized near the edges of the Hall bar while states with positive momenta $\hbar k$ have no significant lateral overlap with states having negative momenta. The spatial separation of edge states with different propagation directions and the resulting reduction of scattering matrix elements is crucial for the occurrence of the quantized Hall plateaus and can obviously be investigated most conveniently by adopting the Landau gauge since the latter ensures translational invariance in the direction of the current. It should be noted however that a full analytical solution cannot be given in terms of the familiar harmonic oscillator functions

(Hermite functions) because of the edge-related boundary condition

$$\chi_k\left(\pm\frac{W}{2}\right) = 0. \tag{7.33}$$

7.4. The temporal gauge

The temporal gauge is given by the condition that the scalar field V vanish identically.

$$V(\mathbf{r}, t) = 0. \tag{7.34}$$

The electric field is then solely represented by the time derivative of the vector potential.

$$\mathbf{E}(\mathbf{r}, t) = -\frac{\partial \mathbf{A}(\mathbf{r}, t)}{\partial t}. \tag{7.35}$$

In particular, this implies that for a static field the vector potential grows unboundedly in time. This gauge has the nice property that from a Lagrangian point of view the electric field is just the canonical momentum conjugated to the vector field variables, i.e.,

$$\mathcal{L} = \frac{1}{2}\varepsilon_0\left(\frac{\partial \mathbf{A}}{\partial t}\right)^2 - \frac{1}{2\mu_0}(\nabla \times \mathbf{A})^2. \tag{7.36}$$

7.5. The axial gauge

The axial gauge is a variation of the theme above. In this gauge one component of the vector potential, e.g., A_z is set identically equal to zero.

$$A_z = 0. \tag{7.37}$$

This gauge may be exploited if a cylindrical symmetry is present. This symmetry can be inserted by setting

$$\mathbf{A}(\rho, \phi, z) = \left(A_\rho(\rho, \phi), A_\phi(\rho, \phi), 0\right) \tag{7.38}$$

in cylindrical coordinates (ρ, ϕ, z). Then

$$\mathbf{B} = \nabla \times \mathbf{A} = \mathbf{e}_z \frac{1}{\rho}\left(\frac{\partial}{\partial \rho}(\rho A_\phi) - \frac{\partial A_\rho}{\partial \phi}\right). \tag{7.39}$$

An infinitely thin solenoid along the z-axis corresponds to a magnetic field distribution like a "needle," i.e., $\mathbf{B} = \Phi\delta(x)\delta(y)\mathbf{e}_z$. Such a field can be represented by the following vector potential:

$$\mathbf{A} = \frac{\Phi}{2\pi\rho}\mathbf{e}_\phi, \tag{7.40}$$

where Φ denotes the magnetic flux generated by the solenoid.

7.6. The 't Hooft gauge

The selection of a gauge should be done by first identifying the problem that one wants to solve. Experience has shown that a proper selection of the gauge condition is essential to handle a particular issue. At all times it should be avoided that in the process of constructing the solution one should jump ad-hoc from one gauge condition to another. There can be found examples in the literature, where this is done, e.g., a sudden jump is taken from the Coulomb gauge to the temporal gauge, without defining the transition function that accompanies such a gauge transformation. Moreover, the demonstration that the physical results are insensitive to such transitions is often neither given. The gauge fixing method due to 'T HOOFT [1971] carefully takes the above considerations into account. It illustrates the freedom in choosing a gauge condition as well as the sliding in going from one gauge condition to another. Whereas 't Hooft's original work deals with the theory of weak interactions, the ideas can also be applied to condensed matter physics. Suppose that the physical system consists of the electromagnetic fields $(V, \mathbf{A})$ and some charged scalar field ϕ. For the latter, there is a Lagrangian density

$$\mathcal{L}_{\text{scalar}} = \frac{1}{2} i\hbar \left(\phi^* \frac{\partial \phi}{\partial t} - \phi \frac{\partial \phi^*}{\partial t} \right) - \frac{\hbar^2}{2m} (\nabla \phi^*) \cdot (\nabla \phi) - W(\phi^* \phi). \tag{7.41}$$

The potential W describes the (massive) mode of this scalar field and possible self-interactions. If this potential has the form

$$W(\phi^* \phi) = c_2 |\phi|^2 + c_3 |\phi|^3 + c_4 |\phi|^4 \tag{7.42}$$

with c_2 a positive number the field ϕ then this Lagrangian density describes massive scalar particles and the vacuum corresponds to $\phi = 0$. On the other hand, if $c_2 < 0$ then the minimum of W occurs at $|\phi| \equiv \phi_0 \neq 0$. In condensed matter physics, the ground state of a superconductor has nonzero expectation value for the presence of Cooper pairs. These Cooper pairs can be considered as a new particle having zero spin, i.e., it is a boson and its charge is $2e$. The corresponding field for these bosons can be given by ϕ as above, and the ground state is characterized by some nonzero value of ϕ. This can be realized by setting $c_2 < 0$. The interaction of this scalar field with the electromagnetic field is provided by the minimal substitution procedure and leads to the following Lagrangian

$$\begin{aligned}
\mathcal{L} &= \mathcal{L}_{\text{EM}} + \mathcal{L}_{\text{scalar}} + \mathcal{L}_{\text{int}}, \\
\mathcal{L}_{\text{int}} &= \mathbf{J} \cdot \mathbf{A} - \rho V - \frac{e}{m} \rho A^2, \\
\rho &= -e\phi^* \phi, \\
\mathbf{J} &= \frac{ie\hbar}{2m} [\phi^* \nabla \phi - (\nabla \phi^*) \phi] + \frac{e}{m} \rho \mathbf{A}.
\end{aligned} \tag{7.43}$$

The complex field $\phi = \phi_1 + i\phi_2$ can now be expanded around the vacuum expectation value $\phi = \phi_0 + \chi + i\phi_2$. The interaction Lagrangian will contain terms being quadratic in the fields that mix the electromagnetic potentials with the scalar fields. Such terms can be eliminated by choosing the gauge condition in such way that these terms cancel, i.e., by properly selecting the constants α_1 and α_2 in

$$C[\mathbf{A}, V, \chi, \phi_2] \equiv \nabla \cdot \mathbf{A} + \frac{1}{c}\frac{\partial V}{\partial t} + \alpha_1\chi + \alpha_2\phi_2 = 0. \tag{7.44}$$

8. The geometry of electrodynamics

Electrodynamics was discovered as a phenomenological theory. Starting from early experiments with amber, permanent magnets, and conducting wires, one finally arrived after much effort at Gauss' law. Biot–Savart's law and Faraday's law of induction. Only Maxwell's laws were obtained by theoretical reasoning being confirmed experimentally later on by Herz. Maxwell's great achievement was later equalized by Einstein who proposed in the general theory of relativity that

gravity = curvature.

Ever since Einstein's achievement of describing gravity in terms of non-Euclidean geometry, theoretical physics has witnessed a stunning development based on geometrical reasoning. Nowadays it is generally accepted that the standard model of matter, based on gauge theories, is the correct description (within present-day experimental accessibility) of matter and its interaction. These gauge theories have a geometrical interpretation very analogous to Einstein's theory of gravity. In fact, we may widen our definition of "geometry" such that gravity (coordinate covariance) and the standard theory (gauge covariance) are two realizations of the same mechanism. Electrodynamics is the low-energy part of the standard model. Being a major aspect of this book, it deserves special attention and in this interpretation. Besides the esthetic beauty that results from these insights, there is also pragmatic benefit. Solving electrodynamic problems on the computer, guided by the geometrical meaning of the variables has been a decisive factor for the success of the calculation. This was already realized by WILSON [1974] when he performed computer calculations of the quantum aspects of gauge theories. In order to perform computer calculations of the classical fields, geometry plays an important role. However, the classical fields $\mathbf{E}$ and $\mathbf{B}$ as well as the sources ρ and $\mathbf{J}$ are invariant under gauge transformations and therefore their deeper geometrical meaning is hidden. In fact, we can identify the proper geometric character for these variables, such as scalars (zero-forms), force fields (one-forms), fluxes (two-forms), or volume densities (three-forms) as can be done for any other fluid dynamic system, but this can be done without making any reference to the geometric nature of electrodynamics in the sense that $\mathbf{E}$ and $\mathbf{B}$ represent the *curvature* in the geometrical interpretation of electrodynamics. Therefore, in this section we will consider the scalar potential and vector potential fields that do depend on gauge transformations and as such will give access to the geometry of electrodynamics.

8.1. Gravity as a gauge theory

The history of the principle of gauge invariance begins with the discovery of the principle of general covariance in general relativity. According to this principle the physical laws should maintain their form for all coordinate systems. In 1918, Hermann Weyl made an attempt to unify electrodynamics with gravity in WEYL [1918]. According to the general theory of relativity, the gravitational field corresponds to curvature of space–time, and therefore, if a vector is parallel transported along a closed loop, the angle between the starting vector and the final vector will differ from zero. Furthermore, this angle is a measure for the curvature in space. Weyl extended the Riemann geometry in such a way that not only the angle changes but also the *length* of the vector. The relative change in length is described by an antisymmetric tensor and this tensor is invariant under changing the "unit of length." This invariance is closely related to charge conservation. Weyl called this "Maszstab Invarianz." The theory turned out to be contradictory and was abandoned, but the term "Maszstab Invarianz" survived (Maszstab = measure = gauge). With the arrival of quantum mechanics the principle of gauge invariance obtained its final interpretation: gauge invariance should refer to the phase transformations that may be applied on the wave functions. In particular, the phase transformation may be applied with different angles for different points in space and time.

$$\psi(\mathbf{r}, t) \rightarrow \psi'(\mathbf{r}, t) = \exp\left(\frac{ie}{\hbar}\chi(\mathbf{r}, t)\right)\psi(\mathbf{r}, t). \tag{8.1}$$

At first sight it looks as if we have lost the geometrical connection and the link is only historical. However, a closer look at gravity shows that the link is still present.

Starting from the idea that all coordinate systems are equivalent, we may consider a general coordinate transformation

$$x^\mu \rightarrow x'^\mu = x'^\mu(x^\nu). \tag{8.2}$$

The transformation rule for coordinate differentials is

$$dx'^\mu = \frac{\partial x'^\mu}{\partial x^\nu} dx^\nu. \tag{8.3}$$

An ordered set of functions transforming under a change of coordinates in the same way as the coordinate differentials is defined to be a *contravariant vector*

$$V'^\mu = \frac{\partial x'^\mu}{\partial x^\nu} V^\nu. \tag{8.4}$$

A *scalar* transforms in an invariant way, i.e.,

$$\phi(x) \rightarrow \phi'(x') = \phi(x). \tag{8.5}$$

The derivatives of a scalar transform as

$$V'_\mu = \frac{\partial x^\nu}{\partial x'^\mu} V_\nu. \tag{8.6}$$

Any ordered set of functions transforming under a change of coordinates as the derivatives of a scalar function is a *covariant vector*. In general, *tensors* transform according to a multiple set of prefactors, i.e.,

$$V'^{\alpha_1\alpha_2\cdots}_{\mu_1\mu_2\cdots} = \frac{\partial x'^{\alpha_1}}{\partial x^{\beta_1}}\frac{\partial x'^{\alpha_2}}{\partial x^{\beta_2}}\frac{\partial x^{\nu_1}}{\partial x'^{\mu_1}}\frac{\partial x^{\nu_2}}{\partial x'^{\mu_2}} \cdots V^{\beta_1\beta_2\cdots}_{\nu_1\nu_2\cdots}. \tag{8.7}$$

The principle of general coordinate covariance can be implemented by claiming that all physical laws should be expressed as tensor equations. Since left- and right-hand sides will transform with equal sets of prefactors, the form invariance is guaranteed.

So far, we have only been concerned with the change from one arbitrary coordinate system to another. One might argue that this will just hide well-known results in a thick shell of notational complexity. In order to peal off these shells and to find the physical implications one must refer to the *intrinsic* properties of the geometric structure. Occasionally, the intrinsic structure is simple, e.g., flat space time, and the familiar relations are recovered. It was Einstein's discovery that space–time is *not* flat in the presence of matter and therefore the physical laws are more involved.

Riemann geometry is a generalization of Euclidean geometry in the sense that locally one can still find coordinate systems $\xi^\mu = (ict, \mathbf{x})$, such that the distance between two near-by points is given by Pythagoras' theorem, i.e.,

$$ds^2 = \delta_{\mu\nu}\, d\xi^\mu\, d\xi^\nu. \tag{8.8}$$

In an arbitrary coordinate system the distance is given by

$$ds^2 = g_{\mu\nu}(x)\, dx^\mu\, dx^\nu, \tag{8.9}$$

where

$$g_{\mu\nu}(x) = \frac{\partial \xi^\alpha}{\partial x^\mu}\frac{\partial \xi^\beta}{\partial x^\nu}\delta_{\alpha\beta} \tag{8.10}$$

is the metric tensor of the coordinate system.

In the local coordinate system, ξ, the equation of motion of a freely falling particle is given by

$$\frac{d^2\xi^\mu}{ds^2} = 0. \tag{8.11}$$

In an arbitrary coordinate system, this equation becomes

$$\frac{d}{ds}\left(\frac{\partial \xi^\mu}{\partial x^\alpha}\frac{dx^\alpha}{ds}\right) = 0. \tag{8.12}$$

This can be evaluated to

$$\frac{d^2x^\alpha}{ds^2} + \Gamma^\alpha_{\mu\nu}\frac{dx^\mu}{ds}\frac{dx^\nu}{ds} = 0, \tag{8.13}$$

where $\Gamma^\alpha_{\mu\nu}$ is the *affine connection*

$$\Gamma^\alpha_{\mu\nu} = \frac{\partial x^\alpha}{\partial \xi^\beta}\frac{\partial^2 \xi^\beta}{\partial x^\mu \partial x^\nu}. \tag{8.14}$$

The affine connection transform under general coordinate transformations as

$$\Gamma'^\alpha_{\mu\nu} = \frac{\partial x'^\alpha}{\partial x^\rho}\frac{\partial x^\tau}{\partial x'^\mu}\frac{\partial x^\sigma}{\partial x'^\nu}\Gamma^\rho_{\tau\sigma} + \frac{\partial x'^\alpha}{\partial x^\rho}\frac{\partial^2 x^\rho}{\partial x'^\mu \partial x'^\nu}. \tag{8.15}$$

The second term destroys the covariance of the affine connection, i.e., the affine connection is *not* a tensor.

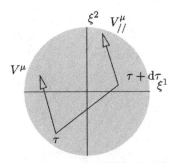

FIG. 8.1. Parallel displacement in the locally Euclidean coordinate system.

The metric tensor $g_{\mu\nu}(x)$ contains information on the local curvature of the Riemann geometry. Now consider a vector $V^\mu(\tau)$ along a curve $x^\mu(\tau)$. In the locally Euclidean coordinate system (ξ), the change of the vector along the curve is $dV^\mu/d\tau$. In another coordinate system (x'), we find from the transformation rule (8.4)

$$\frac{dV'^\mu}{d\tau} = \frac{\partial x'^\mu}{\partial x^\nu}\frac{dV^\nu}{d\tau} + \frac{\partial^2 x'^\mu}{\partial x^\nu \partial x^\lambda}\frac{\partial x^\lambda}{\partial \tau}V^\nu(\tau). \tag{8.16}$$

The second derivative in the second term is an inhomogeneous term in the transformation rule that prevents $dV^\mu/d\tau$ from being a vector and contains the key to curvature. This term is directly related to the affine connection. The combination

$$\frac{DV^\mu}{D\tau} = \frac{dV^\mu}{d\tau} + \Gamma^\mu_{\nu\lambda}\frac{dx^\lambda}{d\tau}V^\nu \tag{8.17}$$

transforms as a vector and is called the *covariant* derivative along the curve. In the restricted region where we can use the Euclidean coordinates, ξ, we may apply Euclidean geometrical methods, and in particular we can shift a vector over an infinitesimal distance from one base point to another and keep the initial and final vector parallel. This is depicted in Fig. 8.1. The component of the vector do not alter by the shift operation: $\delta V^\mu = 0$. Furthermore, in the local frame $x^\mu = \xi^\mu_{x(\tau)}$, the affine connection vanishes, i.e., $\Gamma^\alpha_{\mu\nu} = 0$. Therefore, the conventional operation of parallelly shifting a vector in the locally Euclidean coordinate system can be expressed by the equation $DV^\mu/D\tau = 0$. Being a tensor equation, this it true in all coordinate systems. A vector, whose covariant derivative along a curve vanishes is said to be *parallel* transported along the curve. The coordinates satisfy the following first-order differential equations:

$$\frac{dV^\mu}{d\tau} = -\Gamma^\mu_{\nu\lambda}\frac{dx^\lambda}{d\tau}V^\nu. \tag{8.18}$$

The parallel transport of a vector V^μ over a small distance dx^ν changes the components of the vector by amounts

$$\delta V^\mu = -\Gamma^\mu_{\nu\lambda}V^\nu \delta x^\lambda. \tag{8.19}$$

In general, if we want to perform the differentiation of a tensor field with respect to the coordinates, we must compare tensors in two nearby points. In fact, the comparison corresponds to subtraction, but a subtraction is only defined if the tensors are

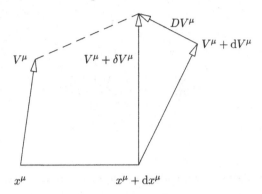

FIG. 8.2. The covariant derivative of a vector field.

anchored to the same point. (In different points, we have different local coordinate systems.) Therefore we must first parallel transport the initial tensor to the nearby point before the subtraction can be performed. This is illustrated in Fig. 8.2. For example, the covariant differential of a vector field is

$$DV^\mu = dV^\mu - \delta V^\mu = \left(\frac{\partial V^\mu}{\partial x^\lambda} + \Gamma^\mu_{\lambda\kappa} V^\kappa \right) dx^\lambda = D_\lambda V^\mu \, dx^\lambda. \tag{8.20}$$

So far, the general coordinate systems include both accelerations originating from nonuniform boosts of the coordinate systems as well as acceleration that may be caused by gravitational field due to the presence of matter. In the first case, space–time is not really curved. In the second case space–time is curved. In order to find out whether gravitation is present one must extract information about the intrinsic properties of space–time. This can be done by the parallel transport of a vector field along a closed loop. If the initial and final vector differ, one can conclude that gravity is present. The difference that a closed loop (see Fig. 8.3) transport generates is given by

$$\Delta V^\mu = V^\mu_{\text{via B}} - V^\mu_{\text{via D}} = R^\mu_{\rho\lambda\sigma} V^\rho \delta x^\lambda \delta x^\sigma, \tag{8.21}$$

where

$$R^\mu_{\rho\lambda\sigma} = \frac{\partial \Gamma^\mu_{\rho\lambda}}{\partial x^\sigma} - \frac{\partial \Gamma^\mu_{\rho\sigma}}{\partial x^\lambda} + \Gamma^\eta_{\rho\lambda} \Gamma^\mu_{\sigma\eta} - \Gamma^\eta_{\rho\sigma} \Gamma^\mu_{\lambda\eta} \tag{8.22}$$

is the *curvature* tensor or Riemann tensor. This tensor describes the intrinsic curvature in a point.

We are now prepared to consider the geometrical basis of electrodynamics and other gauge theories but we will first summarize a few important facts:
- in each space–time point a local frame may be erected,
- the affine connection is a path-dependent quantity,
- the affine connection does not transform as a tensor,
- the field strength (curvature) may be obtained by performing a parallel transport along a closed loop.

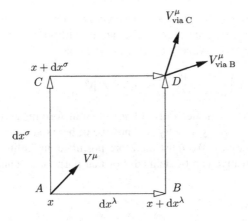

FIG. 8.3. Determination of the curvature from a round trip along a closed loop.

8.2. *The geometrical interpretation of electrodynamics*

As for the local Euclidean coordinate systems, we will consider the possibility of setting up in each space–time point a local frame for fixing the phase of the complex wave function $\psi(\mathbf{r}, t)$ (see Fig. 8.4). Since the choice of such a local frame (gauge) is not unique we may rotate the frame without altering the physical content of a frame fixing.

We can guarantee the latter by demanding appropriate transformation properties (see the above section about tensors) of the variables. Changing the local frame for the phase of a wave function amounts to

$$
\psi'(\mathbf{r}, t) = \exp\left(\frac{ie}{\hbar}\chi(\mathbf{r}, t)\right)\psi(\mathbf{r}, t),
$$

$$
\psi'^*(\mathbf{r}, t) = \exp\left(-\frac{ie}{\hbar}\chi(\mathbf{r}, t)\right)\psi^*(\mathbf{r}, t).
$$

(8.23)

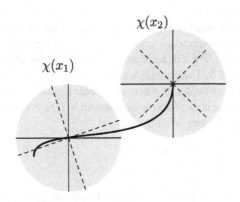

FIG. 8.4. Local frames for the phase of a wave function.

These transformation rules are similar to the contravariant and covariant transformation rules for vectors in the foregoing section. We can similarly construct a "scalar" by taking $\psi^*\psi$. The derivative of the wave function transforms as

$$\frac{\partial \psi'}{\partial x^\mu} = \exp\left(\frac{ie}{\hbar}\chi\right)\frac{\partial \psi}{\partial x^\mu} + \frac{ie}{\hbar}\frac{\partial \chi}{\partial x^\mu}\exp\left(\frac{ie}{\hbar}\chi\right)\psi. \tag{8.24}$$

The second term prevents the derivative of ψ from transforming as a "vector" under the change of gauge. However, geometry will now be of help to construct gauge covariant variables from derivatives. We must therefore postulate an "affine connection," such that a covariant derivative can be defined. For that purpose a connection, A_μ, is proposed that transforms as

$$A_\mu = A_\mu + \frac{\partial \chi}{\partial x^\mu}. \tag{8.25}$$

The covariant derivative is

$$D_\mu = \frac{\partial}{\partial x^\mu} + \frac{ie}{\hbar}A_\mu. \tag{8.26}$$

Similar to the gravitational affine connection, the field A_μ can be used to construct "parallel" transport. Therefore, the field A_μ must be assigned to the *paths* along which the transport takes place. The curvature of the connection can also be constructed by making a complete turn around a closed loop. The result is

$$F_{\mu\nu}\delta x^\mu \delta x^\nu = \oint dx^\mu A_\mu, \tag{8.27}$$

where

$$F_{\mu\nu} = \frac{\partial A_\mu}{\partial x^\nu} - \frac{\partial A_\nu}{\partial x^\mu} \tag{8.28}$$

is the electromagnetic field tensor.

In order to perform numerical computations starting from the fields A_μ it is necessary to introduce a discretization grid. The simulation of a finite space or space–time domain requires that each grid point be separated by a finite distance from its neighboring points. The differential operators that appear in the continuous field equations must be translated to the discretization grid by properly referencing to the geometrical meaning of the variables. The connections A_μ should be assigned to the links of the grid, as depicted in Fig. 8.5. The geometrical interpretation suggests that this is the only correct scheme for solving field and potential problems on the computer.

The numerical consequences of above assignment will be considered in the following example. We will solve the steady-state equation

$$\mathbf{\nabla} \times \mathbf{B} = \mu_0 \mathbf{J}, \qquad \mathbf{B} = \mathbf{\nabla} \times \mathbf{A},$$
$$\mathbf{J} = \sigma \mathbf{E}, \qquad \mathbf{E} = -\mathbf{\nabla}V, \tag{8.29}$$

by discretizing the set of equations on a regular Cartesian grid having N nodes in each direction. The total number of nodes in D dimensions is $M_{\text{nodes}} = N^D$. To each node

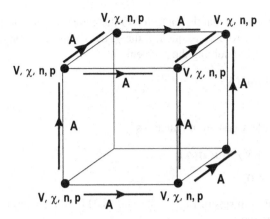

Fɪɢ. 8.5. The fundamental variables on the Cartesian grid.

we may associate D links along the positive directions, and therefore the grid has approximately DN^D links. There are $2D$ sides with each a number of $N^{(D-1)}$ nodes. Half the fraction of side nodes will not contribute a link in the positive direction. Therefore, the precise number of links in the lattice is $M_{\text{links}} = DN^D(1 - \frac{1}{N})$.

As far as the description of the electromagnetic field is concerned, the counting of unknowns for the full lattice results into M_{links} variables (A_{ij}) for the links, and M_{nodes} variables (V_i) for the nodes. Since each link (node) gives rise to one equation, the naive counting is consistent. However, we have not yet implemented the gauge condition. The conventional Coulomb gauge $\nabla \cdot \mathbf{A} = 0$, constraints the link degrees of freedom and therefore not all link fields are independent. There are $3N^3(1 - \frac{1}{N})$ link variables and $3N^3(1 - \frac{1}{N}) + N^3$ equations, including the constraints. As a consequence, at first sight it seems that we are confronted with an overdetermined system of equations, since each node provides an extra equation for $\mathbf{A}$. However, the translation of the Maxwell–Ampère equation on the lattice leads to a singular matrix, i.e., not all rows are independent. The rank of the corresponding matrix is $3N^3(1 - \frac{1}{N})$, whereas there are $3N^3(1 - \frac{1}{N}) + N^3$ rows and $3N^3(1 - \frac{1}{N})$ columns. Such a situation is highly inconvenient for solving nonlinear systems of equations, where the nonlinearity stems from the source terms being explicitly dependent on the fields. The application of the Newton–Raphson method requires that the matrices in the related Newton equation be nonsingular and square. In fact, the nonsingular and square form of the Newton–Raphson matrix can be recovered by introducing the more general gauge $\nabla \cdot \mathbf{A} + \nabla^2 \chi = 0$, where an additional field χ, i.e., one unknown per node, is introduced. In this way the number of unknowns and the number of equations match again. In the continuum limit ($N \to \infty$), the field χ and one component of $\mathbf{A}$ can be eliminated. Though being irrelevant for theoretical understanding, the auxiliary field χ is essential for obtaining numerical stability on a discrete, finite lattice. In other words, our specific gauge solely serves as a tool to obtain a discretization scheme that generates a regular Newton–Raphson matrix, as explained in Mᴇᴜʀɪs, Sᴄʜᴏᴇɴᴍᴀᴋᴇʀ, and Mᴀɢɴᴜs [2001].

It should be emphasized that the inclusion of the gauge-fixing field χ should not lead to unphysical currents. As a consequence, the χ-field should be a solution of $\nabla\chi = 0$. To summarize, instead of solving the problem

$$\nabla \times \nabla \times \mathbf{A} = \mu_0 \mathbf{J}(\mathbf{A}),$$
$$\nabla \cdot \mathbf{A} = 0,$$
(8.30)

we solve the equivalent system of equations

$$\nabla \times \nabla \times \mathbf{A} - \gamma \nabla \chi = \mu_0 \mathbf{J}(\mathbf{A}),$$
$$\nabla \cdot \mathbf{A} + \nabla^2 \chi = 0.$$
(8.31)

The equivalence of both sets of Eqs. (8.30) and (8.31) can be demonstrated by considering the action integral

$$S = -\frac{1}{2\mu_0} \int d\tau \, |\nabla \times \mathbf{A}|^2 + \int d\tau \, \mathbf{J} \cdot \mathbf{A}.$$
(8.32)

Functional differentiation with respect to $\mathbf{A}$ yields the field equations

$$\frac{\delta S}{\delta \mathbf{A}} = -\frac{1}{\mu_0} \nabla \times \nabla \times \mathbf{A} + \mathbf{J} = 0.$$
(8.33)

The constraint corresponding to the Coulomb gauge can be taken into account by adding a Lagrange multiplier term to the action integral

$$S = -\frac{1}{2\mu_0} \int d\tau \, |\nabla \times \mathbf{A}|^2 + \int d\tau \, \mathbf{J} \cdot \mathbf{A} + \gamma \int d\tau \, \chi \nabla \cdot \mathbf{A}$$
(8.34)

and perform the functional differentiation with respect to χ

$$\frac{\delta S}{\delta \chi} = \nabla \cdot \mathbf{A} = 0.$$
(8.35)

Finally, the Lagrange multiplier field χ becomes a dynamical variable by adding a free-field part to the action integral

$$S = -\frac{1}{2\mu_0} \int d\tau \, |\nabla \times \mathbf{A}|^2 + \int d\tau \, \mathbf{J} \cdot \mathbf{A} + \gamma \int d\tau \, \chi \nabla \cdot \mathbf{A} - \frac{1}{2}\gamma \int d\tau \, |\nabla \chi|^2$$
(8.36)

and functional differentiation with respect to $\mathbf{A}$ and χ results into the new system of equations. Physical equivalence is guaranteed provided that $\nabla\chi$ does not lead to an additional current source. Therefore, it is required that $\nabla\chi = 0$. In fact, acting with the divergence operator on the first equation of (8.31) gives Laplace's equation for χ. The solution of the Laplace equation is identically zero if the solution vanishes at the boundary.

We achieved to implement the gauge condition resulting into a unique solution and simultaneously to arrive at a system containing the same number of equations and unknowns. Hence a square Newton–Raphson matrix is guaranteed while solving the full set of nonlinear equations.

8.3. Differential operators in Cartesian grids

Integrated over a test volume ΔV_i surrounding a node i, the divergence operator, acting on vector potential $\mathbf{A}$, can be discretized as a combination of 6 neighboring links

$$\int_{\Delta V_i} \nabla \cdot \mathbf{A} \, d\tau = \int_{\partial(\Delta V_i)} \mathbf{A} \cdot d\mathbf{S} \sim \sum_k^6 S_{ik} A_{ik}. \tag{8.37}$$

The symbol $\sim$ represents the conversion to the grid formulation and $\partial(\Delta V_i)$ denotes the boundary of ΔV_i.

Similarly, the gradient operator acting on the ghost field χ or any scalar field V, can be discretized for a link ij using the nodes i and j. Integration over a surface S_{ij} perpendicular to the link ij gives

$$\int_{\Delta S_{ij}} \nabla \chi \cdot d\mathbf{S} \sim \frac{\chi_j - \chi_i}{h_{ij}} S_{ij}, \tag{8.38}$$

where h_{ij} denotes the length of the link between the nodes i and j.

The gradient operator for a link ij, integrated along the link ij, is given by

$$\int_{\Delta L_{ij}} \nabla \chi \cdot d\mathbf{r} \sim \chi_j - \chi_i. \tag{8.39}$$

The *curl–curl* operator can be discretized for a link ij using a combination of 12 neighboring links and the link ij itself. As indicated in Fig. 8.6, the field $\mathbf{B}_i$ in the center of the "wing" i, can be constructed by taking the circulation of the vector potential $\mathbf{A}$

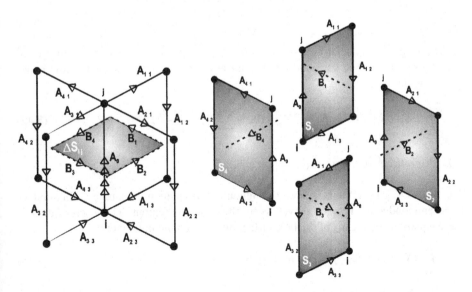

FIG. 8.6. The assembly of the $\nabla \times \nabla \times$-operator using 12 contributions of neighboring links.

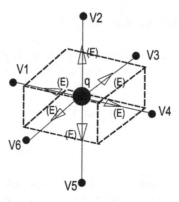

FIG. 8.7. The assembly of the $\nabla \cdot \nabla$-operator using 6 contributions of neighboring nodes.

around the wing i $(i = 1, 4)$

$$\mathbf{B}_i S_i = \sum_{j=1}^{3} A_{ij} h_{ij} + A_0 h_0, \tag{8.40}$$

where h_α is the length of the corresponding link α. Integration over a surface S_{ij} perpendicular to the link ij yields a linear combination of different A_{ij}'s, the coefficients of which are denoted by Λ_{ij}.

$$\int_{\Delta S_{ij}} \nabla \times \nabla \times \mathbf{A} \cdot d\mathbf{S} = \int_{\partial(\Delta S_{ij})} \nabla \times \mathbf{A} \cdot d\mathbf{r} = \int_{\partial(\Delta S_{ij})} \mathbf{B} \cdot d\mathbf{r}$$

$$\sim \Lambda_{ij} A_{ij} + \sum_{kl}^{12} \Lambda_{ij}^{kl} A_{kl}. \tag{8.41}$$

The div-grad (Laplacian) operator can be discretized (see Fig. 8.7) being integrated over a test volume ΔV_i surrounding a node i as a combination of 6 neighboring nodes and the node i itself.

$$\int_{\Delta V_i} \nabla \cdot (\nabla \chi) \, d\tau = \int_{\partial(\Delta V_i)} \nabla \chi \cdot d\mathbf{S} \sim \sum_k^{6} S_{ik} \frac{\chi_k - \chi_i}{h_{ik}}. \tag{8.42}$$

8.4. Discretized equations

The fields $(\mathbf{A}, \chi)$ need to be solved throughout the simulation domain, i.e., for conductors, semiconducting regions as well as for the dielectric regions. The discretization of these equations by means of the box/surface-integration method gives

$$\int_{\Delta S} (\nabla \times \nabla \times \mathbf{A} - \gamma \nabla \chi - \mu_0 \mathbf{J}) \cdot d\mathbf{S} = 0, \tag{8.43}$$

$$\int_{\Delta V} \nabla \cdot \mathbf{J} \, d\tau = 0, \tag{8.44}$$

$$\int_{\Delta V} (\nabla \cdot \mathbf{A} + \nabla^2 \chi) \, d\tau = 0 \qquad (8.45)$$

leading for the independent variables $\mathbf{A}$, χ to

$$\Lambda_{ij} A_{ij} + \sum_{kl}^{12} \Lambda_{ij}^{kl} A_{kl} - \mu_0 S_{ij} J_{ij} - \gamma S_{ij} \frac{\chi_j - \chi_i}{h_{ij}} = 0, \qquad (8.46)$$

$$\sum_{k}^{6} S_{ik} J_{ik} = 0, \qquad (8.47)$$

$$\sum_{k}^{6} S_{ik} \left(A_{ik} + \frac{\chi_k - \chi_i}{h_{ik}} \right) = 0. \qquad (8.48)$$

Depending on the region under consideration, the source terms $(Q_i, \mathbf{J}_{ij})$ differ. In a conductor we implement Ohm's law, $\mathbf{J} = \sigma \mathbf{E}$ on a link ij:

$$J_{ij} = -\sigma_{ij} \left(\frac{V_j - V_i}{h_{ij}} \right) \qquad (8.49)$$

and Q_i is determined by charge conservation.

For the semiconductor environment we follow the Scharfetter–Gummel scheme (SCHARFETTER and GUMMEL [1969]). In this approach, the diffusion equations

$$\mathbf{J} = q \mu c \mathbf{E} \pm kT \mu \nabla c, \qquad (8.50)$$

where the plus (minus) sign refers to negatively (positively) charged particles and c denotes the corresponding carrier density. It is assumed that both the current $\mathbf{J}$ and vector potential $\mathbf{A}$ are constant along a link and that the potential V and the gauge field χ vary linearly along the link. Adopting a local coordinate axis u with $u = 0$ corresponding to node i, and $u = h_{ij}$ corresponding to node j, we may integrate Eq. (8.50) along the link ij to obtain

$$J_{ij} = q \mu_{ij} c \left(\frac{V_i - V_j}{h_{ij}} \right) \pm k_B T \mu_{ij} \frac{dc}{du} \qquad (8.51)$$

which is a first-order differential equation in c. The latter is solved using the aforementioned boundary conditions and gives rise to a nonlinear carrier profile. The current J_{ij} can then be rewritten as

$$\frac{J_{ij}}{\mu_{ij}} = -\frac{\alpha}{h_{ij}} B \left(\frac{-\beta_{ij}}{\alpha} \right) c_i + \frac{\alpha}{h_{ij}} B \left(\frac{\beta_{ij}}{\alpha} \right) c_j, \qquad (8.52)$$

where $B(x)$ is the Bernoulli function

$$B(x) = \frac{x}{e^x - 1} \qquad (8.53)$$

and

$$\alpha = \pm k_B T, \qquad (8.54)$$

$$\beta_{ij} = q(V_i - V_j). \qquad (8.55)$$

8.5. Examples

We present a few examples demonstrating that the proposed potential formulation in terms of the Poisson scalar field V, the vector potential field $\mathbf{A}$, and the ghost field χ, is a viable method to solve the Maxwell field problem. All subtleties related to that formulation, i.e., the positioning of the vector potential on links, and the introduction of the ghost field χ, are already encountered in constructing the solutions of the static equations (SCHOENMAKER and MEURIS [2002]).

8.5.1. Crossing wires

The first example concerns two crossing wires and thereby addresses the three-dimensional features of the solver. The structure is depicted in Fig. 8.8 and has four

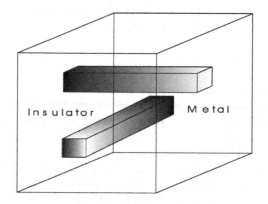

Insulator   Metal

FIG. 8.8. Layout of two crossing wires in insulating environment.

TABLE 8.1
Some characteristic results for two crossing wires

Electric energy (J)		Magnetic energy (J)	
$\frac{1}{2}\varepsilon_0 \int_\Omega d\tau E^2$	1.03984×10^{-18}	$\frac{1}{2\mu_0} \int_\Omega d\tau B^2$	2.89503×10^{-11}
$\frac{1}{2} \int_\Omega d\tau \rho\phi$	1.08573×10^{-18}	$\frac{1}{2} \int_\Omega d\tau \mathbf{J} \cdot \mathbf{A}$	2.92924×10^{-11}

TABLE 8.2
Some characteristic results for a square coaxial cable

a	b	b/a	L	
μm	μm		(cylindrical) (nH)	(square) (nH)
2	6	3	220	255
1	5	5	322	329
1	7	7	389	390
1	10	10	461	458

ports. In the simulation we put one port at 0.1 V and kept the other ports grounded. The current is 4 A. The simulation domain is $10 \times 10 \times 14$ μm³. The metal lines have a perpendicular cross section of 2×2 μm². The resistivity is 10^{-8} Ω m. In Tables 8.1–8.3, some typical results are presented. The energies have been calculated in two different ways and good agreement is observed. This confirms that the new methods underlying the field solver are trustworthy. The χ-field is zero within the numerical accuracy, i.e., $\chi \sim O(10^{-14})$.

8.5.2. *Square coaxial cable*

To show that also inductance calculations are adequately addressed, we calculate the inductance per unit length (L) of a square coaxial cable as depicted in Fig. 8.9. The inductance of such a system with inner dimension a and outer dimension b, was calculated from

$$l \times \frac{1}{2}LI^2 = \frac{1}{2\mu_0} \int_\Omega B^2 \, d\tau = \frac{1}{2} \int_\Omega d\tau \, \mathbf{J} \cdot \mathbf{A} \tag{8.56}$$

with l denoting the length of the cable. As expected, for large values of the ratio $r = b/a$, the numerical result for the square cable approaches the analytical result for a cylindrical cable, $L = (\mu_0/2\pi) \ln(b/a)$.

TABLE 8.3
Some characteristic results for the spiral inductor

Electric energy (J)		Magnetic energy (J)	
$\frac{1}{2}\varepsilon_0 \int_\Omega d\tau \, E^2$	2.2202×10^{-18}	$\frac{1}{2\mu_0} \int_\Omega d\tau \, B^2$	3.8077×10^{-13}
$\frac{1}{2} \int_\Omega d\tau \, \rho\phi$	2.3538×10^{-18}	$\frac{1}{2} \int_\Omega d\tau \, \mathbf{J} \cdot \mathbf{A}$	3.9072×10^{-13}

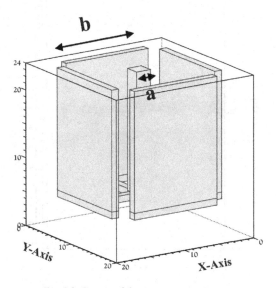

FIG. 8.9. Layout of the square coax structure.

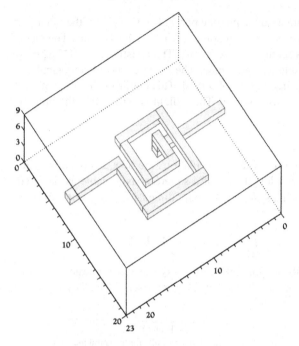

FIG. 8.10. Layout of the spiral inductor structure.

B–field at z= 4.5 (mT)

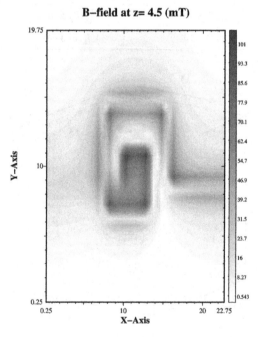

FIG. 8.11. Magnetic field strength in the plane of the spiral inductor.

8.5.3. Spiral inductor

A spiral inductor, as shown in Fig. 8.10 was simulated. This structure also addresses the three-dimensional features of the solver. The cross-section of the different lines is 1 μm $\times$ 1 μm. The overall size of the structure is 8 μm $\times$ 8 μm and the simulation domain is $23 \times 20 \times 9$ μm^3. The resistance is evaluated as $R = V/I$ and equals 0.54 Ω. In Fig. 8.11, the intensity of the magnetic field is shown at height 4.5 μm. From the results in Tables 8.3 we obtain that the inductance of the spiral inductor is 4.23×10^{-11} Henry.

9. Outlook

The preceding sections have been meant to offer the reader a glimpse of the achievements and the present activities in the field of numerical modeling of electromagnetic problems within the framework of 19th century classical electromagnetism that was physically founded by MAXWELL [1954a], MAXWELL [1954b], Faraday, Lenz, Lorentz and many others, and mathematically shaped by the upcoming vector calculus of those days (MORSE and FESHBACH [1953]).

The enormous predictive power of the resulting, "classical" electromagnetic theory and the impressive technological achievements that have emerged from it, may create the false impression that, from the physics point of view, electromagnetism has come to a dead end where no new discoveries should be expected and all remaining questions are reduced to the numerical solubility of the underlying mathematical problems.

Truly, after the inevitable compatibility of electromagnetism with the theory of relativity (EINSTEIN, LORENTZ, MINKOWSKI, and WEYL [1952]) had been achieved and the theory of quantum electrodynamics (QED) (SCHWINGER [1958]) had been successfully established in the first half of the 20th century, neither new fundamental laws nor extensions of the old Maxwell theory have been proposed ever since.

Nevertheless, as was pointed out already in Section 8, modern concepts borrowed from the theory of differential geometry turn out to provide exciting alternatives to formulate the laws of electromagnetism and may gain new insights similar to the understanding of the intimate link between gravity and geometrical curvature of the Minkowski space. Moreover, recent technological developments in the fabrication of nanometer-sized semiconductor structures and mesoscopic devices (DATTA [1995]) have raised new as well as unanswered old questions concerning the basic quantum mechanical features of carrier transport in solids and its relation to both externally applied and induced electromagnetic fields. The topology of electric circuits such as mesoscopic rings carrying persistent currents, mesoscopic devices with macroscopic leads, including quantum wires, quantum dots, quantum point contacts, Hall bars, etc. appears to be a major component determining the transport properties. In particular, the spatial localization of both the electromagnetic fields and the carrier energy dissipation plays an essential role in the quantum theory that governs carrier transport.

This section addresses just a few topics of the above mentioned research domain in order to illustrate that quantum mechanics is invoked not only to provide a correct description of the particles participating in the electric current but also to extend the theory of the electromagnetic field beyond the framework of Maxwell's equations and QED. As the corresponding research area is still being established and theoretical understanding is often still premature, several statements presented in the remainder of this chapter, should be regarded as possible but not final answers to existing problems, thereby mainly reflecting the personal view of the authors. A more detailed treatment of the topics considered below can be found in MAGNUS and SCHOENMAKER [2000c] and MAGNUS and SCHOENMAKER [2002].

9.1. Quantum mechanics, electric circuits and topology

Quantization of the electric conductance of quantum point contacts (QPC) is a striking example of a transport phenomenon that cannot be accounted for by combining classical electrodynamics with conventional transport theory that inherently neglects preservation of phase coherence. A typical QPC consists of a two-dimensional electron gas (2DEG) residing at in a high-mobility semiconductor structure near the interface of, say an AlGaAs/GaAs heterojunction, whereas a negatively biased gate provides a narrow constriction hampering the electron flow in the direction perpendicular to the gate arms (see Fig. 9.1). While the length of the gate arms (along the propagation direction may be of the order of 1 μm, the width is usually smaller than 250 nm. Experimentally, conductance quantization was originally observed by the groups of WHARAM, THORNTON, NEWBURY, PEPPER, AHMED, FROST, HASKO, PEACOCK, RITCHIE, and JONES [1988] and VAN WEES, VAN HOUTEN, BEENAKKER, WILLIAMSON, KOUWENHOVEN, VAN DER MAREL, and FOXON [1988] by connecting the QPC to an external power source (V) through a couple of conducting leads as sketched in Fig. 9.2. While the total resistance R of the circuit was determined by measuring its ohmic response to a given bias voltage V, the resistance R_Q associated with the very QPC was obtained by subtracting the resistance R_L of the two leads:

$$R_Q = R - 2R_L. \tag{9.1}$$

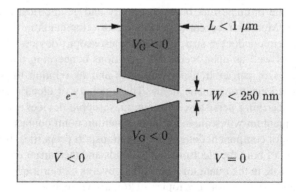

FIG. 9.1. Quantum point contact with length L and width W, considered as a two-terminal device. The source contact (left) is kept on a negative potential V with respect to the drain contact (right).

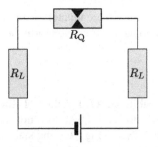

FIG. 9.2. Closed electric circuit containing a QPC connected to a DC power supply through two resistive leads.

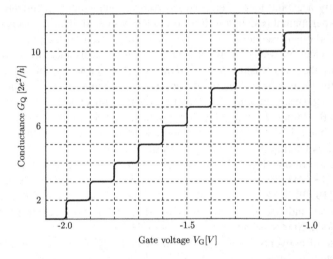

FIG. 9.3. Quantized conductance of a quantum point contact under cryogenic conditions.

After they had cooled down the QPC below 4 K, the experimentalists of both the Delft and Cambridge groups measured the circuit current I as a function of the gate voltage V_G for a fixed bias voltage. As a result, they obtained a staircase-like pattern in the profile of the electric conductance $G_Q = R_Q^{-1}$ associated with the QPC, as indicated schematically in Fig. 9.3. From this observation it follows that the conductance G_Q is quantized in units of $R_K^{-1} = 2e^2/h$ where $R_K = h/e^2 = 25128\ \Omega$ denotes von Klitzing's resistance. A quantitative description is provided by the well-known Landauer–Büttiker formula

$$G_Q = \frac{2e^2}{h}N, \tag{9.2}$$

where N is the number of "conduction channels" that are open for ballistic electron transport through the QPC, given a particular value of the gate voltage V_G. Eq. (9.2) is a special case of a formula that was proposed by LANDAUER [1957], LANDAUER [1970] to describe electron propagation through disordered materials, while it was recovered by BUETTIKER [1986] to cope with semiconductors with mesoscopic active areas. For a two-terminal device the generalized conductance formula reads

$$G = \frac{2e^2}{h} \sum_{n=1}^{N} T_n,$$ (9.3)

where the transmission probabilities $\{T_n\}$ reduce 1 for purely ballistic transport. Although conductance quantization in a QPC does not reach the degree of exactness suggested by the idealized drawing of Fig. 9.3, the stair-case profile has been repeatedly observed by many other researchers in the field and, consequently one should be confident in the experimental verification of this phenomenon. On the other hand, it is the strong belief of the authors that, presently – when this manuscript is being written – the commonly accepted theoretical explanation of conductance quantization runs far behind its experimental realization. It is commonly accepted that the absence of energy dissipation and other decoherence effects and, correspondingly, the preservation of the phase of the electron wave functions over a mesoscopic distance are major keys for understanding the mechanism of quantum transport. Nevertheless, numerous questions concerning the localization of energy dissipation are left unanswered by the underlying theories and a generalized, unifying transport theory connecting the macroscopic models based on the Drude–Lorenz model on one hand and the Landauer–Büttiker picture on the other hand, is still lacking. For a more extensive discussion on common models and theories leading to the Landauer–Büttiker formula, we refer to DATTA [1995], BUTCHER, MARCH, and TOSI [1993], STONE and SZAFER [1988], LENSTRA and SMOKERS [1988], LENSTRA, VAN HAERINGEN, and SMOKERS [1990], STONE [1992], IMRY and LANDAUER [1999]. Here we would like to summarize briefly the main results of conventional theory and discuss an alternative approach which has been proposed recently by the authors in MAGNUS and SCHOENMAKER [2000c].

In the case of conventional conductors one can easily trace back the macroscopic, electric resistance to dissipation of energy and decoherence effects that are due to various elastic and inelastic scattering mechanisms. On the other hand, the question arises why a mesoscopic, ballistic conductor the active region of which is supposed to be free of scattering, can still have a nonzero resistance. Moreover, as one may conclude from Eq. (9.3), this resistance merely depends on the fundamental constants e and $\hbar$ and a set of quantum mechanical transmission coefficients. The latter are usually extracted from the single-electron Schrödinger equation, i.e., under the assumption that many-particle interactions such as electron–electron and even electron–phonon scattering can be neglected. Consequently, the resistance of a ballistic conductor appears to be expressible in quantities that are not referring to neither decoherence nor energy dissipation. As discussed extensively in the above references, a common explanation for this phenomenon is provided by the concept of so-called contact resistance. The underlying picture considers the ballistic conductor as being connected on the "left" and the "right" to two huge reservoirs that are kept on two different chemical potentials μ_L, μ_R so as to maintain between the reservoirs a net current of electrons propagating through one or more channels of the ballistic conductor (such as a QPC or a quantum dot). Due to the mismatch of the huge, macroscopic leads and the mesoscopic active area, two interface regions separating the active area from the "bulk" of the leads. Assuming further that electrons are entering and leaving the active area without undergoing any quantum mechanical reflections in the interface regions, the latter emerge as the missing spots where the phase coherence characterizing the transport in the ballistic region is broken. In other words, the resistance associated with a mesoscopic active areas should be considered localized, being realized in the interface or "contact" regions while the main electrostatic potential drop is still falling over the active area. Even when the notion of nonlocal resistance is rather conceivable in a medium where phase-coherent transport along nanometer-sized paths may demand that Ohm's law $\mathbf{J}(\mathbf{r}) = \sigma \mathbf{E}(\mathbf{r})$ be generalized to $\mathbf{J}(\mathbf{r}) = \int d^3 r' \sigma(\mathbf{r}, \mathbf{r}') \mathbf{E}(\mathbf{r}')$, we feel that the reservoir picture does not satisfactorily explain the phenomenon of quantized conductance. First, to the best of our knowledge, there is neither an unambiguous way of defining the contact regions interfacing between an active area and a reservoir nor a trace of experimental evidence for it. Next, invoking the chemical potentials μ_L and μ_R and the corresponding local thermal equilibria states for the two reservoirs already silently postulates the existence of a finite current without providing explicitly a current limiting mechanism. Moreover, the equation $e V_{app} = \mu_L - \mu_R$ relating the applied bias to the chemical potential difference as a crucial step in conventional treatments, is simply taken for granted (sometimes even taken as a definition of bias voltage!) while FENTON [1992], FENTON [1994] already pointed out that it should be rigorously derived from quantum mechanical first principles. Finally, the topology of an electric circuit containing a ballistic conductor or any mesoscopic device is not reflected in the reservoir concept that treats the circuit as a simply connected, open-ended region. The latter has severe consequences for the description of the driving electric field existing in the active area as will be discussed in the following lines.

For the sake of simplicity, we will consider a DC power source providing the electric circuit with the energy required to maintain a steady current of electrons flowing through a toroidal (doughnut-shaped, torus-like) circuit Ω. In addition, we will assume that no external magnetic field is applied in the circuit region so that the only magnetic field existing in the torus is the self-induced one which is constant in time. According to the third Maxwell equation, the total electric field acting on the electrons in the circuit, should therefore be irrotational, i.e.,

$$\nabla \times \mathbf{E} = 0. \tag{9.4}$$

In spite of Eq. (9.4), the electric field $\mathbf{E}$ is *not* conservative. Indeed, the electromotive force or EMF characterizing the strength of the DC power source, is nothing but the nonvanishing loop integral of $\mathbf{E}$ around any closed curve Γ lying in the interior of the torus and encircling the hole of the torus once and only once (winding number = 1). According to Stokes' theorem for multiply connected regions the curve Γ is arbitrary as long as it is located in a region where $\nabla \times \mathbf{E}$ vanishes, so any internal curve of Ω will do. Physically, the EMF represents the work done by the electric field on a unit charge that makes one complete turn around the circuit (moving along Γ). As an immediate consequence, we need to be most careful when dealing with innocent looking quantities such as electrostatic potential and the notion of potential difference. While an irrotational field $\mathbf{E}(\mathbf{r})$ can always be derived from a scalar potential $V(\mathbf{r})$ in any *simply-connected* subset of the torus (see the Helmholtz theorem), there exists no such scalar potential doing the job along the entire circuit. Mathematically speaking, one could of course imagine a brute force definition for such a potential anyway, namely the line integral of the field $\mathbf{E}$ along a subset of Γ connecting some reference point $\mathbf{r_0}$ with the field point $\mathbf{r}$. However, since the circulation of $\mathbf{E}$ is nonzero when $\mathbf{r}$ travels all around Γ, the value of such a potential would unlimitedly increase (or decrease) when $\mathbf{r}$ keeps on traveling around the circuit. This would give rise to a potential function $V(x_1, x_2, x_3)$ that would be multivalued in the cyclic coordinate, say x_3. Such a function would clearly be unacceptable from the physical point of view which requires all physically meaningful functions to be periodic in x_3. It goes without saying that the concept of EMF is hardly conceivable in a theory describing the electric circuit as an open-ended region. Such a simply-connected region exclusively leads to conservative, irrotational electric fields that cannot give rise to a steady energy supply. The latter is therefore emulated by introducing position dependent chemical potentials artificially keeping the lead reservoirs on different levels of electron supply.

It should be noted at this point that the above topology considerations have already given rise to at least two major conceptual differences between open-ended conductors and closed electric circuits.

First, electrons entering the active area coming from one lead and moving to the other are never seen to return to their "origin" except when they are reflected.[7] As such, the open-ended conductor is very similar to a system of two large water buckets, one of them being emptied into the other through a narrow tube. Although the water flow resembles a steady flow after the initial and before the final transient regime, the water is not being pumped back into the first bucket and the flow trivially stops when the first bucket is empty. On the contrary, although quantum mechanics does not allow an accurate localization of electrons in the transport direction when they reside in delocalized, current carrying states, the electrons are confined to the interior of the circuit region and will make a huge number of turns when a steady-state current is maintained on a macroscopic time scale. Next, in most cases the open-ended conductor model leads to an artificial, spatial division of the circuit into a finite active area and two infinite lead regions. Indeed, position dependence is not only inferred for the chemical potential, in various treatments such as DATTA [1995] one also assigns separate sets of energy spectra and their corresponding quantum states to the three distinct regions: two continuous energy spectra representing the huge and wide leads and a discrete spectrum providing a small number of conduction channels (referred to as N in the Landauer–Büttiker formula). Moreover, at both interfaces emerges a mismatch between the enumerable discrete spectrum and the two continuous spectra and this very mismatch is even considered the origin of the so-called "contact resistance" explaining the phenomenon of conductance quantization.

However, it is known from elementary quantum mechanics that energy and position, being represented by noncommuting operators cannot be simultaneously measured. In other words, there is no physical ground for setting up different quantum mechanical treatments of distinct spatial areas (unless they are completely isolated from each other thereby preventing any exchange of particles, which is obviously not the case). Treating

---

[7]In principle electrons may undergo quantum mechanical reflections at the interfaces between the lead and the active part of the device, but these reflections are explicitly ignored in most of the conventional theories.

the complete circuit – including power source, conducting leads and mesoscopic active area – as a single quantum mechanical entity, a single spectrum of allowed energies and corresponding eigenstates is to be assigned to the entire circuit, not to parts of it. Clearly, unless we are discussing isolated microcircuits such as mesoscopic rings carrying persistent currents, the circuit inevitably becomes huge, due to the presence of the huge leads. Consequently, the single energy spectrum turns out to be a continuous one, consisting virtually of all energies that are accessible by the circuit system. On the other hand, the influence of the active area with either its narrow spatial confinement (QPC) or its huge potential barriers is reflected in the occurrence of a discrete set of sharply peaked resonances emerging in the quantum mechanical transmission coefficient as a function of energy. The corresponding states are genuine "conduction channel states" allowing an appreciable transmission of electrons, while the latter is negligible for any other state. In this picture however, there is no "mismatch" between quantum states, since all states simply pertain to the entire system and only the wave functions (not the energies) depend on position. Consequently, the notion of contact resistance relying on the existence of a mismatch of states, looses its meaning and the basis question remains: what causes the resistance of a mesoscopic active area embedded in a closed electric circuit and why does it take the form of Eq. (9.3)?

Being inspired by the experimental setup, we propose to consider the simplest possible, closed circuit, i.e., a torus-shaped region Ω consisting of a DC power source ("battery" region Ω_B), two ideally conducting leads Ω_{1L} and Ω_{2L} connecting the active area Ω_A, as depicted in Fig. 9.4. In general, the electric field in the circuit region may be decomposed into a conservative and nonconservative part:

$$\mathbf{E}(\mathbf{r}) = \mathbf{E}_C(\mathbf{r}) + \mathbf{E}_{NC}(\mathbf{r}), \tag{9.5}$$

where the conservative component $\mathbf{E}_C$ is derived from an appropriate scalar potential which is periodic along any interior, closed loop Γ (see Fig. 9.5),

$$\mathbf{E}_C(\mathbf{r}) = -\nabla V(\mathbf{r}) \tag{9.6}$$

with

$$\oint_\Gamma \mathbf{E}_C(\mathbf{r}) \cdot d\mathbf{r} = 0, \tag{9.7}$$

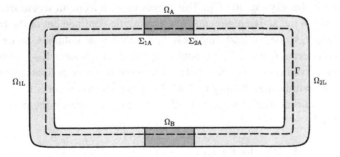

FIG. 9.4. Toroidal electric circuit. (Figure reproduced by permission of the American Physical Society and Springer Verlag.)

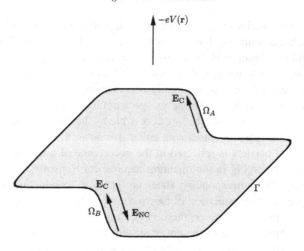

FIG. 9.5. Electrostatic potential energy profile along Γ. (Figure reproduced by permission of the American Physical Society and Springer Verlag.)

whereas the EMF is entirely due to the nonconservative component $\mathbf{E}_{NC}$:

$$V_\varepsilon = \oint_\Gamma \mathbf{E}_{NC}(\mathbf{r}) \cdot d\mathbf{r}. \tag{9.8}$$

Taking the leads to be ideal, dissipationless conductors (which corresponds to the subtraction of the lead resistances of the experimental result setup), we implicitly require that the total electric field vanishes in the leads:

$$\mathbf{E}(\mathbf{r}) = \mathbf{0} \quad \text{for } \mathbf{r} \in \Omega_{1L} \text{ or } \Omega_{2L}. \tag{9.9}$$

Furthermore, as we are looking for a universal mechanism that is able to limit the current in a mesoscopic circuit, we have explicitly omitted any source of incidental inelastic scattering and hence neglected all energy dissipation in the circuit, including the internal resistance of the power source. For the sake of simplicity we have also assumed that the nonconservative electric field component $\mathbf{E}_{NC}$ is strictly localized in the seat of the EMF, i.e., in the "battery region" Ω_B. This leaves us with a circuit where free electrons can pile up only in the active region due to electrostatic confinement or the presence of a potential barrier, while the leads appear to be equipotential volumes. Since the power source has no internal resistance, the nonconservative component $\mathbf{E}_{NC}$ is pumping all electrons that arrived on the positive pole back to the negative pole at no energy cost. In other words, within the "battery region" $\mathbf{E}_{NC}$ exactly counteracts the effect of the conservative field that would decelerate all electrons climbing up the potential hill in Ω_B (see Fig. 9.5):

$$\mathbf{E}_{NC}(\mathbf{r}) = \begin{cases} -\mathbf{E}_C(\mathbf{r}) & \text{for } \mathbf{r} \in \Omega_B, \\ \mathbf{0} & \text{elsewhere.} \end{cases} \tag{9.10}$$

From Eqs. (9.5)–(9.10) it follows that

$$V_\varepsilon = \oint_\Gamma \mathbf{E} \cdot \mathbf{dr} = \int_{\Sigma_{1A}}^{\Sigma_{2A}} \mathbf{E}_C(\mathbf{r}) \cdot \mathbf{dr} = V_1 - V_2. \tag{9.11}$$

In view of the permanently available power supply and the absence of energy dissipation, one would expect the circuit current to grow unlimitedly. Indeed, the counteracting electromotive force arising from the self-induced magnetic field generated by the current, though initially delaying the current increase because of Lenz' law, would not be capable of slowing down the electron flow in the long term. The latter of course follows directly from elementary, classical mechanics but also from the equation of an L–R-circuit where the circuit resistance R tends to zero:

$$I(t) = \frac{V_\varepsilon}{R}\left(1 - e^{-Rt/L}\right) \xrightarrow{R \to 0} \frac{V_\varepsilon}{L} t. \tag{9.12}$$

Clearly, this simple result does not hold if the current should become so large that radiation losses can no longer be neglected. However, the corresponding radiation resistance is typically of the order of the vacuum impedance (see Jackson [1975]) $Z_0 = \mu_0 c \approx 120\pi \ \Omega$, which is not only smaller than von Klitzing's resistance by roughly two orders of magnitude, but also does not inherently contain the constants e and h. We therefore believe that radiation resistance is not the appropriate mechanism to explain conductance quantization.

Although the idealized circuit under consideration should not be regarded as a superconductor, we might be inspired by the phenomenon of flux quantization governing the electromagnetic response of type-I superconductors, as explained in various textbooks by many authors, such as Kittel [1976], Kittel [1963], and Feynman, Leighton, and Sands [1964b]. In type-I superconducting rings with an appreciable thickness (exceeding the coherence length), flux quantization emerges from the Meissner effect according to which all magnetic field lines are expelled from the interior of the ring, and the requirement that the wave function describing Cooper pairs in the superconducting state be single-valued when a virtual turn along an interior closed curve is made. More precisely, as stated in Sakurai [1976], the deflection of the magnetic field causes the vector potential to be irrotational inside the ring which, in turn, allows one to fully absorb the vector potential into the phase of the wave function:[8]

$$\psi(\mathbf{r}) = \psi_0(\mathbf{r}) \exp\left(\frac{2ie}{\hbar} \int_P \mathbf{A} \cdot \mathbf{dr}\right). \tag{9.13}$$

The fields $\psi(\mathbf{r})$ and $\psi_0(\mathbf{r})$ respectively denote the wave function in the presence and absence of an irrotational vector potential and P represents an internal path connecting an arbitrary reference point with the point $\mathbf{r}$. Moving $\mathbf{r}$ all around the ring turns the line integral of $\mathbf{A}$ into the magnetic flux $\Phi = \oint \mathbf{A} \cdot \mathbf{dr}$ trapped by some closed loop Γ. Obviously, $\psi(\mathbf{r})$ becomes multivalued unless the flux Φ equals an integer multiple of the London flux quantum $\Phi_L = h/2e$. In the case of our circuit however, we do not consider external magnetic fields and the only magnetic field that may pierce the circuit region Ω is the self-induced magnetic field $\mathbf{B} = \nabla \times \mathbf{A}$ generated by the current

---

[8]The factor 2 in the phase factor reflects the charge $-2e$ of a Cooper pair.

flowing through the circuit. Though not vanishing everywhere inside Ω, **B** is circulating around the current density vector **J** representing the current distribution in the circuit. As a consequence, the azimuthal component of **B** (along **J**) will generally vanish, while each transverse component changes sign in the region where **J** is nonzero, i.e., inside the circuit region. In other words, there exists a closed, internal curve Γ_0 along which **B** = **0** and **A** is irrotational. Hence, provided the point **r** is close enough to the curve Γ_0, we may repeat the above argument and approximately absorb **A** into the phase of the electron wave functions. Similarly, approximate flux quantization may be invoked, provided that the flux is now strictly defined as the loop integral of **A** around Γ_0 and the flux quantum is taken to be the double of the previous one, i.e., the Dirac flux quantum $\Phi_0 = h/e$. Complying with the flux quantization constraint means that any increase of the induced magnetic flux caused by an increase of the circuit current should be step-wise. Within the scope of a semiclassical picture, one could propose that an electron cannot extract energy from the power supply, unless the time slot during which it is exposed to the external electric field, is large enough to generate one quantum of induced magnetic flux. Indeed, if the energy extraction were continuous, the induced magnetic flux could be raised by an arbitrary small amount, thereby violating the (approximate) flux quantization constraint. The characteristic time τ_0 required to add one flux quantum, can easily be estimated by comparing the electron energy $\Delta E_{\text{MECH},n}$ gained from the external field during a time interval $[t_n - \frac{1}{2}\tau_0, t_n + \frac{1}{2}\tau_0]$ with the corresponding magnetic energy increase ΔU_M of the circuit, where a flux jump occurs at $t = t_n$. Integrating the energy rate equation (2.20) from $t_n - \frac{1}{2}\tau_0$ to $t_n + \frac{1}{2}\tau_0$, we may express $\Delta E_{\text{MECH},n}$ as follows:

$$\Delta E_{\text{MECH},n} = \int_{t_n - \frac{1}{2}\tau_0}^{t_n + \frac{1}{2}\tau_0} dt \int_{\Omega} d\tau\, \mathbf{J}(\mathbf{r}, t) \cdot \mathbf{E}(\mathbf{r}, t). \tag{9.14}$$

During $[t_n - \frac{1}{2}\tau_0, t_n + \frac{1}{2}\tau_0]$, the charge density remains unchanged before and after the jump at $t = t_n$ and consequently, the current density is solenoidal, while the external electric field is irrotational. Hence, according to the recently derived $\mathbf{J} \cdot \mathbf{E}$ integral theorem for multiply connected regions (see Appendix A.1 and MAGNUS and SCHOEN-MAKER [1998]), we may disentangle the right-hand side of Eq. (9.14):

$$\int_{t_n - \frac{1}{2}\tau_0}^{t_n + \frac{1}{2}\tau_0} dt \int_{\Omega} d\tau\, \mathbf{J}(\mathbf{r}, t) \cdot \mathbf{E}(\mathbf{r}, t) = \frac{1}{2}[I_{n-1} + I_n]V_{\varepsilon}\tau_0, \tag{9.15}$$

where $I_n = \int_{\Sigma_{1A}} \mathbf{J}(\mathbf{r}, t_n) \cdot \mathbf{dS}$ is the net current entering the cross section Σ_{1A} at a time t_n. On the other hand the flux change $\Delta \Phi_n$ associated with the jump $\Delta I_n \equiv I_n - I_{n-1}$, reads

$$\Delta \Phi_n = L \Delta I_n, \tag{9.16}$$

where L is the inductance of the circuit. Since $\Delta \Phi_n$ is to be taken equal to Φ_0, we obtain the increased magnetic energy of the circuit:

$$\Delta U_M = \frac{1}{2}LI_n^2 - \frac{1}{2}LI_{n-1}^2 = \frac{1}{2}(I_{n-1} + I_n)\Phi_0. \tag{9.17}$$

Combining Eqs. (9.14), (9.15), and (9.17) and putting $\Delta U_M = \Delta E_{MECH,n}$, we derive the following result:

$$\tau_0 = \frac{\Phi_0}{V_\varepsilon}. \tag{9.18}$$

If an electron has been sufficiently accelerated such that the time it is exposed to the localized electric field becomes smaller than τ_0, energy extraction is stopped and the one-electron current will never exceed e/τ_0. For an electron ensemble carrying spin and being distributed over N ballistic channels, the total current predicted by the Landauer–Büttiker formula (9.2) is therefore recovered:

$$I = 2N\frac{e}{\tau_0} = \frac{2e^2}{h}NV_\varepsilon. \tag{9.19}$$

In spite of the naive calculation leading to Eq. (9.19), it is shown that the interplay between circuit topology, flux quantization, and the localized electric field may lead to a kind of "selection rule" prohibiting the unlimited extraction of energy from a power supply, even if all dissipative mechanisms are (artificially) turned off.

9.2. Quantum circuit theory

On the other hand, it goes without saying that a sound theory is required not only to support and to refine the concept of flux quantization for nonsuperconducting circuits, but also to bridge the gap between the rigorous, microscopic transport description and the global circuit model that is to reflect the quantum mechanical features of coherent transport through the electric circuit or part of it. Such a theory which could be called "quantum circuit theory" (QCT) might emerge as an extension of the good old theory of QED that would generalize the quantization of the electromagnetic field on two levels: not only should one address nontrivial topologies such as toroidal regions in which finite currents may flow and finite charges may be induced, but also an appropriate set of conjugate observables describing the global circuit properties should be defined. In view of the previous considerations regarding the magnetic flux trapped by the circuit, a natural pair of variables could be the flux of the electric displacement field $\mathbf{D}$ through a cross section Σ_0 crossing the circuit in the interior of the active region and the magnetic flux threaded by the loop Γ_0:

$$\Phi_D = \int_{\Sigma_0} \mathbf{D} \cdot \mathbf{dS}, \tag{9.20}$$

$$\Phi_M = \oint_{\Gamma_0} \mathbf{A} \cdot \mathbf{dr}. \tag{9.21}$$

Taking the electric displacement field instead of the electric field itself to construct a "partner" for Φ_M has mainly to do with the requirement that the product of two conjugate variables have the dimension of an action ($\propto \hbar$). Assuming that $\mathbf{D}$ vanishes outside the active region Ω_A, one may consider the latter as a leaky capacitor the plates of which are separated by Σ_0 such that, according to Gauss' law, Φ_D would equal the charge accumulated on one plate, say Q_A (see Fig. 9.6). Canonical quantization would

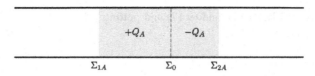

FIG. 9.6. Cross section Σ_0 separating positive and negative charges in the active region Ω_A.

then impose

$$[\Phi_D, \Phi_M] = i\hbar,$$
$$[\Phi_D, \Phi_D] = [\Phi_M, \Phi_M] = 0. \qquad (9.22)$$

It is now tempting to propose a phenomenological expression like

$$H = \frac{\Phi_D^2}{2C} + \frac{\Phi_M^2}{2L} + \Phi_D V_\varepsilon - \Phi_M I \qquad (9.23)$$

for a circuit Hamiltonian describing the interaction between the electromagnetic field variables $\{\Phi_D, \Phi_M\}$ and the electron current operator $I = \int_{\Sigma_0} \mathbf{J} \cdot d\mathbf{S}$ under the constraint $Q_A = \langle \Phi_D \rangle$, and to derive the corresponding Heisenberg equations of motion with the help of the commutation relations (9.22):

$$\frac{d\Phi_D(t)}{dt} = -\frac{i}{\hbar}\left[\Phi_D(t), H\right] = \frac{\Phi_M(t)}{L} - I(t), \qquad (9.24)$$

$$\frac{d\Phi_M(t)}{dt} = -\frac{i}{\hbar}\left[\Phi_M(t), H\right] = -\frac{\Phi_D(t)}{C} - V_\varepsilon. \qquad (9.25)$$

At first sight, the above equations are satisfied by meaningful steady-state solutions that may be obtained by setting the long-time averages $\langle \ldots \rangle = \lim_{t \to \infty} \langle \ldots \rangle_t$ of $d\Phi_D(t)/dt$ and $d\Phi_M(t)/dt$ equal to zero. Indeed, the resulting equations

$$\langle I \rangle = \frac{\langle \Phi_M \rangle}{L}, \qquad (9.26)$$

$$\frac{Q_A}{C} = \frac{\langle \Phi_D \rangle}{C} = -V_\varepsilon \qquad (9.27)$$

are restating the familiar result that the steady-state of the circuit is determined by a current that is proportional to the magnetic flux, while the capacitor voltage tends to the externally applied electromotive force.

However, in order to investigate whether the quantum dynamics generated by the proposed Hamiltonian eventually leads to the Landauer–Büttiker formula or not, would require us to give a meaningful definition of the inductance and capacitance coefficients L and C as well as a recipe to calculate the statistical averages in a straightforward manner. Clearly, this can only be accomplished if a full microscopic investigation of the circuit is performed including both the self-consistent solution of the one-electron Schrödinger equation and the fourth Maxwell equation, and a rigorous evaluation of the dynamical, quantum-statistical ensemble averages. As such, this is quite an elaborate task which, however, may open new perspectives in the boundary region between electromagnetism and quantum mechanics.

Acknowledgements

We gratefully acknowledge Herman Maes (former Director Silicon Technology and Device Integration Division at IMEC) for giving us the opportunity to contribute to this book. We owe special thanks to our colleagues at IMEC for their willingness to discuss issues of electromagnetism and device physics and to give valuable comments and remarks. In particular, we would like to thank Peter Meuris as a coworker realizing the geometrical picture of the magnetic vector potential and its related ghost field including its numerical implementation. Stefan Kubicek is gratefully acknowledged for stimulating discussions on differential geometry and its applications to engineering. Finally, we would like to thank the editors of this special volume to give us the opportunity to present our views concerning the simulation of electromagnetic fields and quantum transport phenomena.

Appendix A.1. Integral theorems

Integral theorems borrowed from the differential geometry of curves, surfaces, and connected regions (MORSE and FESHBACH [1953], MAGNUS and SCHOENMAKER [1998]) turn out to be useful and perhaps even indispensible for a thorough understanding of elementary electromagnetic theory. Not only are they quite helpful in converting the differential form of Maxwell's equations into their equivalent integral form, but they also offer a convenient tool to define a discretized version of the field variables in the framework of numerical simulation. Moreover, they naturally bridge the gap between the microscopic interaction of the electromagnetic fields and charges in a solid-state conductor and the global circuit models envisaged on the macroscopic level.

The first three integral theorems that are summarized below, are extensively referred to in Section 2. The fourth one is the Helmholtz theorem, which allows one to decompose any well-behaved vector field into a longitudinal and a transverse part.

THEOREM A.1 (Stokes' theorem). *Let Σ be an open, orientable, multiply connected surface in $\mathbb{R}^3$ bounded by an outer, closed curve $\partial \Sigma_0$ and n inner, closed curves $\partial \Sigma_1$, ..., $\partial \Sigma_n$ defining n holes. If Σ is oriented by a surface element $\mathbf{dS}$ and if $\mathbf{A}$ is a differentiable vector field defined on Σ, then*

$$\int_{\Sigma} \nabla \times \mathbf{A} \cdot \mathbf{dS} = \oint_{\partial \Sigma_0} \mathbf{A} \cdot \mathbf{dr} - \sum_{j=1}^{n} \oint_{\partial \Sigma_j} \mathbf{A} \cdot \mathbf{dr}, \quad (A.1)$$

where the orientation of all boundary curves is uniquely determined by the orientation of $\mathbf{dS}$.

THEOREM A.2 (Gauss' theorem). *Let Ω be a closed, orientable, multiply connected subset of $\mathbb{R}^3$ bounded by an outer, closed surface $\partial \Omega_0$ and n inner, closed surfaces defining n holes. If $\mathbf{E}$ is a differentiable vector field defined on Ω then*

$$\int_{\Omega} \nabla \cdot \mathbf{E} \, d\tau = \int_{\partial \Omega_0} \mathbf{E} \cdot \mathbf{dS} - \sum_{j=1}^{n} \int_{\partial \Omega_j} \mathbf{E} \cdot \mathbf{dS} \quad (A.2)$$

and

$$\int_\Omega \mathbf{\nabla} \times \mathbf{E}\, d\tau = \int_{\partial \Omega_0} d\mathbf{S} \times \mathbf{E} - \sum_{j=1}^n \int_{\partial \Omega_j} d\mathbf{S} \times \mathbf{E}, \qquad (A.3)$$

where all boundary surfaces have the same orientation as the outward pointing surface element of the outer surface.

The scalar Gauss theorem (A.2) reduces to *Green's Theorem* when the vector field takes the form $\mathbf{E} = f \mathbf{\nabla} g - g \mathbf{\nabla} f$

$$\int_\Omega (f\nabla^2 g - g\nabla^2 f)\, d\tau = \int_{\partial \Omega_0} (f \mathbf{\nabla} g - g \mathbf{\nabla} f) \cdot d\mathbf{S}$$
$$- \sum_{j=1}^n \int_{\partial \Omega_j} (f \mathbf{\nabla} g - g \mathbf{\nabla} f) \cdot d\mathbf{S}, \qquad (A.4)$$

where the scalar fields f and g are differentiable on Ω.

THEOREM A.3 ($\mathbf{J} \cdot \mathbf{E}$ theorem). *Let Ω be a closed, multiply connected, bounded subset of $\mathbb{R}^3$ with one hole and boundary surface $\partial \Omega$. If $\mathbf{J}$ and $\mathbf{E}$ are two differentiable vector fields on Ω, circulating around the hole and satisfying the conditions*

$$\mathbf{\nabla} \cdot \mathbf{J} = 0, \qquad (A.5)$$

$$\mathbf{\nabla} \times \mathbf{E} = \mathbf{0}, \qquad (A.6)$$

$$\mathbf{J} \parallel \partial \Omega \quad or \quad \mathbf{J} = \mathbf{0} \quad in\ each\ point\ of\ \partial \Omega, \qquad (A.7)$$

then

$$\int_\Omega \mathbf{J} \cdot \mathbf{E}\, d\tau = \left(\int_\Sigma \mathbf{J} \cdot d\mathbf{S} \right) \left(\oint_\Gamma \mathbf{E} \cdot d\mathbf{r} \right), \qquad (A.8)$$

where Σ is an arbitrary cross section, intersecting Ω only once and Γ is a simple closed curve, encircling the hole and lying within Ω but not intersecting $\partial \Omega$. The orientation of Σ is uniquely determined by the positive orientation of Γ.

PROOF. Without any loss of generality one may define curvilinear coordinates (x^1, x^2, x^3) and a corresponding set of covariant basis vectors $(\mathbf{a}_1, \mathbf{a}_2, \mathbf{a}_3)$ and its contravariant counterpart, which are compatible with the topology of the toroidal (torus-like) region Ω. More precisely, x^1, x^2, and x^3 may be chosen such that the boundary surface $\partial \Omega$ coincides with one of the coordinate surfaces $dx^1 = 0$ while the curves $dx^1 = dx^2 = 0$ are closed paths encircling the hole only once and x^3 is a cyclic coordinate. Then the inner volume contained within Ω may be conveniently parametrized by restricting the range of (x^1, x^2, x^3) to some rectangular interval $[c^1, d^1] \times [c^2, d^2] \times [c^3, d^3]$. Since Ω is multiply connected, the irrotational vector field $\mathbf{E}$ cannot generally be derived from a scalar potential for the whole region Ω. However, for the given topology of Ω, it is always possible to assign such a potential to the "transverse" components of $\mathbf{E}$ only:

$$E_1(x^1, x^2, x^3) = -\frac{\partial V(x^1, x^2, x^3)}{\partial x^1}, \tag{A.9}$$

$$E_2(x^1, x^2, x^3) = -\frac{\partial V(x^1, x^2, x^3)}{\partial x^2}, \tag{A.10}$$

but

$$E_3(x^1, x^2, x^3) \neq -\frac{\partial V(x^1, x^2, x^3)}{\partial x^3}, \tag{A.11}$$

where $V(x^1, x^2, x^3)$ can be constructed straightaway by invoking the first two components of $\mathbf{V} \times \mathbf{E} = \mathbf{0}$:

$$V(x^1, x^2, x^3) = V(c^1, c^2, x^3) - \int_{c^1}^{x^1} ds \, E_1(s, x^2, x^3) - \int_{c^2}^{x^2} dt \, E_2(c^1, t, x^3). \tag{A.12}$$

The potential term $V(c^1, c^2, x^3)$ naturally arises as an integration constant which, depending on x^3 only, may be absorbed in the definition of $V(x^1, x^2, x^3)$ and will therefore be omitted. Eqs. (A.9) and (A.10) are now easily recovered by taking the derivative of (A.12) with respect to x^1 and x^2, and inserting the third component of $\mathbf{V} \times \mathbf{E} = \mathbf{0}$.

Finally, taking also the derivative with respect to x^3, one obtains:

$$E_3(x^1, x^2, x^3) = -\frac{\partial V(x^1, x^2, x^3)}{\partial x^3} + E_3(c^1, c^2, x^3). \tag{A.13}$$

From Eqs. (A.9), (A.10), and (A.13) arises a natural decomposition of $\mathbf{E}$ into a conservative vector field $\mathbf{E}_C$ and a nonconservative field $\mathbf{E}_{NC}$ that is oriented along a_3, thereby depending only on the cyclic coordinate x^3:

$$\mathbf{E} = \mathbf{E}_C + \mathbf{E}_{NC} \tag{A.14}$$

with

$$\mathbf{E}_C(x^1, x^2, x^3) = -\mathbf{V}V(x^1, x^2, x^3), \tag{A.15}$$

$$\mathbf{E}_{NC}(x^1, x^2, x^3) = E_3(c^1, c^2, x^3)a^3. \tag{A.16}$$

The conservative part of $\mathbf{E}$ does not contribute to the volume integral of $\mathbf{J} \cdot \mathbf{E}$. Indeed, from (A.15) it follows

$$\int_\Omega \mathbf{J} \cdot \mathbf{E}_C \, d\tau = \int_\Omega \mathbf{J} \cdot \mathbf{V}V \, d\tau = \int_\Omega \mathbf{V} \cdot (V\mathbf{J}) \, d\tau - \int_\Omega V\mathbf{V} \cdot \mathbf{J} \, d\tau. \tag{A.17}$$

With the help of Gauss' theorem – which is also valid for multiply connected regions – the first term of the right-hand side of Eq. (A.17) can be rewritten as a surface integral of $V\mathbf{J}$ which is seen to vanish as $\mathbf{J}$ is assumed to be tangential to the surface $\partial\Omega$ in all of its points. Clearly, the second integral in the right-hand side of (A.17) is identically zero due to $\mathbf{V} \cdot \mathbf{J} = 0$ and one is therefore lead to the conclusion

$$\int_\Omega \mathbf{J} \cdot \mathbf{E}_C \, d\tau = 0. \tag{A.18}$$

On the other hand, the contribution of $\mathbf{E}_{NC}$ can readily be evaluated in terms of the curvilinear coordinates. Denoting the Jacobian determinant by $g(x^1, x^2, x^3)$ one may express the volume integral as a threefold integral over the basic interval $[c^1, d^1] \times [c^2, d^2] \times [c^3, d^3]$, thereby exploiting the fact that the nonconservative contribution merely depends on x^3:

$$
\int_\Omega \mathbf{J} \cdot \mathbf{E} \, d\tau
$$

$$
= \int_\Omega \mathbf{J} \cdot \mathbf{E}_{NC} \, d\tau
$$

$$
= \int_{c^3}^{d^3} dx^3 \, E_3(c^1, c^2, x^3) \int_{c^1}^{d^1} dx^1 \int_{c^2}^{d^2} dx^2 \, g(x^1, x^2, x^3) J^3(x^1, x^2, x^3). \quad \text{(A.19)}
$$

The last integral can conveniently be interpreted as the flux of $\mathbf{J}$ through the single cross section $\Sigma(x^3)$ defined by

$$
\Sigma(x^3) = \{(x^1, x^2, x^3) \mid c^1 \leqslant x^1 \leqslant d^1; \ c^2 \leqslant x^2 \leqslant d^2; \ x^3 \text{ fixed}\}. \quad \text{(A.20)}
$$

Indeed, expanding the Jacobian determinant as a mixed product of the three basis vectors, i.e.,

$$
g = \mathbf{a}_1 \times \mathbf{a}_2 \cdot \mathbf{a}_3 \quad \text{(A.21)}
$$

and identifying the two-form $\mathbf{a}_1 \times \mathbf{a}_2 \, dx^1 \, dx^2$ as a generic surface element $\mathbf{dS}$ perpendicular to $\Sigma(x^3)$, one easily arrives at

$$
\int_{c^1}^{d^1} dx^1 \int_{c^2}^{d^2} dx^2 \, g(x^1, x^2, x^3) J^3(x^1, x^2, x^3)
$$

$$
= \int_{c^1}^{d^1} dx^1 \int_{c^2}^{d^2} dx^2 \, \mathbf{a}_1 \times \mathbf{a}_2 \cdot \mathbf{J}(x^1, x^2, x^3) = \int_{\Sigma(x^3)} \mathbf{dS} \cdot \mathbf{J} \equiv I(x^3) \quad \text{(A.22)}
$$

and

$$
\int_\Omega \mathbf{J} \cdot \mathbf{E} \, d\tau = \int_{c^3}^{d^3} dx^3 \, E_3(c^1, c^2, x^3) I(x^3). \quad \text{(A.23)}
$$

The sign of the flux $I(x^3)$ obviously depends on the orientation of $\Sigma(x^3)$, which is unequivocally determined by the surface element $\mathbf{dS} = \mathbf{a}_1 \times \mathbf{a}_2 \, dx^1 \, dx^2$. As long as only positive body volumes are concerned, one may equally require that each infinitesimal volume element $d\tau = g \, dx^1 \, dx^2 \, dx^3$ be positive for positive incremental values dx^1, dx^2, and dx^3. Moreover, since $\mathbf{dr} = dx^3 \mathbf{a}_3$ is the elementary tangent vector of the coordinate curve $\Gamma(x^1, x^2) = \{(x^1, x^2, x^3) \mid x^1, x^2 \text{ fixed}; c^3 \leqslant x^3 \leqslant d^3\}$ orienting $\Gamma(x^1, x^2)$ in a positive traversal sense through increasing x^3, one easily arrives at

$$
d\tau = \mathbf{dS} \cdot \mathbf{dr} > 0. \quad \text{(A.24)}
$$

In other words, the orientation of $\Sigma(x^3)$ is completely fixed by the positive traversal sense of $\Gamma(x^1, x^2)$. However, since $\mathbf{J}$ is solenoidal within Ω as well as tangential to $\partial\Omega$, one may conclude from Gauss' theorem that the value of the flux $I(x^3)$ does not depend on the particular choice of the cross section $\Sigma(x^3)$ which may thus be replaced by any other single cross section Σ provided that the orientation is preserved. Consequently, $I(x^3)$ reduces to a constant value I and may be taken out of the integral of Eq. (A.23) which now simplifies to:

$$\int_\Omega \mathbf{J} \cdot \mathbf{E} \, d\tau = I \int_{c^3}^{d^3} dx^3 \, E_3(c^1, c^2, x^3). \tag{A.25}$$

The remaining integral turns out to be the line integral of $\mathbf{E}$ along the coordinate curve $\Gamma(c^1, c^2)$:

$$\int_\Omega \mathbf{J} \cdot \mathbf{E} \, d\tau = I V_\varepsilon(c^1, c^2) \tag{A.26}$$

with

$$V_\varepsilon(c^1, c^2) = \oint_{\Gamma(c^1, c^2)} \mathbf{E} \cdot d\mathbf{r}. \tag{A.27}$$

Since $\mathbf{E}$ is irrotational, according to Stokes' theorem its circulation does not depend on the particular choice of the circulation curve as was already discussed in more detail in the previous section. Consequently, $\Gamma(c^1, c^2)$ may be replaced by any other interior closed curve Γ encircling the hole region and sharing the traversal sense with $\Gamma(c^1, c^2)$:

$$V_\varepsilon(c^1, c^2) = V_\varepsilon \equiv \oint_\Gamma \mathbf{E} \cdot d\mathbf{r}. \tag{A.28}$$

Hence,

$$\int_\Omega \mathbf{J} \cdot \mathbf{E} \, d\tau = I V_\varepsilon. \tag{A.29}$$

This completes the proof. $\qquad\square$

THEOREM A.4 (Helmholtz' theorem). *Let Ω be a simply connected, bounded subset of $\mathbb{R}^3$. Then, any finite, continuous vector field $\mathbf{F}$ defined on Ω can be derived from a differentiable vector potential $\mathbf{A}$ and a differentiable scalar potential χ such that*

$$\mathbf{F} = \mathbf{F}_L + \mathbf{F}_T, \tag{A.30}$$

$$\mathbf{F}_L = \nabla\chi, \tag{A.31}$$

$$\mathbf{F}_T = \nabla \times \mathbf{A}. \tag{A.32}$$

Due to the obvious properties

$$\begin{aligned} \nabla \times \mathbf{F}_L &= \mathbf{0}, \\ \nabla \cdot \mathbf{F}_T &= 0. \end{aligned} \tag{A.33}$$

$\mathbf{F}_L$ and $\mathbf{F}_T$ are respectively called the longitudinal and transverse components of $\mathbf{F}$.

Appendix A.2. Vector identities

Let f, $\mathbf{A}$ and $\mathbf{B}$ represent a scalar field and two vector fields defined on a connected subset Ω of $\mathbb{R}^3$, all being differentiable on Ω. Then the following (nonexhaustive) list of identities may be derived using familiar vector calculus:

$$\nabla \cdot (\nabla \times \mathbf{A}) \equiv 0, \tag{A.34}$$

$$\nabla \times (\nabla f) \equiv \mathbf{0}, \tag{A.35}$$

$$\nabla(\mathbf{A} \cdot \mathbf{B}) = \mathbf{A}(\nabla \cdot \mathbf{B}) + \mathbf{B}(\nabla \cdot \mathbf{A}) + (\mathbf{A} \cdot \nabla)\mathbf{B}$$
$$+ (\mathbf{B} \cdot \nabla)\mathbf{A} + \mathbf{A} \times (\nabla \times \mathbf{B}) + \mathbf{A} \times (\nabla \times \mathbf{B}), \tag{A.36}$$

$$\nabla \times (\mathbf{A} \times \mathbf{B}) = -\mathbf{A}(\nabla \cdot \mathbf{B}) + \mathbf{B}(\nabla \cdot \mathbf{A}) - (\mathbf{A} \cdot \nabla)\mathbf{B}$$
$$+ (\mathbf{B} \cdot \nabla)\mathbf{A} - \mathbf{A} \times (\nabla \times \mathbf{B}) + \mathbf{A} \times (\nabla \times \mathbf{B}), \tag{A.37}$$

$$\nabla \times (f\mathbf{A}) = f\nabla \times \mathbf{A} + \nabla f \times \mathbf{A}, \tag{A.38}$$

$$\nabla \cdot (\mathbf{A} \times \mathbf{B}) = \mathbf{B} \cdot \nabla \times \mathbf{A} - \mathbf{A} \cdot \nabla \times \mathbf{B}, \tag{A.39}$$

$$\nabla \cdot (f\mathbf{A}) = f\nabla \cdot \mathbf{A} + \nabla f \cdot \mathbf{A}, \tag{A.40}$$

$$\nabla \times (\nabla \times \mathbf{A}) = \nabla(\nabla \cdot \mathbf{A}) - \nabla^2 \mathbf{A}. \tag{A.41}$$

It should be noted that Eq. (A.41) should be considered as a definition of the vectorial Laplace operator ("Laplacian"), rather than a vector identity. Clearly, if one expands the left-hand side of Eq. (A.41) in Cartesian coordinates, one may straightforwardly obtain

$$\left[\nabla \times (\nabla \times \mathbf{A})\right]_x = \frac{\partial}{\partial x}\nabla \cdot \mathbf{A} - \nabla^2 A_x, \tag{A.42}$$

etc., which does indeed justify the identification $\nabla^2\mathbf{A} = (\nabla^2 A_x, \nabla^2 A_y, \nabla^2 A_z)$ for Cartesian coordinates, but not for an arbitrary system of curvilinear coordinates.

References

BARDEEN, J. (1961). Tunneling from a many-particle point of view. *Phys. Rev. Lett.* **6**, 57–59.

BRAR, B., WILK, G., SEABAUGH, A. (1996). Direct extraction of the electron tunneling effective mass in ultrathin SiO_2. *Appl. Phys. Lett.* **69**, 2728–2730.

BREIT, G., WIGNER, E.P. (1936). Capture of slow neutrons. *Phys. Rev.* **49**, 519–531.

BUETTIKER, M. (1986). Role of quantum coherence in series resistors. *Phys. Rev. B* **33**, 3020–3026.

BUTCHER, P., MARCH, N.H., TOSI, M.P. (eds.) (1993). *Physics of Low-Dimensional Semiconductor Structures* (Plenum Press, New York).

COLLIN, R. (1960). *Field Theory of Guided Waves* (Mc-Graw Hill, New York).

DATTA, S. (1995). *Electronic Transport in Mesoscopic Systems* (Cambridge University Press, UK).

DEPAS, M., VANMEIRHAEGHE, R., LAFLERE, W., CARDON, F. (1994). Electrical characteristics of $Al/SiO_2/n$-Si tunnel-diodes with an oxide layer grown by rapid thermal-oxidation. *Solid-State Electron.* **37**.

DITTRICH, T., HAENGGI, P., INGOLD, G.-L., KRAMER, B., SCHOEN, G., ZWERGER, W. (1997). *Quantum Transport and Dissipation* (Wiley-VCH, Weinheim, Germany).

DRUDE, P. (1900a). Zur Elektronentheorie der Metalle, I. Teil. *Ann. Phys.* **1**, 566–613.

DRUDE, P. (1900b). Zur Elektronentheorie der Metalle, II. Teil. *Ann. Phys.* **3**, 369–402.

EINSTEIN, A., LORENTZ, H.A., MINKOWSKI, H., WEYL, H. (1952). *The Principle of Relativity, Collected Papers* (Dover, New York).

EZAWA, Z.F. (2000). *Quantum Hall Effects — Field Theoretical Approach and Related Topics* (World Scientific Publishing, Singapore).

FENTON, E.W. (1992). Electric-field conditions for Landauer and Boltzmann-Drude conductance equations. *Phys. Rev. B* **46**, 3754–3770.

FENTON, E.W. (1994). Electrical and chemical potentials in a quantum-mechanical conductor. *Superlattices Microstruct.* **16**, 87–91.

FEYNMAN, R.P., LEIGHTON, R.B., SANDS, M. (1964a). *The Feynman Lectures on Physics, vol. 2* (Addison-Wesley, New York).

FEYNMAN, R.P., LEIGHTON, R.B., SANDS, M. (1964b). *The Feynman Lectures on Physics, vol. 3* (Addison-Wesley, New York).

FLUEGGE, S. (1974). *Practical Quantum Mechanics* (Springer, New York).

FORGHIERI, A., GUERRRI, R., CIAMPOLINI, P., GNUDI, A., RUDAN, M. (1988). A new discretization strategy of the semiconductor equations comprising momentum and energy balance. *IEEE Trans. Computer-Aided Design* **7**, 231–242.

FOWLER, R.H. (1936). *Statistical Mechanics* (MacMillan, New York).

HUANG, K. (1963). *Statistical Mechanics* (John Wiley & Sons, New York).

IMRY, Y., LANDAUER, R. (1999). Conductance viewed as transmission. *Rev. Mod. Phys.* **71**, S306-S312.

JACKSON, J.D. (1975). *Classical Electrodynamics* (John Wiley & Sons, New York).

JOOSTEN, H., NOTEBORN, H., LENSTRA, D. (1990). Numerical study of coherent tunneling in a doublebarrier structure. *Thin Solid Films* **184**, 199–206.

KITTEL, C. (1963). *Quantum Theory of Solids* (John Wiley, New York).

KITTEL, C. (1976). *Introduction to Solid State Physics* (John Wiley, New York).

KUBO, R. (1957). Statistical-mechanical theory of irreversible processes. I. General theory and simple applications to magnetic and conduction problems. *J. Phys. Soc. Japan* **12**, 570.

LANDAU, L.D., LIFSHITZ, E.M. (1958). *Quantum Mechanics (Non-Relativistic Theory)* (Pergamon Press, London).

LANDAU, L., LIFSHITZ, E.M. (1962). *The Classical Theory of Fields* (Addison-Wesley, Reading, MA).

LANDAUER, R. (1957). Spatial variation of currents and fields due to localized scatterers in metallic conduction. *IBM J. Res. Dev.* **1**, 223–231.

LANDAUER, R. (1970). Electrical resistance of disordered one-dimensional lattices. *Philos. Mag.* **21–25**, 863.

LENSTRA, D., SMOKERS, T.M. (1988). Theory of nonlinear quantum tunneling resistance in one-dimensional disordered systems. *Phys. Rev. B* **38**, 6452–6460.

LENSTRA, D., VAN HAERINGEN, W., SMOKERS, T.M. (1990). Carrier dynamics in a ring, Landauer resistance and localization in a periodic system. *Physica A* **162**, 405–413.

LUNDSTROM, M. (1999). *Fundamentals of Carrier Transport*, second ed. (Cambridge University Press, Cambridge).

MAGNUS, W., SCHOENMAKER, W. (1993). Dissipative motion of an electron-phonon system in a uniform electric field: an exact solution. *Phys. Rev. B* **47**, 1276–1281.

MAGNUS, W., SCHOENMAKER, W. (1998). On the use of a new integral theorem for the quantum mechanical treatment of electric circuits. *J. Math. Phys.* **39**, 6715–6719.

MAGNUS, W., SCHOENMAKER, W. (1999). Full quantum mechanical treatment of charge leakage in MOS capacitors with ultra-thin oxide layers. In: *Proc. 29th European Solid-State Device Research Conference (ESSDERC'99), Leuven, Editions Frontières, 13–15 September 1999*, pp. 248–251.

MAGNUS, W., SCHOENMAKER, W. (2000a). Full quantum mechanical model for the charge distribution and the leakage currents in ultrathin metal-insulator-semiconductor capacitors. *J. Appl. Phys.* **88**, 5833–5842.

MAGNUS, W., SCHOENMAKER, W. (2000b). On the calculation of gate tunneling currents in ultra-thin metalinsulator- semiconductor capacitors. *Microelectronics and Reliability* **41**, 31–35.

MAGNUS, W., SCHOENMAKER, W. (2000c). Quantized conductance, circuit topology, and flux quantization. *Phys. Rev. B* **61**, 10883–10889.

MAGNUS, W., SCHOENMAKER, W. (2002). *Quantum Transport in Submicron Devices: A Theoretical Introduction* (Springer, Berlin, Heidelberg).

MAHAN, G.D. (1981). *Many-Particle Physics* (Plenum Press, New York).

MAXWELL, J.C. (1954a). *A Treatise on Electricity and Magnetism, vol. 1* (Dover Publications, New York).

MAXWELL, J.C. (1954b). *A Treatise on Electricity and Magnetism, vol. 2* (Dover Publications, New York).

MERZBACHER, E. (1970). *Quantum Mechanics* (John Wiley & Sons, New York).

MEURIS, P., SCHOENMAKER, W., MAGNUS, W. (2001). Strategy for electromagnetic interconnect modeling. *IEEE Trans. Computer-Aided Design of Circuits and Integrated Systems* **20** (6), 753–762.

MORSE, P.M., FESHBACH, H. (1953). *Methods of Theoretical Physics, Part I* (McGraw-Hill Book Company, Inc).

NOTEBORN, H., JOOSTEN, H., LENSTRA, D., KASKI, K. (1990). Selfconsistent study of coherent tunneling through a double-barrier structure. *Phys. Scripta T* **33**, 219–226.

SAKURAI, J.J. (1976). *Advanced Quantum Mechanics* (Addison-Wesley, Reading, MA).

SCHARFETTER, D., GUMMEL, H. (1969). Large-signal analysis of a silicon read diode oscillator. *IEEE Trans. Electron Devices* **ED-16**, 64–77.

SCHOENMAKER, W., MAGNUS, W., MEURIS, P. (2002). Ghost fields in classical gauge theories. *Phys. Rev. Lett.* **88** (18), 181602-01-181602-04.

SCHOENMAKER, W., MEURIS, P. (2002). Electromagnetic interconnects and passives modeling: software implementation issues. *IEEE Trans. Computer-Aided Design of Circuits and Integrated Systems* **21** (5), 534–543.

SCHWINGER, J. (1958). *Selected Papers on Quantum Electrodynamics* (Dover Publications, New York).

STONE, A.D., SZAFER, A. (1988).What is measured when you measure a resistance – the Landauer formula revisited. *IBM J. Res. Dev.* **32**, 384–413.

STONE, D. (1992). Physics of nanostructures. In: Davies, J.H., Long, A.R. (eds.), *Proceedings of the 38th Scottish Universities Summer School in Physics, St Andrews, July–August 1991* (IOP Publishing Ltd., London), pp. 65–76.

STRATTON, R. (1962). Diffusion of hot and cold electrons in semiconductor devices. *Phys. Rev.* **126**, 2002–2014.

SUNE, J., OLIVIO, P., RICCO, B. (1991). Self-consistent solution of the Poisson and Schrödinger-equations in accumulated semiconductor-insulator interfaces. *J. Appl. Phys.* **70**, 337–345.

'T HOOFT, G. (1971). Renormalizable Lagrangians for massive Yang–Mills fields. *Nucl. Phys.* B **35**, 167–188.

VAN WEES, B.J., VAN HOUTEN, H., BEENAKKER, C.W.J., WILLIAMSON, J.G., KOUWENHOVEN, L.P., VAN DER MAREL, D., FOXON, C.T. (1988). Quantized conductance of point contacts in a two-dimensional electron-gas. *Phys. Rev. Lett.* **60**, 848–850.

VON KLITZING, K., DORDA, G., PEPPER, M. (1980). New method for high-accuracy determination of the fine-structure constant based on quantized Hall resistance. *Phys. Rev. Lett.* **45**, 494–497.

WEYL, H. (1918). Gravitation und Elektrizität. *Sitzungsber. Preussischen Akad. Wiss.* **26**, 465–480.

WHARAM, D.A., THORNTON, T.J., NEWBURY, R., PEPPER, M., AHMED, H., FROST, J.E.F., HASKO, D.G., PEACOCK, D.C., RITCHIE, D.A., JONES, G.A.C. (1988). One-dimensional transport and the quantization of the ballistic resistance. *J. Phys.* C **21**, L209–L214.

WILSON, K. (1974). Confinement of quarks. *Phys. Rev.* D **10**, 2445–2459.

Finite-Difference Time-Domain Methods

Susan C. Hagness

University of Wisconsin-Madison, College of Engineering, Department of Electrical and Computer Engineering, 3419 Engineering Hall, 1415 Engineering Drive, Madison, WI 53706, USA
E-mail address: hagness@engr.wisc.edu

Allen Taflove

Northwestern University, Computational Electromagnetics Lab (NUEML), Department of Electrical and Computer Engineering, 2145 Sheridan Road, Evanston, IL 60208-3118, USA
E-mail address: taflove@ece.northwestern.edu

Stephen D. Gedney

University of Kentucky, College of Engineering, Department of Electrical and Computer Engineering, 687C F. Paul Anderson Tower Lexington, KY 40506-0046, USA
E-mail address: gedney@engr.uky.edu

1. Introduction

1.1. Background

Prior to about 1990, the modeling of electromagnetic engineering systems was primarily implemented using solution techniques for the sinusoidal steady-state Maxwell's equations. Before about 1960, the principal approaches in this area involved closed-form and infinite-series analytical solutions, with numerical results from these analyses obtained using mechanical calculators. After 1960, the increasing availability of programmable electronic digital computers permitted such frequency-domain approaches to rise

Essential Numerical Methods in Electromagnetics
Special Volume (W.H.A. Schilders and E.J.W. ter Maten, Guest Editors) of
HANDBOOK OF NUMERICAL ANALYSIS, VOL. XIII
P.G. Ciarlet (Editor)

markedly in sophistication. Researchers were able to take advantage of the capabilities afforded by powerful new high-level programming languages such as Fortran, rapid random-access storage of large arrays of numbers, and computational speeds orders of magnitude faster than possible with mechanical calculators. In this period, the principal computational approaches for Maxwell's equations included the high-frequency asymptotic methods of KELLER [1962] and KOUYOUMJIAN and PATHAK [1974] and the integral-equation techniques of HARRINGTON [1968].

However, these frequency-domain techniques have difficulties and trades-off. For example, while asymptotic analyses are well suited for modeling the scattering properties of electrically large complex shapes, such analyses have difficulty treating nonmetallic material composition and volumetric complexity of a structure. While integral equation methods can deal with material and structural complexity, their need to construct and solve systems of linear equations limits the electrical size of possible models, especially those requiring detailed treatment of geometric details within a volume, as opposed to just the surface shape.

While significant progress has been made in solving the ultralarge systems of equations generated by frequency-domain integral equations (see, for example, SONG and CHEW [1998]), the capabilities of even the latest such technologies are exhausted by many volumetrically complex structures of engineering interest. This also holds for frequency-domain finite-element techniques, which generate sparse rather than dense matrices. Further, the very difficult incorporation of material and device nonlinearities into frequency-domain solutions of Maxwell's equations poses a significant problem as engineers seek to design active electromagnetic/electronic and electromagnetic/quantum-optical systems such as high-speed digital circuits, microwave and millimeter-wave amplifiers, and lasers.

1.2. Rise of finite-difference time-domain methods

During the 1970s and 1980s, a number of researchers realized the limitations of frequency-domain integral-equation solutions of Maxwell's equations. This led to early explorations of a novel alternative approach: direct time-domain solutions of Maxwell's differential (curl) equations on spatial grids or lattices. The finite-difference time-domain (FDTD) method, introduced by YEE [1966], was the first technique in this class, and has remained the subject of continuous development (see TAFLOVE and HAGNESS [2000]).

There are seven primary reasons for the expansion of interest in FDTD and related computational solution approaches for Maxwell's equations:

(1) FDTD uses no linear algebra. Being a fully explicit computation, FDTD avoids the difficulties with linear algebra that limit the size of frequency-domain integral-equation and finite-element electromagnetics models to generally fewer than 10^6 field unknowns. FDTD models with as many as 10^9 field unknowns have been run. There is no intrinsic upper bound to this number.

(2) FDTD is accurate and robust. The sources of error in FDTD calculations are well understood and can be bounded to permit accurate models for a very large variety of electromagnetic wave interaction problems.

(3) FDTD treats impulsive behavior naturally. Being a time-domain technique, FDTD directly calculates the impulse response of an electromagnetic system. Therefore, a single FDTD simulation can provide either ultrawideband temporal waveforms or the sinusoidal steady-state response at any frequency within the excitation spectrum.

(4) FDTD treats nonlinear behavior naturally. Being a time-domain technique, FDTD directly calculates the nonlinear response of an electromagnetic system.

(5) FDTD is a systematic approach. With FDTD, specifying a new structure to be modeled is reduced to a problem of mesh generation rather than the potentially complex reformulation of an integral equation. For example, FDTD requires no calculation of structure-dependent Green's functions.

(6) Computer memory capacities are increasing rapidly. While this trend positively influences all numerical techniques, it is of particular advantage to FDTD methods which are founded on discretizing space over a volume, and therefore inherently require a large random access memory.

(7) Computer visualization capabilities are increasing rapidly. While this trend positively influences all numerical techniques, it is of particular advantage to FDTD methods which generate time-marched arrays of field quantities suitable for use in color videos to illustrate the field dynamics.

An indication of the expanding level of interest in FDTD Maxwell's equations solvers is the hundreds of papers currently published in this area worldwide each year, as opposed to fewer than ten as recently as 1985 (see SHLAGER and SCHNEIDER [1998]). This expansion continues as engineers and scientists in nontraditional electromagnetics-related areas such as digital systems and integrated optics become aware of the power of such direct solution techniques for Maxwell's equations.

1.3. Characteristics of FDTD and related space-grid time-domain techniques

FDTD and related space-grid time-domain techniques are direct solution methods for Maxwell's curl equations. These methods employ no potentials. Rather, they are based upon volumetric sampling of the unknown electric and magnetic fields within and surrounding the structure of interest, and over a period of time. The sampling in space is at subwavelength resolution set by the user to properly sample the highest near-field spatial frequencies thought to be important in the physics of the problem. Typically, 10–20 samples per wavelength are needed. The sampling in time is selected to ensure numerical stability of the algorithm.

Overall, FDTD and related techniques are marching-in-time procedures that simulate the continuous actual electromagnetic waves in a finite spatial region by sampled-data numerical analogs propagating in a computer data space. Time-stepping continues as the numerical wave analogs propagate in the space lattice to causally connect the physics of the modeled region. For simulations where the modeled region must extend to infinity, absorbing boundary conditions (ABCs) are employed at the outer lattice truncation planes which ideally permit all outgoing wave analogs to exit the region with negligible reflection. Phenomena such as induction of surface currents, scattering and multiple scattering, aperture penetration, and cavity excitation are modeled time-step by time-step by the action of the numerical analog to the curl equations. Self-consistency of these modeled phenomena is generally assured if their spatial and temporal variations are well resolved by the space and time sampling process. In fact, the goal is to provide a self-consistent model of the mutual coupling of all of the electrically small volume cells constituting the structure and its near field, even if the structure spans tens of wavelengths in three dimensions and there are hundreds of millions of space cells.

Time-stepping is continued until the desired late-time pulse response is observed at the field points of interest. For linear wave interaction problems, the sinusoidal response at these field points can be obtained over a wide band of frequencies by discrete Fourier transformation of the computed field-versus-time waveforms at these points. Prolonged "ringing" of the computed field waveforms due to a high Q-factor or large electrical size of the structure being modeled requires a combination of extending the computational window in time and extrapolation of the windowed data before Fourier transformation.

1.4. Classes of algorithms

Current FDTD and related space-grid time-domain algorithms are fully explicit solvers employing highly vectorizable and parallel schemes for time-marching the six components of the electric and magnetic field vectors at each of the space cells. The explicit nature of the solvers is usually maintained by employing a leapfrog time-stepping scheme. Current methods differ primarily in how the space lattice is set up. In fact, gridding methods can be categorized according to the degree of structure or regularity in the mesh cells:

(1) Almost completely structured. In this case, the space lattice is organized so that its unit cells are congruent wherever possible. The most basic example of such a mesh is the pioneering work of YEE [1966], who employed a uniform Cartesian grid having rectangular cells. Staircasing was used to approximate the surface of structural features not parallel to the grid coordinate axes. Later work showed that it is possible to modify the size and shape of the space cells located immediately adjacent to a structural feature to conformally fit its surface (see, for example, JURGENS, TAFLOVE, UMASHANKAR, and MOORE [1992] and DEY and MITTRA [1997]). This is accurate and computationally efficient for large structures because the number of modified cells is proportional to the surface area of the structure. Thus, the number of modified cells becomes progressively smaller relative to the number of regular cells filling the structure volume as its size increases. As a result, the computer resources needed to implement a fully conformal model approximate those required for a staircased model. However, a key disadvantage of this technique is that special mesh-generation software must be constructed.

(2) Surface-fitted. In this case, the space lattice is globally distorted to fit the shape of the structure of interest. The lattice can be divided into multiple zones to accommodate a set of distinct surface features (see, for example, SHANKAR, MOHAMMADIAN, and HALL [1990]). The major advantage of this approach is

that well-developed mesh-generation software of this type is available. The major disadvantage is that, relative to the Yee algorithm, there is substantial added computer burden due to:
(a) memory allocations for the position and stretching factors of each cell;
(b) extra computer operations to implement Maxwell's equations at each cell and to enforce field continuity at the interfaces of adjacent cells.
Another disadvantage is the possible presence of numerical dissipation in the time-stepping algorithm used for such meshes. This can limit the range of electrical size of the structure being modeled due to numerical wave-attenuation artifacts.
(3) Completely unstructured. In this case, the space containing the structure of interest is completely filled with a collection of lattice cells of varying sizes and shapes, but conforming to the structure surface (see, for example, MADSEN and ZIOLKOWSKI [1990]). As for the case of surface-fitted lattices, mesh-generation software is available and capable of modeling complicated three-dimensional shapes possibly having volumetric inhomogeneities. A key disadvantage of this approach is its potential for numerical inaccuracy and instability due to the unwanted generation of highly skewed space cells at random points within the lattice. A second disadvantage is the difficulty in mapping the unstructured mesh computations onto the architecture of either parallel vector computers or massively parallel machines. The structure-specific irregularity of the mesh mandates a robust preprocessing algorithm that optimally assigns specific mesh cells to specific processors.

At present, the best choice of computational algorithm and mesh remains unclear. For the next several years, we expect continued progress in this area as various groups develop their favored approaches and perform validations.

1.5. Predictive dynamic range

For computational modeling of electromagnetic wave interaction structures using FDTD and related space-grid time-domain techniques, it is useful to consider the concept of predictive dynamic range. Let the power density of the primary (incident) wave in the space grid be P_0 W/m². Further, let the minimum observable power density of a secondary (scattered) wave be P_S W/m², where "minimum observable" means that the accuracy of the field computation degrades due to numerical artifacts to poorer than n dB (some desired figure of merit) at lower levels than P_S. Then, we can define the predictive dynamic range as $10 \log_{10}(P_0/P_S)$ dB.

This definition is well suited for FDTD and other space-grid time-domain codes for two reasons:
- It squares nicely with the concept of a "quiet zone" in an experimental anechoic chamber, which is intuitive to most electromagnetics engineers;
- It succinctly quantifies the fact that the desired numerical wave analogs propagating in the lattice exist in an additive noise environment due to nonphysical propagating wave analogs caused by the imperfect ABCs.

In addition to additive noise, the desired physical wave analogs undergo gradual progressive deterioration while propagating due to accumulating numerical dispersion artifacts, including phase velocity anisotropies and inhomogeneities within the mesh.

In the 1980s, researchers accumulated solid evidence for a predictive dynamic range
on the order of 40–50 dB for FDTD codes. This value is reasonable if one considers
the additive noise due to imperfect ABCs to be the primary limiting factor, since the
analytical ABCs of this era (see, for example, MUR [1981]) provided outer-boundary
reflection coefficients in the range of about 0.3–3% (-30 to -50 dB).

The 1990s saw the emergence of powerful, entirely new classes of ABCs including
the perfectly matched layer (PML) of BERENGER [1994]; the uniaxial anisotropic PML
(UPML) of SACKS, KINGSLAND, LEE, and LEE [1995] and GEDNEY [1996]; and the
complementary operator methods (COM) of RAMAHI [1997], RAMAHI [1998]. These
ABCs were shown to have effective outer-boundary reflection coefficients of better than
-80 dB for impinging pulsed electromagnetic waves having ultrawideband spectra.
Solid capabilities were demonstrated to terminate free-space lattices, multimoding and
dispersive waveguiding structures, and lossy and dispersive materials.

However, for electrically large problems, the overall dynamic range may not reach
the maximum permitted by these new ABCs because of inaccuracies due to accumulating numerical-dispersion artifacts generated by the basic grid-based solution of
the curl equations. Fortunately, by the end of the 1990s, this problem was being attacked
by a new generation of low-dispersion algorithms. Examples include the wavelet-
based multiresolution time-domain (MRTD) technique introduced by KRUMPHOLZ and
KATEHI [1996] and the pseudo-spectral time-domain (PSTD) technique introduced by
LIU, Q.H. [1996], LIU, Q.H. [1997]. As a result of these advances, there is emerging
the possibility of FDTD and related space-grid time-domain methods demonstrating
predictive dynamic ranges of 80 dB or more in the first decade of the 21st century.

1.6. Scaling to very large problem sizes

Using FDTD and related methods, we can model electromagnetic wave interaction
problems requiring the solution of considerably more than 10^8 field-vector unknowns.
At this level of complexity, it is possible to develop detailed, three-dimensional models
of complete engineering systems, including the following:

- Entire aircraft and missiles illuminated by radar at 1 GHz and above;
- Entire multilayer circuit boards and multichip modules for digital signal propagation, crosstalk, and radiation;
- Entire microwave and millimeter-wave amplifiers, including the active and passive
  circuit components and packaging;
- Entire integrated-optical structures, including lasers, waveguides, couplers, and
  resonators.

A key goal for such large models is to achieve algorithm/computer-architecture scaling such that for N field unknowns to be solved on M processors, we approach an
order(N/M) scaling of the required computational resources.

We now consider the factors involved in determining the computational burden for
the class of FDTD and related space-grid time-domain solvers.

(1) *Number of volumetric grid cells, N.* The six vector electromagnetic field components located at each lattice cell must be updated at every time step. This yields by itself an order(N) scaling.

(2) *Number of time steps, n_{max}.* A self-consistent solution in the time domain mandates that the numerical wave analogs propagate over time scales sufficient to causally connect each portion of the structure of interest. Therefore, n_{max} must increase as the maximum electrical size of the structure. In three dimensions, it can be argued that n_{max} is a fractional power function of N such as $N^{1/3}$. Further, n_{max} must be adequate to step through "ring-up" and "ring-down" times of energy storage features such as cavities. These features vary from problem to problem and cannot be ascribed a dependence relative to N.

(3) *Cumulative propagation errors.* Additional computational burdens may arise due to the need for either progressive mesh refinement or progressively higher-accuracy algorithms to bound cumulative positional or phase errors for propagating numerical modes in progressively enlarged meshes. Any need for progressive mesh refinement would feed back to factor 1.

For most free-space problems, factors 2 and 3 are weaker functions of the size of the modeled structure than factor 1. This is because geometrical features at increasing electrical distances from each other become decoupled due to radiative losses by the electromagnetic waves propagating between these features. Further, it can be shown that replacing second-order accurate algorithms by higher-order versions sufficiently reduces numerical dispersion error to avoid the need for progressive mesh refinement for object sizes up to the order of 100 wavelengths. Overall, a computational burden of order($N \cdot n_{max}$) = order($N^{4/3}$) is estimated for very large FDTD and related models.

2. Maxwell's equations

In this section, we establish the fundamental equations and notation for the electromagnetic fields used in the remainder of this chapter.

2.1. Three-dimensional case

Using MKS units, the time-dependent Maxwell's equations in three dimensions are given in differential and integral form by

Faraday's Law:

$$\frac{\partial \vec{B}}{\partial t} = -\nabla \times \vec{E} - \vec{M}, \tag{2.1a}$$

$$\frac{\partial}{\partial t} \iint_A \vec{B} \cdot d\vec{A} = -\oint_\ell \vec{E} \cdot d\vec{\ell} - \iint_A \vec{M} \cdot d\vec{A}. \tag{2.1b}$$

Ampere's Law:

$$\frac{\partial \vec{D}}{\partial t} = \nabla \times \vec{H} - \vec{J}, \tag{2.2a}$$

$$\frac{\partial}{\partial t} \iint_A \vec{D} \cdot d\vec{A} = \oint_\ell \vec{H} \cdot d\vec{\ell} - \iint_A \vec{J} \cdot d\vec{A}. \tag{2.2b}$$

Gauss' Law for the electric field:

$$\nabla \cdot \vec{D} = 0, \tag{2.3a}$$

$$\oiint_A \vec{D} \cdot d\vec{A} = 0. \tag{2.3b}$$

Gauss' Law for the magnetic field:

$$\nabla \cdot \vec{B} = 0, \tag{2.4a}$$

$$\oiint_A \vec{B} \cdot d\vec{A} = 0. \tag{2.4b}$$

In (2.1)–(2.4), the following symbols (and their MKS units) are defined:

$\vec{E}$:   electric field (volts/meter)

$\vec{D}$:   electric flux density (coulombs/meter^2)

$\vec{H}$:   magnetic field (amperes/meter)

$\vec{B}$:   magnetic flux density (webers/meter^2)

A: arbitrary three-dimensional surface

$d\vec{A}$:   differential normal vector that characterizes surface A (meter^2)

ℓ: closed contour that bounds surface A

$d\vec{\ell}$:   differential length vector that characterizes contour ℓ (meters)

$\vec{J}$:   electric current density (amperes/meter^2)

$\vec{M}$:   equivalent magnetic current density (volts/meter^2)

In linear, isotropic, nondispersive materials (i.e., materials having field-independent, direction-independent, and frequency-independent electric and magnetic properties), we can relate $\vec{D}$ to $\vec{E}$ and $\vec{B}$ to $\vec{H}$ using simple proportions:

$$\vec{D} = \varepsilon \vec{E} = \varepsilon_r \varepsilon_0 \vec{E}; \qquad \vec{B} = \mu \vec{H} = \mu_r \mu_0 \vec{H}, \tag{2.5}$$

where

ε: electrical permittivity (farads/meter)

ε_r: relative permittivity (dimensionless scalar)

ε_0:   free-space permittivity (8.854×10^{-12} farads/meter)

μ: magnetic permeability (henrys/meter)

μ_r: relative permeability (dimensionless scalar)

μ_0:   free-space permeability ($4\pi \times 10^{-7}$ henrys/meter)

Note that $\vec{J}$ and $\vec{M}$ can act as *independent sources* of E- and H-field energy, $\vec{J}_{\text{source}}$ and $\vec{M}_{\text{source}}$. We also allow for materials with isotropic, nondispersive electric and magnetic losses that attenuate E- and H-fields via conversion to heat energy. This yields:

$$\vec{J} = \vec{J}_{\text{source}} + \sigma\vec{E}; \qquad \vec{M} = \vec{M}_{\text{source}} + \sigma^*\vec{H}, \tag{2.6}$$

where

σ: electric conductivity (siemens/meter)

σ^*: equivalent magnetic loss (ohms/meter)

Finally, we substitute (2.5) and (2.6) into (2.1a) and (2.2a). This yields Maxwell's curl equations in linear, isotropic, nondispersive, lossy materials:

$$\frac{\partial\vec{H}}{\partial t} = -\frac{1}{\mu}\nabla\times\vec{E} - \frac{1}{\mu}\left(\vec{M}_{\text{source}} + \sigma^*\vec{H}\right), \tag{2.7}$$

$$\frac{\partial\vec{E}}{\partial t} = \frac{1}{\varepsilon}\nabla\times\vec{H} - \frac{1}{\varepsilon}\left(\vec{J}_{\text{source}} + \sigma\vec{E}\right). \tag{2.8}$$

We now write out the vector components of the curl operators of (2.7) and (2.8) in Cartesian coordinates. This yields the following system of six coupled scalar equations:

$$\frac{\partial H_x}{\partial t} = \frac{1}{\mu}\left[\frac{\partial E_y}{\partial z} - \frac{\partial E_z}{\partial y} - \left(M_{\text{source}_x} + \sigma^*H_x\right)\right], \tag{2.9a}$$

$$\frac{\partial H_y}{\partial t} = \frac{1}{\mu}\left[\frac{\partial E_z}{\partial x} - \frac{\partial E_x}{\partial z} - \left(M_{\text{source}_y} + \sigma^*H_y\right)\right], \tag{2.9b}$$

$$\frac{\partial H_z}{\partial t} = \frac{1}{\mu}\left[\frac{\partial E_x}{\partial y} - \frac{\partial E_y}{\partial x} - \left(M_{\text{source}_z} + \sigma^*H_z\right)\right], \tag{2.9c}$$

$$\frac{\partial E_x}{\partial t} = \frac{1}{\varepsilon}\left[\frac{\partial H_z}{\partial y} - \frac{\partial H_y}{\partial z} - \left(J_{\text{source}_x} + \sigma E_x\right)\right], \tag{2.10a}$$

$$\frac{\partial E_y}{\partial t} = \frac{1}{\varepsilon}\left[\frac{\partial H_x}{\partial z} - \frac{\partial H_z}{\partial x} - \left(J_{\text{source}_y} + \sigma E_y\right)\right], \tag{2.10b}$$

$$\frac{\partial E_z}{\partial t} = \frac{1}{\varepsilon}\left[\frac{\partial H_y}{\partial x} - \frac{\partial H_x}{\partial y} - \left(J_{\text{source}_z} + \sigma E_z\right)\right]. \tag{2.10c}$$

The system of six coupled partial differential equations of (2.9) and (2.10) forms the basis of the FDTD numerical algorithm for electromagnetic wave interactions with general three-dimensional objects. The FDTD algorithm need not explicitly enforce the Gauss' Law relations indicating zero free electric and magnetic charge, (2.3) and (2.4). This is because these relations are theoretically a direct consequence of the curl equations, as can be readily shown. However, the FDTD space grid must be structured so that the Gauss' Law relations are *implicit* in the positions of the E- and H-field vector components in the grid, and in the numerical space-derivative operations upon these components that model the action of the curl operator. This will be discussed later in the context of the Yee mesh.

Before proceeding with the introduction of the Yee algorithm, it is instructive to consider simplified two-dimensional cases for Maxwell's equations. These cases demonstrate important electromagnetic wave phenomena and can yield insight into the analytical and algorithmic features of the general three-dimensional case.

2.2. Reduction to two dimensions

Let us assume that the structure being modeled extends to infinity in the z-direction with no change in the shape or position of its transverse cross section. If the incident wave is also uniform in the z-direction, then all partial derivatives of the fields with respect to z must equal zero. Under these conditions, the full set of Maxwell's curl equations given by (2.9) and (2.10) reduces to two modes, the *transverse-magnetic mode with respect to z* (TM$_z$) and the *transverse-electric mode with respect to z* (TE$_z$). The reduced sets of Maxwell's equations for these modes are as follows.

TM$_z$ mode (involving only H_x, H_y, and E_z)

$$\frac{\partial H_x}{\partial t} = \frac{1}{\mu}\left[-\frac{\partial E_z}{\partial y} - \left(M_{source_x} + \sigma^* H_x\right)\right], \tag{2.11a}$$

$$\frac{\partial H_y}{\partial t} = \frac{1}{\mu}\left[\frac{\partial E_z}{\partial x} - \left(M_{source_y} + \sigma^* H_y\right)\right], \tag{2.11b}$$

$$\frac{\partial E_z}{\partial t} = \frac{1}{\varepsilon}\left[\frac{\partial H_y}{\partial x} - \frac{\partial H_x}{\partial y} - \left(J_{source_z} + \sigma E_z\right)\right]. \tag{2.11c}$$

TE$_z$ mode (involving only E_x, E_y, and H_z)

$$\frac{\partial E_x}{\partial t} = \frac{1}{\varepsilon}\left[\frac{\partial H_z}{\partial y} - \left(J_{source_x} + \sigma E_x\right)\right], \tag{2.12a}$$

$$\frac{\partial E_y}{\partial t} = \frac{1}{\varepsilon}\left[-\frac{\partial H_z}{\partial x} - \left(J_{source_y} + \sigma E_y\right)\right], \tag{2.12b}$$

$$\frac{\partial H_z}{\partial t} = \frac{1}{\mu}\left[\frac{\partial E_x}{\partial y} - \frac{\partial E_y}{\partial x} - \left(M_{source_z} + \sigma^* H_z\right)\right]. \tag{2.12c}$$

The TM$_z$ and TE$_z$ modes contain no common field vector components. Thus, these modes can exist simultaneously with *no* mutual interactions for structures composed of isotropic materials or anisotropic materials having no off-diagonal components in the constitutive tensors.

Physical phenomena associated with these two modes can be very different. The TE$_z$ mode can support propagating electromagnetic fields bound closely to, or guided by, the surface of a metal structure (the "creeping wave" being a classic example for curved metal surfaces). On the other hand, the TM$_z$ mode sets up an E-field which must be negligible at a metal surface. This diminishes or eliminates bound or guided near-surface propagating waves for metal surfaces. The presence or absence of surface-type waves can have important implications for scattering and radiation problems.

3. The Yee algorithm

3.1. Basic ideas

YEE [1966] originated a set of finite-difference equations for the time-dependent Maxwell's curl equations of (2.9) and (2.10) for the lossless materials case $\sigma = 0$ and

$\sigma^* = 0$. This section summarizes Yee's algorithm, which forms the basis of the FDTD technique. Key ideas underlying the robust nature of the Yee algorithm are as follows:

(1) The Yee algorithm solves for both electric and magnetic fields in time and space using the coupled Maxwell's curl equations rather than solving for the electric field alone (or the magnetic field alone) with a wave equation.

- This is analogous to the combined-field integral equation formulation of the method of moments, wherein both $\vec{E}$ and $\vec{H}$ boundary conditions are enforced on the surface of a material structure.
- Using both $\vec{E}$ and $\vec{H}$ information, the solution is more robust than using either alone (i.e., it is accurate for a wider class of structures). Both electric and magnetic material properties can be modeled in a straightforward manner. This is especially important when modeling radar cross section mitigation.
- Features unique to each field such as tangential $\vec{H}$ singularities near edges and corners, azimuthal (looping) $\vec{H}$ singularities near thin wires, and radial $\vec{E}$ singularities near points, edges, and thin wires can be individually modeled if both electric and magnetic fields are available.

(2) As illustrated in Fig. 3.1, the Yee algorithm centers its $\vec{E}$ and $\vec{H}$ components in three-dimensional space so that every $\vec{E}$ component is surrounded by four circulating $\vec{H}$ components, and every $\vec{H}$ component is surrounded by four circulating $\vec{E}$ components.

This provides a beautifully simple picture of three-dimensional space being filled by an interlinked array of Faraday's Law and Ampere's Law contours. For example, it is possible to identify Yee $\vec{E}$ components associated with displacement current flux linking $\vec{H}$ loops, as well as $\vec{H}$ components associated with magnetic flux linking $\vec{E}$ loops. In effect, the Yee algorithm simultaneously simulates the pointwise differential form *and* the macroscopic integral form of

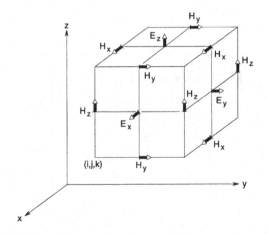

FIG. 3.1. Position of the electric and magnetic field vector components about a cubic unit cell of the Yee space lattice. *After*: K.S. Yee, *IEEE Trans. Antennas and Propagation*, Vol. 14, 1966, pp. 302–307, © 1966 IEEE.

Maxwell's equations. The latter is extremely useful in specifying field boundary conditions and singularities.

In addition, we have the following attributes of the Yee space lattice:

- The finite-difference expressions for the space derivatives used in the curl operators are central-difference in nature and second-order accurate.
- Continuity of tangential $\vec{E}$ and $\vec{H}$ is naturally maintained across an interface of dissimilar materials if the interface is parallel to one of the lattice coordinate axes. For this case, there is no need to specially enforce field boundary conditions at the interface. At the beginning of the problem, we simply specify the material permittivity and permeability at each field component location. This yields a stepped or "staircase" approximation of the surface and internal geometry of the structure, with a space resolution set by the size of the lattice unit cell.
- The location of the $\vec{E}$ and $\vec{H}$ components in the Yee space lattice and the central-difference operations on these components implicitly enforce the two Gauss' Law relations. Thus, the Yee mesh is divergence-free with respect to its E- and H-fields in the absence of free electric and magnetic charge.

(3) As illustrated in Fig. 3.2, the Yee algorithm also centers its $\vec{E}$ and $\vec{H}$ components in time in what is termed a leapfrog arrangement. All of the $\vec{E}$ computations in the modeled space are completed and stored in memory for a particular time point using previously stored $\vec{H}$ data. Then all of the $\vec{H}$ computations in the space are completed and stored in memory using the $\vec{E}$ data just computed. The cycle begins again with the recomputation of the $\vec{E}$ components based on the newly obtained $\vec{H}$. This process continues until time-stepping is concluded.

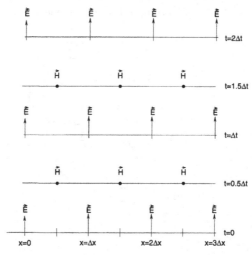

FIG. 3.2. Space–time chart of the Yee algorithm for a one-dimensional wave propagation example showing the use of central differences for the space derivatives and leapfrog for the time derivatives. Initial conditions for both electric and magnetic fields are zero everywhere in the grid.

- Leapfrog time-stepping is fully explicit, thereby avoiding problems involved with simultaneous equations and matrix inversion.
- The finite-difference expressions for the time derivatives are central-difference in nature and second-order accurate.
- The time-stepping algorithm is nondissipative. That is, numerical wave modes propagating in the mesh do not spuriously decay due to a nonphysical artifact of the time-stepping algorithm.

3.2. Finite differences and notation

YEE [1966] introduced the following notation for space points and functions of space and time. A space point in a uniform, rectangular lattice is denoted as

$$(i, j, k) = (i\Delta x, j\Delta y, k\Delta z). \tag{3.1}$$

Here, Δx, Δy, and Δz are, respectively, the lattice space increments in the x, y, and z coordinate directions, and i, j, and k are integers. Further, we denote any function u of space and time evaluated at a discrete point in the grid and at a discrete point in time as

$$u(i\Delta x, j\Delta y, k\Delta z, n\Delta t) = u_{i,j,k}^n, \tag{3.2}$$

where Δt is the time increment, assumed uniform over the observation interval, and n is an integer.

Yee used centered finite-difference (central-difference) expressions for the space and time derivatives that are both simply programmed and second-order accurate in the space and time increments. Consider his expression for the first partial space derivative of u in the x-direction, evaluated at the fixed time $t_n = n\Delta t$:

$$\frac{\partial u}{\partial x}(i\Delta x, j\Delta y, k\Delta z, n\Delta t) = \frac{u_{i+1/2,j,k}^n - u_{i-1/2,j,k}^n}{\Delta x} + \mathrm{O}[(\Delta x)^2]. \tag{3.3}$$

We note the $\pm 1/2$ increment in the i subscript (x-coordinate) of u, denoting a space finite-difference over $\pm 1/2\Delta x$. Yee's goal was second-order accurate central differencing, but it is apparent that he desired to take data for his central differences to the right and left of his observation point by only $\Delta x/2$, rather than a full Δx.

Yee chose this notation because he wished to interleave his $\vec{E}$ and $\vec{H}$ components in the space lattice at intervals of $\Delta x/2$. For example, the difference of two adjacent $\vec{E}$ components, separated by Δx and located $\pm 1/2\Delta x$ on either side of an $\vec{H}$ component, would be used to provide a numerical approximation for $\partial E/\partial x$ to permit stepping the $\vec{H}$ component in time. For completeness, it should be added that a numerical approximation analogous to (3.3) for $\partial u/\partial y$ or $\partial u/\partial z$ can be written simply by incrementing the j or k subscript of u by $\pm 1/2\Delta y$ or $\pm 1/2\Delta z$, respectively.

Yee's expression for the first time partial derivative of u, evaluated at the fixed space point (i, j, k), follows by analogy:

$$\frac{\partial u}{\partial t}(i\Delta x, j\Delta y, k\Delta z, n\Delta t) = \frac{u_{i,j,k}^{n+1/2} - u_{i,j,k}^{n-1/2}}{\Delta t} + \mathrm{O}[(\Delta t)^2]. \tag{3.4}$$

Now the $\pm 1/2$ increment is in the n superscript (time coordinate) of u, denoting a time finite-difference over $\pm 1/2\Delta t$. Yee chose this notation because he wished to interleave his $\vec{E}$ and $\vec{H}$ components in time at intervals of $1/2\Delta t$ for purposes of implementing a leapfrog algorithm.

3.3. Finite-difference expressions for Maxwell's equations in three dimensions

We now apply the above ideas and notation to achieve a finite-difference numerical approximation of the Maxwell's curl equations in three dimensions given by (2.9) and (2.10). We begin by considering as an example the E_x field-component equation (2.10a). Referring to Figs. 3.1 and 3.2, a typical substitution of central differences for the time and space derivatives in (2.10a) at $E_x(i, j+1/2, k+1/2, n)$ yields the following expression:

$$
\frac{E_x|_{i,j+1/2,k+1/2}^{n+1/2} - E_x|_{i,j+1/2,k+1/2}^{n-1/2}}{\Delta t}
$$

$$
= \frac{1}{\varepsilon_{i,j+1/2,k+1/2}} \cdot \left(\frac{\frac{H_z|_{i,j+1,k+1/2}^{n}-H_z|_{i,j,k+1/2}^{n}}{\Delta y} - \frac{H_y|_{i,j+1/2,k+1}^{n}-H_y|_{i,j+1/2,k}^{n}}{\Delta z}}{- J_{\text{source}x}|_{i,j+1/2,k+1/2}^{n} - \sigma_{i,j+1/2,k+1/2} E_x|_{i,j+1/2,k+1/2}^{n}} \right).
$$

(3.5)

Note that all field quantities on the right-hand side are evaluated at time-step n, including the electric field E_x appearing due to the material conductivity σ. Since E_x values at time-step n are not assumed to be stored in the computer's memory (only the previous values of E_x at time-step $n-1/2$ are assumed to be in memory), we need some way to estimate such terms. A very good way is as follows, using what we call a *semi-implicit approximation*:

$$
E_x|_{i,j+1/2,k+1/2}^{n} = \frac{E_x|_{i,j+1/2,k+1/2}^{n+1/2} + E_x|_{i,j+1/2,k+1/2}^{n-1/2}}{2}.
$$

(3.6)

Here E_x values at time-step n are assumed to be simply the arithmetic average of the stored values of E_x at time-step $n-1/2$ and the yet-to-be computed new values of E_x at time-step $n+1/2$. Substituting (3.6) into (3.5) and collecting terms yields the following explicit time-stepping relation for E_x (which is numerically stable for values of σ from zero to infinity):

$$
E_x|_{i,j+1/2,k+1/2}^{n+1/2} = \left(\frac{1 - \frac{\sigma_{i,j+1/2,k+1/2}\Delta t}{2\varepsilon_{i,j+1/2,k+1/2}}}{1 + \frac{\sigma_{i,j+1/2,k+1/2}\Delta t}{2\varepsilon_{i,j+1/2,k+1/2}}} \right) E_x|_{i,j+1/2,k+1/2}^{n-1/2}
$$

$$
+ \left(\frac{\frac{\Delta t}{\varepsilon_{i,j+1/2,k+1/2}}}{1 + \frac{\sigma_{i,j+1/2,k+1/2}\Delta t}{2\varepsilon_{i,j+1/2,k+1/2}}} \right) \cdot \left(\frac{\frac{H_z|_{i,j+1,k+1/2}^{n}-H_z|_{i,j,k+1/2}^{n}}{\Delta y}}{- \frac{H_y|_{i,j+1/2,k+1}^{n}-H_y|_{i,j+1/2,k}^{n}}{\Delta z}}{- J_{\text{source}x}|_{i,j+1/2,k+1/2}^{n}} \right).
$$

(3.7a)

Similarly, we can derive finite-difference expressions based on Yee's algorithm for the E_y and E_z field components given by Maxwell's equations (2.10b) and (2.10c).

Referring again to Fig. 3.1, we have:

$$E_y\Big|_{i-1/2,j+1,k+1/2}^{n+1/2}$$

$$= \left(\frac{1 - \frac{\sigma_{i-1/2,j+1,k+1/2}\Delta t}{2\varepsilon_{i-1/2,j+1,k+1/2}}}{1 + \frac{\sigma_{i-1/2,j+1,k+1/2}\Delta t}{2\varepsilon_{i-1/2,j+1,k+1/2}}}\right) E_y\Big|_{i-1/2,j+1,k+1/2}^{n-1/2}$$

$$+ \left(\frac{\frac{\Delta t}{\varepsilon_{i-1/2,j+1,k+1/2}}}{1 + \frac{\sigma_{i-1/2,j+1,k+1/2}\Delta t}{2\varepsilon_{i-1/2,j+1,k+1/2}}}\right) \cdot \left(\begin{array}{c} \frac{H_x|_{i-1/2,j+1,k+1}^n - H_x|_{i-1/2,j+1,k}^n}{\Delta z} \\ - \frac{H_z|_{i,j+1,k+1/2}^n - H_z|_{i-1,j+1,k+1/2}^n}{\Delta x} \\ - J_{\text{source}_y}|_{i-1/2,j+1,k+1/2}^n \end{array}\right), \quad (3.7b)$$

$$E_z\Big|_{i-1/2,j+1/2,k+1}^{n+1/2}$$

$$= \left(\frac{1 - \frac{\sigma_{i-1/2,j+1/2,k+1}\Delta t}{2\varepsilon_{i-1/2,j+1/2,k+1}}}{1 + \frac{\sigma_{i-1/2,j+1/2,k+1}\Delta t}{2\varepsilon_{i-1/2,j+1/2,k+1}}}\right) E_z\Big|_{i-1/2,j+1/2,k+1}^{n-1/2}$$

$$+ \left(\frac{\frac{\Delta t}{\varepsilon_{i-1/2,j+1/2,k+1}}}{1 + \frac{\sigma_{i-1/2,j+1/2,k+1}\Delta t}{2\varepsilon_{i-1/2,j+1/2,k+1}}}\right) \cdot \left(\begin{array}{c} \frac{H_y|_{i,j+1/2,k+1}^n - H_y|_{i-1,j+1/2,k+1}^n}{\Delta x} \\ - \frac{H_x|_{i-1/2,j+1,k+1}^n - H_x|_{i-1/2,j,k+1}^n}{\Delta y} \\ - J_{\text{source}_z}|_{i-1/2,j+1/2,k+1}^n \end{array}\right). \quad (3.7c)$$

By analogy we can derive finite-difference equations for (2.9a)–(2.9c) to time-step H_x, H_y, and H_z. Here σ^*H represents a magnetic loss term on the right-hand side of each equation, which is estimated using a semi-implicit procedure analogous to (3.6). Referring again to Figs. 3.1 and 3.2, we have for example the following time-stepping expressions for the H components located about the unit cell:

$$H_x\Big|_{i-1/2,j+1,k+1}^{n+1}$$

$$= \left(\frac{1 - \frac{\sigma_{i-1/2,j+1,k+1}^*\Delta t}{2\mu_{i-1/2,j+1,k+1}}}{1 + \frac{\sigma_{i-1/2,j+1,k+1}^*\Delta t}{2\mu_{i-1/2,j+1,k+1}}}\right) H_x\Big|_{i-1/2,j+1,k+1}^{n}$$

$$+ \left(\frac{\frac{\Delta t}{\mu_{i-1/2,j+1,k+1}}}{1 + \frac{\sigma_{i-1/2,j+1,k+1}^*\Delta t}{2\mu_{i-1/2,j+1,k+1}}}\right) \cdot \left(\begin{array}{c} \frac{E_y|_{i-1/2,j+1,k+3/2}^{n+1/2} - E_y|_{i-1/2,j+1,k+1/2}^{n+1/2}}{\Delta z} \\ - \frac{E_z|_{i-1/2,j+3/2,k+1}^{n+1/2} - E_z|_{i-1/2,j+1/2,k+1}^{n+1/2}}{\Delta y} \\ - M_{\text{source}_x}|_{i-1/2,j+1,k+1}^{n+1/2} \end{array}\right), \quad (3.8a)$$

$$H_y\Big|_{i,j+1/2,k+1}^{n+1}$$

$$= \left(\frac{1 - \frac{\sigma_{i,j+1/2,k+1}^*\Delta t}{2\mu_{i,j+1/2,k+1}}}{1 + \frac{\sigma_{i,j+1/2,k+1}^*\Delta t}{2\mu_{i,j+1/2,k+1}}}\right) H_y\Big|_{i,j+1/2,k+1}^{n}$$

$$+ \left(\frac{\frac{\Delta t}{\mu_{i,j+1/2,k+1}}}{1 + \frac{\sigma_{i,j+1/2,k+1}^*\Delta t}{2\mu_{i,j+1/2,k+1}}}\right) \cdot \left(\begin{array}{c} \frac{E_z|_{i+1/2,j+1/2,k+1}^{n+1/2} - E_z|_{i-1/2,j+1/2,k+1}^{n+1/2}}{\Delta x} \\ - \frac{E_x|_{i,j+1/2,k+3/2}^{n+1/2} - E_x|_{i,j+1/2,k+1/2}^{n+1/2}}{\Delta z} \\ - M_{\text{source}_y}|_{i,j+1/2,k+1}^{n+1/2} \end{array}\right), \quad (3.8b)$$

$$H_z\big|_{i,j+1,k+1/2}^{n+1}$$

$$= \left(\frac{1 - \frac{\sigma_{i,j+1,k+1/2}^* \Delta t}{2\mu_{i,j+1,k+1/2}}}{1 + \frac{\sigma_{i,j+1,k+1/2}^* \Delta t}{2\mu_{i,j+1,k+1/2}}}\right) H_z\big|_{i,j+1,k+1/2}^{n}$$

$$+ \left(\frac{\frac{\Delta t}{\mu_{i,j+1,k+1/2}}}{1 + \frac{\sigma_{i,j+1,k+1/2}^* \Delta t}{2\mu_{i,j+1,k+1/2}}}\right) \cdot \left(\begin{array}{c} \frac{E_x\big|_{i,j+3/2,k+1/2}^{n+1/2} - E_x\big|_{i,j+1/2,k+1/2}^{n+1/2}}{\Delta y} \\[2mm] - \frac{E_y\big|_{i+1/2,j+1,k+1/2}^{n+1/2} - E_y\big|_{i-1/2,j+1,k+1/2}^{n+1/2}}{\Delta x} \\[2mm] - M_{\text{source}z}\big|_{i,j+1,k+1/2}^{n+1/2} \end{array}\right). \qquad (3.8c)$$

With the systems of finite-difference expressions of (3.7) and (3.8) , the new value of an electromagnetic field vector component at any space lattice point depends only on its previous value, the previous values of the components of the other field vector at adjacent points, and the known electric and magnetic current sources. Therefore, at any given time step, the computation of a field vector can proceed either one point at a time, or, if p parallel processors are employed concurrently, p points at a time.

3.4. Field updating coefficients

To implement the finite-difference systems of (3.7) and (3.8) for a region having a continuous variation of material properties with spatial position, it is desirable to define and store the following updating coefficients for each field vector component:

Updating coefficients at the general E-field component location (i, j, k):

$$C_a|_{i,j,k} = \left(1 - \frac{\sigma_{i,j,k} \Delta t}{2\varepsilon_{i,j,k}}\right) \Big/ \left(1 + \frac{\sigma_{i,j,k} \Delta t}{2\varepsilon_{i,j,k}}\right), \qquad (3.9a)$$

$$C_{b_1}|_{i,j,k} = \left(\frac{\Delta t}{\varepsilon_{i,j,k}\Delta_1}\right) \Big/ \left(1 + \frac{\sigma_{i,j,k} \Delta t}{2\varepsilon_{i,j,k}}\right), \qquad (3.9b)$$

$$C_{b_2}|_{i,j,k} = \left(\frac{\Delta t}{\varepsilon_{i,j,k}\Delta_2}\right) \Big/ \left(1 + \frac{\sigma_{i,j,k} \Delta t}{2\varepsilon_{i,j,k}}\right). \qquad (3.9c)$$

Updating coefficients at the general H-field component location (i, j, k):

$$D_a|_{i,j,k} = \left(1 - \frac{\sigma_{i,j,k}^* \Delta t}{2\mu_{i,j,k}}\right) \Big/ \left(1 + \frac{\sigma_{i,j,k}^* \Delta t}{2\mu_{i,j,k}}\right), \qquad (3.10a)$$

$$D_{b_1}|_{i,j,k} = \left(\frac{\Delta t}{\mu_{i,j,k}\Delta_1}\right) \Big/ \left(1 + \frac{\sigma_{i,j,k}^* \Delta t}{2\mu_{i,j,k}}\right), \qquad (3.10b)$$

$$D_{b_2}|_{i,j,k} = \left(\frac{\Delta t}{\mu_{i,j,k}\Delta_2}\right) \Big/ \left(1 + \frac{\sigma_{i,j,k}^* \Delta t}{2\mu_{i,j,k}}\right). \qquad (3.10c)$$

In (3.9) and (3.10), Δ_1 and Δ_2 denote the two possible lattice space increments used for the finite differences in each field-component calculation. For a cubic lattice, $\Delta x = \Delta y = \Delta z = \Delta$ and thus $\Delta_1 = \Delta_2 = \Delta$. For this case, $C_{b_1} = C_{b_2}$ and

$D_{b_1} = D_{b_2}$, reducing the storage requirement to two updating coefficients per field vector component. Here, the approximate total computer storage needed is $18N$, where N is the number of space cells in the FDTD lattice. The finite-difference expressions of (3.7) and (3.8) can now be rewritten more simply. For example, to update E_x we have:

$$
\begin{aligned}
E_x\Big|_{i,j+1/2,k+1/2}^{n+1/2} &= C_{a,E_x}\big|_{i,j+1/2,k+1/2}\, E_x\Big|_{i,j+1/2,k+1/2}^{n-1/2} \\
&\quad + C_{b,E_x}\big|_{i,j+1/2,k+1/2}\cdot
\begin{pmatrix}
H_z\big|_{i,j+1,k+1/2}^{n} - H_z\big|_{i,j,k+1/2}^{n} + H_y\big|_{i,j+1/2,k}^{n} \\
- H_y\big|_{i,j+1/2,k+1}^{n} - J_{\text{source}x}\big|_{i,j+1/2,k+1/2}^{n}\Delta
\end{pmatrix}.
\end{aligned}
\tag{3.11}
$$

Similarly, to update H_x we have:

$$
\begin{aligned}
H_x\Big|_{i-1/2,j+1,k+1}^{n+1} &= D_{a,H_x}\big|_{i-1/2,j+1,k+1}\, H_x\Big|_{i-1/2,j+1,k+1}^{n} \\
&\quad + D_{b,H_x}\big|_{i-1/2,j+1,k+1}\cdot
\begin{pmatrix}
E_y\big|_{i-1/2,j+1,k+3/2}^{n+1/2} - E_y\big|_{i-1/2,j+1,k+1/2}^{n+1/2} \\
+ E_z\big|_{i-1/2,j+1/2,k+1}^{n+1/2} - E_z\big|_{i-1/2,j+3/2,k+1}^{n+1/2} \\
- M_{\text{source}x}\big|_{i-1/2,j+1,k+1}^{n+1/2}\Delta
\end{pmatrix}.
\end{aligned}
\tag{3.12}
$$

For a space region with a finite number of media having distinct electrical properties, the computer storage requirement can be further reduced. This can be done by defining an integer array, $\text{MEDIA}(i, j, k)$, for each set of field vector components. This array stores an integer "pointer" at each location of such a field component in the space lattice, enabling the proper algorithm coefficients to be extracted. For example, to update E_x we have:

$$
m = \text{MEDIA}_{E_x}\big|_{i,j+1/2,k+1/2},
$$
$$
\begin{aligned}
E_x\Big|_{i,j+1/2,k+1/2}^{n+1/2} &= C_a(m)\, E_x\Big|_{i,j+1/2,k+1/2}^{n-1/2} + C_b(m)\cdot\Big(H_z\big|_{i,j+1,k+1/2}^{n} - H_z\big|_{i,j,k+1/2}^{n} \\
&\quad + H_y\big|_{i,j+1/2,k}^{n} - H_y\big|_{i,j+1/2,k+1}^{n} - J_{\text{source}x}\big|_{i,j+1/2,k+1/2}^{n}\Delta\Big).
\end{aligned}
\tag{3.13}
$$

Similarly, to update H_x we have:

$$
m = \text{MEDIA}_{H_x}\big|_{i-1/2,j+1,k+1},
$$
$$
\begin{aligned}
H_x\Big|_{i-1/2,j+1,k+1}^{n+1} &= D_a(m)\, H_x\Big|_{i-1/2,j+1,k+1}^{n} + D_b(m)\cdot\Big(E_y\big|_{i-1/2,j+1,k+3/2}^{n+1/2} - E_y\big|_{i-1/2,j+1,k+1/2}^{n+1/2} \\
&\quad + E_z\big|_{i-1/2,j+1/2,k+1}^{n+1/2} - E_z\big|_{i-1/2,j+3/2,k+1}^{n+1/2} - M_{\text{source}x}\big|_{i-1/2,j+1,k+1}^{n+1/2}\Delta\Big).
\end{aligned}
\tag{3.14}
$$

We note that the coefficient arrays $C_a(m)$, $C_b(m)$, $D_a(m)$, and $D_b(m)$ each contain only M elements, where M is the number of distinct material media in the FDTD space lattice. Thus, if separate MEDIA(i, j, k) integer pointer arrays are provided for each field vector component, the approximate total computer storage needed is reduced to $12N$, where N is the number of space cells in the FDTD lattice. This reduction in computer storage comes at some cost, however, since additional computer instructions must be executed at each field vector location to obtain the pointer integer m from the associated MEDIA array and then extract the $C(m)$ or $D(m)$ updating coefficients.

Taking advantage of the integer nature of the MEDIA arrays, further reduction in computer storage can be achieved by bitwise packing of these integers. For example, a 64-bit word can be divided into sixteen 4-bit pointers. Such a composite pointer could specify up to $2^4 = 16$ distinct media at each of 16 locations in the grid. This provides the means to reduce the overall computer storage for the MEDIA arrays by a factor of $15/16$ (94%).

3.5. Space region with nonpermeable media

Many electromagnetic wave interaction problems involve nonpermeable media ($\mu = \mu_0$, $\sigma^* = 0$) and can be implemented on a uniform cubic-cell FDTD space lattice. For such problems, the field updating expressions can be further simplified by defining the proportional $\hat{\vec{E}}$ and $\hat{\vec{M}}$ vectors:

$$\hat{\vec{E}} = (\Delta t/\mu_0 \Delta)\vec{E};\tag{3.15a}$$

$$\hat{\vec{M}} = (\Delta t/\mu_0)\vec{M},\tag{3.15b}$$

where $\Delta = \Delta x = \Delta y = \Delta z$ is the cell size of the space lattice. Assuming that $\hat{E}_x$, $\hat{E}_y$, and $\hat{E}_z$ are stored in the computer memory, and further defining a scaled E-field updating coefficient $\hat{C}_b(m)$ as

$$\hat{C}_b(m) = (\Delta t/\mu_0 \Delta)C_b(m)\tag{3.16}$$

we can rewrite (3.13) as:

$$m = \mathrm{MEDIA}_{E_x}|_{i,j+1/2,k+1/2},$$

$$\hat{E}_x\big|_{i,j+1/2,k+1/2}^{n+1/2}$$
$$= C_a(m)\hat{E}_x\big|_{i,j+1/2,k+1/2}^{n-1/2} + \hat{C}_b(m)\cdot\left(H_z\big|_{i,j+1,k+1/2}^{n} - H_z\big|_{i,j,k+1/2}^{n}\right.$$
$$\left. + H_y\big|_{i,j+1/2,k}^{n} - H_y\big|_{i,j+1/2,k+1}^{n} - J_{\text{source}x}\big|_{i,j+1/2,k+1/2}^{n}\Delta\right).\tag{3.17}$$

Finite-difference expression (3.14) can now be rewritten very simply as:

$$H_x\big|_{i-1/2,j+1,k+1}^{n+1}$$
$$= H_x\big|_{i-1/2,j+1,k+1}^{n} + \hat{E}_y\big|_{i-1/2,j+1,k+3/2}^{n+1/2} - \hat{E}_y\big|_{i-1/2,j+1,k+1/2}^{n+1/2}$$
$$+ \hat{E}_z\big|_{i-1/2,j+1/2,k+1}^{n+1/2} - \hat{E}_z\big|_{i-1/2,j+3/2,k+1}^{n+1/2} - \hat{M}_{\text{source}x}\big|_{i-1/2,j+1,k+1}^{n+1/2}.\tag{3.18}$$

This technique eliminates the multiplications previously needed to update the H components, and requires storage of MEDIA arrays only for the E components. At the end of the run, the desired values of the unscaled E-fields can be obtained simply by multiplying the stored E values by the reciprocal of the scaling factor of (3.15a).

3.6. Reduction to the two-dimensional TM_z and TE_z modes

The finite-difference systems of (3.7) and (3.8) can be reduced for the decoupled, two-dimensional TM_z and TE_z modes summarized in Section 2.2. For convenience and consistency, we again consider the field vector components in the space lattice represented by the unit cell of Fig. 3.1. Assuming now that all partial derivatives of the fields with respect to z are equal to zero, the following conditions hold:

(1) The sets of (E_z, H_x, H_y) components located in each lattice cut plane k, $k + 1$, etc. are identical and can be completely represented by any one of these sets, which we designate as the TM_z mode.

(2) The sets of (H_z, E_x, E_y) components located in each lattice cut plane $k + 1/2$, $k + 3/2$, etc. are identical and can be completely represented by any one of these sets, which we designate as the TE_z mode.

The resulting finite-difference systems for the TM_z and TE_z modes are as follows:

TM_z *mode, corresponding to the system of* (2.11)

$$H_x\Big|_{i-1/2,j+1}^{n+1} = \left(\frac{1 - \frac{\sigma_{i-1/2,j+1}^* \Delta t}{2\mu_{i-1/2,j+1}}}{1 + \frac{\sigma_{i-1/2,j+1}^* \Delta t}{2\mu_{i-1/2,j+1}}}\right) H_x\Big|_{i-1/2,j+1}^{n} + \left(\frac{\frac{\Delta t}{\mu_{i-1/2,j+1}}}{1 + \frac{\sigma_{i-1/2,j+1}^* \Delta t}{2\mu_{i-1/2,j+1}}}\right)$$

$$\times \left(\frac{E_z\Big|_{i-1/2,j+1/2}^{n+1/2} - E_z\Big|_{i-1/2,j+3/2}^{n+1/2}}{\Delta y} - M_{\text{source}x}\Big|_{i-1/2,j+1}^{n+1/2}\right),$$

$$(3.19a)$$

$$H_y\Big|_{i,j+1/2}^{n+1} = \left(\frac{1 - \frac{\sigma_{i,j+1/2}^* \Delta t}{2\mu_{i,j+1/2}}}{1 + \frac{\sigma_{i,j+1/2}^* \Delta t}{2\mu_{i,j+1/2}}}\right) H_y\Big|_{i,j+1/2}^{n} + \left(\frac{\frac{\Delta t}{\mu_{i,j+1/2}}}{1 + \frac{\sigma_{i,j+1/2}^* \Delta t}{2\mu_{i,j+1/2}}}\right)$$

$$\times \left(\frac{E_z\Big|_{i+1/2,j+1/2}^{n+1/2} - E_z\Big|_{i-1/2,j+1/2}^{n+1/2}}{\Delta x} - M_{\text{source}y}\Big|_{i,j+1/2}^{n+1/2}\right), \quad (3.19b)$$

$$E_z\Big|_{i-1/2,j+1/2}^{n+1/2} = \left(\frac{1 - \frac{\sigma_{i-1/2,j+1/2} \Delta t}{2\varepsilon_{i-1/2,j+1/2}}}{1 + \frac{\sigma_{i-1/2,j+1/2} \Delta t}{2\varepsilon_{i-1/2,j+1/2}}}\right) E_z\Big|_{i-1/2,j+1/2}^{n-1/2} + \left(\frac{\frac{\Delta t}{\varepsilon_{i-1/2,j+1/2}}}{1 + \frac{\sigma_{i-1/2,j+1/2} \Delta t}{2\varepsilon_{i-1/2,j+1/2}}}\right)$$

$$\times \left(\frac{H_y\Big|_{i,j+1/2}^{n} - H_y\Big|_{i-1,j+1/2}^{n}}{\Delta x} + \frac{H_x\Big|_{i-1/2,j}^{n} - H_x\Big|_{i-1/2,j+1}^{n}}{\Delta y}\right). \quad (3.19c)$$

$$- J_{\text{source}z}\Big|_{i-1/2,j+1/2}^{n}$$

TE_z mode, corresponding to the system of (2.12)

$$E_x\big|_{i,j+1/2}^{n+1/2} = \left(\frac{1 - \frac{\sigma_{i,j+1/2}\Delta t}{2\varepsilon_{i,j+1/2}}}{1 + \frac{\sigma_{i,j+1/2}\Delta t}{2\varepsilon_{i,j+1/2}}}\right) E_x\big|_{i,j+1/2}^{n-1/2} + \left(\frac{\frac{\Delta t}{\varepsilon_{i,j+1/2}}}{1 + \frac{\sigma_{i,j+1/2}\Delta t}{2\varepsilon_{i,j+1/2}}}\right)$$

$$\times \left(\frac{H_z\big|_{i,j+1}^{n} - H_z\big|_{i,j}^{n}}{\Delta y} - J_{\text{source}_x}\big|_{i,j+1/2}^{n}\right), \qquad (3.20\text{a})$$

$$E_y\big|_{i-1/2,j+1}^{n+1/2} = \left(\frac{1 - \frac{\sigma_{i-1/2,j+1}\Delta t}{2\varepsilon_{i-1/2,j+1}}}{1 + \frac{\sigma_{i-1/2,j+1}\Delta t}{2\varepsilon_{i-1/2,j+1}}}\right) E_y\big|_{i-1/2,j+1}^{n-1/2} + \left(\frac{\frac{\Delta t}{\varepsilon_{i-1/2,j+1}}}{1 + \frac{\sigma_{i-1/2,j+1}\Delta t}{2\varepsilon_{i-1/2,j+1}}}\right)$$

$$\times \left(\frac{H_z\big|_{i-1,j+1}^{n} - H_z\big|_{i,j+1}^{n}}{\Delta x} - J_{\text{source}_y}\big|_{i-1/2,j+1}^{n}\right), \qquad (3.20\text{b})$$

$$H_z\big|_{i,j+1}^{n+1} = \left(\frac{1 - \frac{\sigma_{i,j+1}^{*}\Delta t}{2\mu_{i,j+1}}}{1 + \frac{\sigma_{i,j+1}^{*}\Delta t}{2\mu_{i,j+1}}}\right) H_z\big|_{i,j+1}^{n} + \left(\frac{\frac{\Delta t}{\mu_{i,j+1}}}{1 + \frac{\sigma_{i,j+1}^{*}\Delta t}{2\mu_{i,j+1}}}\right)$$

$$\times \left(\begin{array}{l}\frac{E_x\big|_{i,j+3/2}^{n+1/2} - E_x\big|_{i,j+1/2}^{n+1/2}}{\Delta y} \\ + \frac{E_y\big|_{i-1/2,j+1}^{n+1/2} - E_y\big|_{i+1/2,j+1}^{n+1/2}}{\Delta x} - M_{\text{source}_z}\big|_{i,j+1}^{n+1/2}\end{array}\right). \qquad (3.20\text{c})$$

3.7. Interpretation as Faraday's and Ampere's Laws in integral form

The Yee algorithm for FDTD was originally interpreted as a direct approximation of the pointwise derivatives of Maxwell's time-dependent curl equations by numerical central differences. Although this interpretation is useful for understanding how FDTD models wave propagation away from material interfaces, it sheds little light on what algorithm modifications are needed to properly model the electromagnetic field physics of fine geometrical features such as wires, slots, and curved surfaces requiring subcell spatial resolution.

The literature indicates that FDTD modeling can be extended to such features by departing from Yee's original pointwise derivative thinking (see, for example, TAFLOVE, UMASHANKAR, BEKER, HARFOUSH, and YEE [1988] and JURGENS, TAFLOVE, UMASHANKAR, and MOORE [1992]). As shown in Fig. 3.3, the idea involves starting with a more macroscopic (but still local) combined-field description based upon Ampere's Law and Faraday's Law in *integral* form, implemented on an array of electrically small, spatially orthogonal contours. These contours mesh (intersect) in the manner of links in a chain, providing a geometrical interpretation of the coupling of these two laws. This meshing results in the filling of the FDTD modeled space by a three-dimensional "chain-link" array of intersecting orthogonal contours. The presence of wires, slots, and curved surfaces can be modeled by incorporating appropriate field behavior into the contour and surface integrals used to implement Ampere's and Faraday's Laws at selected meshes, and by deforming contour paths as required to conform with surface curvature.

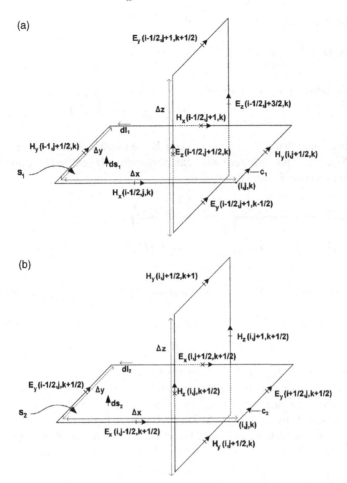

FIG. 3.3. Examples of chain-linked orthogonal contours in the free-space Yee mesh. (a) Ampere's Law for time-stepping E_z; (b) Faraday's Law for time-stepping H_z. *Adapted from*: A. Taflove et al., *IEEE Trans. Antennas and Propagation*, 1988, pp. 247–257, © 1988 IEEE.

This approach is intuitively satisfying to an electrical engineer since it permits the FDTD numerical model to deal with physical quantities such as:

- Electromotive forces (EMFs) and magnetomotive forces (MMFs) developed when completing one circuit about a Faraday's or Ampere's Law contour path;
- Magnetic flux and electric displacement current when performing the surface integrations on the patches bounded by the respective contours.

In this section, we demonstrate the equivalence of the Yee and contour-path interpretations for the free-space case. For simplicity, FDTD time-stepping expressions are developed for only one E and one H field component. Extension to all the rest is straightforward. We further assume lossless free space with no electric or magnetic current sources. Applying Ampere's Law along contour C_1 in Fig. 3.3(a), and assuming that the field value at a midpoint of one side of the contour equals the average value of that field component along that side, we obtain

$$\frac{\partial}{\partial t} \int S_1 \vec{D} \cdot d\vec{S}_1 = \oint_{C_1} \vec{H} \cdot d\vec{\ell}_1, \tag{3.21a}$$

$$\frac{\partial}{\partial t} \int S_1 \varepsilon_0 E_z|_{i-1/2,j+1/2,k}\, dS_1 \cong H_x|_{i-1/2,j,k}\Delta x + H_y|_{i,j+1/2,k}\Delta y$$
$$- H_x|_{i-1/2,j+1,k}\Delta x - H_y|_{i-1,j+1/2,k}\Delta y. \tag{3.21b}$$

Now further assume that $E_z|_{i-1/2,j+1/2,k}$ equals the average value of E_z over the surface patch S_1 and that the time derivative can be numerically realized by using a central-difference expression. Then (3.21b) yields

$$\varepsilon_0 \Delta x \Delta y \left(\frac{E_z|_{i-1/2,j+1/2,k}^{n+1/2} - E_z|_{i-1/2,j+1/2,k}^{n-1/2}}{\Delta t} \right)$$
$$= \left(H_x|_{i-1/2,j,k}^n - H_x|_{i-1/2,j+1,k}^n \right)\Delta x + \left(H_y|_{i,j+1/2,k}^n - H_y|_{i-1,j+1/2,k}^n \right)\Delta y. \tag{3.21c}$$

Multiplying both sides by $\Delta t/(\varepsilon_0 \Delta x \Delta y)$ and solving for $E_z|_{i-1/2,j+1/2,k}^{n+1/2}$ provides

$$E_z|_{i-1/2,j+1/2,k}^{n+1/2} = E_z|_{i-1/2,j+1/2,k}^{n-1/2} + \left(H_x|_{i-1/2,j,k}^n - H_x|_{i-1/2,j+1,k}^n \right)\Delta t/(\varepsilon_0 \Delta y)$$
$$+ \left(H_y|_{i,j+1/2,k}^n - H_y|_{i-1,j+1/2,k}^n \right)\Delta t/(\varepsilon_0 \Delta x). \tag{3.22}$$

(Eq. 3.22) is simply the free-space version of (3.7c), the Yee time-stepping equation for E_z that was obtained directly from implementing the curl $\vec{H}$ equation with finite differences. The only difference is that (3.22) is evaluated at $(i-1/2, j+1/2, k)$ whereas (3.7c) is evaluated at $(i-1/2, j+1/2, k+1)$ shown in Fig. 3.1.

In an analogous manner, we can apply Faraday's Law along contour C_2 in Fig. 3.3(b) to obtain

$$\frac{\partial}{\partial t} \int S_2 \vec{B} \cdot d\vec{S}_2 = -\oint_{C_2} \vec{E} \cdot d\vec{\ell}_2, \tag{3.23a}$$

$$\frac{\partial}{\partial t} \int S_2 \mu_0 H_z|_{i,j,k+1/2}\, dS_2 \cong -E_x|_{i,j-1/2,k+1/2}\Delta x - E_y|_{i+1/2,j,k+1/2}\Delta y$$
$$+ E_x|_{i,j+1/2,k+1/2}\Delta x + E_y|_{i-1/2,j,k+1/2}\Delta y, \tag{3.23b}$$

$$\mu_0 \Delta x \Delta y \left(\frac{H_z|_{i,j,k+1/2}^{n+1} - H_z|_{i,j,k+1/2}^n}{\Delta t} \right)$$
$$= \left(E_x|_{i,j+1/2,k+1/2}^{n+1/2} - E_x|_{i,j-1/2,k+1/2}^{n+1/2} \right)\Delta x$$
$$+ \left(E_y|_{i-1/2,j,k+1/2}^{n+1/2} - E_y|_{i+1/2,j,k+1/2}^{n+1/2} \right)\Delta y. \tag{3.23c}$$

Multiplying both sides by $\Delta t/(\mu_0 \Delta x \Delta y)$ and solving for $H_z|_{i,j,k+1/2}^{n+1/2}$ provides

$$H_z|_{i,j,k+1/2}^{n+1} = H_z|_{i,j,k+1/2}^n + \left(E_x|_{i,j+1/2,k+1/2}^{n+1/2} - E_x|_{i,j-1/2,k+1/2}^{n+1/2} \right)\Delta t/(\mu_0 \Delta y)$$
$$+ \left(E_y|_{i-1/2,j,k+1/2}^{n+1/2} - E_y|_{i+1/2,j,k+1/2}^{n+1/2} \right)\Delta t/(\mu_0 \Delta x). \tag{3.24}$$

Eq. (3.24) is simply the free-space version of (3.8c), the Yee time-stepping expression for H_z that was obtained directly from implementing the curl $\vec{E}$ equation with finite differences. The only difference is that (3.24) is evaluated at $(i, j, k + 1/2)$ whereas (3.8c) is evaluated at $(i, j + 1, k + 1/2)$ shown in Fig. 3.1.

3.8. *Divergence-free nature*

We now demonstrate that the Yee algorithm satisfies Gauss' Law for the electric field, Eq. (2.3), and hence is divergence-free in source-free space. We first form the time derivative of the total electric flux over the surface of a single Yee cell of Fig. 3.1:

$$\frac{\partial}{\partial t} \oiint_{\text{Yee cell}} \vec{D} \cdot d\vec{S}$$

$$= \varepsilon_0 \frac{\partial}{\partial t} \underbrace{(E_x|_{i,j+1/2,k+1/2} - E_x|_{i-1,j+1/2,k+1/2})}_{\text{Term 1}} \Delta y \Delta z$$

$$+ \varepsilon_0 \frac{\partial}{\partial t} \underbrace{(E_y|_{i-1/2,j+1,k+1/2} - E_y|_{i-1/2,j,k+1/2})}_{\text{Term 2}} \Delta x \Delta z$$

$$+ \varepsilon_0 \frac{\partial}{\partial t} \underbrace{(E_z|_{i-1/2,j+1/2,k+1} - E_z|_{i-1/2,j+1/2,k})}_{\text{Term 3}} \Delta x \Delta y. \qquad (3.25)$$

Using the Yee algorithm time-stepping relations for the E-field components according to (3.7), we substitute appropriate H-field spatial finite differences for the E-field time derivatives in each term:

Term 1

$$= \left(\frac{H_z|_{i,j+1,k+1/2} - H_z|_{i,j,k+1/2}}{\Delta y} - \frac{H_y|_{i,j+1/2,k+1} - H_y|_{i,j+1/2,k}}{\Delta z} \right)$$
$$- \left(\frac{H_z|_{i-1,j+1,k+1/2} - H_z|_{i-1,j,k+1/2}}{\Delta y} - \frac{H_y|_{i-1,j+1/2,k+1} - H_y|_{i-1,j+1/2,k}}{\Delta z} \right),$$
$$\qquad (3.26a)$$

Term 2

$$= \left(\frac{H_x|_{i-1/2,j+1,k+1} - H_x|_{i-1/2,j+1,k}}{\Delta z} - \frac{H_z|_{i,j+1,k+1/2} - H_z|_{i-1,j+1,k+1/2}}{\Delta x} \right)$$
$$- \left(\frac{H_x|_{i-1/2,j,k+1} - H_x|_{i-1/2,j,k}}{\Delta z} - \frac{H_z|_{i,j,k+1/2} - H_z|_{i-1,j,k+1/2}}{\Delta x} \right),$$
$$\qquad (3.26b)$$

Term 3

$$= \left(\frac{H_y|_{i,j+1/2,k+1} - H_y|_{i-1,j+1/2,k+1}}{\Delta x} - \frac{H_x|_{i-1/2,j+1,k+1} - H_x|_{i-1/2,j,k+1}}{\Delta y} \right)$$
$$- \left(\frac{H_y|_{i,j+1/2,k} - H_y|_{i-1,j+1/2,k}}{\Delta x} - \frac{H_x|_{i-1/2,j+1,k} - H_x|_{i-1/2,j,k}}{\Delta y} \right).$$
$$\qquad (3.26c)$$

For all time steps, this results in

$$\frac{\partial}{\partial t} \oiint_{\text{Yee cell}} \vec{D} \cdot d\vec{S} = (\text{Term } 1)\Delta y \Delta z + (\text{Term } 2)\Delta x \Delta z + (\text{Term } 3)\Delta x \Delta y$$
$$= 0. \tag{3.27}$$

Assuming zero initial conditions, the constant zero value of the time derivative of the net electric flux leaving the Yee cell means that this flux never departs from zero:

$$\oiint_{\text{Yee cell}} \vec{D}(t) \cdot d\vec{S} = \oiint_{\text{Yee cell}} \vec{D}(t=0) \cdot d\vec{S} = 0. \tag{3.28}$$

Therefore, the Yee cell satisfies Gauss' Law for the E-field in charge-free space and thus is divergence-free with respect to its E-field computations. The proof of the satisfaction of Gauss' Law for the magnetic field, Eq. (2.4), is by analogy.

4. Nonuniform Yee grid

4.1. Introduction

The FDTD algorithm is second-order-accurate by nature of the central-difference approximations used to realize the first-order spatial and temporal derivatives. This leads to a discrete approximation for the fields based on a uniform space lattice. Unfortunately, structures with fine geometrical features cannot always conform to the edges of a uniform lattice. Further, it is often desirable to have a refined lattice in localized regions, such as near sharp edges or corners, to accurately model the local field phenomena.

A quasi-nonuniform grid FDTD algorithm was introduced by SHEEN [1991]. This method is based on reducing the grid size by exactly one-third. By choosing the subgrid to be exactly one-third, the spatial derivatives of the fields at the interface between the two regions can be expressed using central-difference approximations, resulting in a second-order-accurate formulation. This technique was successfully applied to a number of microwave circuit and antenna problems (see, for example, SHEEN [1991] and TULINTSEFF [1992]). However, this method is limited to specific geometries that conform to this specialized grid.

It is clear that more general geometries could be handled by a grid with arbitrary spacing. Unfortunately, central differences can no longer be used to evaluate the spatial derivatives of the fields for such a grid, leading to first-order error. However, it was demonstrated by MONK and SULI [1994] and MONK [1994] that, while this formulation does lead to first-order error locally, it results in second-order error globally. This is known as *supraconvergence* (see also MANTEUFFEL and WHITE [1986] and KREISS, MANTEUFFEL, SCHWARTZ, WENDROFF, and WHITE [1986]).

4.2. Supraconvergent FDTD algorithm

This section presents the supraconvergent FDTD algorithm based on nonuniform meshing that was discussed by GEDNEY and LANSING [1995]. Following their notation, a three-dimensional nonuniform space lattice is introduced. The vertices of the lattice are defined by the general one-dimensional coordinates:

$$\{x_i; i = 1, N_x\}; \qquad \{y_j; j = 1, N_y\}; \qquad \{z_k; k = 1, N_z\}. \tag{4.1}$$

The edge lengths between vertices are also defined as

$$\begin{aligned}
&\{\Delta x_i = x_{i+1} - x_i; \; i = 1, \, N_x - 1\}; \\
&\{\Delta y_j = y_{j+1} - y_j; \; j = 1, N_y - 1\}; \\
&\{\Delta z_k = z_{k+1} - z_k; \; k = 1, N_z - 1\}.
\end{aligned} \tag{4.2}$$

Within the nonuniform space, a reduced notation is introduced, defining the cell and edge centers:

$$x_{i+1/2} = x_i + \Delta x_i/2; \qquad y_{j+1/2} = y_j + \Delta y_j/2; \qquad z_{k+1/2} = z_k + \Delta z_k/2. \tag{4.3}$$

A set of dual edge lengths representing the distances between the edge centers is then introduced:

$$\begin{aligned}
&\{h_i^x = (\Delta x_i + \Delta x_{i-1})/2; \; i = 2, N_x\}; \\
&\{h_j^y = (\Delta y_j + \Delta y_{j-1})/2; \; j = 2, N_y\}; \\
&\{h_k^z = (\Delta z_k + \Delta z_{k-1})/2; \; k = 2, N_z\}.
\end{aligned} \tag{4.4}$$

Finally, the E- and H-fields in the discrete nonuniform grid are denoted as in the following examples:

$$E_x \big|_{i+1/2,j,k}^n \equiv E_x(x_{i+1/2}, y_j, z_k, n\Delta t), \tag{4.5a}$$

$$H_x \big|_{i,j+1/2,k+1/2}^{n+1/2} \equiv H_x\big(x_i, y_{j+1/2}, z_{k+1/2}, (n + 1/2)\Delta t\big). \tag{4.5b}$$

The nonuniform FDTD algorithm is based on a discretization of Maxwell's equations in their integral form, specifically, Faraday's Law and Ampere's Law:

$$\oint_C \vec{E} \cdot d\vec{\ell} = -\frac{\partial}{\partial t} \iint_S \vec{B} \cdot d\vec{s} - \iint_S \vec{M} \cdot d\vec{s}, \tag{4.6}$$

$$\oint_{C'} \vec{H} \cdot d\vec{\ell} = \frac{\partial}{\partial t} \iint_{S'} \vec{D} \cdot d\vec{s} + \iint_{S'} \sigma \vec{E} \cdot d\vec{s} + \iint_{S'} \vec{J} \cdot d\vec{s}. \tag{4.7}$$

The surface integral in (4.6) is performed over a lattice cell face, and the contour integral is performed over the edges bounding the face, as illustrated in Fig. 4.1(a). Similarly, the surface integral in (4.7) is performed over a dual-lattice cell face.

Evaluating (4.6) and (4.7) over the cell faces using (4.5), and evaluating the time derivatives using central-differencing leads to

$$\begin{aligned}
E_x \big|_{i+1/2,j+1,k}^n \Delta x_i &- E_x \big|_{i+1/2,j,k}^n \Delta x_i - E_y \big|_{i+1,j+1/2,k}^n \Delta y_j + E_y \big|_{i,j+1/2,k}^n \Delta y_j \\
&= -\left[\mu_{i+1/2,j+1/2,k} \left(\frac{H_z \big|_{i+1/2,j+1/2,k}^{n+1/2} - H_z \big|_{i+1/2,j+1/2,k}^{n-1/2}}{\Delta t} \right) + M_z \big|_{i+1/2,j+1/2,k}^{n+1/2} \right] \\
&\quad \times \Delta x_i \Delta y_j,
\end{aligned} \tag{4.8}$$

S.C. Hagness et al.

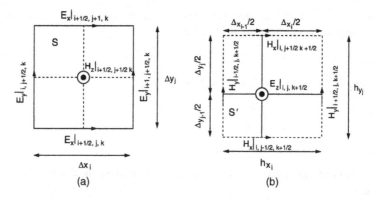

FIG. 4.1. Lattice faces bounded by lattice edges defining surfaces of integration bounded by closed contours.
(a) Lattice cell face bounded by grid edges, showing a dual-lattice edge passing through its center. (b) The
dual-lattice face bounded by dual edges. *Source*: GEDNEY and LANSING [1995].

$$
H_x\Big|_{i,j+1/2,k+1/2}^{n+1/2} h_{x_i} - H_x\Big|_{i,j-1/2,k+1/2}^{n+1/2} h_{x_i} - H_y\Big|_{i+1/2,j,k+1/2}^{n+1/2} h_{y_j} + H_y\Big|_{i-1/2,j,k+1/2}^{n+1/2} h_{y_j}
$$

$$
= \left[\varepsilon_{i,j,k+1/2}\left(\frac{E_z\big|_{i,j,k+1/2}^{n+1} - E_z\big|_{i,j,k+1/2}^{n}}{\Delta t} \right) + \frac{\sigma_{i,j,k+1/2}}{2}\left(\frac{E_z\big|_{i,j,k+1/2}^{n+1} + E_z\big|_{i,j,k+1/2}^{n}}{\Delta t} \right) + J_z\big|_{i,j,k+1/2}^{n+1/2} \right] h_{x_i} h_{y_j}, \qquad (4.9)
$$

where $\varepsilon_{i,j,k+1/2}$, $\sigma_{i,j,k+1/2}$, and $\mu_{i+1/2,j+1/2,k}$ are the averaged permittivity, conductivity, and permeability, respectively, about the grid edges. Subsequently, this leads to an explicit update scheme:

$$
H_z\Big|_{i+1/2,j+1/2,k}^{n+1/2}
$$

$$
= H_z\Big|_{i+1/2,j+1/2,k}^{n-1/2} - \frac{\Delta t}{\mu_{i+1/2,j+1/2,k}}
$$

$$
\times \left[\frac{1}{\Delta y_j}\left(E_x\big|_{i+1/2,j+1,k}^{n} - E_x\big|_{i+1/2,j,k}^{n} \right) - \frac{1}{\Delta x_i}\left(E_y\big|_{i+1,j+1/2,k}^{n} - E_y\big|_{i,j+1/2,k}^{n} \right) + M_z\big|_{i+1/2,j+1/2,k}^{n+1/2} \right], \qquad (4.10)
$$

$$
E_z\Big|_{i,j,k+1/2}^{n+1}
$$

$$
= \left(\frac{2\varepsilon_{i,j,k+1/2} - \sigma_{i,j,k+1/2}\Delta t}{2\varepsilon_{i,j,k+1/2} + \sigma_{i,j,k+1/2}\Delta t} \right) E_z\big|_{i,j,k+1/2}^{n} + \left(\frac{2\Delta t}{2\varepsilon_{i,j,k+1/2} + \sigma_{i,j,k+1/2}\Delta t} \right)
$$

$$
\times \left[\frac{1}{h_{y_j}}\left(H_x\big|_{i,j+1/2,k+1/2}^{n+1/2} - H_x\big|_{i,j-1/2,k+1/2}^{n+1/2} \right) - \frac{1}{h_{x_i}}\left(H_y\big|_{i+1/2,j,k+1/2}^{n+1/2} - H_y\big|_{i-1/2,j,k+1/2}^{n+1/2} \right) - J_z\big|_{i,j,k+1/2}^{n+1/2} \right]. \qquad (4.11)
$$

Similar updates for the remaining field components are easily derived by permuting the indices in (4.10) and (4.11) in a right-handed manner.

4.3. Demonstration of supraconvergence

The explicit updates for the H-fields in (4.10) are second-order-accurate in both space and time, since the vertices of the dual lattice are assumed to be located at the cell centers of the primary lattice. On the other hand, the explicit updates for the E-fields in (4.11) are only first-order-accurate in space. This results in local first-order error in regions where the grid is nonuniform.

However, via a numerical example, GEDNEY and LANSING [1995] showed that this method is supraconvergent, i.e., it converges with a higher order accuracy than the local error mandates. They considered calculation of the resonant frequencies of a fixed-size rectangular cavity having perfect electric conductor (PEC) walls. A random, nonuniform grid spacing for x_i, y_j, and z_k was assumed within the cavity such that

$$\{x_i = (i - 1)\Delta x + 0.5\Re\Delta x; \ i = 1, N_x\}, \tag{4.12a}$$

$$\{y_j = (j - 1)\Delta y + 0.5\Re\Delta y; \ j = 1, N_y\}, \tag{4.12b}$$

$$\{z_k = (k - 1)\Delta z + 0.5\Re\Delta z; \ k = 1, N_z\}, \tag{4.12c}$$

where $-1/2 \leqslant \Re \leqslant 1/2$ denotes a random number. The interior of the cavity was excited with a Gaussian-pulsed, z-directed magnetic dipole placed off a center axis:

$$\vec{M}(t) = \hat{z}e^{-(t-t_0)^2/T^2}. \tag{4.13}$$

The calculated time-varying E-field was probed off a center axis (to avoid the nulls of odd resonant modes), and the cavity resonant frequencies were extracted using a fast Fourier transform (FFT). Spectral peaks resulting from this procedure corresponded to the resonant frequencies of the cavity modes. Subsequently, the average grid cell size h was reduced, and the entire simulation was run again.

Fig. 4.2 graphs the results of four such runs for the error of the nonuniform grid FDTD model in calculating the resonant frequency of the TE_{110} mode relative to the exact solution, as well as a generic order(h^2) accuracy slope. We see that the convergence of the resonant frequency is indeed second-order.

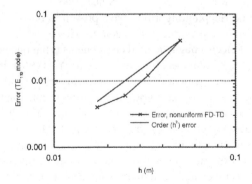

FIG. 4.2. Error convergence of the resonant frequency of the TE_{110} mode of a rectangular PEC cavity computed using the nonuniform FDTD algorithm. *Source:* GEDNEY and LANSING [1995].

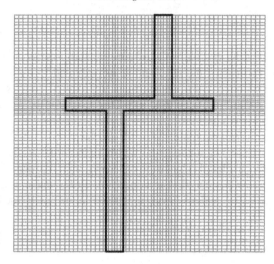

FIG. 4.3. Typical nonuniform grid used for a planar microwave circuit.

The nonuniform FDTD method is well suited for the analysis of planar microwave circuits. The geometrical details of such circuits are typically electrically small, leading to small cell sizes. Further, microwave circuits are often located in an unbounded medium, requiring absorbing boundaries to be placed a sufficient distance from the circuit to avoid nonphysical reflections. For uniform meshing, these two characteristics can combine to produce very large space lattices. With nonuniform meshing, the local cell size can be refined such that the circuit trace size, shape, and field behavior are accurately modeled, while coarser cells are used in regions further from the metal traces. Fig. 4.3 illustrates a typical nonuniform grid used for a microstrip circuit.

5. Alternative finite-difference grids

Thus far, this chapter has considered several fundamental aspects of the uniform Cartesian Yee space lattice for Maxwell's equations. Since 1966, this lattice and its associated staggered leapfrog time-stepping algorithm have proven to be very flexible, accurate, and robust for a wide variety of engineering problems. However, Yee's staggered, uncollocated arrangement of electromagnetic field components is but one possible alternative in a Cartesian coordinate system (see, for example, Liu, Y. [1996]). In turn, a Cartesian grid is but one possible arrangement of field components in two and three dimensions. Other possibilities include hexagonal grids in two dimensions and tetradecahedron/dual-tetrahedron meshes in three dimensions (see again Liu, Y. [1996]).

It is important to develop criteria for the use of a particular space lattice and time-stepping algorithm to allow optimum selection for a given problem. A key consideration is the capability of rendering the geometry of the structure of interest within the space lattice with sufficient accuracy and detail to obtain meaningful results. A second fundamental consideration is the accuracy by which the algorithm simulates the propagation of electromagnetic waves as they interact with the structure.

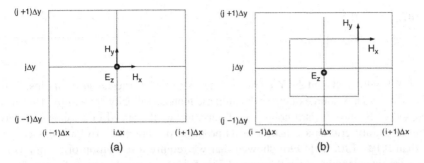

FIG. 5.1. Two Cartesian grids that are alternatives to Yee's arrangement (illustrated in two dimensions for the TM$_z$ case). (a) Unstaggered, collocated grid. (b) Staggered, collocated grid. *Source*: Y. Liu, *J. Computational Physics*, 1996, pp. 396–416.

5.1. Cartesian grids

Fig. 5.1 illustrates two Cartesian grids that are alternatives to Yee's arrangement in two dimensions for the TM$_z$ case, as discussed by LIU, Y. [1996]: (a) the unstaggered, collocated grid, in which all E- and H-components are collocated at a single set of grid-cell vertices; and (b) the staggered, collocated grid, in which all E-components are collocated at a distinct set of grid-cell vertices that are spatially interleaved with a second distinct set of vertices where all H-components are collocated.

Upon applying second-order-accurate central space differences to the TM$_z$ mode equations of (2.11) for the unstaggered, collocated grid of Fig. 5.1(a) (with a lossless material background assumed for simplicity), we obtain as per LIU, Y. [1996]:

$$\frac{\partial H_x|_{i,j}}{\partial t} = -\frac{1}{\mu_{i,j}} \cdot \left(\frac{E_z|_{i,j+1} - E_z|_{i,j-1}}{2\Delta y} \right), \tag{5.1a}$$

$$\frac{\partial H_y|_{i,j}}{\partial t} = \frac{1}{\mu_{i,j}} \cdot \left(\frac{E_z|_{i+1,j} - E_z|_{i-1,j}}{2\Delta x} \right), \tag{5.1b}$$

$$\frac{\partial E_z|_{i,j}}{\partial t} = \frac{1}{\varepsilon_{i,j}} \cdot \left(\frac{H_y|_{i+1,j} - H_y|_{i-1,j}}{2\Delta x} - \frac{H_x|_{i,j+1} - H_x|_{i,j-1}}{2\Delta y} \right). \tag{5.1c}$$

Similarly, applying second-order-accurate central space differences to the TM$_z$ mode equations of (2.11) for the staggered, collocated grid of Fig. 5.1(b) yields:

$$\frac{\partial H_x|_{i+1/2,j+1/2}}{\partial t}$$
$$= -\frac{0.5}{\mu_{i+1/2,j+1/2}} \cdot \left[\frac{(E_z|_{i,j+1} + E_z|_{i+1,j+1}) - (E_z|_{i,j} + E_z|_{i+1,j})}{\Delta y} \right], \tag{5.2a}$$

$$\frac{\partial H_y|_{i+1/2,j+1/2}}{\partial t}$$
$$= \frac{0.5}{\mu_{i+1/2,j+1/2}} \cdot \left[\frac{(E_z|_{i+1,j} + E_z|_{i+1,j+1}) - (E_z|_{i,j} + E_z|_{i,j+1})}{\Delta x} \right], \tag{5.2b}$$

$$\frac{\partial E_z|_{i,j}}{\partial t} = \frac{0.5}{\varepsilon_{i,j}} \cdot \left[\begin{array}{c} \frac{(H_y|_{i+1/2,j-1/2}+H_y|_{i+1/2,j+1/2})-(H_y|_{i-1/2,j-1/2}+H_y|_{i-1/2,j+1/2})}{\Delta x} \\ -\frac{(H_x|_{i-1/2,j+1/2}+H_x|_{i+1/2,j+1/2})-(H_x|_{i-1/2,j-1/2}+H_x|_{i+1/2,j-1/2})}{\Delta y} \end{array} \right] \cdot$$

$$(5.2c)$$

LIU, Y. [1996] analyzed the Yee grid and the alternative Cartesian grids of Figs. 5.1(a) and 5.1(b) for a key source of error: the numerical phase-velocity anisotropy. This error, discussed in Section 6, is a nonphysical variation of the speed of a numerical wave within an empty grid as a function of its propagation direction. To limit this error to less than 0.1%, LIU, Y. [1996] showed that we require a resolution of 58 points per free-space wavelength λ_0 for the grid of Fig. 5.1(a), 41 points per λ_0 for the grid of Fig. 5.1(b), and only 29 points per λ_0 for the Yee grid. Thus, Yee's grid provides more accurate modeling results than the two alternatives of Fig. 5.1.

5.2. Hexagonal grids

LIU, Y. [1996] proposed using regular hexagonal grids in two dimensions to reduce the numerical phase-velocity anisotropy well below that of Yee's Cartesian mesh. Here, the primary grid is composed of equilateral hexagons of edge length Δs. Each hexagon can be considered to be the union of six equilateral triangles. Connecting the centroids of these triangles yields a second set of regular hexagons that comprises a dual grid.

Fig. 5.2 illustrates for the TM_z case in two dimensions the two principal ways of arranging E and H components in hexagonal grids. Fig. 5.2(a) shows the unstaggered, collocated grid in which Cartesian E_z, H_x, and H_y components are collocated at the vertices of the equilateral triangles. No dual grid is used. Fig. 5.2(b) shows the field arrangement for the staggered, uncollocated grid and its associated dual grid, the latter indicated by the dashed line segments. Here, only E_z components are defined at the vertices of the equilateral triangles, which are the centroids of the hexagonal faces of the dual grid. Magnetic field components H_1, H_2, H_3, etc. are defined to be tangential to, and centered on, the edges of the dual-grid hexagons. These magnetic components

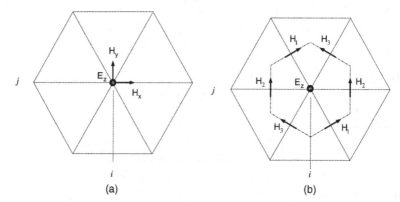

FIG. 5.2. Two central-difference hexagonal grids that are alternatives to Yee's arrangement (illustrated in two dimensions for the TM_z case). (a) Unstaggered, collocated grid, with no dual grid. (b) Staggered, uncollocated grid and its associated dual grid. *Source*: Y. Liu, *J. Computational Physics*, 1996, pp. 396–416.

are also perpendicular to, and centered on, the edges of the primary-grid triangles. We note that the grid of Fig. 5.2(b) is a direct extension of Yee's interleaved E and H component topology from rectangular to hexagonal cells.

Upon applying second-order-accurate central space differences to the TM$_z$ mode equations of (2.11) for the unstaggered, collocated hexagonal grid of Fig. 5.2(a) (with a lossless material background assumed for simplicity), we obtain as per LIU, Y. [1996]:

$$\frac{\partial H_x|_{i,j}}{\partial t} = -\frac{\sqrt{3}}{\mu_{i,j}6\Delta s}\left(\begin{array}{c} E_z|_{i-1/2,j+0.5\sqrt{3}} + E_z|_{i+1/2,j+0.5\sqrt{3}} \\ - E_z|_{i-1/2,j-0.5\sqrt{3}} - E_z|_{i+1/2,j-0.5\sqrt{3}} \end{array}\right), \qquad (5.3a)$$

$$\frac{\partial H_y|_{i,j}}{\partial t}$$

$$= \frac{1}{\mu_{i,j}6\Delta s}\left(\begin{array}{c} 2E_z|_{i+1,j} - 2E_z|_{i-1,j} + E_z|_{i+1/2,j|0.5\sqrt{3}} \\ - E_z|_{i-1/2,j+0.5\sqrt{3}} + E_z|_{i+1/2,j-0.5\sqrt{3}} - E_z|_{i-1/2,j-0.5\sqrt{3}} \end{array}\right), \qquad (5.3b)$$

$$\frac{\partial E_z|_{i,j}}{\partial t}$$

$$= \frac{1}{\varepsilon_{i,j}6\Delta s}\left(\begin{array}{c} 2H_y|_{i+1,j} - 2H_y|_{i-1,j} + H_y|_{i+1/2,j+0.5\sqrt{3}} \\ - H_y|_{i-1/2,j+0.5\sqrt{3}} + H_y|_{i+1/2,j-0.5\sqrt{3}} - H_y|_{i-1/2,j-0.5\sqrt{3}} \\ - \sqrt{3}H_x|_{i+1/2,j+0.5\sqrt{3}} + \sqrt{3}H_x|_{i+1/2,j-0.5\sqrt{3}} \\ - \sqrt{3}H_x|_{i-1/2,j+0.5\sqrt{3}} + \sqrt{3}H_x|_{i-1/2,j-0.5\sqrt{3}} \end{array}\right). \qquad (5.3c)$$

Similarly, applying second-order-accurate central space differences to the TM$_z$ mode equations for the staggered, uncollocated grid of Fig. 5.2(b) yields:

$$\frac{\partial H_1|_{i+1/4,j-0.25\sqrt{3}}}{\partial t} = \frac{1}{\mu_{i+1/4,j-0.25\sqrt{3}}\Delta s}(E_z|_{i+1/2,j-0.5\sqrt{3}} - E_z|_{i,j}), \qquad (5.4a)$$

$$\frac{\partial H_2|_{i+1/2,j}}{\partial t} = \frac{1}{\mu_{i+1/2,j}\Delta s}(E_z|_{i+1,j} - E_z|_{i,j}), \qquad (5.4b)$$

$$\frac{\partial H_3|_{i+1/4,j+0.25\sqrt{3}}}{\partial t} = \frac{1}{\mu_{i+1/4,j+0.25\sqrt{3}}\Delta s}(E_z|_{i+1/2,j+0.5\sqrt{3}} - E_z|_{i,j}), \qquad (5.4c)$$

$$\frac{\partial E_z|_{i,j}}{\partial t} = \frac{2}{\varepsilon_{i,j}3\Delta s}$$

$$\times \left(\begin{array}{c} H_1|_{i+1/4,j-0.25\sqrt{3}} + H_2|_{i+1/2,j} + H_3|_{i+1/4,j+0.25\sqrt{3}} \\ - H_1|_{i-1/4,j+0.25\sqrt{3}} - H_2|_{i-1/2,j} - H_3|_{i-1/4,j-0.25\sqrt{3}} \end{array}\right). \qquad (5.4d)$$

We note that the total number of field unknowns for the staggered, uncollocated grid of Fig. 5.2(b) is 33% more than that for the unstaggered grid of Fig. 5.2(a), but the discretization is simpler and the number of total operations is less by about 50%.

LIU, Y. [1996] showed that the numerical velocity anisotropy errors of the hexagonal grids of Figs. 5.2(a) and 5.2(b) are 1/200th and 1/1200th, respectively, that of the rectangular Yee grid for a grid sampling density of 20 points per free-space wavelength. This represents a large potential advantage in computational accuracy for the hexagonal grids. Additional details are provided in Section 6.

5.3. Tetradecahedron/dual-tetrahedron mesh in three dimensions

In three dimensions, the uniform Cartesian Yee mesh consists of an ordered array of hexahedral unit cells ("bricks"), as shown in Fig. 3.1. This simple arrangement is attractive since the location of every field component in the mesh is easily and compactly specified, and geometry generation can be performed in many cases with paper and pencil.

However, from the discussion of Section 5.2, it is clear that constructing a uniform mesh in three dimensions using shapes other than rectangular "bricks" may lead to superior computational accuracy with respect to the reduction of the velocity-anisotropy error. Candidate shapes for unit cells must be capable of assembly in a regular mesh to completely fill space. In addition to the hexahedron, space-filling shapes include the tetradecahedron (truncated octahedron), hexagonal prism, rhombic dodecahedron, and elongated rhombic dodecahedron (see LIU, Y. [1996]). We note that the three-dimensional lattice corresponding to the two-dimensional, staggered, uncollocated hexagonal grid of Fig. 5.2(b) is the tetradecahedron/dual-tetrahedron configuration shown in Fig. 5.3. Here, the primary mesh is comprised of tetradecahedral units cells having 6 square faces and 8 regular hexagonal faces. The dual mesh is comprised of tetrahedral cells having isosceles-triangle faces with sides in the ratio of $\sqrt{3}$ to 2.

LIU, Y. [1996] reports a study of the extension of Yee's method to the staggered tetradecahedron/dual-tetrahedron mesh of Fig. 5.3. The algorithm uses a centered finite-difference scheme involving 19 independent unknown field components, wherein 12

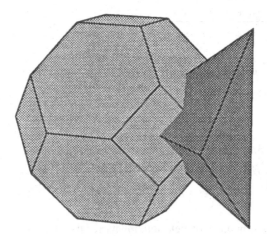

FIG. 5.3. Tetradecahedron and dual-tetrahedron unit cells for the extension of Yee's method to a regular non-Cartesian mesh in three dimensions. This mesh has very favorable numerical wave-velocity anisotropy characteristics relative to the Cartesian arrangement of Fig. 3.1. *Source*: Y. Liu, *J. Computational Physics*, 1996, pp. 396–416.

are defined on the edges of tetradecahedra, and 7 are defined on the edges of the dual tetrahedra. Similar to the staggered, uncollocated hexagonal grid of Fig. 5.2(b), this mesh has very favorable numerical wave-velocity anisotropy characteristics relative to its Yee Cartesian counterpart, shown in Fig. 3.1.

Despite this advantage, the usage of the tetradecahedron/dual-tetrahedron mesh by the FDTD community has been very limited. This is due to the additional complexity in its mesh generation relative to Yee's Cartesian space lattice.

6. Numerical dispersion

6.1. Introduction

The FDTD algorithms for Maxwell's curl equations reviewed in Sections 3–5 cause nonphysical dispersion of the simulated waves in a free-space computational lattice. That is, the phase velocity of numerical wave modes can differ from c by an amount varying with the wavelength, direction of propagation in the grid, and grid discretization. An intuitive way to view this phenomenon is that the FDTD algorithm embeds the electromagnetic wave interaction structure of interest in a tenuous "numerical aether" having properties very close to vacuum, but not quite. This "aether" causes propagating numerical waves to accumulate delay or phase errors that can lead to nonphysical results such as broadening and ringing of pulsed waveforms, imprecise cancellation of multiple scattered waves, anisotropy, and pseudorefraction. Numerical dispersion is a factor that must be accounted to understand the operation of FDTD algorithms and their accuracy limits, especially for electrically large structures.

This section reviews the numerical dispersion characteristics of Yee's FDTD formulation. Section 7 will review proposed low-dispersion FDTD methods, not necessarily based on Yee's space grid and/or the use of explicit central differences.

6.2. Two-dimensional wave propagation, Cartesian Yee grid

We begin our discussion of numerical dispersion with an analysis of the Yee algorithm for the two-dimensional TM_z mode, (3.19a)–(3.19c), assuming for simplicity no electric or magnetic loss. It can be easily shown that the dispersion relation obtained is valid for any two-dimensional TM or TE mode in a Cartesian Yee grid. The analysis procedure involves substitution of a plane, monochromatic, sinusoidal traveling-wave mode into (3.19a)–(3.19c). After algebraic manipulation, an equation is derived that relates the numerical wavevector components, the wave frequency, the time step, and the grid space increments. This equation, the numerical dispersion relation, can be solved for a variety of grid discretizations, wavevectors, and wave frequencies to illustrate the principal nonphysical results associated with numerical dispersion.

Initiating this procedure, we assume the following plane, monochromatic, sinusoidal traveling wave for the TM_z mode:

$$E_z|_{I,J}^n = E_{z0} e^{j(\omega n \Delta t - \tilde{k}_x I \Delta x - \tilde{k}_y J \Delta y)}, \tag{6.1a}$$

$$H_x|_{I,J}^n = H_{x0} e^{j(\omega n \Delta t - \tilde{k}_x I \Delta x - \tilde{k}_y J \Delta y)}, \tag{6.1b}$$

$$H_y|_{I,J}^n = H_{y0} e^{j(\omega n \Delta t - \tilde{k}_x I \Delta x - \tilde{k}_y J \Delta y)}, \tag{6.1c}$$

where $\tilde{k}_x$ and $\tilde{k}_y$ are the x- and y-components of the numerical wavevector and ω is the wave angular frequency. Substituting the traveling-wave expressions of (6.1) into the finite-difference equations of (3.19) yields, after simplification, the following relations for the lossless material case:

$$H_{x0} = \frac{\Delta t E_{z0}}{\mu \Delta y} \cdot \frac{\sin(\tilde{k}_y \Delta y/2)}{\sin(\omega \Delta t/2)}, \tag{6.2a}$$

$$H_{y0} = -\frac{\Delta t E_{z0}}{\mu \Delta x} \cdot \frac{\sin(\tilde{k}_x \Delta x/2)}{\sin(\omega \Delta t/2)}, \tag{6.2b}$$

$$E_{z0} \sin\left(\frac{\omega \Delta t}{2}\right) = \frac{\Delta t}{\varepsilon} \left[\frac{H_{x0}}{\Delta y} \sin\left(\frac{\tilde{k}_y \Delta y}{2}\right) - \frac{H_{y0}}{\Delta x} \sin\left(\frac{\tilde{k}_x \Delta x}{2}\right) \right]. \tag{6.2c}$$

Upon substituting H_{x0} of (6.2a) and H_{y0} of (6.2b) into (6.2c), we obtain

$$\left[\frac{1}{c\Delta t} \sin\left(\frac{\omega \Delta t}{2}\right) \right]^2 = \left[\frac{1}{\Delta x} \sin\left(\frac{\tilde{k}_x \Delta x}{2}\right) \right]^2 + \left[\frac{1}{\Delta y} \sin\left(\frac{\tilde{k}_y \Delta y}{2}\right) \right]^2, \tag{6.3}$$

where $c = 1/\sqrt{\mu\varepsilon}$ is the speed of light in the material being modeled. (Eq. 6.3) is the general numerical dispersion relation of the Yee algorithm for the TM_z mode.

We shall consider the important special case of a square-cell grid having $\Delta x = \Delta y = \Delta$. Then, defining the *Courant stability factor* $S = c\Delta t/\Delta$ and the *grid sampling density* $N_\lambda = \lambda_0/\Delta$, we rewrite (6.3) in a more useful form:

$$\frac{1}{S^2} \sin^2\left(\frac{\pi S}{N_\lambda}\right) = \sin^2\left(\frac{\Delta \cdot \tilde{k} \cos\phi}{2}\right) + \sin^2\left(\frac{\Delta \cdot \tilde{k} \sin\phi}{2}\right), \tag{6.4}$$

where ϕ is the propagation direction of the numerical wave with respect to the grid's x-axis. To obtain the numerical dispersion relation for the one-dimensional wave-propagation case, we can assume without loss of generality that $\phi = 0$ in (6.4), yielding

$$\frac{1}{S} \sin\left(\frac{\pi S}{N_\lambda}\right) = \sin\left(\frac{\tilde{k}\Delta}{2}\right) \tag{6.5a}$$

or equivalently

$$\tilde{k} = \frac{2}{\Delta} \sin^{-1}\left[\frac{1}{S} \sin\left(\frac{\pi S}{N_\lambda}\right) \right]. \tag{6.5b}$$

6.3. Extension to three dimensions, Cartesian Yee lattice

The dispersion analysis presented above is now extended to the full three-dimensional case, following the analysis presented by TAFLOVE and TAFLOVE and BRODWIN [1975]. We consider a normalized, lossless region of space with $\mu = 1$, $\varepsilon = 1$, $\sigma = 0$, $\sigma^* = 0$, and $c = 1$. Letting $j = \sqrt{-1}$, we rewrite Maxwell's equations in compact form as

$$j\nabla \times (\vec{H} + j\vec{E}) = \frac{\partial}{\partial t}(\vec{H} + j\vec{E}) \tag{6.6a}$$

or more simply as

$$j\nabla \times \vec{V} = \frac{\partial \vec{V}}{\partial t},$$ (6.6b)

where $\vec{V} = \vec{H} + j\vec{E}$. Substituting the vector-field traveling-wave expression

$$\vec{V}\big|_{I,J,K}^n = \vec{V}_0 \, e^{j(\omega n \Delta t - \tilde{k}_x I \Delta x - \tilde{k}_y J \Delta y - \tilde{k}_z K \Delta z)}$$ (6.7)

into the Yee space–time central-differencing realization of (6.6b), we obtain

$$\left[\frac{\hat{x}}{\Delta x} \sin\left(\frac{\tilde{k}_x \Delta x}{2}\right) + \frac{\hat{y}}{\Delta y} \sin\left(\frac{\tilde{k}_y \Delta y}{2}\right) + \frac{\hat{z}}{\Delta z} \sin\left(\frac{\tilde{k}_z \Delta z}{2}\right) \right] \times \vec{V}\big|_{I,J,K}^n$$
$$= \frac{-j}{\Delta t} \vec{V}\big|_{I,J,K}^n \sin\left(\frac{\omega \Delta t}{2}\right),$$ (6.8)

where $\hat{x}$, $\hat{y}$, and $\hat{z}$ are unit vectors in the x-, y-, and z-coordinate directions. After performing the vector cross product in (6.8) and writing out the x, y, and z vector component equations, we obtain a homogeneous system (zero right-hand side) of three equations in the unknowns V_x, V_y, and V_z. Setting the determinant of this system equal to zero results in

$$\left[\frac{1}{\Delta t} \sin\left(\frac{\omega \Delta t}{2}\right) \right]^2 = \left[\frac{1}{\Delta x} \sin\left(\frac{\tilde{k}_x \Delta x}{2}\right) \right]^2 + \left[\frac{1}{\Delta y} \sin\left(\frac{\tilde{k}_y \Delta y}{2}\right) \right]^2$$
$$+ \left[\frac{1}{\Delta z} \sin\left(\frac{\tilde{k}_z \Delta z}{2}\right) \right]^2.$$ (6.9)

Finally, we denormalize to a nonunity c and obtain the general form of the numerical dispersion relation for the full-vector-field Yee algorithm in three dimensions:

$$\left[\frac{1}{c\Delta t} \sin\left(\frac{\omega \Delta t}{2}\right) \right]^2 = \left[\frac{1}{\Delta x} \sin\left(\frac{\tilde{k}_x \Delta x}{2}\right) \right]^2 + \left[\frac{1}{\Delta y} \sin\left(\frac{\tilde{k}_y \Delta y}{2}\right) \right]^2$$
$$+ \left[\frac{1}{\Delta z} \sin\left(\frac{\tilde{k}_z \Delta z}{2}\right) \right]^2.$$ (6.10)

This equation is seen to reduce to (6.3), the numerical dispersion relation for the two-dimensional TM_z mode, simply by letting $\tilde{k}_z = 0$.

6.4. Comparison with the ideal dispersion case

In contrast to (6.10), the analytical (ideal) dispersion relation for a physical plane wave propagating in three dimensions in a homogeneous lossless medium is simply

$$\left(\frac{\omega}{c} \right)^2 = (k_x)^2 + (k_y)^2 + (k_z)^2.$$ (6.11)

Although at first glance (6.10) bears little resemblance to the ideal case of (6.11), we can easily show that the two dispersion relations are identical in the limit as Δx, Δy,

Δz, and Δt approach zero. Qualitatively, this suggests that numerical dispersion can be reduced to any degree that is desired if we only use fine enough FDTD gridding.

It can also be shown that (6.10) reduces to (6.11) if the Courant factor S and the wave-propagation direction are suitably chosen. For example, reduction to the ideal dispersion case can be demonstrated for a numerical plane wave propagating along a diagonal of a three-dimensional cubic lattice ($\tilde{k}_x = \tilde{k}_y = \tilde{k}_z = \tilde{k}/\sqrt{3}$) if $S = 1/\sqrt{3}$. Similarly, ideal dispersion results for a numerical plane wave propagating along a diagonal of a two-dimensional square grid ($\tilde{k}_x = \tilde{k}_y = \tilde{k}/\sqrt{2}$) if $S = 1/\sqrt{2}$. Finally, ideal dispersion results for any numerical wave in a one-dimensional grid if $S = 1$. These reductions to the ideal case have little practical value for two- and three-dimensional simulations, occurring only for diagonal propagation. However, the reduction to ideal dispersion in one dimension is very interesting, since it implies that the Yee algorithm (based upon numerical finite-difference approximations) yields an *exact* solution for wave propagation.

6.5. Anisotropy of the numerical phase velocity

This section probes a key implication of numerical dispersion relations (6.3) and (6.10). Namely, numerical waves in a two- or three-dimensional Yee space lattice have a propagation velocity that is dependent upon the direction of wave propagation. The space lattice thus represents an anisotropic medium.

Our strategy in developing an understanding of this phenomenon is to first calculate sample values of the numerical phase velocity $\tilde{v}_p$ versus wave-propagation direction ϕ in order to estimate the magnitude of the problem. Then, we will conduct an appropriate analysis to examine the issue more deeply.

6.5.1. Sample values of numerical phase velocity

For simplicity, we start with the simplest possible situation where numerical phase-velocity anisotropy arises: two-dimensional TM_z modes propagating in a square-cell grid. Dispersion relation (6.4) can be solved directly for $\tilde{k}$ for propagation along the major axes of the grid: $\phi = 0°, 90°, 180°$, and $270°$. For this case, the solution for $\tilde{k}$ is given by (6.5b), which is repeated here for convenience:

$$\tilde{k} = \frac{2}{\Delta} \sin^{-1}\left[\frac{1}{S}\sin\left(\frac{\pi S}{N_\lambda}\right)\right]. \tag{6.12a}$$

The corresponding numerical phase velocity is given by

$$\tilde{v}_p = \frac{\omega}{\tilde{k}} = \frac{\pi}{N_\lambda \sin^{-1}\left[\frac{1}{S}\sin\left(\frac{\pi S}{N_\lambda}\right)\right]}c. \tag{6.12b}$$

Dispersion relation (6.4) can also be solved directly for $\tilde{k}$ for propagation along the diagonals of the grid $\phi = 45°, 135°, 225°$, and $315°$, yielding

$$\tilde{k} = \frac{2\sqrt{2}}{\Delta} \sin^{-1}\left[\frac{1}{S\sqrt{2}}\sin\left(\frac{\pi S}{N_\lambda}\right)\right], \tag{6.13a}$$

$$\tilde{v}_p = \frac{\pi}{N_\lambda \sqrt{2}\sin^{-1}\left[\frac{1}{S\sqrt{2}}\sin\left(\frac{\pi S}{N_\lambda}\right)\right]}c. \tag{6.13b}$$

As an example, assume a grid having $S = 0.5$ and $N_\lambda = 20$. Then (6.12b) and (6.13b) provide unequal $\tilde{v}_p$ values of $0.996892c$ and $0.998968c$, respectively. The implication is that a sinusoidal numerical wave propagating obliquely within this grid has a speed that is $0.998968/0.996892 = 1.00208$ times that of a wave propagating along the major grid axes. This represents a velocity anisotropy of about 0.2% between oblique and along-axis numerical wave propagation.

TAFLOVE and HAGNESS [2000, pp. 115–117] demonstrated that this theoretical anisotropy of the numerical phase velocity appears in FDTD simulations. Fig. 6.1 presents their modeling results for a radially outward-propagating sinusoidal cylindrical wave in a two-dimensional TM$_z$ grid. Their grid was configured with 360×360 square cells with $\Delta x = \Delta y = \Delta = 1.0$. A unity-amplitude sinusoidal excitation was provided

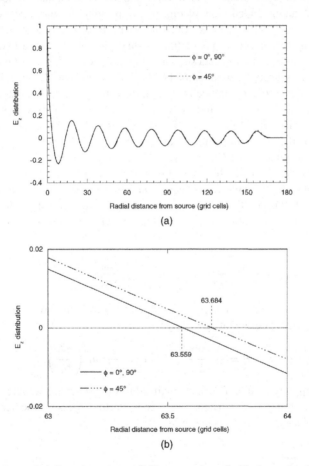

FIG. 6.1. Effect of numerical dispersion upon a radially propagating cylindrical wave in a 2D TM Yee grid. The grid is excited at its center by applying a unity-amplitude sinusoidal time function to a single E_z field component. $S = 0.5$ and the grid sampling density is $N_\lambda = 20$. (a) Comparison of calculated wave propagation along the grid axes and along a grid diagonal. (b) Expanded view of (a) at distances between 63 and 64 grid cells from the source.

to a single E_z component at the center of the grid. Choosing a grid-sampling density of $N_\lambda = 20$ and a Courant factor $S = 0.5$ permitted direct comparison of the FDTD modeling results with the theoretical results for the anisotropy of $\tilde{v}_p$, discussed immediately above.

Fig. 6.1(a) illustrates snapshots of the E_z field distribution vs. radial distance from the source at the center of the grid. Here, field observations are made along cuts through the grid passing through the source point and either parallel to the principal grid axes $\phi = 0°, 90°$ or parallel to the grid diagonal $\phi = 45°$. (Note that, by the 90° rotational symmetry of the Cartesian grid geometry, identical field distributions are obtained along $\phi = 0°$ and $\phi = 90°$.) The snapshots are taken $328\Delta t$ after the beginning of time-stepping. At this time, the wave has not yet reached the outer grid boundary, and the calculated E_z field distribution is free of error due to outer-boundary reflections.

Fig. 6.1(b) is an expanded view of Fig. 6.1(a) at radial distances between 63Δ and 64Δ from the source. This enables evaluation (with three-decimal-place precision) of the locations of the zero-crossings of the E_z distributions along the two observation cuts through the grid. From the data shown in Fig. 6.1(b), the sinusoidal wave along the $\phi = 45°$ cut passes through zero at 63.684 cells, whereas the wave along the $\phi = 0°, 90°$ cut passes through zero at 63.559 cells. Taking the difference, we see that the obliquely propagating wave "leads" the on-axis wave by 0.125 cells. This yields a numerical phase-velocity anisotropy $\Delta\tilde{v}_p/\tilde{v}_p \cong 0.125/63.6 = 0.197\%$. This number is only about 5% less than the 0.208% value obtained using (6.12b) and (6.13b).

To permit determination of $\tilde{k}$ and $\tilde{v}_p$ for any wave-propagation direction ϕ, it would be very useful to derive closed-form equations analogous to (6.12) and (6.13). However, for this general case, the underlying dispersion relation (6.4) is a transcendental equation. TAFLOVE [1995, pp. 97–98] provided a useful alternative approach for obtaining sample values of $\tilde{v}_p$ by applying the following Newton's method iterative procedure to (6.4):

$$\tilde{k}_{\text{icount}+1} = \tilde{k}_{\text{icount}} - \frac{\sin^2(A\tilde{k}_{\text{icount}}) + \sin^2(B\tilde{k}_{\text{icount}}) - C}{A \sin(2A\tilde{k}_{\text{icount}}) + B \sin(2B\tilde{k}_{\text{icount}})}. \tag{6.14a}$$

Here, $\tilde{k}_{\text{icount}+1}$ is the improved estimate of $\tilde{k}$, and $\tilde{k}_{\text{icount}}$ is the previous estimate of $\tilde{k}$. The A, B, and C are coefficients given by

$$A = \frac{\Delta \cdot \cos\phi}{2}, \qquad B = \frac{\Delta \cdot \sin\phi}{2}, \qquad C = \frac{1}{S^2} \sin^2\left(\frac{\pi S}{N_\lambda}\right). \tag{6.14b}$$

Additional simplicity results if Δ is normalized to the free-space wavelength, λ_0. This is equivalent to setting $\lambda_0 = 1$. Then, a very good starting guess for the iterative process is simply 2π. For this case, $\tilde{v}_p$ is given by

$$\frac{\tilde{v}_p}{c} = \frac{2\pi}{\tilde{k}_{\text{final icount}}}. \tag{6.15}$$

Usually, only two or three iterations are required for convergence.

Fig. 6.2 graphs results obtained using this procedure that illustrate the variation of $\tilde{v}_p$ with propagation direction ϕ. Here, for the Courant factor fixed at $S = 0.5$, three

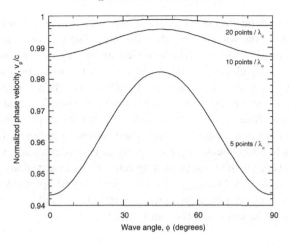

FIG. 6.2. Variation of the numerical phase velocity with wave propagation angle in a 2D FDTD grid for three sampling densities of the square unit cells. $S = c\Delta t/\Delta = 0.5$ for all cases.

different grid sampling densities N_λ are examined: $N_\lambda = 5$ points per λ_0, $N_\lambda = 10$, and $N_\lambda = 20$. We see that $\tilde{v}_p < c$ and is a function of both ϕ and N_λ. $\tilde{v}_p$ is maximum for waves propagating obliquely within the grid ($\phi = 45°$), and is minimum for waves propagating along either grid axis ($\phi = 0°, 90°$).

It is useful to summarize the algorithmic dispersive-error performance by defining two normalized error measures: (1) the physical phase-velocity error $\Delta\tilde{v}_{\text{physical}}$, and (2) the velocity-anisotropy error $\Delta\tilde{v}_{\text{aniso}}$. These are given by

$$\Delta\tilde{v}_{\text{physical}}|_{N_\lambda} = \frac{\min[\tilde{v}_p(\phi)] - c}{c} \times 100\%, \tag{6.16}$$

$$\Delta\tilde{v}_{\text{aniso}}|_{N_\lambda} = \frac{\max[\tilde{v}_p(\phi)] - \min[\tilde{v}_p(\phi)]}{\min[\tilde{v}_p(\phi)]} \times 100\%. \tag{6.17}$$

$\Delta\tilde{v}_{\text{physical}}$ is useful in quantifying the phase lead or lag that numerical modes suffer relative to physical modes propagating at c. For example, from Fig. 6.2 and (6.12b), $\Delta\tilde{v}_{\text{physical}} = -0.31\%$ for $N_\lambda = 20$. This means that a sinusoidal numerical wave traveling over a $10\lambda_0$ distance in the grid (200 cells) could develop a lagging phase error up to $11°$. We note that $\Delta\tilde{v}_{\text{physical}}$ is a function of N_λ. Since the grid cell size Δ is fixed, for an impulsive wave-propagation problem there exists a spread of effective N_λ values for the spectral components comprising the pulse. This causes a spread of $\Delta\tilde{v}_{\text{physical}}$ over the pulse spectrum, which in turn yields a temporal dispersion of the pulse evidenced in the spreading and distortion of its waveform as it propagates.

$\Delta\tilde{v}_{\text{aniso}}$ is useful in quantifying wavefront distortion. For example, a circular cylindrical wave would suffer progressive distortion of its wavefront since the portions propagating along the grid diagonals would travel slightly faster than the portions traveling along the major grid axes. For example, from Fig. 6.2 and (6.12b) and (6.13b), $\Delta\tilde{v}_{\text{aniso}} = 0.208\%$ for $N_\lambda = 20$. The wavefront distortion due to this anisotropy would total about 2.1 cells for each 1000 cells of propagation distance.

It is clear that errors due to inaccurate numerical velocities are cumulative, i.e., they increase linearly with the wave-propagation distance. These errors represent a fundamental limitation of *all* grid-based Maxwell's equations' algorithms, and can be troublesome when modeling electrically large structures. A positive aspect seen in Fig. 6.2 is that both $\Delta\tilde{v}_{\mathrm{physical}}$ and $\Delta\tilde{v}_{\mathrm{aniso}}$ decrease by approximately a 4:1 factor each time the grid-sampling density doubles, indicative of the second-order accuracy of the Yee algorithm. Therefore, finer meshing is one way to control the dispersion error.

As discussed in Section 7, there are proposed means to improve the accuracy of FDTD algorithms to allow much larger structures to be modeled. Specifically, $\Delta\tilde{v}_{\mathrm{aniso}}$ can be reduced to very low levels approaching zero. In this case, residual errors involve primarily the dispersion of $\Delta\tilde{v}_{\mathrm{physical}}$ with N_λ, which can be optimized by the proper choice of Δt. However, the new approaches presently have limitations regarding their ability to model material discontinuities, and require more research.

6.5.2. Intrinsic grid velocity anisotropy

Following TAFLOVE and HAGNESS [2000, pp. 120–123], this section provides a deeper discussion of the numerical phase-velocity errors of the Yee algorithm. We show that the nature of the grid discretization, in a manner virtually independent of the time-stepping scheme, determines the velocity anisotropy $\Delta\tilde{v}_{\mathrm{aniso}}$.

Relation of the time and space discretizations in generating numerical velocity error. In Section 6.5.1, we determined that $\Delta\tilde{v}_{\mathrm{aniso}} = 0.208\%$ for a two-dimensional Yee algorithm having $N_\lambda = 20$ and $S = 0.5$. An important and revealing question is: How is $\Delta\tilde{v}_{\mathrm{aniso}}$ affected by the choice of S, assuming that N_λ is fixed at 20?

To begin to answer this question, we first choose (what will later be shown to be) the largest possible value of S for numerical stability in two dimensions, $S = 1/\sqrt{2}$. Substituting this value of S into (6.12b) and (6.13b) yields

$$\left.\begin{array}{l} \tilde{v}_p(\phi = 0°) = 0.997926c \\ \tilde{v}_p(\phi = 45°) = c \end{array}\right\} \qquad \Delta\tilde{v}_{\mathrm{aniso}} = \frac{c - 0.997926c}{0.997926c} \times 100\% = 0.208\%.$$

To three decimal places, there is no change in $\Delta\tilde{v}_{\mathrm{aniso}}$ from the previous value, $S = 0.5$. We next choose a very small value $S = 0.01$ for substitution into (6.12b) and (6.13b):

$$\left.\begin{array}{l} \tilde{v}_p(\phi = 0°) = 0.995859c \\ \tilde{v}_p(\phi = 45°) = 0.997937c \end{array}\right\}$$

$$\Delta\tilde{v}_{\mathrm{aniso}} = \frac{0.997937c - 0.995859c}{0.995859c} \times 100\% = 0.208\%.$$

Again, there is no change in $\Delta\tilde{v}_{\mathrm{aniso}}$ to three decimal places.

We now suspect that, for a given N_λ, $\Delta\tilde{v}_{\mathrm{aniso}}$ is at most a weak function of S, and therefore is only weakly dependent on Δt. In fact, this is the case. More generally, LIU, Y. [1996] has shown that $\Delta\tilde{v}_{\mathrm{aniso}}$ is only weakly dependent on the specific type of time-marching scheme used, whether leapfrog, Runge–Kutta, etc. Thus, we can say that $\Delta\tilde{v}_{\mathrm{aniso}}$ is virtually an intrinsic characteristic of the space-lattice discretization. Following LIU, Y. [1996], three key points should be made in this regard:

- Numerical-dispersion errors associated with the time discretization are isotropic relative to the propagation direction of the wave.
- The choice of time discretization has little effect upon the phase-velocity anisotropy $\Delta\tilde{v}_{\text{aniso}}$ for $N_\lambda > 10$.
- The choice of time discretization does influence $\Delta\tilde{v}_{\text{physical}}$. However, it is not always true that higher-order time-marching schemes, such as Runge–Kutta, yield less $\Delta\tilde{v}_{\text{physical}}$ than simple Yee leapfrogging. Errors in $\Delta\tilde{v}_{\text{physical}}$ are caused separately by the space and time discretizations, and can either partially reinforce or cancel each other. Thus, the use of fourth-order Runge–Kutta may actually shift the $\tilde{v}_p(\phi)$ profile away from c, representing an increased $\Delta\tilde{v}_{\text{physical}}$ relative to ordinary leapfrogging.

The associated eigenvalue problem. LIU, Y. [1996] has shown that, to determine the relative velocity anisotropy characteristic intrinsic to a space grid, it is useful to set up an eigenvalue problem for the matrix that delineates the spatial derivatives used in the numerical algorithm. Consider as an example the finite-difference system of (3.19) for the case of two-dimensional TM_z electromagnetic wave propagation. The associated eigenvalue problem for the lossless-medium case is written as:

$$-\frac{1}{\mu}\left(\frac{E_z|_{i,j+1/2} - E_z|_{i,j-1/2}}{\Delta y}\right) = \Lambda H_x|_{i,j}, \tag{6.18a}$$

$$\frac{1}{\mu}\left(\frac{E_z|_{i+1/2,j} - E_z|_{i-1/2,j}}{\Delta x}\right) = \Lambda H_y|_{i,j}, \tag{6.18b}$$

$$\frac{1}{\varepsilon}\left(\frac{H_y|_{i+1/2,j} - H_y|_{i-1/2,j}}{\Delta x} - \frac{H_x|_{i,j+1/2} - H_x|_{i,j-1/2}}{\Delta y}\right) = \Lambda E_z|_{i,j}. \tag{6.18c}$$

We note that, at any time step n, the instantaneous values of the E- and H-fields distributed in space across the grid can be Fourier-transformed with respect to the i and j grid coordinates to provide a spectrum of sinusoidal modes. The result is often called the two-dimensional spatial-frequency spectrum, or the plane-wave eigenmodes of the grid. Let the following specify a typical mode of this spectrum having $\tilde{k}_x$ and $\tilde{k}_y$ as, respectively, the x- and y-components of its numerical wavevector:

$$E_z|_{I,J} = E_{z0}e^{j(\tilde{k}_x I\Delta x + \tilde{k}_y J\Delta y)};$$

$$H_x|_{I,J} = H_{x0}e^{j(\tilde{k}_x I\Delta x + \tilde{k}_y J\Delta y)}; \tag{6.19}$$

$$H_y|_{I,J} = H_{y0}e^{j(\tilde{k}_x I\Delta x + \tilde{k}_y J\Delta y)}$$

Upon substituting the eigenmode expressions of (6.19) into (6.18a), we obtain

$$-\frac{1}{\mu}\left(\frac{E_{z0}e^{j[\tilde{k}_x I\Delta x + \tilde{k}_y(J+1/2)\Delta y]} - E_{z0}e^{j[\tilde{k}_x I\Delta x + \tilde{k}_y(J-1/2)\Delta y]}}{\Delta y}\right)$$

$$= \Lambda H_{x0}e^{j(\tilde{k}_x I\Delta x + \tilde{k}_y J\Delta y)}. \tag{6.20}$$

Factoring out the $e^{j(\tilde{k}_x I \Delta x + \tilde{k}_y J \Delta y)}$ term that is common to both sides and then applying Euler's identity yields

$$H_{x0} = -\frac{2jE_{z0}}{\Lambda \mu \Delta y} \sin\left(\frac{\tilde{k}_y \Delta y}{2}\right). \tag{6.21a}$$

In a similar manner, substituting the eigenmode expressions of (6.19) into (6.18b) and (6.18c) yields

$$H_{y0} = \frac{2jE_{z0}}{\Lambda \mu \Delta x} \sin\left(\frac{\tilde{k}_x \Delta x}{2}\right), \tag{6.21b}$$

$$E_{z0} = \frac{2j}{\Lambda \varepsilon}\left[\frac{H_{y0}}{\Delta x} \sin\left(\frac{\tilde{k}_x \Delta x}{2}\right) - \frac{H_{x0}}{\Delta y} \sin\left(\frac{\tilde{k}_y \Delta y}{2}\right)\right]. \tag{6.21c}$$

Substituting H_{x0} of (6.21a) and H_{y0} of (6.21b) into (6.21c) yields

$$E_{z0} = \frac{2j}{\Lambda \varepsilon}\left[\begin{array}{l} \frac{1}{\Delta x} \cdot \frac{2jE_{z0}}{\Lambda \mu \Delta x} \cdot \sin\left(\frac{\tilde{k}_x \Delta x}{2}\right) \cdot \sin\left(\frac{\tilde{k}_x \Delta x}{2}\right) \\ -\frac{1}{\Delta y} \cdot \frac{-2jE_{z0}}{\Lambda \mu \Delta y} \cdot \sin\left(\frac{\tilde{k}_y \Delta y}{2}\right) \cdot \sin\left(\frac{\tilde{k}_y \Delta y}{2}\right) \end{array}\right]. \tag{6.22}$$

Now factoring out the common E_{z0} term, simplifying, and solving for Λ^2, we obtain

$$\Lambda^2 = -\frac{4}{\mu \varepsilon}\left[\frac{1}{(\Delta x)^2} \sin^2\left(\frac{\tilde{k}_x \Delta x}{2}\right) + \frac{1}{(\Delta y)^2} \sin^2\left(\frac{\tilde{k}_y \Delta y}{2}\right)\right]. \tag{6.23}$$

From the elementary properties of the sine function (assuming that $\tilde{k}_x$ and $\tilde{k}_y$ are real numbers for propagating numerical waves), the right-hand side of (6.23) is negative. Hence, Λ is a pure imaginary number given by

$$\Lambda = j2c\left[\frac{1}{(\Delta x)^2} \sin^2\left(\frac{\tilde{k}_x \Delta x}{2}\right) + \frac{1}{(\Delta y)^2} \sin^2\left(\frac{\tilde{k}_y \Delta y}{2}\right)\right]^{1/2}, \tag{6.24}$$

where $c = 1/\sqrt{\mu \varepsilon}$ is the speed of light in the homogeneous material being modeled. Finally, following the definition provided by LIU, Y. [1996], we obtain the "normalized numerical phase speed" c^*/c intrinsic to the grid discretization, given by

$$\frac{c^*}{c} = \frac{\Lambda_{\text{imag}}}{c\tilde{k}} = \frac{2}{\tilde{k}}\left[\frac{1}{(\Delta x)^2} \sin^2\left(\frac{\tilde{k}_x \Delta x}{2}\right) + \frac{1}{(\Delta y)^2} \sin^2\left(\frac{\tilde{k}_y \Delta y}{2}\right)\right]^{1/2}. \tag{6.25}$$

A convenient closed-form expression for c^*/c can be written by using the approximation $\tilde{k} \cong k$. Then, assuming a uniform square-cell grid, we obtain

$$\frac{c^*}{c} \cong \frac{N_\lambda}{\pi}\left[\sin^2\left(\frac{\pi \cos \phi}{N_\lambda}\right) + \sin^2\left(\frac{\pi \sin \phi}{N_\lambda}\right)\right]^{1/2}, \quad N_\lambda > 10. \tag{6.26}$$

The meaning of c^*/c. The reader is cautioned that c^*/c is *not* the same as $\tilde{v}_p/c$. This is because the derivation of c^*/c utilizes no information regarding the time-stepping process. Thus, c^*/c cannot be used to determine $\Delta\tilde{v}_{\text{physical}}$ defined in (6.16). However, c^*/c does provide information regarding $\Delta\tilde{v}_{\text{aniso}}$ defined in (6.17). Following LIU, Y. [1996], we can expand (6.26) to isolate the leading-order velocity-anisotropy term. This yields a simple expression for $\Delta\tilde{v}_{\text{aniso}}$ that is useful for $N_\lambda > 10$:

$$\Delta\tilde{v}_{\text{aniso}}|_{\text{Yee}} \cong \frac{\max\left[\frac{c^*(\phi)}{c}\right] - \min\left[\frac{c^*(\phi)}{c}\right]}{\min\left[\frac{c^*(\phi)}{c}\right]} \times 100\%$$

$$\cong \frac{\pi^2}{12(N_\lambda)^2} \times 100\%. \tag{6.27}$$

For example, (6.27) provides $\Delta\tilde{v}_{\text{aniso}} \cong 0.206\%$ for $N_\lambda = 20$. This is very close to the 0.208% value previously obtained using (6.12b) and (6.13b), the exact solutions of the full numerical dispersion relation for $\phi = 0°$ and $\phi = 45°$, respectively.

In summary, we can use (6.27) to estimate the numerical phase-velocity anisotropy $\Delta\tilde{v}_{\text{aniso}}$ of the Yee algorithm applied to a square-cell grid without having to resort to the Newton's method solution (6.14). This approach provides a convenient means to compare the relative anisotropy of alternative space-gridding techniques, including the higher-order methods and non-Cartesian meshes to be discussed in Section 7.

6.6. Complex-valued numerical wavenumbers

SCHNEIDER and WAGNER [1999] found that the Yee algorithm has a low-sampling-density regime that allows complex-valued numerical wavenumbers. In this regime, spatially decaying numerical waves can propagate faster than light, causing a weak, nonphysical signal to appear ahead of the nominal leading edges of sharply defined pulses. This section reviews the theory underlying this phenomenon.

6.6.1. Case 1: Numerical wave propagation along the principal lattice axes
Consider again numerical wave propagation along the major axes of a Yee space grid. For convenience, we rewrite 6.12a), the corresponding numerical dispersion relation:

$$\tilde{k} = \frac{2}{\Delta}\sin^{-1}\left[\frac{1}{S}\sin\left(\frac{\pi S}{N_\lambda}\right)\right] \equiv \frac{2}{\Delta}\sin^{-1}(\zeta), \tag{6.28}$$

where

$$\zeta = \frac{1}{S}\sin\left(\frac{\pi S}{N_\lambda}\right). \tag{6.29}$$

SCHNEIDER and WAGNER [1999] realized that, in evaluating numerical dispersion relations such as (6.28), it is possible to choose S and N_λ such that $\tilde{k}$ is complex. In the case of (6.28), it can be shown that the transition between real and complex values of $\tilde{k}$ occurs when $\zeta = 1$. Solving for N_λ at this transition results in

$$N_\lambda|_{\text{transition}} = \frac{\pi S}{\sin^{-1}(S)}. \tag{6.30}$$

For a grid sampling density greater than this value, i.e., $N_\lambda > N_\lambda|_{\text{transition}}$, $\tilde{k}$ is a real number and the numerical wave undergoes no attenuation while propagating in the grid. Here, $\tilde{v}_p < c$. For a coarser grid-sampling density $N_\lambda < N_\lambda|_{\text{transition}}$, $\tilde{k}$ is a complex number and the numerical wave undergoes a nonphysical exponential decay while propagating. Further, in this coarse-resolution regime, $\tilde{v}_p$ can exceed c.

Following SCHNEIDER and WAGNER [1999], we now discuss how $\tilde{k}$ and $\tilde{v}_p$ vary with grid sampling N_λ, both above and below the transition between real and complex numerical wavenumbers.

Real-numerical-wavenumber regime. For $N_\lambda > N_\lambda|_{\text{transition}}$ we have from (6.28)

$$\tilde{k}_{\text{real}} = \frac{2}{\Delta} \sin^{-1}\left[\frac{1}{S} \sin\left(\frac{\pi S}{N_\lambda}\right)\right];$$
(6.31a)

$$\tilde{k}_{\text{imag}} = 0.$$
(6.31b)

The numerical phase velocity is given by

$$\tilde{v}_p = \frac{\omega}{\tilde{k}_{\text{real}}} = \frac{\pi}{N_\lambda \sin^{-1}\left[\frac{1}{S}\sin\left(\frac{\pi S}{N_\lambda}\right)\right]} c.$$
(6.32)

This is exactly expression (6.12b). The wave-amplitude multiplier per grid cell of propagation is given by

$$e^{\tilde{k}_{\text{imag}}\Delta} \equiv e^{-\alpha\Delta} = e^0 = 1.$$
(6.33)

Thus, there is a constant wave amplitude with spatial position for this range of N_λ.

Complex-numerical-wavenumber regime. For $N_\lambda < N_\lambda|_{\text{transition}}$, we observe that $\zeta > 1$ in (6.28). Here, the following relation for the complex-valued arc-sine function given by CHURCHILL, BROWN, and VERHEY [1976] is useful:

$$\sin^{-1}(\zeta) = -j\ln\left(j\zeta + \sqrt{1 - \zeta^2}\right).$$
(6.34)

Substituting (6.34) into (6.28) yields after some algebraic manipulation

$$\tilde{k}_{\text{real}} = \frac{\pi}{\Delta};$$
(6.35a)

$$\tilde{k}_{\text{imag}} = -\frac{2}{\Delta} \ln\left(\zeta + \sqrt{\zeta^2 - 1}\right).$$
(6.35b)

The numerical phase velocity is then

$$\tilde{v}_p = \frac{\omega}{\tilde{k}_{\text{real}}} = \frac{\omega}{(\pi/\Delta)} = \frac{2\pi f \Delta}{\pi} = \frac{2f\lambda_0}{N_\lambda} = \frac{2}{N_\lambda} c$$
(6.36)

and the wave-amplitude multiplier per grid cell of propagation is

$$e^{\tilde{k}_{\text{imag}}\Delta} \equiv e^{-\alpha\Delta} = e^{-2\ln(\zeta+\sqrt{\zeta^2-1})} = \frac{1}{(\zeta + \sqrt{\zeta^2 - 1})^2}.$$
(6.37)

Since $\zeta > 1$, the numerical wave amplitude decays exponentially with spatial position.

We now consider the possibility of $\tilde{v}_p$ exceeding c in this situation. Nyquist theory states that any physical or numerical process that obtains samples of a time waveform every Δt seconds can reproduce the original waveform without aliasing for spectral content up to $f_{max} = 1/(2\Delta t)$. In the present case, the corresponding minimum free-space wavelength that can be sampled without aliasing is therefore

$$\lambda_{0,min} = c/f_{max} = 2c\Delta t. \tag{6.38a}$$

The corresponding minimum spatial-sampling density is

$$N_{\lambda,min} = \lambda_{0,min}/\Delta = 2c\Delta t/\Delta = 2S. \tag{6.38b}$$

Then from (6.36), the maximum numerical phase velocity is given by

$$\tilde{v}_{p,max} = \frac{2}{N_{\lambda,min}}c = \frac{2}{2S}c = \frac{c}{S}. \tag{6.39a}$$

From the definition of S, this maximum phase velocity can also be expressed as

$$\tilde{v}_{p,max} = \frac{1}{S}c = \left(\frac{\Delta}{c\Delta t}\right)c = \frac{\Delta}{\Delta t}. \tag{6.39b}$$

This relation tells us that in one Δt, a numerical value can propagate at most one Δ. This is intuitively correct given the local nature of the spatial differences used in the Yee algorithm. That is, a field point more than one Δ away from a source point that undergoes a sudden change cannot possibly "feel" the effect of that change during the next Δt. Note that $\tilde{v}_{p,max}$ is independent of material parameters and is an inherent property of the grid and its method of obtaining space derivatives.

6.6.2. *Case 2: Numerical wave propagation along a grid diagonal*
We next explore the possibility of complex-valued wavenumbers arising for oblique numerical wave propagation in a square-cell grid. For convenience, we rewrite (6.13a), the corresponding numerical dispersion relation:

$$\tilde{k} = \frac{2\sqrt{2}}{\Delta}\sin^{-1}\left[\frac{1}{S\sqrt{2}}\sin\left(\frac{\pi S}{N_\lambda}\right)\right] \equiv \frac{2\sqrt{2}}{\Delta}\sin^{-1}(\zeta), \tag{6.40}$$

where

$$\zeta = \frac{1}{S\sqrt{2}}\sin\left(\frac{\pi S}{N_\lambda}\right). \tag{6.41}$$

Similar to the previous case of numerical wave propagation along the principal lattice axes, it is possible to choose S and N_λ such that $\tilde{k}$ is complex. In the specific case of (6.40), the transition between real and complex values of $\tilde{k}$ occurs when $\zeta = 1$. Solving for N_λ at this transition results in

$$N_\lambda|_{transition} = \frac{\pi S}{\sin^{-1}(S\sqrt{2})}. \tag{6.42}$$

We now discuss how $\tilde{k}$ and $\tilde{v}_p$ vary with grid sampling N_λ, both above and below the transition between real and complex numerical wavenumbers.

Real-numerical-wavenumber regime. For $N_\lambda \geqslant N_\lambda|_{\text{transition}}$ we have from (6.40)

$$\tilde{k}_{\text{real}} = \frac{2\sqrt{2}}{\Delta} \sin^{-1}\left[\frac{1}{S\sqrt{2}} \sin\left(\frac{\pi S}{N_\lambda}\right)\right];$$

(6.43a)

$$\tilde{k}_{\text{imag}} = 0.$$

(6.43b)

The numerical phase velocity is given by

$$\tilde{v}_p = \frac{\omega}{\tilde{k}_{\text{real}}} = \frac{\pi}{N_\lambda\sqrt{2}\sin^{-1}\left[\frac{1}{S\sqrt{2}}\sin\left(\frac{\pi S}{N_\lambda}\right)\right]} c.$$

(6.44)

This is exactly expression (6.13b). The wave-amplitude multiplier per grid cell of propagation is given by

$$e^{\tilde{k}_{\text{imag}}\Delta} \equiv e^{-\alpha\Delta} = e^0 = 1.$$

(6.45)

Thus, there is a constant wave amplitude with spatial position for this range of N_λ.

Complex-numerical-wavenumber regime. For $N_\lambda < N_\lambda|_{\text{transition}}$, we observe that $\zeta > 1$ in (6.40). Substituting the complex-valued arc-sine function of (6.34) into (6.40) yields after some algebraic manipulation

$$\tilde{k}_{\text{real}} = \frac{\pi\sqrt{2}}{\Delta};$$

(6.46a)

$$\tilde{k}_{\text{imag}} = -\frac{2\sqrt{2}}{\Delta} \ln\left(\zeta + \sqrt{\zeta^2 - 1}\right).$$

(6.46b)

The numerical phase velocity for this case is

$$\tilde{v}_p = \frac{\omega}{\tilde{k}_{\text{real}}} = \frac{\omega}{(\pi\sqrt{2}/\Delta)} = \frac{\sqrt{2}f\lambda_0}{N_\lambda} = \frac{\sqrt{2}}{N_\lambda}c$$

(6.47)

and the wave-amplitude multiplier per grid cell of propagation is

$$e^{\tilde{k}_{\text{imag}}\Delta} \equiv e^{-\alpha\Delta} = e^{-2\sqrt{2}\ln(\zeta+\sqrt{\zeta^2-1})} = \frac{1}{(\zeta+\sqrt{\zeta^2-1})^{2\sqrt{2}}}.$$

(6.48)

Since $\zeta > 1$, the numerical wave amplitude decays exponentially with spatial position.

We again consider the possibility of $\tilde{v}_p$ exceeding c. From our previous discussion of (6.38a) and (6.38b), the minimum free-space wavelength that can be sampled without aliasing is $\lambda_{0,\text{min}} = c/f_{\text{max}} = 2c\Delta t$, and the corresponding minimum spatial-sampling density is $N_{\lambda,\text{min}} = \lambda_{0,\text{min}}/\Delta = 2S$. Then from (6.47), the maximum numerical phase velocity is given by

$$\tilde{v}_{p,\text{max}} = \frac{\sqrt{2}}{N_{\lambda,\text{min}}} c = \frac{\sqrt{2}}{2S} c.$$

(6.49a)

From the definition of S, this maximum phase velocity can also be expressed as

$$\tilde{v}_{p,\text{max}} = \frac{\sqrt{2}}{2}\left(\frac{\Delta}{c\Delta t}\right) c = \frac{\sqrt{2}\Delta}{2\Delta t}.$$

(6.49b)

This relation tells us that in $2\Delta t$, a numerical value can propagate at most $\sqrt{2}\,\Delta$ along the grid diagonal. We can show that this upper bound on $\tilde{v}_p$ is intuitively correct given the local nature of the spatial differences used in the Yee algorithm. Consider two nearest neighbor field points $P_{i,j}$ and $P_{i+1,j+1}$ along a grid diagonal, and how a sudden change at $P_{i,j}$ could be communicated to $P_{i+1,j+1}$. Now, a basic principle is that the Yee algorithm can communicate field data only along Cartesian (x and y) grid lines, and not along grid diagonals. Thus, at the minimum, $1\Delta t$ would be needed to transfer any part of the field perturbation at $P_{i,j}$ over a distance of 1Δ in the x-direction to $P_{i+1,j}$. Then, a second Δt would be needed, at the minimum, to transfer any part of the resulting field perturbation at $P_{i+1,j}$ over a distance of 1Δ in the y-direction to reach $P_{i+1,j+1}$. Because the distance between $P_{i,j}$ and $P_{i+1,j+1}$ is $\sqrt{2}\,\Delta$, the maximum effective velocity of signal transmission between the two points is $\sqrt{2}\,\Delta/2\Delta t$. By this reasoning, we see that $\tilde{v}_{p,\max}$ is independent of material parameters modeled in the grid. It is an inherent property of the FDTD grid and its method of obtaining space derivatives.

6.6.3. Example of calculation of numerical phase velocity and attenuation

This section provides sample calculations of values of the numerical phase velocity and the exponential attenuation constant for the case of a two-dimensional square-cell Yee grid. These calculations are based upon the numerical dispersion analyses of Sections 6.6.1 and 6.6.2.

Fig. 6.3 graphs the normalized numerical phase velocity and the exponential attenuation constant per grid cell as a function of grid sampling density N_λ. A Courant factor $S = 0.5$ is assumed. From this figure, we note that:

- For propagation along the principal grid axes $\phi = 0°$, $90°$, a minimum value of $\tilde{v}_p = (2/3)c$ is reached at $N_\lambda = 3$. This sampling density is also the onset of attenuation. As N_λ is reduced below 3, $\tilde{v}_p$ increases inversely with N_λ. Eventually, $\tilde{v}_p$ exceeds c for $N_\lambda < 2$, and reaches a limiting velocity of $2c$ as $N_\lambda \to 1$. In this limit, as well, the attenuation constant approaches a value of 2.634 nepers/cell.

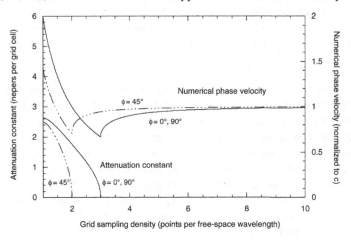

FIG. 6.3. Normalized numerical phase velocity and exponential attenuation constant per grid cell versus grid sampling density for on-axis and oblique wave propagation. $S = 0.5$.

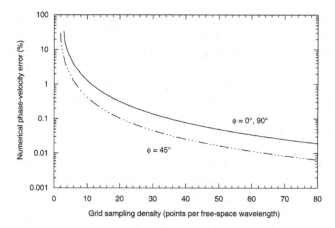

FIG. 6.4. Percent numerical phase-velocity error relative to the free-space speed of light as a function of the grid sampling density for on-axis and oblique wave propagation. $S = 0.5$.

- For propagation along the grid diagonal at $\phi = 45°$, a minimum value of $\tilde{v}_p = (\sqrt{2}/2)c$ is reached at $N_\lambda = 2$. This point is also the onset of exponential attenuation. As N_λ is reduced below 2, $\tilde{v}_p$ increases inversely with N_λ. Eventually, $\tilde{v}_p$ exceeds c for $N_\lambda < \sqrt{2}$, and reaches a limiting velocity of $\sqrt{2}c$ as $N_\lambda \to 1$. In this limit, as well, the attenuation constant approaches a value of 2.493 nepers/cell.

Overall, for both the on-axis and oblique cases of numerical wave propagation, we see that very coarsely resolved wave modes in the grid can propagate at superluminal speeds, but are rapidly attenuated.

Fig. 6.4 graphs the percent error in the numerical phase velocity relative to c for lossless wave propagation along the principal grid axes $\phi = 0°$, $90°$. In the present example wherein $S = 0.5$, this lossless propagation regime exists for $N_\lambda \geqslant 3$. Fig. 6.4 also graphs the percent velocity error for lossless wave propagation along the grid diagonal $\phi = 45°$. This lossless regime exists for $N_\lambda \geqslant 2$ for $S = 0.5$. As $N_\lambda \gg 10$, we see that the numerical phase-velocity error at each wave-propagation angle diminishes as the inverse square of N_λ. This is indicative of the second-order-accurate nature of the Yee algorithm.

6.6.4. Examples of calculations of pulse propagation in a one-dimensional grid

Fig. 6.5(a) graphs examples of the calculated propagation of a 40-cell-wide rectangular pulse in free space for two cases of the Courant factor: $S = 1$ (i.e., Δt is equal to the value for dispersionless propagation in a one-dimensional grid); and $S = 0.99$. To permit a direct comparison of these results, both "snapshots" are taken at the same absolute time after the onset of time-stepping. There are three key observations:

(1) When $S = 1$, the rectangular shape and spatial width of the pulse are completely preserved. For this case, the abrupt step discontinuities of the propagating pulse are modeled perfectly. In fact, this is expected since $\tilde{v}_p \equiv c$ for all numerical modes in the grid.

(2) When $S = 0.99$, there is appreciable "ringing" located behind the leading and trailing edges of the pulse. This is due to short-wavelength numerical modes in

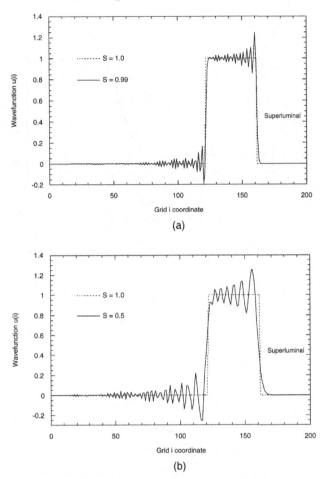

FIG. 6.5. Effect of numerical dispersion upon a rectangular pulse propagating in free space in a one-dimensional grid for three different Courant factors: $S = 1$, $S = 0.99$, and $S = 0.5$. (a) Comparison of calculated pulse propagation for $S = 1$ and $S = 0.99$. (b) Comparison of calculated pulse propagation for $S = 1$ and $S = 0.5$.

the grid generated at the step discontinuities of the wave. These numerical modes are poorly sampled in space and hence travel slower than c, thereby lagging behind the causative discontinuities.

(3) When $S = 0.99$, a weak superluminal response propagates just ahead of the leading edge of the pulse. This is again due to short-wavelength numerical modes in the grid generated at the step-function wavefront. However, these modes have spatial wavelengths even shorter than those noted in point (2), in fact so short that their grid sampling density drops below the upper bound for complex wavenumbers, and the modes appear in the superluminal, exponentially decaying regime.

Fig. 6.5(b) repeats the examples of Fig. 6.5(a), but for the Courant factors $S = 1$ and $S = 0.5$. We see that the duration and periodicity of the ringing is greater than that

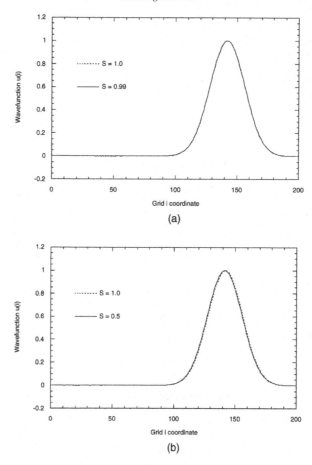

FIG. 6.6. Effect of numerical dispersion upon a Gaussian pulse propagating in free space in a one-dimensional grid for three different Courant factors: $S = 1$, $S = 0.99$, and $S = 0.5$. (a) Comparison of calculated pulse propagation for $S = 1$ and $S = 0.99$. (b) Comparison of calculated pulse propagation for $S = 1$ and $S = 0.5$.

for the $S = 0.99$ case. Further, the superluminal response is more pronounced and less damped.

Figs. 6.6(a) and 6.6(b) repeat the above examples, but for a Gaussian pulse having a 40-grid-cell spatial width between its $1/e$ points. We see that this pulse undergoes much less distortion than the rectangular pulse. The calculated propagation for $S = 0.99$ shows no observable difference (at the scale of Fig. 6.6(a)) relative to the perfect propagation case of $S = 1$. Even for $S = 0.5$, the calculated pulse propagation shows only a slight retardation relative to the exact solution, as expected because $\tilde{v}_p < c$ for virtually all modes in the grid. Further, there is no observable superluminal precursor. All of these phenomena are due to the fact that, for this case, virtually the entire spatial spectrum of propagating wavelengths within the grid is well resolved by the grid's sampling process. As a result, almost all numerical phase-velocity errors relative to c are well below 1%.

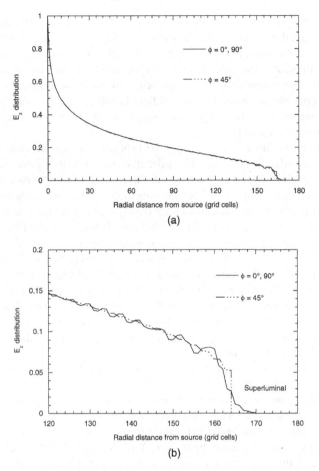

FIG. 6.7. Effect of numerical dispersion upon a radially propagating cylindrical wave in a 2D TM$_z$ Yee grid. The grid is excited at its center point by applying a unit-step time-function to a single E_z field component. The Courant factor is $S = \sqrt{2}/2$. (a) Comparison of calculated wave propagation along the grid axes and along a grid diagonal. (b) Expanded view of (a) at distances between 120 and 180 grid cells from the source.

This allows the Gaussian pulse to "hold together" while propagating over significant distances within the grid.

6.6.5. *Example of calculation of pulse propagation in a two-dimensional grid*

Fig. 6.7 presents an example of the calculation of a radially outward-propagating cylindrical wave in a two-dimensional TM$_z$ Yee grid. A 360×360-cell square grid with $\Delta x = \Delta y = \Delta = 1.0$ is used in this example. The grid is numerically excited at its center point by applying a unit-step time-function to a single E_z field component. We assume the Courant factor $S = \sqrt{2}/2$, which yields dispersionless propagation for numerical plane-wave modes propagating along the grid diagonals $\phi = 45°$, $135°$, $225°$, and $315°$. In Fig. 6.7(a), we graph snapshots of the E_z distribution vs. radial distance from the source. Here, field observations are made along cuts through the grid passing through the source and either parallel to the principal grid axes $\phi = 0°$, $90°$ or parallel to the grid diagonal $\phi = 45°$. The snapshots are taken $232\Delta t$ after the beginning of time-stepping. At this time, the wave has not yet reached the outer grid boundary.

Fig. 6.7(a) illustrates two nonphysical artifacts arising from numerical dispersion. First, for both observation cuts, the leading edge of the wave exhibits an oscillatory spatial jitter superimposed upon the normal field falloff profile. Second, for the observation cuts along the grid axes, the leading edge of the wave exhibits a small, spatially decaying, superluminal component.

To more easily see these artifacts, Fig. 6.7(b) shows an expanded view in the vicinity of the leading edge of the wave. Consider first the oscillatory jitter. Similar to the results shown in Fig. 6.5, this is due to short-wavelength numerical modes in the grid generated at the leading edge of the propagating step-function wave. According to our dispersion theory, these numerical modes are poorly sampled in space and hence travel slower than c, thereby lagging behind the actual wavefront. While the jitter is most pronounced along the grid axes $\phi = 0°$, $90°$, it is nonetheless finite along $\phi = 45°$ despite our choice of $S = \sqrt{2}/2$ (which implies dispersionless propagation along grid diagonals). This apparent conflict between theory and numerical experiment is resolved by noting that numerical dispersion introduces a slightly anisotropic propagation characteristic of the background "free space" within the grid versus azimuth angle ϕ. The resulting inhomogeneity of the free-space background scatters part of the radially propagating numerical wave into the ϕ-direction. Thus, no point behind the wavefront can avoid the short-wavelength numerical jitter.

Consider next the superluminal artifact present at the leading edge of the wave shown in Fig. 6.7(b) for $\phi = 0°$, $90°$ but not for $\phi = 45°$. This is again due to short-wavelength numerical modes in the grid generated at the leading edge of the outgoing step-function wave. However, these modes have spatial wavelengths so short that their grid sampling density drops below the threshold delineated in (6.30), and the modes appear in the superluminal, exponentially decaying regime. With $S = \sqrt{2}/2$ in the present example, we conclude that the lack of a superluminal artifact along the $\phi = 45°$ cut (and the consequent exact modeling of the step discontinuity at the leading edge of the wave) is due to dispersionless numerical wave propagation along grid diagonals.

7. Algorithms for improved numerical dispersion

7.1. Introduction

The numerical algorithm for Maxwell's equations introduced by YEE [1966] is very robust. Evidence of this claim is provided by the existence of thousands of successful electromagnetic engineering applications and refereed papers derived from the basic Yee algorithm. However, it is clear from Section 6 that Yee's approach is not perfect. For certain modeling problems, numerical dispersion can cause significant errors to arise in the calculated field.

This section reviews a small set of representative strategies aimed at mitigating the effects of numerical dispersion. No attempt is made to provide a comprehensive summary because such an effort would require several sections. The intent here is to provide the flavor of what may be possible in this area.

7.2. Strategy 1: Center a specific numerical phase-velocity curve about c

We have seen from Fig. 6.2 that the Yee algorithm yields a family of numerical phase-velocity curves contained in the range $\tilde{v}_p < c$. We also observe that each velocity curve is centered about the value

$$\tilde{v}_{avg} = \frac{\tilde{v}_p(\phi = 0°) + \tilde{v}_p(\phi = 45°)}{2}, \tag{7.1}$$

where $\tilde{v}_p(\phi = 0°)$ and $\tilde{v}_p(\phi = 45°)$ are given by (6.12b) and (6.13b), respectively. This symmetry can be exploited if a narrowband grid excitation is used such that a specific phase-velocity curve accurately characterizes the propagation of most of the numerical modes in the grid. Then, it is possible to shift the phase-velocity curve of interest so that it is centered about the free-space speed of light c, thereby cutting $\Delta\tilde{v}_{physical}$ by almost 3:1. Centering is implemented by simply scaling the free-space values of ε_0 and μ_0 used in the finite-difference system of (3.19):

$$\varepsilon_0' = \left(\frac{\tilde{v}_{avg}}{c}\right)\varepsilon_0; \qquad \mu_0' = \left(\frac{\tilde{v}_{avg}}{c}\right)\mu_0. \tag{7.2}$$

This scaling increases the baseline value of the model's "free-space" speed of light to compensate for the too-slow value of $\tilde{v}_{avg}$. By scaling both ε_0 and μ_0, the required shift in $\tilde{v}_{avg}$ is achieved without introducing any changes in wave impedance.

There are three primary difficulties with this approach: (1) The phase-velocity anisotropy error $\Delta\tilde{v}_{aniso}$ remains unmitigated. (2) The velocity compensation is only in the average sense over all possible directions in the grid. Hence, important numerical modes can still have phase velocities not equal to c. (3) Propagating wave pulses having broad spectral content cannot be compensated over their entire frequency range. Nevertheless, this approach is so easy to implement that its use can be almost routine.

7.3. Strategy 2: Use fourth-order-accurate spatial differences

It is possible to substantially reduce the phase-velocity anisotropy error $\Delta\tilde{v}_{aniso}$ for the Yee algorithm by incorporating a fourth-order-accurate finite-difference scheme for the spatial first-derivatives needed to implement the curl operator. This section reviews two such approaches. The first, by FANG [1989], is an explicit method wherein a fourth-order-accurate spatial central-difference is calculated at one observation point at a time from two pairs of field values: a pair on each side of the observation point at distances of $\Delta/2$ and $3\Delta/2$. The second approach, by TURKEL [1998], is an implicit method wherein a tridiagonal matrix is solved to obtain fourth-order-accurate spatial derivatives simultaneously at all observation points along a linear cut through the grid.

7.3.1. Explicit method

Assuming that Yee leapfrog time-stepping is used, the fourth-order-accurate spatial-difference scheme of FANG [1989] results in the following set of finite-difference expressions for the TM$_z$ mode:

$$\frac{H_x|_{i,j+1/2}^{n+1/2} - H_x|_{i,j+1/2}^{n-1/2}}{\Delta t}$$
$$= -\frac{1}{\mu_{i,j+1/2}}\left(\frac{-E_z|_{i,j+2}^n + 27E_z|_{i,j+1}^n - 27E_z|_{i,j}^n + E_z|_{i,j-1}^n}{24\Delta y}\right), \tag{7.3a}$$

$$\frac{H_y|_{i+1/2,j}^{n+1/2} - H_y|_{i+1/2,j}^{n-1/2}}{\Delta t}$$
$$= \frac{1}{\mu_{i+1/2,j}}\left(\frac{-E_z|_{i+2,j}^n + 27E_z|_{i+1,j}^n - 27E_z|_{i,j}^n + E_z|_{i-1,j}^n}{24\Delta x}\right), \tag{7.3b}$$

$$\frac{E_z|_{i,j}^{n+1} - E_z|_{i,j}^n}{\Delta t}$$
$$= \frac{1}{\varepsilon_{i,j}}\left(\begin{array}{c}\frac{-H_y|_{i+3/2,j}^{n+1/2} + 27H_y|_{i+1/2,j}^{n+1/2} - 27H_y|_{i-1/2,j}^{n+1/2} + H_y|_{i-3/2,j}^{n+1/2}}{24\Delta x} \\ -\frac{-H_x|_{i,j+3/2}^{n+1/2} + 27H_x|_{i,j+1/2}^{n+1/2} - 27H_x|_{i,j-1/2}^{n+1/2} + H_x|_{i,j-3/2}^{n+1/2}}{24\Delta y}\end{array}\right). \tag{7.3c}$$

The numerical dispersion relation for this algorithm analogous to (6.3) is given by

$$\left[\frac{1}{c\Delta t}\sin\left(\frac{\omega\Delta t}{2}\right)\right]^2 = \frac{1}{(\Delta x)^2}\left[\frac{27}{24}\sin\left(\frac{\tilde{k}_x\Delta x}{2}\right) - \frac{1}{24}\sin\left(\frac{3\tilde{k}_x\Delta x}{2}\right)\right]^2$$
$$+ \frac{1}{(\Delta y)^2}\left[\frac{27}{24}\sin\left(\frac{\tilde{k}_y\Delta y}{2}\right) - \frac{1}{24}\sin\left(\frac{3\tilde{k}_y\Delta y}{2}\right)\right]^2. \tag{7.4}$$

By analogy with the development in Section 6.5.2 culminating in (6.26), it can be shown that the intrinsic numerical phase-velocity anisotropy for a square-cell grid of this type is given by

$$\frac{c^*}{c} \cong \frac{N_\lambda}{\pi}\sqrt{\frac{\left[\frac{27}{24}\sin\left(\frac{\pi\cos\phi}{N_\lambda}\right) - \frac{1}{24}\sin\left(\frac{3\pi\cos\phi}{N_\lambda}\right)\right]^2}{+\left[\frac{27}{24}\sin\left(\frac{\pi\sin\phi}{N_\lambda}\right) - \frac{1}{24}\sin\left(\frac{3\pi\sin\phi}{N_\lambda}\right)\right]^2}} \tag{7.5}$$

and the numerical phase-velocity anisotropy error (by analogy with (6.27)) is given by

$$\Delta\tilde{v}_{\text{aniso}}|_{\text{explicit 4th-order}} \cong \frac{\pi^4}{18(N_\lambda)^4} \times 100\%. \tag{7.6}$$

7.3.2. Implicit method

TURKEL [1998] reported the Ty operator, an implicit fourth-order-accurate finite-difference scheme defined on the Yee space lattice for calculating the spatial first-derivatives involved in the curl. To see how the Ty operator is constructed, consider an x-directed cut through the Yee lattice. At every sample point along this cut, we wish to compute with fourth-order accuracy the x-derivatives of the general field component V. By manipulating Taylor's series expansions for V along this line, it was shown in TURKEL [1998, (Eq. (2.23)] that

$$\frac{1}{24}\left(\frac{\partial V}{\partial x}\bigg|_{i+1} + \frac{\partial V}{\partial x}\bigg|_{i-1}\right) + \frac{11}{12}\frac{\partial V}{\partial x}\bigg|_{i} = \frac{V_{i+1/2} - V_{i-1/2}}{\Delta x}. \tag{7.7}$$

Here, $\{(\partial V/\partial x)_i\}$ represents the set of initially unknown fourth-order-accurate x-derivatives of V at all grid-points along the observation cut; and $\{V_i\}$ represents the set of known values of V at the same grid-points. Upon writing (7.7) at each grid-point along the cut, a system of simultaneous equations for the unknowns $\{(\partial V/\partial x)_i\}$ is obtained. From the subscripts in (7.7), we see that this linear system has a tridiagonal matrix. This can be efficiently solved to yield $\{(\partial V/\partial x)_i\}$ in one step.

It was shown in TURKEL [1998, Eq. (2.70b)] that the Ty operator results in the following intrinsic grid-velocity anisotropy for a two-dimensional square-cell grid:

$$\frac{c^*}{c} = \frac{12 N_\lambda}{\pi} \sqrt{\left[\frac{\sin\left(\frac{\pi \cos\phi}{N_\lambda}\right)}{11 + \cos\left(\frac{2\pi \cos\phi}{N_\lambda}\right)}\right]^2 + \left[\frac{\sin\left(\frac{\pi \sin\phi}{N_\lambda}\right)}{11 + \cos\left(\frac{2\pi \sin\phi}{N_\lambda}\right)}\right]^2}. \tag{7.8}$$

From TURKEL [1998, (Eq. 2.71b)], the phase-velocity anisotropy error is

$$\Delta \tilde{v}_{aniso}|_{\text{Ty 4th-order}} \cong \frac{17 \cdot 3 \cdot 2 \cdot \pi^4}{2880 (N_\lambda)^4} \times 100\% \cong \frac{\pi^4}{28 (N_\lambda)^4} \times 100\%. \tag{7.9}$$

Fig. 7.1 compares the accuracy of the Ty method to the Yee algorithm for a generic two-dimensional wave-propagation problem, a sinusoidal line source radiating in free

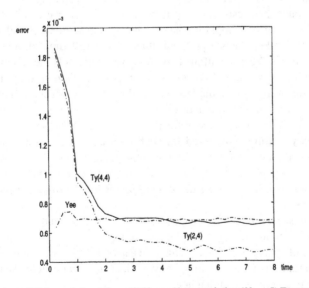

FIG. 7.1. Comparison of high-resolution ($N_\lambda = 40$) Yee and low-resolution ($N_\lambda = 5$) Ty errors in the L_2 norm for a radially propagating sinusoidal wave as a function of the simulated time. *Source*: E. Turkel, Chapter 2 in *Advances in Computational Electrodynamics: The Finite-Difference Time-Domain Method*, A. Taflove, ed., © 1998 Artech House, Inc.

space after being switched on at $t = 0$. Here, Ty(2, 4) and Ty(4, 4) refer to the implicit spatial-differentiation scheme of (7.7) used in conjunction with second-order Yee and fourth-order Runge–Kutta time-stepping, respectively. The two Ty approaches are implemented on square-cell grids of sampling-density N_λ with separate, accuracy-optimized Courant factors: $S = 1/18$ for Ty(2, 4) and $S = 1/4$ for Ty(4, 4).

From Fig. 7.1, we see that both Ty schemes run with $N_\lambda = 5$ achieve accuracy comparable to that of the Yee algorithm run with $N_\lambda = 40$ and $S = 2/3$. Under these conditions, both Ty methods are much more efficient than Yee's algorithm, requiring $(40/5)^2 : 1 = 8^2 : 1$ less computer storage and 23:1 less running-time. The permissible coarseness of the Ty grid is decisive in reducing its running-time, more than compensating for the extra operations required by its tridiagonal matrix inversions. These advantages in computer resources scale to the order of $8^3 : 1$ in three dimensions.

7.3.3. Discussion

Consider comparing (6.27) with (7.6) and (7.9). This allows us to form approximate ratios of the numerical phase-velocity anisotropy errors of the fourth-order-accurate spatial-differencing schemes discussed above relative to the Yee algorithm:

$$\frac{\Delta \tilde{v}_{\text{aniso}}|_{\text{explicit 4th-order}}}{\Delta \tilde{v}_{\text{aniso}}|_{\text{Yee}}} \cong \frac{2\pi^2}{3(N_\lambda)^2}; \qquad \frac{\Delta \tilde{v}_{\text{aniso}}|_{\text{Ty 4th-order}}}{\Delta \tilde{v}_{\text{aniso}}|_{\text{Yee}}} \cong \frac{3\pi^2}{7(N_\lambda)^2}. \qquad (7.10)$$

With the reminder that these ratios were derived based upon assuming $N_\lambda > 10$, we see that greatly reduced $\Delta \tilde{v}_{\text{aniso}}$ error is possible for both fourth-order spatial-differencing schemes. In addition, optimally choosing the Courant number S for each fourth-order spatial technique permits minimizing the overall error, including $\Delta \tilde{v}_{\text{physical}}$. From a growing set of published results similar to those of Fig. 7.1, we conclude that fourth-order-accurate explicit and implicit spatial schemes allow modeling electromagnetic wave-propagation and interaction problems that are at least 8 times the electrical size of those permitted by the Yee algorithm. This is a very worthwhile increase in capability.

However, this improvement is not without cost. Although easy to set up in homogeneous-material regions, the larger stencil needed to calculate fourth-order spatial differences is troublesome when dealing with material interfaces. Metal boundaries are especially challenging since they effectively cause field discontinuities in the grid. Special boundary conditions required for such interfaces significantly complicate the computer software used to render structures in the grid.

7.3.4. Fourth-order-accurate approximation of jumps in material parameters at interfaces

As stated above, special boundary conditions must be derived and programmed to deal with material discontinuities when implementing high-order-accuracy finite-difference approximations of spatial derivatives. This is because the nonlocal nature of the numerical space-differentiation process may convey electromagnetic field data across such discontinuities in a nonphysical manner.

TURKEL [1998] reported a means to markedly reduce error due to abrupt dielectric interfaces. This approach replaces the discontinuous permittivity function ε by a fourth-order-accurate smooth implicit approximation. (A similar strategy can be applied to jumps in μ.) Relative to the use of a polynomial approximation to ε, this strategy avoids the overshoot artifact. We note that, with an implicit interpolation, ε varies in the entire domain and not just near the interface. However, far from the interface, the variation is small.

Consider a dielectric interface separating two regions defined along the x-axis of the space lattice. Following TURKEL [1998, Eq. (2.82)], a fourth-order-accurate interpolation of the permittivity distribution with grid position i can be achieved using

$$\frac{1}{8}\begin{bmatrix} 10 & -5 & 4 & -1 & \cdot & \cdot & 0 \\ 1 & 6 & 1 & 0 & \cdot & \cdot & 0 \\ 0 & 1 & 6 & 1 & 0 & \cdot & 0 \\ \cdot & \cdot & \cdot & \cdot & \cdot & \cdot & \cdot \\ 0 & \cdot & \cdot & 0 & 1 & 6 & 1 \\ 0 & \cdot & \cdot & -1 & 4 & -5 & 10 \end{bmatrix}\begin{bmatrix} \varepsilon_1 \\ \varepsilon_2 \\ \cdot \\ \cdot \\ \cdot \\ \varepsilon_{p-1} \end{bmatrix}$$

$$= \frac{1}{2}\left(\begin{bmatrix} \varepsilon_{3/2} \\ \varepsilon_{5/2} \\ \cdot \\ \varepsilon_{p-3/2} \\ \varepsilon_{p-1/2} \end{bmatrix} + \begin{bmatrix} \varepsilon_{1/2} \\ \varepsilon_{3/2} \\ \cdot \\ \varepsilon_{p-5/2} \\ \varepsilon_{p-3/2} \end{bmatrix}\right). \tag{7.11}$$

Here, $[\varepsilon_1, \varepsilon_2, \ldots, \varepsilon_{p-1}]$ is the initially unknown set of values of the smooth approximation to the abrupt dielectric interface; and $[\varepsilon_{1/2}, \varepsilon_{3/2}, \ldots, \varepsilon_{p-1/2}]$ is the known set of permittivities for the original dielectric interface geometry. Inversion of the linear system of (7.11) yields the desired smooth approximation, $[\varepsilon_1, \varepsilon_2, \ldots, \varepsilon_{p-1}]$.

TURKEL [1998] presented a comparative example (shown in Fig. 7.2) of the use of this dielectric interface smoothing technique for both the Yee and Ty algorithms. His example modeled the standing wave within a two-dimensional rectangular cavity comprised of a block of lossless dielectric of $\varepsilon_r = 4$ surrounded by free space. An available exact solution for the sinusoidal space–time variation of the standing-wave mode was used to specify the computational domain's initial conditions and boundary conditions for both the Yee and Ty simulations. It was also used to develop L_2-normed errors of the calculated Yee and Ty fields as a function of the simulated time.

Fig. 7.2(a) compares the error of the Ty method with that of the Yee algorithm for the case where ε at the dielectric interfaces of the cavity is simply set to the arithmetic average of the values on both sides. Here, both the Yee and Ty grids use square unit cells wherein $N_\lambda = 30$ within the dielectric material. The Courant factors are selected as $S_{\text{Yee}} = 2/3$ and $S_{\text{Ty}(2,4)} = 1/18$. We see from this figure that, while the Ty results show less error than the Yee data, the error performance of the Ty scheme is clearly hurt by the treatment of the interfaces, which gives only second-order accuracy.

Fig. 7.2(b) shows the corresponding numerical errors for the case where the permittivity is smoothed as per (7.11). While there is a modest reduction in the error of the Yee data, there is a much greater reduction in the error of the Ty results. In additional studies discussed by TURKEL [1998], it was demonstrated that the Yee error can be reduced to that of Ty by increasing the Yee-grid-sampling density to a level eight times that of Ty, just what was observed for the free-space propagation example discussed previously in the context of Fig. 7.1. Thus, we see that fourth-order-accurate smoothing of abrupt permittivity jumps succeeds in preserving the fourth-order-accuracy advantage of Ty versus Yee observed for the homogeneous-permittivity case.

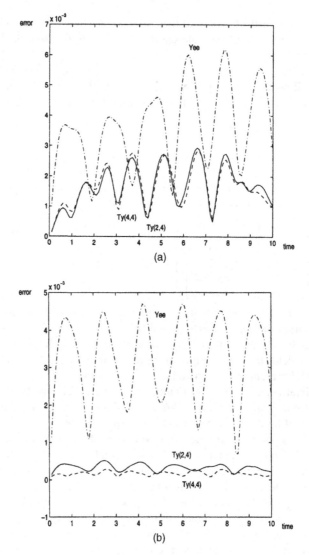

FIG. 7.2. Comparison of Yee and Ty errors (L_2 norm) for the standing-wave fields within a rectangular dielectric cavity. The same grid-sampling density ($N_\lambda = 30$) is used for both algorithms. (a) Second-order-accurate
arithmetic averaging of the permittivity at the dielectric interfaces. (b) Fourth-order-accurate smoothing of
the permittivity at the dielectric interfaces as per (4.86). *Source*: E. Turkel, Chapter 2 in *Advances in Computational Electrodynamics: The Finite-Difference Time-Domain Method*, A. Taflove, ed., © 1998 Artech
House, Inc.

7.4. Strategy 3: Use hexagonal grids

Regular hexagonal grids in two dimensions have been proposed to reduce the numerical
phase-velocity anisotropy well below that of Yee's Cartesian mesh. Here, the primary
grid is composed of equilateral hexagons having edge length Δs. Each hexagon can be

considered to be the union of six equilateral triangles. Connecting the centroids of these triangles yields a second set of regular hexagons that comprises a dual grid.

Fig. 5.2 illustrated for the TM$_z$ case in two dimensions the two principal ways of arranging E and H vector components about hexagonal grids, as discussed by LIU, Y. [1996] and reviewed in Section 5.2. There, the finite-difference equations for the TM$_z$ mode for the unstaggered, collocated hexagonal grid of Fig. 5.2(a) were given by (5.3), and the finite-difference equations for the TM$_z$ mode for the staggered, uncollocated hexagonal grid of Fig. 5.2(b) were given by (5.4).

Using the analysis method of Section 6.5.2, LIU, Y. [1996] obtained the following expressions for the numerical phase-velocity anisotropy error for the hexagonal grids of Fig. 5.2:

$$\Delta \tilde{v}_{aniso}\Big|_{\substack{\text{hex.grid,}\\ \text{Fig. 5.2(a)}}} \cong \frac{1 \cdot 2 \cdot \pi^4}{120(N_\lambda)^4} \times 100\% = \frac{\pi^4}{60(N_\lambda)^4} \times 100\%, \qquad (7.12)$$

$$\Delta \tilde{v}_{aniso}\Big|_{\substack{\text{hex.grid,}\\ \text{Fig. 5.2(b)}}} \cong \frac{1 \cdot 2 \cdot \pi^4}{720(N_\lambda)^4} \times 100\% = \frac{\pi^4}{360(N_\lambda)^4} \times 100\%. \qquad (7.13)$$

Interestingly, we note that $\Delta \tilde{v}_{aniso}$ for both hexagonal grids exhibits a *fourth-order* dependence on the grid-sampling density N_λ despite the second-order accuracy of each spatial difference used. As shown by LIU, Y. [1996], this is because the leading second-order error term becomes isotropic for the hexagonal gridding case, with a value exactly equal to the average of its ϕ-dependent Cartesian counterpart.

Comparison of $\Delta \tilde{v}_{aniso}$ of the Yee algorithm given by (6.27) with $\Delta \tilde{v}_{aniso}$ of the hexagonal gridding given by (7.12) and (7.13) yields the following error ratios:

$$\frac{\Delta \tilde{v}_{aniso}\big|_{\text{hex.grid, Fig. 5.2(a)}}}{\Delta \tilde{v}_{aniso}\big|_{\text{Yee}}}(N_\lambda) \cong \frac{\pi^2}{5(N_\lambda)^2} \qquad (7.14)$$

and

$$\frac{\Delta \tilde{v}_{aniso}\big|_{\text{hex.grid, Fig. 5.2(b)}}}{\Delta \tilde{v}_{aniso}\big|_{\text{Yee}}}(N_\lambda) \cong \frac{\pi^2}{30(N_\lambda)^2}. \qquad (7.15)$$

Hexagonal gridding is seen to yield velocity-anisotropy errors as little as $1/300$th that of the Yee grid at a sampling density of 10 points per wavelength.

We can also compare the velocity-anisotropy errors of hexagonal gridding with those of the fourth-order gridding schemes discussed previously:

$$\frac{\Delta \tilde{v}_{aniso}\big|_{\text{hex.grid, Fig. 5.2(a)}}}{\Delta \tilde{v}_{aniso}\big|_{\text{explicit 4th-order}}}(N_\lambda) \cong \frac{3}{10}; \qquad \frac{\Delta \tilde{v}_{aniso}\big|_{\text{hex.grid, Fig. 5.2(a)}}}{\Delta \tilde{v}_{aniso}\big|_{\text{Ty 4th-order}}}(N_\lambda) \cong \frac{7}{15}; \qquad (7.16)$$

$$\frac{\Delta \tilde{v}_{aniso}\big|_{\text{hex.grid, Fig. 5.2(b)}}}{\Delta \tilde{v}_{aniso}\big|_{\text{explicit 4th-order}}}(N_\lambda) \cong \frac{1}{20}; \qquad \frac{\Delta \tilde{v}_{aniso}\big|_{\text{hex.grid, Fig. 5.2(b)}}}{\Delta \tilde{v}_{aniso}\big|_{\text{Ty 4th-order}}}(N_\lambda) \cong \frac{7}{90}. \qquad (7.17)$$

We see that hexagonal gridding yields less velocity-anisotropy error than the two fourth-order-accurate Cartesian spatial-differencing techniques reviewed previously. In the case of the hexagonal grid of Fig. 5.2(b), this dispersion error is lower by more than one order-of-magnitude.

Hexagonal gridding has a second advantage relative to the fourth-order spatial algorithms: it uses only nearest-neighbor field data. Therefore, hexagonal grids can model material discontinuities including metal boundaries as easily as the Yee algorithm. There is no need to develop special boundary conditions.

What, if any, are the limitations in using hexagonal grid algorithms relative to Yee's method? The answer is: none very significant in two dimensions. This is because it is only moderately more complicated to generate (even manually) uniform hexagonal grids than it is to generate uniform Cartesian grids. The difficulty arises in attempting to extend hexagonal gridding to three dimensions. As shown in Fig. 5.3, such an extension involves filling space with tetradecahedron and dual-tetrahedron unit cells. This increases the complexity of the computational mesh to the point where sophisticated computer-based mesh-generation techniques are mandatory.

7.5. Strategy 4: Use discrete Fourier transforms to calculate the spatial derivatives

The fourth and final approach reviewed here for reduction of the numerical dispersion artifact is the *pseudospectral time-domain* (PSTD) method of LIU, Q.H. [1996], LIU, Q.H. [1997]. This technique uses a discrete Fourier transform (DFT) algorithm to represent the spatial derivatives in the Maxwell's equations' computational lattice. The fast Fourier transform (FFT) can also be applied to increase numerical efficiency.

7.5.1. Formulation
The PSTD method works on unstaggered, collocated Cartesian space lattices wherein all field components are located at the same points. An example of such an arrangement is the two-dimensional TM$_z$ grid of Fig. 5.1(a). To see how PSTD works, consider an x-directed cut through this grid. At every sample point along this cut, we wish to compute the x-derivatives of the general field component V. Let $\{V_i\}$ denote the set of initially known values of V at all grid-points along the observation cut, and let $\{(\partial V/\partial x)_i\}$ denote the set of initially unknown x-derivatives of V at the same grid-points. Then, using the differentiation theorem for Fourier transforms, we can write:

$$\left\{ \left. \frac{\partial V}{\partial x} \right|_i \right\} = -\mathcal{F}^{-1}\left(j\tilde{k}_x \mathcal{F}\{V_i\} \right), \tag{7.18}$$

where $\mathcal{F}$ and $\mathcal{F}^{-1}$ denote respectively the forward and inverse DFTs, and $\tilde{k}_x$ is the Fourier transform variable representing the x-component of the numerical wavevector. In this manner, the entire set of spatial derivatives of V along the observation cut can be calculated in one step. In multiple dimensions, this process is repeated for each observation cut parallel to the major axes of the space lattice.

According to the Nyquist sampling theorem, the representation in (7.18) is *exact* for $|\tilde{k}_x| \leqslant \pi/\Delta x$, i.e., $\Delta x \leqslant \tilde{\lambda}/2$. Thus, the spatial-differencing process here can be said to be of "infinite order" for grid-sampling densities of two or more points per wavelength. The wraparound effect, a potentially major limitation caused by the periodicity assumed in the FFT, is eliminated by using the perfectly matched layer absorbing boundary condition (to be discussed in Section 10). Finally, the time-differencing for PSTD uses conventional second-order-accurate Yee leapfrogging.

LIU, Q.H. [1996], LIU, Q.H. [1997], derived the following expressions for the wavenumber and phase velocity of a sinusoidal numerical wave of temporal period $T = 2\pi/\omega$ propagating in an arbitrary direction within a three-dimensional PSTD space lattice:

$$|\vec{\tilde{k}}| = \frac{2}{c\Delta t} \sin\left(\frac{\omega\Delta t}{2}\right), \tag{7.19}$$

$$\tilde{v}_p = \frac{\omega}{\tilde{k}} = \frac{\omega}{\frac{2}{c\Delta t}\sin\left(\frac{\omega\Delta t}{2}\right)} = \frac{\omega\Delta t/2}{\sin(\omega\Delta t/2)}c. \tag{7.20}$$

Eq. (7.20) implies that the numerical phase velocity is *independent* of the propagation direction of the wave, unlike any of the methods considered previously. Applying our definitions of numerical phase-velocity error, we therefore have the following figures of merit for the PSTD method:

$$\Delta\tilde{v}_{\text{physical}}|_{\text{PSTD}} = \left[\frac{\omega\Delta t/2}{\sin(\omega\Delta t/2)} - 1\right] \times 100\% = \left[\frac{\pi/N_T}{\sin(\pi/N_T)} - 1\right] \times 100\%, \tag{7.21}$$

$$\Delta\tilde{v}_{\text{aniso}}|_{\text{PSTD}} = 0, \tag{7.22}$$

where we define the temporal sampling density $N_T = T/\Delta t$ time samples per wave-oscillation period.

7.5.2. Discussion

Remarkably, $\Delta\tilde{v}_{\text{aniso}}$ is zero for the PSTD method for *all* propagating sinusoidal waves sampled at $N_\lambda \geqslant 2$. Therefore, to specify the gridding density of the PSTD simulation, we need only a reliable estimate of λ_{min}, the fastest oscillating spectral component of significance. This estimate is based upon the wavelength spectrum of the exciting pulse and the size of significant structural details such as material inhomogeneities. Then, the space-cell dimension is set at $\Delta = \lambda_{\text{min}}/2$, regardless of the problem's overall electrical size. This is because our choice of Δ assures zero $\Delta\tilde{v}_{\text{aniso}}$ error, and thus, zero accumulation of this error even if the number of space cells increases without bound. Consequently, we conclude that:

- The density of the PSTD mesh-sampling is independent of the electrical size of the modeling problem.

However, the fact that $\Delta\tilde{v}_{\text{aniso}} = 0$ does *not* mean that PSTD yields perfect results. In fact, (7.21) shows that there remains a numerical phase-velocity error relative to c. This residual velocity error is not a function of the wave-propagation direction ϕ, and is therefore isotropic within the space grid. The residual velocity error arises from the Yee-type leapfrog time-stepping used in the algorithm, and is a function only of N_T. Table 7.1 provides representative values of this residual velocity error.

The key point from Table 7.1 is that N_T limits the accuracy of the PSTD technique when modeling impulsive propagation. This is not an issue for a monochromatic wave where there is only a single value of N_T, and $\Delta\tilde{v}_{\text{physical}}$ can be nulled in the manner of Strategy 1. For a pulse, however, with Δt a fixed algorithm parameter, there exists a spread of equivalent N_T values for the spectral components of the pulse which possess a range of temporal periods T. This causes a spread of $\tilde{v}_p$ over the pulse spectrum, which

TABLE 7.1
Residual numerical phase-velocity error of the PSTD method versus its time-sampling

N_T	$\Delta \tilde{v}_{\text{physical}}$	N_T	$\Delta \tilde{v}_{\text{physical}}$
2	+57%	15	+0.73%
4	+11%	20	+0.41%
8	+2.6%	25	+0.26%
10	+1.7%	30	+0.18%

in turn results in an isotropic progressive broadening and distortion of the pulse wave-form as it propagates in the grid. To bound such dispersion, it is important to choose Δt small enough so that it adequately resolves the period T_{min} of the fastest oscillating spectral component λ_{min}. Because this dispersion is cumulative with the wave-propagation distance, we have a second key point:

- The density of the PSTD time sampling must increase with the electrical size of the modeling problem if we apply a fixed upper bound on the maximum total phase error of propagating waves within the mesh.

Despite this need for a small Δt, PSTD can provide a large reduction in computer resources relative to the Yee algorithm for electrically large problems not having spatial details or material inhomogeneities smaller than $\lambda_{\text{min}}/2$. Increased efficiency is expected even relative to the fourth-order-accurate spatial algorithms reviewed previously. Liu, Q.H. [1996], Liu, Q.H. [1997] reported that, within the range of problem sizes from 16–64 wavelengths, the use of PSTD permits an $8^D : 1$ reduction in computer storage and running time relative to the Yee algorithm, where D is the problem dimensionality. While this savings is comparable to that shown for the fourth-order spatial techniques, we expect the PSTD advantage to increase for even larger problems. In fact, the computational benefit of PSTD theoretically increases without limit as the electrical size of the modeling problem expands.

The second topic in our discussion is whether PSTD's global calculation of space derivatives along observation cuts through the lattice (similar to the Ty method) has difficulties at material interfaces. An initial concern is that PSTD might yield nonphysical results for problems having abrupt jumps in ε unless an ε-smoothing technique such as (7.11) is used. However, this is not the case. As reported by Liu, Q.H. [1996], Liu, Q.H. [1997], the PSTD method is successful for dielectric interfaces because, at such discontinuities, the normal derivatives that it calculates via DFT or FFT are implemented on continuous tangential-field data across the interface.

However, structures having metal surfaces comprise a very important set of problems where the required continuity of tangential fields within the PSTD space lattice is effectively violated. Depending upon the orientation and thickness of the metal surfaces, tangential-field discontinuities could appear for two reasons:

(1) A space-cell boundary lies at a metal surface or within a metal layer. The tangential H-field within the space cell drops abruptly to zero at the metal surface, and remains at zero for the remainder of the space cell.

(2) A metal sheet splits the space lattice so there exist distinct lit and shadow regions within the lattice. Here, the tangential H-field on the far (shadowed or shielded)

side of the metal sheet may be physically isolated from the field immediately across the metal sheet on its near (lit) side. Gross error is caused by the global nature of PSTD's spatial-derivative calculation which nonphysically transports field information directly across the shielding metal barrier from the lit to the shadow sides.

Until these problems are solved by the prescription of special boundary conditions, PSTD will likely find its primary applications in Cartesian-mesh modeling of structures comprised entirely of dielectrics. We note that the PSTD usage of collocated field components simplifies rendering such structures within the mesh. It is also ideal for modeling nonlinear optical problems where the local index of refraction is dependent upon a power of the magnitude of the local $\vec{E}$. Here, collocation of the vector components of $\vec{E}$ avoids the need for error-causing spatial interpolations of nearby electric field vector components staggered in space.

8. Numerical stability

8.1. Introduction

In Section 6, we saw that the choice of Δ and Δt can affect the propagation characteristics of numerical waves in the Yee space lattice, and therefore the numerical error. In this section, we show that, in addition, Δt must be bounded to ensure numerical stability. Our approach to determine the upper bound on Δt is based upon the complex-frequency analysis reported by TAFLOVE and HAGNESS [2000, pp. 133–140]. As noted there, the complex-frequency approach is conceptually simple, yet rigorous. It also allows straightforward estimates of the exponential growth rate of unstable numerical solutions.

Subsequently, Section 9 will review a representative approach of a new class of algorithms proposed to eliminate the need to bound Δt. This class replaces Yee's leapfrog time-stepping with an implicit alternating-direction technique.

8.2. Complex-frequency analysis

We first postulate a sinusoidal traveling wave present in the three-dimensional FDTD space lattice and discretely sampled at (x_I, y_J, z_K, t_n), allowing for the possibility of a complex-valued numerical angular frequency, $\tilde{\omega} = \tilde{\omega}_{\text{real}} + j\tilde{\omega}_{\text{imag}}$. A field vector in this wave can be written as

$$
\begin{aligned}
\vec{V}\big|_{I,J,K}^{n} &= \vec{V}_0 e^{j[(\tilde{\omega}_{\text{real}}+j\tilde{\omega}_{\text{imag}})n\Delta t - \tilde{k}_x I\Delta x - \tilde{k}_y J\Delta y - \tilde{k}_z K\Delta z]} \\
&= \vec{V}_0 e^{-\tilde{\omega}_{\text{imag}}n\Delta t} e^{j(\tilde{\omega}_{\text{real}}n\Delta t - \tilde{k}_x I\Delta x - \tilde{k}_y J\Delta y - \tilde{k}_z K\Delta z)},
\end{aligned}
\tag{8.1}
$$

where $\tilde{k}$ is the wavenumber of the numerical sinusoidal traveling wave. We note that (8.1) permits either a constant wave amplitude with time ($\tilde{\omega}_{\text{imag}} = 0$), an exponentially decreasing amplitude with time ($\tilde{\omega}_{\text{imag}} > 0$), or an exponentially increasing amplitude with time ($\tilde{\omega}_{\text{imag}} < 0$).

Given this basis, we proceed to analyze numerical dispersion relation (6.10) allowing for a complex-valued angular frequency:

$$\left[\frac{1}{c\Delta t}\sin\left(\frac{\tilde{\omega}\Delta t}{2}\right)\right]^2 = \left[\frac{1}{\Delta x}\sin\left(\frac{\tilde{k}_x\Delta x}{2}\right)\right]^2 + \left[\frac{1}{\Delta y}\sin\left(\frac{\tilde{k}_y\Delta y}{2}\right)\right]^2$$
$$+ \left[\frac{1}{\Delta z}\sin\left(\frac{\tilde{k}_z\Delta z}{2}\right)\right]^2. \tag{8.2}$$

We first solve (8.2) for $\tilde{\omega}$. This yields

$$\tilde{\omega} = \frac{2}{\Delta t}\sin^{-1}(\xi), \tag{8.3}$$

where

$$\xi = c\Delta t\sqrt{\frac{1}{(\Delta x)^2}\sin^2\left(\frac{\tilde{k}_x\Delta x}{2}\right) + \frac{1}{(\Delta y)^2}\sin^2\left(\frac{\tilde{k}_y\Delta y}{2}\right) + \frac{1}{(\Delta z)^2}\sin^2\left(\frac{\tilde{k}_z\Delta z}{2}\right)}. \tag{8.4}$$

We observe from (8.4) that

$$0 \leqslant \xi \leqslant c\Delta t\sqrt{\frac{1}{(\Delta x)^2} + \frac{1}{(\Delta y)^2} + \frac{1}{(\Delta z)^2}} \equiv \xi_{\text{upper bound}} \tag{8.5}$$

for all possible real values of $\tilde{k}$, that is, those numerical waves having zero exponential attenuation per grid space cell. $\xi_{\text{upper bound}}$ is obtained when each $\sin^2$ term under the square root of (8.4) simultaneously reaches a value of 1. This occurs for the propagating numerical wave having the wavevector components

$$\tilde{k}_x = \pm\frac{\pi}{\Delta x}; \tag{8.6a}$$

$$\tilde{k}_y = \pm\frac{\pi}{\Delta y}; \tag{8.6b}$$

$$\tilde{k}_z = \pm\frac{\pi}{\Delta z}. \tag{8.6c}$$

It is clear that $\xi_{\text{upper bound}}$ can exceed 1 depending upon the choice of Δt. This yields complex values of $\sin^{-1}(\xi)$ in (8.3), and therefore complex values of $\tilde{\omega}$. To investigate further, we divide the range of ξ given in (8.5) into two subranges.

8.2.1. Stable range: $0 \leqslant \xi \leqslant 1$
Here, $\sin^{-1}(\xi)$ is real-valued and hence, real values of $\tilde{\omega}$ are obtained in (8.3). With $\tilde{\omega}_{\text{imag}} = 0$, (8.1) yields a constant wave amplitude with time.

8.2.2. Unstable range: $1 < \xi < \xi_{\text{upper bound}}$
This subrange exists only if

$$\xi_{\text{upper bound}} = c\Delta t\sqrt{\frac{1}{(\Delta x)^2} + \frac{1}{(\Delta y)^2} + \frac{1}{(\Delta z)^2}} > 1. \tag{8.7}$$

The unstable range is defined in an equivalent manner by

$$\Delta t > \frac{1}{c\sqrt{\frac{1}{(\Delta x)^2} + \frac{1}{(\Delta y)^2} + \frac{1}{(\Delta z)^2}}} \equiv \Delta t_{\substack{\text{stable} \\ \text{limit-3D}}} . \tag{8.8}$$

To prove the claim of instability for the range $\xi > 1$, we substitute the complex-valued $\sin^{-1}(\xi)$ function of (6.34) into (8.3) and solve for $\tilde{\omega}$. This yields

$$\tilde{\omega} = \frac{-j^2}{\Delta t} \ln\left(jxi + \sqrt{1 - \xi^2}\right). \tag{8.9}$$

Upon taking the natural logarithm, we obtain

$$\tilde{\omega}_{\text{real}} = \frac{\pi}{\Delta t}; \qquad \tilde{\omega}_{\text{imag}} = -\frac{2}{\Delta t} \ln\left(\xi + \sqrt{\xi^2 - 1}\right). \tag{8.10}$$

Substituting (8.10) into (8.1) yields

$$\begin{aligned}
\vec{V}\big|_{I,J,K}^{n} &= \vec{V}_0 e^{2n \ln(\xi + \sqrt{\xi^2 - 1})} e^{j[(\pi/\Delta t)(n\Delta t) - \tilde{k}_x I \Delta x - \tilde{k}_y J \Delta y - \tilde{k}_z K \Delta z]} \\
&= \vec{V}_0 \left(\xi + \sqrt{\xi^2 - 1}\right)^{**2n} e^{j[(\pi/\Delta t)(n\Delta t) - \tilde{k}_x I \Delta x - \tilde{k}_y J \Delta y - \tilde{k}_z K \Delta z]},
\end{aligned} \tag{8.11}$$

where **$2n$ denotes the $2n$th power. From (8.11), we define the following multiplicative factor greater than 1 that amplifies the numerical wave every time step:

$$q_{\text{growth}} \equiv \left(\xi + \sqrt{\xi^2 - 1}\right)^2. \tag{8.12}$$

Eqs. (8.11) and (8.12) define an exponential growth of the numerical wave with time-step number n. We see that the dominant exponential growth occurs for the most positive possible value of ξ, i.e., $\xi_{\text{upper bound}}$ defined in (8.5).

8.2.3. Example of calculating a stability bound: 3D cubic-cell lattice

Consider the practical case of a three-dimensional cubic-cell space lattice with $\Delta x = \Delta y = \Delta z = \Delta$. From (8.8), numerical instability arises when

$$\Delta t > \frac{1}{c\sqrt{\frac{1}{(\Delta)^2} + \frac{1}{(\Delta)^2} + \frac{1}{(\Delta)^2}}} = \frac{1}{c\sqrt{\frac{3}{(\Delta)^2}}} = \frac{\Delta}{c\sqrt{3}}. \tag{8.13}$$

We define an equivalent Courant stability limit for the cubic-cell lattice case:

$$S_{\substack{\text{stability} \\ \text{limit-3D}}} = \frac{1}{\sqrt{3}}. \tag{8.14}$$

From (8.6), the dominant exponential growth is seen to occur for numerical waves propagating along the lattice diagonals. The relevant wavevectors are

$$\tilde{k} = \frac{\pi}{\Delta}(\pm\hat{x} \pm \hat{y} \pm \hat{z}) \rightarrow |\tilde{k}| = \frac{\pi\sqrt{3}}{\Delta} \rightarrow \tilde{\lambda} = \left(\frac{2\sqrt{3}}{3}\right)\Delta, \tag{8.15}$$

where $\hat{x}$, $\hat{y}$, and $\hat{z}$ are unit vectors defining the major lattice axes. Further, (8.5) yields

$$\xi_{\substack{\text{upper} \\ \text{bound}}} = c\Delta t \sqrt{\frac{1}{(\Delta)^2} + \frac{1}{(\Delta)^2} + \frac{1}{(\Delta)^2}} = \left(\frac{c\Delta t}{\Delta}\right)\sqrt{3} = S\sqrt{3}. \tag{8.16}$$

From (8.12), this implies the following maximum possible growth factor per time step under conditions of numerical instability:

$$q_{\text{growth}} \equiv \left[S\sqrt{3} + \sqrt{\left(S\sqrt{3} \right)^2 - 1} \right]^2 \quad \text{for } S \geqslant \frac{1}{\sqrt{3}}. \tag{8.17}$$

8.2.4. Courant factor normalization and extension to 2D and 1D grids
It is instructive to use the results of (8.14) to normalize the Courant factor S in (8.17). This will permit us to generalize the three-dimensional results for the maximum growth factor q to two-dimensional and one-dimensional Yee grids. In this spirit, we define

$$S_{\text{norm-3D}} \equiv \frac{S}{\underset{\text{limit-3D}}{S_{\text{stability}}}} = \frac{S}{(1/\sqrt{3})} = S\sqrt{3}. \tag{8.18}$$

Then, (8.17) can be written as

$$q_{\text{growth}} = \left[S_{\text{norm-3D}} + \sqrt{(S_{\text{norm-3D}})^2 - 1} \right]^2 \quad \text{for } S_{\text{norm-3D}} \geqslant 1. \tag{8.19}$$

Given this notation, it can be shown that analogous expressions for the Courant stability limit and the growth-factor under conditions of numerical instability are given by:

Two-dimensional square Yee grid:

$$\underset{\text{limit-2D}}{S_{\text{stability}}} = \frac{1}{\sqrt{2}}, \tag{8.20}$$

$$S_{\text{norm-2D}} \equiv \frac{S}{\underset{\text{limit-2D}}{S_{\text{stability}}}} = \frac{S}{(1/\sqrt{2})} = S\sqrt{2}. \tag{8.21}$$

Here, dominant exponential growth occurs for numerical waves propagating along the grid diagonals. The relevant wavevectors are

$$\tilde{\vec{k}} = \frac{\pi}{\Delta}(\pm\hat{x} \pm \hat{y}) \rightarrow |\tilde{\vec{k}}| = \frac{\pi\sqrt{2}}{\Delta} \rightarrow \tilde{\lambda} = \sqrt{2}\,\Delta. \tag{8.22}$$

This yields the following solution growth factor per time step:

$$q_{\text{growth}} = \left[S_{\text{norm-2D}} + \sqrt{(S_{\text{norm-2D}})^2 - 1} \right]^2 \quad \text{for } S_{\text{norm-2D}} \geqslant 1. \tag{8.23}$$

One-dimensional uniform Yee grid:

$$\underset{\text{limit-1D}}{S_{\text{stability}}} = 1, \tag{8.24}$$

$$S_{\text{norm-1D}} \equiv \frac{S}{\underset{\text{limit-1D}}{S_{\text{stability}}}} = \frac{S}{1} = S. \tag{8.25}$$

Dominant exponential growth occurs for the wavevectors

$$\tilde{\vec{k}} = \pm\frac{\pi}{\Delta}\hat{x} \rightarrow |\tilde{\vec{k}}| = \frac{\pi}{\Delta} \rightarrow \tilde{\lambda} = 2\Delta. \tag{8.26}$$

This yields the following solution growth factor per time step:

$$q_{\text{growth}} = \left(S + \sqrt{S^2 - 1}\right)^2 \quad \text{for } S \geqslant 1. \tag{8.27}$$

We see from the above discussion that the solution growth factor q under conditions of numerical instability is the same, regardless of the dimensionality of the FDTD space lattice, if the same normalized Courant factor is used. A normalized Courant factor equal to one yields no exponential solution growth for any dimensionality grid. However, a normalized Courant factor only 0.05% larger, i.e.,

$S = 1.0005$ for a uniform, one-dimensional grid;

$S = 1.0005 \times (1/\sqrt{2}) = 0.707460$ for a uniform, square, two-dimensional grid;

$S = 1.0005 \times (1/\sqrt{3}) = 0.577639$ for a uniform, cubic, three-dimensional grid

yields a multiplicative solution growth of 1.0653 every time step for each dimensionality grid. This is equivalent to a solution growth of 1.8822 every 10 time steps, 558.7 every 100 time steps, and 2.96×10^{27} every 1000 time steps.

8.3. Examples of calculations involving numerical instability in a 1D grid

We first consider an example of the beginning of a numerical instability arising because the Courant stability condition is violated equally at *every* point in a uniform one-dimensional grid. Fig. 8.1(a) graphs three snapshots of the free-space propagation of a Gaussian pulse within a grid having the Courant factor $S = 1.0005$. The exciting pulse waveform has a $40\Delta t$ temporal width between its $1/e$ points, and reaches its peak value of 1.0 at time step $n = 60$. Graphs of the wavefunction $u(i)$ versus the grid coordinate i are shown at time steps $n = 200$, $n = 210$, and $n = 220$. We see that the trailing edge of the Gaussian pulse is contaminated by a rapidly oscillating and growing noise component that does not exist in Fig. 6.6(a), which shows the same Gaussian pulse at the same time but with $S \leqslant 1.0$. In fact, the noise component in Fig. 8.1(a) results from the onset of numerical instability within the grid due to $S = 1.0005 > 1.0$. Because this noise grows exponentially with time-step number n, it quickly overwhelms the desired numerical results for the propagating Gaussian pulse. Shortly thereafter, the exponential growth of the noise increases the calculated field values beyond the dynamic range of the computer being used, resulting in run-time floating-point overflows and errors.

Fig. 8.1(b) is an expanded view of Fig. 8.1(a) between grid points $i = 1$ and $i = 20$, showing a segment of the numerical noise on the trailing edge of the Gaussian pulse. We see that the noise oscillates with a spatial period of 2 grid cells, i.e., $\tilde{\lambda} = 2\Delta x$, in accordance with (8.26). In addition, upon analyzing the raw data underlying Fig. 8.1(b), it is observed that the growth factor q is in the range 1.058–1.072 per time step. This compares favorably with the theoretical value of 1.0653 determined using (8.27).

We next consider an example of the beginning of a numerical instability arising because the Courant stability condition is violated at only a *single* point in a uniform one-dimensional grid. Fig. 8.2(a) graphs two snapshots of the free-space propagation of a narrow Gaussian pulse within a grid having the Courant factor $S = 1.0$ at all points except at $i = 90$, where $S = 1.2075$. The exciting pulse has a $10\Delta t$ temporal width between its $1/e$ points, and reaches its peak value of 1.0 at time step $n = 60$. Graphs of

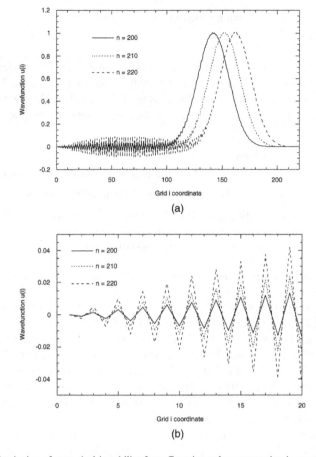

FIG. 8.1. The beginning of numerical instability for a Gaussian pulse propagating in a uniform, free-space 1D grid. The Courant factor is $S = 1.0005$ at each grid point. (a) Comparison of calculated pulse propagation at $n = 200, 210$, and 220 time steps over grid coordinates $i = 1$ through $i = 220$. (b) Expanded view of (a) over grid coordinates $i = 1$ through $i = 20$.

the wavefunction $u(i)$ versus the grid coordinate i are shown at time-steps $n = 190$ and $n = 200$. In contrast to Fig. 8.1(a), the rapidly oscillating and growing noise component due to numerical instability originates at just a single grid point along the trailing edge of the Gaussian pulse ($i = 90$) where S exceeds 1.0, rather than along the entirety of the trailing edge. Despite this localization of the source of the instability, the noise again grows exponentially with time step number n. In this case, the noise propagates symmetrically in both directions from the unstable point. Ultimately, the noise again fills the entire grid, overwhelms the desired numerical results for the propagating Gaussian pulse, and causes run-time floating-point overflows.

Fig. 8.2(b) is an expanded view of Fig. 8.2(a) between grid points $i = 70$ and $i = 110$, showing how the calculated noise due to the numerical instability originates at grid point $i = 90$. Again, in accordance with (8.26), the noise oscillates with a spatial

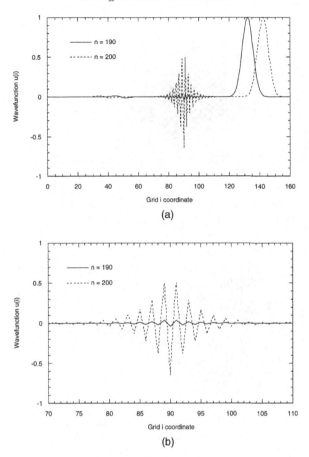

Fig. 8.2. The beginning of numerical instability for a Gaussian pulse propagating in a uniform, free-space 1D grid. The Courant factor is $S = 1$ at all grid points but $i = 90$, where $S = 1.2075$. (a) Comparison of calculated pulse propagation at $n = 190$ and $n = 200$ time steps over grid coordinates $i = 1$ through $i = 160$. (b) Expanded view of (a) over grid coordinates $i = 70$ through $i = 110$.

period of 2 grid cells, i.e., $\tilde{\lambda} = 2\Delta x$. However, the rate of exponential growth here is much less than that predicted by (8.27), wherein *all* grid points were assumed to violate Courant stability. Upon analyzing the raw data underlying Fig. 8.2(b), a growth factor of $q \cong 1.31$ is observed per time step. This compares to $q \cong 3.55$ per time step determined by substituting $S = 1.2075$ into (8.27). Thus, it is clear that a grid having one or just a few localized points of numerical instability can "blow up" much more slowly than a uniformly unstable grid having a comparable or even smaller Courant factor S.

8.4. *Example of calculation involving numerical instability in a 2D grid*

We next consider an FDTD modeling example where the Courant stability condition is violated equally at every point in a uniform two-dimensional TM_z grid. To allow

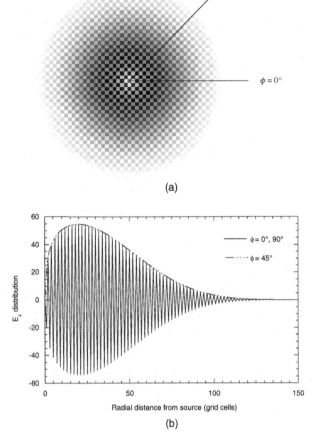

FIG. 8.3. Effect of numerical instability upon a two-dimensional pulse-propagation model. (a) Visualization of the two-dimensional E_z distribution at $n = 40$ for $S = 1.005 \times (1/\sqrt{2})$. (b) E_z distributions along the grid axes and grid diagonal at $n = 200$ for $S = 1.0005 \times (1/\sqrt{2})$. The theoretical and measured growth factor is
$q_{\text{growth}} \cong 1.065$ per time step.

direct comparison with a previous example of stable pulse propagation, the same grid discussed in Section 6.6.5 and Fig. 6.7 is used. The overall grid size is again 360×360 square cells with $\Delta x = \Delta y = 1.0 \equiv \Delta$. Numerical excitation to the grid is again provided by specifying a unit-step time-function for the center E_z component. The only condition that differs from those assumed in Section 6.6.5 is that the Courant stability factor S is increased just above the threshold for numerical instability given by (8.20).

Fig. 8.3(a) visualizes the two-dimensional E_z distribution at $n = 40$ time steps for $S = 1.005 \times (1/\sqrt{2})$. This value of S quickly generates a region of numerical instability spreading out radially from the source, where the field amplitudes are large enough to mask the normal wave propagation. This permits a high-resolution visualization of individual E_z components in the grid which are depicted as square pixels. We see that the unstable field pattern has the form of a checkerboard wherein the dark and gray pixels denote positive and negative E_z field values, respectively. Here, the pixel saturation denotes the relative amplitude of its positive or negative value.

Fig. 8.3(b) graphs the variation of E_z versus radial distance from the source at $n = 200$ time steps for $S = 1.0005 \times (1/\sqrt{2})$. Two distinct plots are shown. The solid line graph exhibits a rapid spatial oscillation with the period 2Δ. This is the E_z behavior along the $\phi = 0°, 90°$ (and similar on-axis) cuts through the grid. The smooth dashed-dotted curve with no spatial oscillation represents the E_z behavior along the $\phi = 45°$ (and similar oblique) cuts through the grid. Analysis of the underlying data reveals growth factors in the range 1.060 to 1.069 per time step along the leading edge of the instability region. This agrees very well with $q_{\text{growth}} = 1.0653$ calculated using (8.23), and is an excellent validation of the Courant-factor-normalization theory.

An interesting observation in Fig. 8.3(b) is that the smooth E_z variation along $\phi = 45°$ forms the envelope of the oscillatory E_z distribution observed along the grid's major axes. This difference in behavior is confirmed in Fig. 8.3(a), which shows that the $\phi = 45°$ cut lies entirely within a diagonal string of dark (positive) pixels, whereas the $\phi = 0°$ cut passes through alternating dark (positive) and gray (negative) pixels. We attribute this behavior to (8.22), which states that the exponential growth along the grid diagonal has $\tilde{\lambda} = \sqrt{2}\,\Delta$. That is, the numerical wavelength along the 45° observation cut for the unstable mode is exactly the diagonal length across one $\Delta \times \Delta$ grid cell. Thus, there exists 2π (or equivalently, 0) phase shift of the unstable mode between adjacent observation points along the $\phi = 45°$ cut. This means that adjacent E_z values along $\phi = 45°$ cannot change sign. In contrast, (8.22) reduces to $\tilde{k} = \pi/\Delta$, i.e., $\tilde{\lambda} = 2\Delta$, for the unstable mode along the $\phi = 0°, 90°$ cuts. Therefore, there is π phase shift of the unstable mode between adjacent observation points along $\phi = 0°, 90°$; yielding the point-by-point sign reversals (rapid spatial oscillations) seen in Fig. 8.3(b).

8.5. *Linear instability when the normalized Courant factor equals 1*

The general field vector postulated in (8.1) permits a numerical wave amplitude that is either constant, exponentially growing, or exponentially decaying as time-stepping progresses. Recently, MIN and TENG [2001] have identified a linear growth mode, i.e., a linear instability, that can occur if the normalized Courant factor equals exactly 1. While this growth mode is much slower than the exponential instability discussed previously, the analyst should proceed with caution when using $S_{\text{norm}} = 1$.

8.6. *Generalized stability problem*

The previous discussion focused on the numerical stability of the Yee algorithm. However, the stability of the entire FDTD procedure depends upon more than this. In fact, a *generalized stability problem* arises due to interactions between the Yee algorithm and any augmenting algorithms used to model boundary conditions, variable meshing, and lossy, dispersive, nonlinear, and gain materials. Factors involved in the generalized stability problem are now reviewed.

8.6.1. Boundary conditions

Numerical realizations of electromagnetic field boundary conditions that require the processing of field data located nonlocally in space or time can lead to instability of the overall time-stepping algorithm. An important example of this possibility arises when implementing *absorbing boundary conditions* (ABCs) at the outermost space-lattice planes to simulate the extension of the lattice to infinity for modeling scattering or radiation phenomena in unbounded regions. ABCs have been the subject of much research since the 1970s, with several distinct physics modeling approaches and numerical implementations emphasized by the FDTD research community.

The nature of the numerical stability problem here is exemplified by one of the most popular ABCs of the early 1990s, that reported by LIAO, WONG, YANG, and YUAN [1984]. This ABC implements a polynomial extrapolation of field data at interior grid points and past time steps to the desired outer-boundary grid point at the latest time step. However, the Liao ABC was found by later workers to be marginally stable. It requires double-precision computer arithmetic and/or perturbation of its algorithm coefficients away from the theoretical optimum to ensure numerical stability during prolonged time stepping. Similar issues had previously arisen with regard to the ABCs of ENGQUIST and MAJDA [1977] and HIGDON [1986]. More recently, the perfectly matched layer ABC of BERENGER [1994] has come under scrutiny for potential numerical instability.

Overall, operational experience with a wide variety of ABCs has shown that numerical stability can be maintained for many thousands of iterations, if not indefinitely, with the proper choice of time step. A similar experience base has been established for the numerical stability of a variety of impedance boundary conditions.

8.6.2. Variable and unstructured meshing

The analysis of numerical instability can become complicated when the FDTD space lattice is generated to conformally fit a specific structure by varying the size, position, and shape of the lattice cells, rather than using the uniform "bricks" postulated by Yee. Groups working in this area have found that even if the mesh construction is so complex that an exact stability criterion cannot be derived, a part-analytical/part-empirical upper bound on the time step can be derived for each gridding approach so that numerical stability is maintained for many thousands of time steps, if not indefinitely. This has permitted numerous successful engineering applications for non-Cartesian and unstructured FDTD meshes.

8.6.3. Lossy, dispersive, nonlinear, and gain materials

Much literature has emerged concerning FDTD modeling of dispersive and nonlinear materials. For linear-dispersion algorithms, it is usually possible to derive precise bounds on numerical stability. However, stability analysis may not be feasible for dispersion models of nonlinear materials. Fortunately, substantial modeling experience has shown that numerical stability can be maintained for thousands of time steps, if not indefinitely, for linear, nonlinear, and gain materials with a properly chosen time step. Again, this has permitted numerous successful engineering applications.

9. Alternating-direction-implicit time-stepping algorithm for operation beyond the Courant limit

9.1. Introduction

Section 8 showed that numerical stability of the Yee algorithm requires placing an upper bound on the time step Δt relative to the space increments Δx, Δy, and Δz. This has allowed the successful application of FDTD methods to a wide variety of electromagnetic wave modeling problems of moderate electrical size and quality factor. Typically, such problems require 10^3–10^4 time-steps to complete a single simulation.

However, there are important potential applications of FDTD modeling where the Courant stability bounds determined in Sections 8.2.3 and 8.2.4 are much too restrictive. Modeling problems that fall into this regime have the following properties:
- The cell size Δ needed to resolve the fine-scale geometric detail of the electromagnetic wave interaction structure is much less than the shortest wavelength λ_{min} of a significant spectral component of the source.
- The simulated time T_{sim} needed to evolve the electromagnetic wave physics to the desired endpoint is related to the cycle time T of λ_{min}.

With Δ fixed by the need to resolve the problem geometry, the requirement for numerical stability in turn specifies the maximum possible time step Δt_{max}. This, in turn, fixes the total number of time steps needed to complete the simulation, $N_{sim} = T_{sim}/\Delta t_{max}$. Table 9.1 lists parameters of two important classes of problems where this decision process results in values of N_{sim} that are so large that standard FDTD modeling in three dimensions is difficult, or even impossible.

If these classes of electromagnetics problems are to be explored using FDTD modeling, we need an advancement of FDTD techniques that permits accurate and numerically stable operation for values of Δt exceeding the Courant limit by much more than 10:1. A candidate, computationally efficient approach for this purpose is to use an *alternating-direction-implicit* (ADI) time-stepping algorithm rather than the usual explicit Yee leapfrogging. In fact, work with ADI FDTD methods in the early 1980s by HOLLAND [1984] and HOLLAND and CHO [1986] achieved promising results for two-dimensional models. However, using these early ADI techniques, it proved difficult to demonstrate numerical stability for the general three-dimensional case, and research in this area was largely discontinued.

Recently, key publications by ZHENG, CHEN, and ZHANG [2000] and NAMIKI [2000] have reported the development of unconditionally stable three-dimensional ADI

TABLE 9.1

Two important classes of three-dimensional FDTD modeling problems made difficult or impossible by the Courant limit on Δt

Problem class	Δ	T_{sim}	Δt_{max}	N_{sim}
Propagation of bioelectric signals	~ 1 mm	~ 100 ms	~ 2 ps	$\sim 5 \times 10^{10}$
Propagation of digital logic signals	$\sim 0.25\,\mu$m	~ 1 ns	~ 0.5 fs	$\sim 2 \times 10^6$

FDTD algorithms. This section reviews the work by ZHENG, CHEN, and ZHANG [2000] wherein for the first time unconditional numerical stability is derived for the full three-dimensional case.

Note that, with any unconditionally stable ADI FDTD algorithm, the upper bound on Δt is relaxed to only that value needed to provide good accuracy in numerically approximating the time derivatives of the electromagnetic field. Thus, in theory, Δt need only be small enough to provide about 20 or more field samples during the cycle time T of the fastest oscillating significant spectral component of the exciting source. For example, in Table 9.1, Δt could be $10\,\mu s$ rather than 2 ps for studies of signal propagation within human muscles, yielding $N_{sim} = 10^4$ rather than $N_{sim} = 5 \times 10^{10}$.

9.2. Formulation of the Zheng et al. algorithm

ZHENG, CHEN, and ZHANG [2000] reported a new ADI time-stepping algorithm for FDTD that has theoretical unconditional numerical stability for the general three-dimensional case. While this technique uses the same Yee space lattice as conventional FDTD, the six field-vector components are collocated rather than staggered in time. In discussing the formulation of this algorithm, we assume that all of the field components are known everywhere in the lattice at time step n and stored in the computer memory.

9.2.1. Unsimplified system of time-stepping equations

The ADI nature of the Zheng et al. algorithm can be best understood by first considering its unsimplified form, and then proceeding to obtain the final simplified system of field update equations. To advance a single time step from n to $n + 1$, we perform two subiterations: the first from n to $n + 1/2$, and the second from $n + 1/2$ to $n + 1$. These subiterations are as follows.

Subiteration 1. Advance the 6 field components from time step n to time step $n + 1/2$

$$E_x\big|_{i+1/2,j,k}^{n+1/2} = E_x\big|_{i+1/2,j,k}^{n} + \frac{\Delta t}{2\varepsilon\Delta y}\Big(H_z\big|_{i+1/2,j+1/2,k}^{n+1/2} - H_z\big|_{i+1/2,j-1/2,k}^{n+1/2}\Big)$$
$$- \frac{\Delta t}{2\varepsilon\Delta z}\Big(H_y\big|_{i+1/2,j,k+1/2}^{n} - H_y\big|_{i+1/2,j,k-1/2}^{n}\Big), \tag{9.1a}$$

$$E_y\big|_{i,j+1/2,k}^{n+1/2} = E_y\big|_{i,j+1/2,k}^{n} + \frac{\Delta t}{2\varepsilon\Delta z}\Big(H_x\big|_{i,j+1/2,k+1/2}^{n+1/2} - H_x\big|_{i,j+1/2,k-1/2}^{n+1/2}\Big)$$
$$- \frac{\Delta t}{2\varepsilon\Delta x}\Big(H_z\big|_{i+1/2,j+1/2,k}^{n} - H_z\big|_{i-1/2,j+1/2,k}^{n}\Big), \tag{9.1b}$$

$$E_z\big|_{i,j,k+1/2}^{n+1/2} = E_z\big|_{i,j,k+1/2}^{n} + \frac{\Delta t}{2\varepsilon\Delta x}\Big(H_y\big|_{i+1/2,j,k+1/2}^{n+1/2} - H_y\big|_{i-1/2,j,k+1/2}^{n+1/2}\Big)$$
$$- \frac{\Delta t}{2\varepsilon\Delta y}\Big(H_x\big|_{i,j+1/2,k+1/2}^{n} - H_x\big|_{i,j-1/2,k+1/2}^{n}\Big), \tag{9.1c}$$

$$H_x\big|_{i,j+1/2,k+1/2}^{n+1/2} = H_x\big|_{i,j+1/2,k+1/2}^{n} + \frac{\Delta t}{2\mu\Delta z}\Big(E_y\big|_{i,j+1/2,k+1}^{n+1/2} - E_y\big|_{i,j+1/2,k}^{n+1/2}\Big)$$
$$- \frac{\Delta t}{2\mu\Delta y}\Big(E_z\big|_{i,j+1,k+1/2}^{n} - E_z\big|_{i,j,k+1/2}^{n}\Big), \tag{9.2a}$$

$$H_y\big|_{i+1/2,j,k+1/2}^{n+1/2} = H_y\big|_{i+1/2,j,k+1/2}^{n} + \frac{\Delta t}{2\mu\Delta x}\left(E_z\big|_{i+1,j,k+1/2}^{n+1/2} - E_z\big|_{i,j,k+1/2}^{n+1/2}\right)$$

$$- \frac{\Delta t}{2\mu\Delta z}\left(E_x\big|_{i+1/2,j,k+1}^{n} - E_x\big|_{i+1/2,j,k}^{n}\right), \tag{9.2b}$$

$$H_z\big|_{i+1/2,j+1/2,k}^{n+1/2} = H_z\big|_{i+1/2,j+1/2,k}^{n} + \frac{\Delta t}{2\mu\Delta y}\left(E_x\big|_{i+1/2,j+1,k}^{n+1/2} - E_x\big|_{i+1/2,j,k}^{n+1/2}\right)$$

$$- \frac{\Delta t}{2\mu\Delta x}\left(E_y\big|_{i+1,j+1/2,k}^{n} - E_y\big|_{i,j+1/2,k}^{n}\right). \tag{9.2c}$$

In each of the above equations, the first finite-difference on the right-hand side is set up to be evaluated implicitly from as-yet unknown field data at time step $n + 1/2$, while the second finite-difference on the right-hand side is evaluated explicitly from known field data at time step n.

Subiteration 2. Advance the 6 field components from time step $n + 1/2$ to $n + 1$

$$E_x\big|_{i+1/2,j,k}^{n+1} = E_x\big|_{i+1/2,j,k}^{n+1/2} + \frac{\Delta t}{2\varepsilon\Delta y}\left(H_z\big|_{i+1/2,j+1/2,k}^{n+1/2} - H_z\big|_{i+1/2,j-1/2,k}^{n+1/2}\right)$$

$$- \frac{\Delta t}{2\varepsilon\Delta z}\left(H_y\big|_{i+1/2,j,k+1/2}^{n+1} - H_y\big|_{i+1/2,j,k-1/2}^{n+1}\right), \tag{9.3a}$$

$$E_y\big|_{i,j+1/2,k}^{n+1} = E_y\big|_{i,j+1/2,k}^{n+1/2} + \frac{\Delta t}{2\varepsilon\Delta z}\left(H_x\big|_{i,j+1/2,k+1/2}^{n+1/2} - H_x\big|_{i,j+1/2,k-1/2}^{n+1/2}\right)$$

$$- \frac{\Delta t}{2\varepsilon\Delta x}\left(H_z\big|_{i+1/2,j+1/2,k}^{n+1} - H_z\big|_{i-1/2,j+1/2,k}^{n+1}\right), \tag{9.3b}$$

$$E_z\big|_{i,j,k+1/2}^{n+1} = E_z\big|_{i,j,k+1/2}^{n+1/2} + \frac{\Delta t}{2\varepsilon\Delta x}\left(H_y\big|_{i+1/2,j,k+1/2}^{n+1/2} - H_y\big|_{i-1/2,j,k+1/2}^{n+1/2}\right)$$

$$- \frac{\Delta t}{2\varepsilon\Delta y}\left(H_x\big|_{i,j+1/2,k+1/2}^{n+1} - H_x\big|_{i,j-1/2,k+1/2}^{n+1}\right), \tag{9.3c}$$

$$H_x\big|_{i,j+1/2,k+1/2}^{n+1} = H_x\big|_{i,j+1/2,k+1/2}^{n+1/2} + \frac{\Delta t}{2\mu\Delta z}\left(E_y\big|_{i,j+1/2,k+1}^{n+1/2} - E_y\big|_{i,j+1/2,k}^{n+1/2}\right)$$

$$- \frac{\Delta t}{2\mu\Delta y}\left(E_z\big|_{i,j+1,k+1/2}^{n+1} - E_z\big|_{i,j,k+1/2}^{n+1}\right), \tag{9.4a}$$

$$H_y\big|_{i+1/2,j,k+1/2}^{n+1} = H_y\big|_{i+1/2,j,k+1/2}^{n+1/2} + \frac{\Delta t}{2\mu\Delta x}\left(E_z\big|_{i+1,j,k+1/2}^{n+1/2} - E_z\big|_{i,j,k+1/2}^{n+1/2}\right)$$

$$- \frac{\Delta t}{2\mu\Delta z}\left(E_x\big|_{i+1/2,j,k+1}^{n+1} - E_x\big|_{i+1/2,j,k}^{n+1}\right), \tag{9.4b}$$

$$H_z\big|_{i+1/2,j+1/2,k}^{n+1} = H_z\big|_{i+1/2,j+1/2,k}^{n+1/2} + \frac{\Delta t}{2\mu\Delta y}\left(E_x\big|_{i+1/2,j+1,k}^{n+1/2} - E_x\big|_{i+1/2,j,k}^{n+1/2}\right)$$

$$- \frac{\Delta t}{2\mu\Delta x}\left(E_y\big|_{i+1,j+1/2,k}^{n+1} - E_y\big|_{i,j+1/2,k}^{n+1}\right). \tag{9.4c}$$

In each of the above equations, the second finite-difference on the right-hand side is set up to be evaluated implicitly from as-yet unknown field data at time step $n + 1$, while the first finite-difference on the right-hand side is evaluated explicitly from known field data at time step $n + 1/2$ previously computed using (9.1) and (9.2).

9.2.2. Simplified system of time-stepping equations

The system of equations summarized above for each subiteration can be greatly simplified. For Subiteration 1, this is done by substituting the expressions of (9.2) for the H-field components evaluated at time step $n + 1/2$ into the E-field updates of (9.1). Similarly, for Subiteration 2, this is done by substituting the expressions of (9.4) for the H-field components evaluated at time step $n + 1$ into the E-field updates of (9.3). This yields the following simplified system of time-stepping equations for the algorithm of ZHENG, CHEN, and ZHANG [2000]:

Subiteration 1.   Advance the 6 field components from time step n to time step $n + 1/2$

$$
\left[1 + \frac{(\Delta t)^2}{2\mu\varepsilon(\Delta y)^2}\right] E_x\big|_{i+1/2,j,k}^{n+1/2} - \left[\frac{(\Delta t)^2}{4\mu\varepsilon(\Delta y)^2}\right]\left(E_x\big|_{i+1/2,j-1,k}^{n+1/2} + E_x\big|_{i+1/2,j+1,k}^{n+1/2}\right)
$$

$$
= E_x\big|_{i+1/2,j,k}^{n} + \frac{\Delta t}{2\varepsilon\Delta y}\left(H_z\big|_{i+1/2,j+1/2,k}^{n} - H_z\big|_{i+1/2,j-1/2,k}^{n}\right)
$$

$$
- \frac{\Delta t}{2\varepsilon\Delta z}\left(H_y\big|_{i+1/2,j,k+1/2}^{n} - H_y\big|_{i+1/2,j,k-1/2}^{n}\right) - \left[\frac{(\Delta t)^2}{4\mu\varepsilon\Delta x\Delta y}\right]
$$

$$
\times \left(E_y\big|_{i+1,j+1/2,k}^{n} - E_y\big|_{i,j+1/2,k}^{n} - E_y\big|_{i+1,j-1/2,k}^{n} + E_y\big|_{i,j-1/2,k}^{n}\right), \quad (9.5a)
$$

$$
\left[1 + \frac{(\Delta t)^2}{2\mu\varepsilon(\Delta z)^2}\right] E_y\big|_{i,j+1/2,k}^{n+1/2} - \left[\frac{(\Delta t)^2}{4\mu\varepsilon(\Delta z)^2}\right]\left(E_y\big|_{i,j+1/2,k-1}^{n+1/2} + E_y\big|_{i,j+1/2,k+1}^{n+1/2}\right)
$$

$$
= E_y\big|_{i,j+1/2,k}^{n} + \frac{\Delta t}{2\varepsilon\Delta z}\left(H_x\big|_{i,j+1/2,k+1/2}^{n} - H_x\big|_{i,j+1/2,k-1/2}^{n}\right)
$$

$$
- \frac{\Delta t}{2\varepsilon\Delta x}\left(H_z\big|_{i+1/2,j+1/2,k}^{n} - H_z\big|_{i-1/2,j+1/2,k}^{n}\right) - \left[\frac{(\Delta t)^2}{4\mu\varepsilon\Delta y\Delta z}\right]
$$

$$
\times \left(E_z\big|_{i,j+1,k+1/2}^{n} - E_z\big|_{i,j,k+1/2}^{n} - E_z\big|_{i,j+1,k-1/2}^{n} + E_z\big|_{i,j,k-1/2}^{n}\right), \quad (9.5b)
$$

$$
\left[1 + \frac{(\Delta t)^2}{2\mu\varepsilon(\Delta x)^2}\right] E_z\big|_{i,j,k+1/2}^{n+1/2} - \left[\frac{(\Delta t)^2}{4\mu\varepsilon(\Delta x)^2}\right]\left(E_z\big|_{i-1,j,k+1/2}^{n+1/2} + E_z\big|_{i+1,j,k+1/2}^{n+1/2}\right)
$$

$$
= E_z\big|_{i,j,k+1/2}^{n} + \frac{\Delta t}{2\varepsilon\Delta x}\left(H_y\big|_{i+1/2,j,k+1/2}^{n} - H_y\big|_{i-1/2,j,k+1/2}^{n}\right)
$$

$$
- \frac{\Delta t}{2\varepsilon\Delta y}\left(H_x\big|_{i,j+1/2,k+1/2}^{n} - H_x\big|_{i,j-1/2,k+1/2}^{n}\right) - \left[\frac{(\Delta t)^2}{4\mu\varepsilon\Delta x\Delta z}\right]
$$

$$
\times \left(E_x\big|_{i+1/2,j,k+1}^{n} - E_x\big|_{i+1/2,j,k}^{n} - E_x\big|_{i-1/2,j,k+1}^{n} + E_x\big|_{i-1/2,j,k}^{n}\right). \quad (9.5c)
$$

We see that (9.5a) yields a set of simultaneous equations for $E_x^{n+1/2}$ when written for each j coordinate along a y-directed line through the space lattice. The matrix associated with this system is tridiagonal, and hence, easily solved. This process is repeated for each y-cut through the lattice where E_x components are located. Similarly, (9.5b) yields a tridiagonal matrix system for each z-cut through the lattice to obtain $E_y^{n+1/2}$, and (9.5c) yields a tridiagonal matrix system for each x-cut through the lattice to obtain $E_z^{n+1/2}$.

To complete Subiteration 1, we next apply (9.2a)–(9.2c). These H-field updating equations are now fully explicit because all of their required E-field component data at time step $n + 1/2$ are available upon solving (9.5a)–(9.5c) in the manner described above.

Subiteration 2. Advance the 6 field components from time step $n + 1/2$ to $n + 1$

$$\left[1 + \frac{(\Delta t)^2}{2\mu\varepsilon(\Delta z)^2}\right] E_x\big|_{i+1/2,j,k}^{n+1} - \left[\frac{(\Delta t)^2}{4\mu\varepsilon(\Delta z)^2}\right]\left(E_x\big|_{i+1/2,j,k-1}^{n+1} + E_x\big|_{i+1/2,j,k+1}^{n+1}\right)$$

$$= E_x\big|_{i+1/2,j,k}^{n+1/2} + \frac{\Delta t}{2\varepsilon\Delta y}\left(H_z\big|_{i+1/2,j+1/2,k}^{n+1/2} - H_z\big|_{i+1/2,j-1/2,k}^{n+1/2}\right)$$

$$- \frac{\Delta t}{2\varepsilon\Delta z}\left(H_y\big|_{i+1/2,j,k+1/2}^{n+1/2} - H_y\big|_{i+1/2,j,k-1/2}^{n+1/2}\right) - \left[\frac{(\Delta t)^2}{4\mu\varepsilon\Delta x\Delta z}\right]$$

$$\times \left(E_z\big|_{i+1,j,k+1/2}^{n+1/2} - E_z\big|_{i,j,k+1/2}^{n+1/2} - E_z\big|_{i+1,j,k-1/2}^{n+1/2} + E_z\big|_{i,j,k-1/2}^{n+1/2}\right), \quad (9.6a)$$

$$\left[1 + \frac{(\Delta t)^2}{2\mu\varepsilon(\Delta x)^2}\right] E_y\big|_{i,j+1/2,k}^{n+1} - \left[\frac{(\Delta t)^2}{4\mu\varepsilon(\Delta x)^2}\right]\left(E_y\big|_{i-1,j+1/2,k}^{n+1} + E_y\big|_{i+1,j+1/2,k}^{n+1}\right)$$

$$= E_y\big|_{i,j+1/2,k}^{n+1/2} + \frac{\Delta t}{2\varepsilon\Delta z}\left(H_x\big|_{i,j+1/2,k+1/2}^{n+1/2} - H_x\big|_{i,j+1/2,k-1/2}^{n+1/2}\right)$$

$$- \frac{\Delta t}{2\varepsilon\Delta x}\left(H_z\big|_{i+1/2,j+1/2,k}^{n+1/2} - H_z\big|_{i-1/2,j+1/2,k}^{n+1/2}\right) - \left[\frac{(\Delta t)^2}{4\mu\varepsilon\Delta x\Delta y}\right]$$

$$\times \left(E_x\big|_{i+1/2,j+1,k}^{n+1/2} - E_x\big|_{i+1/2,j,k}^{n+1/2} - E_x\big|_{i-1/2,j+1,k}^{n+1/2} + E_x\big|_{i-1/2,j,k}^{n+1/2}\right), \quad (9.6b)$$

$$\left[1 + \frac{(\Delta t)^2}{2\mu\varepsilon(\Delta y)^2}\right] E_z\big|_{i,j,k+1/2}^{n+1} - \left[\frac{(\Delta t)^2}{4\mu\varepsilon(\Delta y)^2}\right]\left(E_z\big|_{i,j-1,k+1/2}^{n+1} + E_z\big|_{i,j+1,k+1/2}^{n+1}\right)$$

$$= E_z\big|_{i,j,k+1/2}^{n+1/2} + \frac{\Delta t}{2\varepsilon\Delta x}\left(H_y\big|_{i+1/2,j,k+1/2}^{n+1/2} - H_y\big|_{i-1/2,j,k+1/2}^{n+1/2}\right)$$

$$- \frac{\Delta t}{2\varepsilon\Delta y}\left(H_x\big|_{i,j+1/2,k+1/2}^{n+1/2} - H_x\big|_{i,j-1/2,k+1/2}^{n+1/2}\right) - \left[\frac{(\Delta t)^2}{4\mu\varepsilon\Delta y\Delta z}\right]$$

$$\times \left(E_y\big|_{i,j+1/2,k+1}^{n+1/2} - E_y\big|_{i,j+1/2,k}^{n+1/2} - E_y\big|_{i,j-1/2,k+1}^{n+1/2} + E_y\big|_{i,j-1/2,k}^{n+1/2}\right). \quad (9.6c)$$

We see that (9.6a) yields a set of simultaneous equations for E_x^{n+1} when written for each k coordinate along a z-directed line through the space lattice. The matrix associated with this system is tridiagonal, and hence, easily solved. This process is repeated for each z-cut through the lattice where E_x components are located. Similarly, (9.6b) yields a tridiagonal matrix system for each x-cut through the lattice to obtain E_y^{n+1}, and (9.6c) yields a tridiagonal matrix system for each y-cut through the lattice to obtain E_z^{n+1}.

To complete Subiteration 2, we next apply (9.4a)–(9.4c). These H-field updating equations are now fully explicit because all of their required E-component data at time step $n + 1$ are available upon solving (9.6a)–(9.6c) in the manner described above. This completes the ADI algorithm.

9.3. Proof of numerical stability

ZHENG, CHEN, and ZHANG [2000] provided the following proof of the numerical stability of their ADI algorithm. Assume that for each time step n, the instantaneous values of the E- and H-fields are Fourier-transformed into the spatial spectral domain with wavenumbers $\tilde{k}_x$, $\tilde{k}_y$, and $\tilde{k}_z$ along the x-, y-, and z-directions, respectively. Denoting the composite field vector in the spatial spectral domain at time step n as

$$
\mathbf{F}^n = \begin{bmatrix} E_x^n \\ E_y^n \\ E_z^n \\ H_x^n \\ H_y^n \\ H_z^n \end{bmatrix}
\tag{9.7}
$$

then Subiteration 1 (consisting of the systems (9.5) and (9.2)) can be written as

$$
\mathbf{F}^{n+1/2} = \overline{\overline{\mathbf{M}}}_1 \mathbf{F}^n,
\tag{9.8}
$$

where

$$
\overline{\overline{\mathbf{M}}}_1 = \begin{bmatrix}
\dfrac{1}{Q_y} & \dfrac{W_x W_y}{\mu\varepsilon Q_y} & 0 & 0 & \dfrac{jW_z}{\varepsilon Q_y} & \dfrac{-jW_y}{\varepsilon Q_y} \\[2.5ex]
0 & \dfrac{1}{Q_z} & \dfrac{W_z W_y}{\mu\varepsilon Q_z} & \dfrac{-jW_z}{\varepsilon Q_z} & 0 & \dfrac{jW_x}{\varepsilon Q_z} \\[2.5ex]
\dfrac{W_x W_z}{\mu\varepsilon Q_x} & 0 & \dfrac{1}{Q_x} & \dfrac{jW_y}{\varepsilon Q_x} & \dfrac{-jW_x}{\varepsilon Q_x} & 0 \\[2.5ex]
0 & \dfrac{-jW_z}{\mu Q_z} & \dfrac{jW_z}{\mu Q_z} & \dfrac{1}{Q_z} & 0 & \dfrac{W_x W_z}{\mu\varepsilon Q_z} \\[2.5ex]
\dfrac{jW_z}{\mu Q_x} & 0 & \dfrac{-jW_x}{\mu Q_x} & \dfrac{W_x W_y}{\mu\varepsilon Q_x} & \dfrac{1}{Q_x} & 0 \\[2.5ex]
\dfrac{-jW_y}{\mu Q_y} & \dfrac{jW_x}{\mu Q_y} & 0 & 0 & \dfrac{W_z W_y}{\mu\varepsilon Q_y} & \dfrac{1}{Q_y}
\end{bmatrix}
\tag{9.9}
$$

and

$$
W_x = \frac{\Delta t}{\Delta x}\sin\left(\frac{\tilde{k}_x \Delta x}{2}\right); \qquad W_y = \frac{\Delta t}{\Delta y}\sin\left(\frac{\tilde{k}_y \Delta y}{2}\right); \qquad W_z = \frac{\Delta t}{\Delta z}\sin\left(\frac{\tilde{k}_z \Delta z}{2}\right),
\tag{9.10}
$$

$$
Q_x = 1 + \frac{(W_x)^2}{\mu\varepsilon}; \qquad Q_y = 1 + \frac{(W_y)^2}{\mu\varepsilon}; \qquad Q_z = 1 + \frac{(W_z)^2}{\mu\varepsilon}.
\tag{9.11}
$$

Similarly, it can be shown that Subiteration 2 (consisting of the systems (9.6) and (9.4)) can be written as

$$
\mathbf{F}^{n+1} = \overline{\overline{\mathbf{M}}}_2 \mathbf{F}^{n+1/2},
\tag{9.12}
$$

where

$$
\overline{\overline{M}}_2 =
\begin{bmatrix}
\dfrac{1}{Q_z} & 0 & \dfrac{W_z W_x}{\mu\varepsilon Q_z} & 0 & \dfrac{jW_z}{\varepsilon Q_z} & \dfrac{-jW_y}{\varepsilon Q_z} \\[8pt]
\dfrac{W_x W_y}{\mu\varepsilon Q_x} & \dfrac{1}{Q_x} & 0 & \dfrac{-jW_z}{\varepsilon Q_x} & 0 & \dfrac{jW_x}{\varepsilon Q_x} \\[8pt]
0 & \dfrac{W_y W_z}{\mu\varepsilon Q_y} & \dfrac{1}{Q_y} & \dfrac{jW_y}{\varepsilon Q_y} & \dfrac{-jW_x}{\varepsilon Q_y} & 0 \\[8pt]
0 & \dfrac{-jW_z}{\mu Q_y} & \dfrac{jW_y}{\mu Q_y} & \dfrac{1}{Q_y} & \dfrac{W_z W_y}{\mu\varepsilon Q_y} & 0 \\[8pt]
\dfrac{jW_z}{\mu Q_z} & 0 & \dfrac{-jW_x}{\mu Q_z} & 0 & \dfrac{1}{Q_z} & \dfrac{W_z W_y}{\mu\varepsilon Q_z} \\[8pt]
\dfrac{-jW_y}{\mu Q_x} & \dfrac{jW_x}{\mu Q_x} & 0 & 0 & \dfrac{W_x W_z}{\mu\varepsilon Q_x} & \dfrac{1}{Q_x}
\end{bmatrix}.
\tag{9.13}
$$

Now, we substitute (9.8) into (9.12) to obtain in matrix form the complete single time-step update expression in the spatial spectral domain:

$$
\mathbf{F}^{n+1} = \overline{\overline{\mathbf{M}}}_2 \overline{\overline{\mathbf{M}}}_1 \mathbf{F}^n.
\tag{9.14}
$$

Using the software package MAPLE™, Zheng, Chen, and Zhang found that the magnitudes of all of the eigenvalues of the composite matrix $\overline{\overline{\mathbf{M}}} = \overline{\overline{\mathbf{M}}}_2\overline{\overline{\mathbf{M}}}_1$ equal unity, regardless of the time-step Δt. Therefore, they concluded that their ADI algorithm is *unconditionally stable* for all Δt, and the Courant stability condition is removed.

9.4. Numerical dispersion

ZHENG and CHEN [2001] derived the following numerical dispersion relation for their ADI algorithm:

$$
\sin^2(\omega t) = \frac{4\mu\varepsilon
\begin{bmatrix}
\mu\varepsilon(W_x)^2 + \mu\varepsilon(W_y)^2 + \mu\varepsilon(W_z)^2 \\
+ (W_x)^2(W_y)^2 + (W_y)^2(W_z)^2 \\
+ (W_z)^2(W_x)^2
\end{bmatrix}
[(\mu\varepsilon)^3 + (W_x)^2(W_y)^2(W_z)^2]}
{[\mu\varepsilon + (W_x)^2]^2[\mu\varepsilon + (W_y)^2]^2[\mu\varepsilon + (W_z)^2]^2}.
\tag{9.15}
$$

For Δt below the usual Courant limit, the numerical dispersion given by (9.15) is quite close to that of Yee's leapfrog time-stepping algorithm. For Δt above the usual Courant limit, the dispersive error given by (9.15) increases steadily.

9.5. Additional accuracy limitations and their implications

GONZALEZ GARCIA, LEE, and HAGNESS [2002] demonstrated additional accuracy limitations of ADI-FDTD not revealed by previously published numerical dispersion analyses such as that given in ZHENG and CHEN [2001]. They showed that some terms of its truncation error grow with Δt^2 multiplied by the spatial derivatives of the fields. These error terms, which are not present in a fully implicit time-stepping method such as the Crank–Nicolson scheme, give rise to potentially large numerical errors as Δt is increased. Excessive error can occur even if Δt is still small enough to highly resolve key temporal features of the modeled electromagnetic field waveform.

As a result, the primary usage of existing ADI-FDTD techniques appears to be for problems involving a fine mesh needed to model a small geometric feature in an overall much-larger structure that is discretized using a coarse mesh; and where, for computational efficiency, it is desirable to use a large time-step satisfying Courant stability for the coarse mesh. While this limits the impact of the excess error introduced locally within the fine mesh, this also limits the usefulness of ADI-FDTD when considering how to model the key problem areas outlined in Table 9.1.

10. Perfectly matched layer absorbing boundary conditions

10.1. Introduction to absorbing boundary conditions

A basic consideration with the FDTD approach to solving electromagnetic wave interaction problems is that many geometries of interest are defined in "open" regions where the spatial domain of the field is ideally unbounded in one or more directions. Clearly, no computer can store an unlimited amount of data, and therefore, the computational domain must be bounded. However, on this domain's outer boundary, only outward numerical wave motion is desired. That is, all outward-propagating numerical waves should exit the domain with negligible spurious reflections returning to the vicinity of the modeled structure. This would permit the FDTD solution to remain valid for all time steps. Depending upon their theoretical basis, outer-boundary conditions of this type have been called either *radiation boundary conditions* (RBCs) or *absorbing boundary conditions* (ABCs). The notation ABC will be used here.

ABCs cannot be directly obtained from the numerical algorithms for Maxwell's equations reviewed earlier. Principally, this is because these algorithms require field data on both sides of an observation point, and hence cannot be implemented at the outermost planes of the space lattice (since by definition there is no information concerning the fields at points outside of these planes). Although backward finite differences could conceivably be used here, these are generally of lower accuracy for a given space discretization and have not been used in major FDTD software.

Research in this area since 1970 has resulted in two principal categories of ABCs for FDTD simulations:

(1) Special analytical boundary conditions imposed upon the electromagnetic field at the outermost planes of the space lattice. This category was recently reviewed by TAFLOVE and HAGNESS [2000, Chapter 6].

(2) Incorporation of impedance-matched electromagnetic wave absorbing layers adjacent to the outer planes of the space lattice (by analogy with the treatment of the walls of an anechoic chamber). ABCs of this type have excellent capabilities for truncation of FDTD lattices in free space, in lossy or dispersive materials, or in metal or dielectric waveguides. Extremely small numerical-wave reflection coefficients in the order of 10^{-4} to 10^{-6} can be attained with an acceptable computational burden, allowing the possibility of achieving FDTD simulations having a dynamic range of 70 dB or more.

This section reviews modern, perfectly matched electromagnetic wave absorbing layers (Category 2 above). The review is based upon the recent publication by GEDNEY and TAFLOVE [2000].

10.2. Introduction to impedance-matched absorbing layers

Consider implementing an ABC by using an impedance-matched electromagnetic wave absorbing layer adjacent to the outer planes of the FDTD space lattice. Ideally, the absorbing medium is only a few lattice cells thick, reflectionless to all impinging waves over their full frequency spectrum, highly absorbing, and effective in the near field of a source or a scatterer. An early attempt at implementing such an absorbing material boundary condition was reported by HOLLAND and WILLIAMS [1983] who utilized a conventional lossy, dispersionless, absorbing medium. The difficulty with this tactic is that such an absorbing layer is matched only to normally incident plane waves.

BERENGER [1994] provided the seminal insight that a nonphysical absorber can be postulated that is matched independent of the frequency, angle of incidence, and polarization of an impinging plane wave by exploiting additional degrees of freedom arising from a novel split-field formulation of Maxwell's equations. Here, each vector field component is split into two orthogonal components, and the 12 resulting field components are then expressed as satisfying a coupled set of first-order partial differential equations. By choosing loss parameters consistent with a dispersionless medium, a perfectly matched planar interface is derived. This strategy allows the construction of what Berenger called a *perfectly matched layer* (PML) adjacent to the outer boundary of the FDTD space lattice for absorption of all outgoing waves.

Following Berenger's work, many papers appeared validating his technique as well as applying FDTD with the PML medium. An important advance was made by CHEW and WEEDON [1994], who restated the original split-field PML concept in a stretched-coordinate form. Subsequently, this allowed TEIXEIRA and CHEW [1997] to extend PML to cylindrical and spherical coordinate systems. A second important advance was made by SACKS, KINGSLAND, LEE, and LEE [1995] and GEDNEY [1995, 1996], who re-posed the split-field PML as a lossy, uniaxial anisotropic medium having both magnetic permeability and electric permittivity tensors. The uniaxial PML, or UPML, is intriguing because it is based on a potentially physically realizable material formulation rather than Berenger's nonphysical mathematical model.

10.3. Berenger's perfectly matched layer

10.3.1. Two-dimensional TE_z case
This section reviews the theoretical basis of Berenger's PML for the case of a TE_z-polarized plane wave incident from Region 1, the lossless material half-space $x < 0$, onto Region 2, the PML half-space $x > 0$.

Field-splitting modification of Maxwell's equations. Within Region 2, Maxwell's curl equations (2.12a)–(2.12c) as modified by Berenger are expressed in their time-dependent form as

$$\varepsilon_2 \frac{\partial E_x}{\partial t} + \sigma_y E_x = \frac{\partial H_z}{\partial y}, \tag{10.1a}$$

$$\varepsilon_2 \frac{\partial E_y}{\partial t} + \sigma_x E_y = -\frac{\partial H_z}{\partial x}, \tag{10.1b}$$

$$\mu_2 \frac{\partial H_{zx}}{\partial t} + \sigma_x^* H_{zx} = -\frac{\partial E_y}{\partial x}, \tag{10.1c}$$

$$\mu_2 \frac{\partial H_{zy}}{\partial t} + \sigma_y^* H_{zy} = \frac{\partial E_x}{\partial y}. \tag{10.1d}$$

Here, H_z is assumed to be split into two additive subcomponents

$$H_z = H_{zx} + H_{zy}. \tag{10.2}$$

Further, the parameters σ_x and σ_y denote electric conductivities, and the parameters σ_x^* and σ_y^* denote magnetic losses.

We see that Berenger's formulation represents a generalization of normally modeled physical media. If $\sigma_x = \sigma_y = 0$ and $\sigma_x^* = \sigma_y^* = 0$, (10.1a)–(10.1d) reduce to Maxwell's equations in a lossless medium. If $\sigma_x = \sigma_y = \sigma$ and $\sigma_x^* = \sigma_y^* = 0$, (10.1a)–(10.1d) describe an electrically conductive medium. And, if $\varepsilon_2 = \varepsilon_1$, $\mu_2 = \mu_1$, $\sigma_x = \sigma_y = \sigma$, $\sigma_x^* = \sigma_y^* = \sigma^*$, and

$$\sigma^*/\mu_1 = \sigma/\varepsilon_1 \quad \rightarrow \quad \sigma^* = \sigma\mu_1/\varepsilon_1 = \sigma(\eta_1)^2 \tag{10.3}$$

then (10.1a)–(10.1d) describe an absorbing medium that is impedance-matched to Region 1 for normally incident plane waves.

Additional possibilities present themselves, however. If $\sigma_y = \sigma_y^* = 0$, the medium can absorb a plane wave having field components (E_y, H_{zx}) propagating along x, but does not absorb a wave having field components (E_x, H_{zy}) propagating along y, since in the first case propagation is governed by (10.1b) and (10.1c) , and in the second case by (10.1a) and (10.1d). The converse situation is true for waves (E_y, H_{zx}) and (E_x, H_{zy}) if $\sigma_x = \sigma_x^* = 0$. These properties of particular Berenger media characterized by the pairwise parameter sets $(\sigma_x, \sigma_x^*, 0, 0)$ and $(0, 0, \sigma_y, \sigma_y^*)$ are closely related to the fundamental premise of this novel ABC, proved later. That is, if the pairwise electric and magnetic losses satisfy (10.3), then at interfaces normal to x and y, respectively, the Berenger media have zero reflection of electromagnetic waves.

Now consider (10.1a)–(10.1d)expressed in their time-harmonic form in the Berenger medium. Letting the hat symbol denote a phasor quantity, we write

$$j\omega\varepsilon_2\left(1 + \frac{\sigma_y}{j\omega\varepsilon_2}\right)\breve{E}_x = \frac{\partial}{\partial y}(\breve{H}_{zx} + \breve{H}_{zy}), \tag{10.4a}$$

$$j\omega\varepsilon_2\left(1 + \frac{\sigma_x}{j\omega\varepsilon_2}\right)\breve{E}_y = -\frac{\partial}{\partial x}(\breve{H}_{zx} + \breve{H}_{zy}), \tag{10.4b}$$

$$j\omega\mu_2\left(1 + \frac{\sigma_x^*}{j\omega\mu_2}\right)\breve{H}_{zx} = -\frac{\partial \breve{E}_y}{\partial x}, \tag{10.4c}$$

$$j\omega\mu_2\left(1 + \frac{\sigma_y^*}{j\omega\mu_2}\right)\breve{H}_{zy} = \frac{\partial \breve{E}_x}{\partial y}. \tag{10.4d}$$

The notation is simplified by introducing the variables

$$s_w = \left(1 + \frac{\sigma_w}{j\omega\varepsilon_2}\right); \qquad s_w^* = \left(1 + \frac{\sigma_w^*}{j\omega\mu_2}\right) : w = x, y. \qquad (10.5)$$

Then, (10.4a) and (10.4b) are rewritten as

$$j\omega\varepsilon_2 s_y \breve{E}_x = \frac{\partial}{\partial y}(\breve{H}_{zx} + \breve{H}_{zy}), \qquad (10.6a)$$

$$j\omega\varepsilon_2 s_x \breve{E}_y = -\frac{\partial}{\partial x}(\breve{H}_{zx} + \breve{H}_{zy}). \qquad (10.6b)$$

Plane-wave solution within the Berenger medium. The next step is to derive the plane-wave solution within the Berenger medium. To this end, (10.6a) is differentiated with respect to y and (10.6b) with respect to x. Substituting the expressions for $\partial\breve{E}_y/\partial x$ and $\partial\breve{E}_x/\partial y$ from (10.4c) and (10.4d) leads to

$$-\omega^2\mu_2\varepsilon_2\breve{H}_{zx} = -\frac{1}{s_x^*}\frac{\partial}{\partial x}\frac{1}{s_x}\frac{\partial}{\partial x}(\breve{H}_{zx} + \breve{H}_{zy}), \qquad (10.7a)$$

$$-\omega^2\mu_2\varepsilon_2\breve{H}_{zy} = -\frac{1}{s_y^*}\frac{\partial}{\partial y}\frac{1}{s_y}\frac{\partial}{\partial y}(\breve{H}_{zx} + \breve{H}_{zy}). \qquad (10.7b)$$

Adding these together and using (10.2) leads to the representative wave equation

$$\frac{1}{s_x^*}\frac{\partial}{\partial x}\frac{1}{s_x}\frac{\partial}{\partial x}\breve{H}_z + \frac{1}{s_y^*}\frac{\partial}{\partial y}\frac{1}{s_y}\frac{\partial}{\partial y}\breve{H}_z + \omega^2\mu_2\varepsilon_2\breve{H}_z = 0. \qquad (10.8)$$

This wave equation supports the solutions

$$\breve{H}_z = H_0\tau e^{-j\sqrt{s_x s_x^*}\beta_{2x}x - j\sqrt{s_y s_y^*}\beta_{2y}y} \qquad (10.9)$$

with the dispersion relationship

$$(\beta_{2x})^2 + (\beta_{2y})^2 = (k_2)^2 \quad \rightarrow \quad \beta_{2x} = \left[(k_2)^2 - (\beta_{2y})^2\right]^{1/2}. \qquad (10.10)$$

Then, from (10.6a), (10.6b), and (10.2), we have

$$\breve{E}_x = -H_0\tau\frac{\beta_{2y}}{\omega\varepsilon_2}\sqrt{\frac{s_y^*}{s_y}}\,e^{-j\sqrt{s_x s_x^*}\beta_{2x}x - j\sqrt{s_y s_y^*}\beta_{2y}y}, \qquad (10.11)$$

$$\breve{E}_y = H_0\tau\frac{\beta_{2x}}{\omega\varepsilon_2}\sqrt{\frac{s_x^*}{s_x}}\,e^{-j\sqrt{s_x s_x^*}\beta_{2x}x - j\sqrt{s_y s_y^*}\beta_{2y}y}. \qquad (10.12)$$

Despite the field splitting, continuity of the tangential electric and magnetic fields must be preserved across the $x = 0$ interface. To enforce this field continuity, we have $s_y = s_y^* = 1$, or equivalently $\sigma_y = 0 = \sigma_y^*$. This yields the phase-matching condition $\beta_{2y} = \beta_{1y} = k_1\sin\theta$. Further, we derive the H-field reflection and transmission coefficients

$$\Gamma = \left(\frac{\beta_{1x}}{\omega\varepsilon_1} - \frac{\beta_{2x}}{\omega\varepsilon_2}\sqrt{\frac{s_x^*}{s_x}}\right) \cdot \left(\frac{\beta_{1x}}{\omega\varepsilon_1} + \frac{\beta_{2x}}{\omega\varepsilon_2}\sqrt{\frac{s_x^*}{s_x}}\right)^{-1}; \qquad (10.13a)$$

$$\tau = 1 + \Gamma. \qquad (10.13b)$$

Reflectionless matching condition. Now, assume $\varepsilon_1 = \varepsilon_2$, $\mu_1 = \mu_2$, and $s_x = s_x^*$. This is equivalent to $k_1 = k_2$, $\eta_1 = \sqrt{\mu_1/\varepsilon_1} = \sqrt{\mu_2/\varepsilon_2}$, and $\sigma_x/\varepsilon_1 = \sigma_x^*/\mu_1$ (i.e., σ_x and σ_x^* satisfying (10.3) in a pairwise manner). With $\beta_{2_y} = \beta_{1_y}$, (10.10) now yields $\beta_{2_x} = \beta_{1_x}$. Substituting into (10.13a) gives the reflectionless condition $\Gamma = 0$ for *all* incident angles regardless of frequency ω. For this case, (10.9), (10.11), (10.12), and (10.13b) specify the following transmitted fields within the Berenger medium:

$$\check{H}_z = H_0 e^{-js_x\beta_{1_x}x - j\beta_{1_y}y} = H_0 e^{-j\beta_{1_x}x - j\beta_{1_y}y} e^{-\sigma_x x \eta_1 \cos\theta}, \tag{10.14}$$

$$\check{E}_x = -H_0 \eta_1 \sin\theta\, e^{-j\beta_{1_x}x - j\beta_{1_y}y} e^{-\sigma_x x \eta_1 \cos\theta}, \tag{10.15}$$

$$\check{E}_y = H_0 \eta_1 \cos\theta\, e^{-j\beta_{1_x}x - j\beta_{1_y}y} e^{-\sigma_x x \eta_1 \cos\theta}. \tag{10.16}$$

Within the matched Berenger medium, the transmitted wave propagates with the same speed and direction as the impinging wave while simultaneously undergoing exponential decay along the x-axis normal to the interface between Regions 1 and 2. Further, the attenuation factor $\sigma_x \eta_1 \cos\theta$ is independent of frequency. These desirable properties apply to all angles of incidence. Hence, Berenger's coining of the term "perfectly matched layer" makes excellent sense.

Structure of an FDTD grid employing Berenger's PML ABC. The above analysis can be repeated for PMLs that are normal to the y-direction. This permitted Berenger to propose the two-dimensional TE$_z$ FDTD grid shown in Fig. 10.1 which uses PMLs to greatly reduce outer-boundary reflections. Here, a free-space computation zone is surrounded by PML backed by perfect electric conductor (PEC) walls. At the left and right sides of the grid (x_1 and x_2), each PML has σ_x and σ_x^* matched according to (10.3) along with $\sigma_y = 0 = \sigma_y^*$ to permit reflectionless transmission across the interface between the

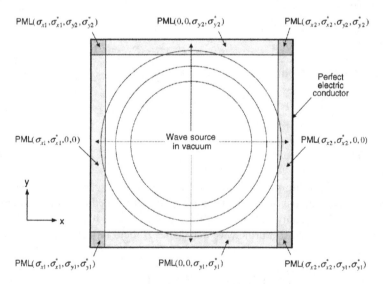

FIG. 10.1. Structure of a two-dimensional TE$_z$ FDTD grid employing the J.P. Berenger PML ABC. *After*: J.P. Berenger, *J. Computational Physics*, 1994, pp. 185–200.

free-space and PML regions. At the lower and upper sides of the grid (y_1 and y_2), each PML has σ_y and σ_y^* matched according to (10.3) along with $\sigma_x = 0 = \sigma_x^*$. At the four corners of the grid where there is overlap of two PMLs, all four losses (σ_x, σ_x^*, σ_y, and σ_y^*) are present and set equal to those of the adjacent PMLs.

10.3.2. Two-dimensional TM$_z$ case

The analysis of Section 10.3 can be repeated for the case of a TM$_z$-polarized incident wave wherein we implement the field splitting $E_z = E_{zx} + E_{zy}$. Analogous to (10.1), Maxwell's curl equations (2.11a)–(2.11c) as modified by Berenger are expressed in their time-dependent form as

$$\mu_2 \frac{\partial H_x}{\partial t} + \sigma_y^* H_x = -\frac{\partial E_z}{\partial y}, \tag{10.17a}$$

$$\mu_2 \frac{\partial H_y}{\partial t} + \sigma_x^* H_y = \frac{\partial E_z}{\partial x}, \tag{10.17b}$$

$$\varepsilon_2 \frac{\partial E_{zx}}{\partial t} + \sigma_x E_{zx} = \frac{\partial H_y}{\partial x}, \tag{10.17c}$$

$$\varepsilon_2 \frac{\partial E_{zy}}{\partial t} + \sigma_y E_{zy} = -\frac{\partial H_x}{\partial y}. \tag{10.17d}$$

A derivation of the PML properties conducted in a manner analogous to that of the TE$_z$ case yields slightly changed results. In most of the equations, the change is only a permutation of ε_2 with μ_2, and of σ with σ^*. However, the PML matching conditions are unchanged. This permits an absorbing reflectionless layer to be constructed adjacent to the outer grid boundary, as in the TE$_z$ case.

10.3.3. Three-dimensional case

KATZ, THIELE, and TAFLOVE [1994] showed that Berenger's PML can be realized in three dimensions by splitting all six Cartesian field vector components. For example, the modified Ampere's Law is given by

$$\left(\varepsilon \frac{\partial}{\partial t} + \sigma_y \right) E_{xy} = \frac{\partial}{\partial y} (H_{zx} + H_{zy}), \tag{10.18a}$$

$$\left(\varepsilon \frac{\partial}{\partial t} + \sigma_z \right) E_{xz} = -\frac{\partial}{\partial z} (H_{yx} + H_{yz}), \tag{10.18b}$$

$$\left(\varepsilon \frac{\partial}{\partial t} + \sigma_z \right) E_{yz} = \frac{\partial}{\partial z} (H_{xy} + H_{xz}), \tag{10.18c}$$

$$\left(\varepsilon \frac{\partial}{\partial t} + \sigma_x \right) E_{yx} = -\frac{\partial}{\partial x} (H_{zx} + H_{zy}), \tag{10.18d}$$

$$\left(\varepsilon \frac{\partial}{\partial t} + \sigma_x \right) E_{zx} = \frac{\partial}{\partial x} (H_{yx} + H_{yz}), \tag{10.18e}$$

$$\left(\varepsilon \frac{\partial}{\partial t} + \sigma_y \right) E_{zy} = -\frac{\partial}{\partial y} (H_{xy} + H_{xz}). \tag{10.18f}$$

Similarly, the modified Faraday's Law is given by

$$\left(\mu\frac{\partial}{\partial t}+\sigma_y^*\right)H_{xy} = -\frac{\partial}{\partial y}(E_{zx}+E_{zy}),\tag{10.19a}$$

$$\left(\mu\frac{\partial}{\partial t}+\sigma_z^*\right)H_{xz} = \frac{\partial}{\partial z}(E_{yx}+E_{yz}),\tag{10.19b}$$

$$\left(\mu\frac{\partial}{\partial t}+\sigma_z^*\right)H_{yz} = -\frac{\partial}{\partial z}(E_{xy}+E_{xz}),\tag{10.19c}$$

$$\left(\mu\frac{\partial}{\partial t}+\sigma_x^*\right)H_{yx} = \frac{\partial}{\partial x}(E_{zx}+E_{zy}),\tag{10.19d}$$

$$\left(\mu\frac{\partial}{\partial t}+\sigma_x^*\right)H_{zx} = -\frac{\partial}{\partial x}(E_{yx}+E_{yz}),\tag{10.19e}$$

$$\left(\mu\frac{\partial}{\partial t}+\sigma_y^*\right)H_{zy} = \frac{\partial}{\partial y}(E_{xy}+E_{xz}).\tag{10.19f}$$

PML matching conditions analogous to the two-dimensional cases discussed previously are used. Specifically, if we denote $w = x, y, z$, the matching condition at a normal-to-w PML interface has the parameter pair (σ_w, σ_w^*) satisfy (10.3). This causes the transmitted wave within the PML to undergo exponential decay in the $\pm w$-directions. All other (σ_w, σ_w^*) pairs within this PML are zero. In a corner region, the PML is provided with each matched (σ_w, σ_w^*) pair that is assigned to the overlapping PMLs forming the corner. Thus, PML media located in dihedral-corner overlapping regions have two nonzero and one zero (σ_w, σ_w^*) pairs. PML media located in trihedral-corner overlapping regions have three nonzero (σ_w, σ_w^*) pairs.

10.4. Stretched-coordinate formulation of Berenger's PML

A more compact form of the split-field equations of (10.18) and (10.19) was introduced by CHEW and WEEDON [1994]. Here, the split-field equations are re-posed in a non-split form that maps Maxwell's equations into a complex coordinate space. To this end, the following coordinate mapping is introduced:

$$\tilde{x} \to \int 0^x s_x(x')\,dx'; \qquad \tilde{y} \to \int 0^y s_y(y')\,dy'; \qquad \tilde{z} \to \int 0^z s_z(z')\,dz'. \tag{10.20}$$

In (10.20), we assume that the PML parameters s_w are continuous functions along the axial directions. The partial derivatives in the stretched coordinate space are then

$$\frac{\partial}{\partial\tilde{x}} = \frac{1}{s_x}\frac{\partial}{\partial x}; \qquad \frac{\partial}{\partial\tilde{y}} = \frac{1}{s_y}\frac{\partial}{\partial y}; \qquad \frac{\partial}{\partial\tilde{z}} = \frac{1}{s_z}\frac{\partial}{\partial z}. \tag{10.21}$$

Thus, the ∇ operator in the mapped space is defined as

$$\tilde{\nabla} = \hat{x}\frac{\partial}{\partial\tilde{x}} + \hat{y}\frac{\partial}{\partial\tilde{y}} + \hat{z}\frac{\partial}{\partial\tilde{z}} = \hat{x}\frac{1}{s_x}\frac{\partial}{\partial x} + \hat{y}\frac{1}{s_y}\frac{\partial}{\partial y} + \hat{z}\frac{1}{s_z}\frac{\partial}{\partial z}. \tag{10.22}$$

The time-harmonic Maxwell's equations in the complex-coordinate stretched space are then expressed as

$$j\omega\varepsilon\breve{\vec{E}} = \tilde{\nabla}\times\breve{\vec{H}}$$

$$= \hat{x}\left(\frac{1}{s_y}\frac{\partial}{\partial y}\breve{H}_z - \frac{1}{s_z}\frac{\partial}{\partial z}\breve{H}_y\right) + \hat{y}\left(\frac{1}{s_z}\frac{\partial}{\partial z}\breve{H}_x - \frac{1}{s_x}\frac{\partial}{\partial x}\breve{H}_z\right)$$

$$+ \hat{z}\left(\frac{1}{s_x}\frac{\partial}{\partial x}\breve{H}_y - \frac{1}{s_y}\frac{\partial}{\partial y}\breve{H}_x\right), \tag{10.23}$$

$$-j\omega\mu\breve{\vec{H}} = \tilde{\nabla}\times\breve{\vec{E}}$$

$$= \hat{x}\left(\frac{1}{s_y}\frac{\partial}{\partial y}\breve{E}_z - \frac{1}{s_z}\frac{\partial}{\partial z}\breve{E}_y\right) + \hat{y}\left(\frac{1}{s_z}\frac{\partial}{\partial z}\breve{E}_x - \frac{1}{s_x}\frac{\partial}{\partial x}\breve{E}_z\right)$$

$$+ \hat{z}\left(\frac{1}{s_x}\frac{\partial}{\partial x}\breve{E}_y - \frac{1}{s_y}\frac{\partial}{\partial y}\breve{E}_x\right). \tag{10.24}$$

A direct relationship can now be shown between the stretched-coordinate form of Maxwell's equations and Berenger's split-field PML. To demonstrate this, we first rewrite the split-field equations of (10.18) for the time-harmonic case:

$$j\omega\varepsilon s_y\breve{E}_{xy} = \frac{\partial}{\partial y}(\breve{H}_{zx} + \breve{H}_{zy}), \tag{10.25a}$$

$$j\omega\varepsilon s_z\breve{E}_{xz} = -\frac{\partial}{\partial z}(\breve{H}_{yx} + \breve{H}_{yz}), \tag{10.25b}$$

$$j\omega\varepsilon s_z\breve{E}_{yz} = \frac{\partial}{\partial z}(\breve{H}_{xy} + \breve{H}_{xz}), \tag{10.25c}$$

$$j\omega\varepsilon s_x\breve{E}_{yx} = -\frac{\partial}{\partial x}(\breve{H}_{zx} + \breve{H}_{zy}), \tag{10.25d}$$

$$j\omega\varepsilon s_x\breve{E}_{zx} = \frac{\partial}{\partial x}(\breve{H}_{yx} + \breve{H}_{yz}), \tag{10.25e}$$

$$j\omega\varepsilon s_y\breve{E}_{zy} = -\frac{\partial}{\partial y}(\breve{H}_{xy} + \breve{H}_{xz}). \tag{10.25f}$$

Then, we add (10.25a) + (10.25b); (10.25c) + (10.25d); and (10.25e) + (10.25f); and use the relationships $E_x = E_{xy} + E_{xz}$, $E_y = E_{yx} + E_{yz}$, and $E_z = E_{zx} + E_{zy}$. This yields

$$j\omega\varepsilon\breve{E}_x = \frac{1}{s_y}\frac{\partial}{\partial y}\breve{H}_z - \frac{1}{s_z}\frac{\partial}{\partial z}\breve{H}_y, \tag{10.26a}$$

$$j\omega\varepsilon\breve{E}_y = \frac{1}{s_z}\frac{\partial}{\partial z}\breve{H}_x - \frac{1}{s_x}\frac{\partial}{\partial x}\breve{H}_z, \tag{10.26b}$$

$$j\omega\varepsilon\breve{E}_z = \frac{1}{s_x}\frac{\partial}{\partial x}\breve{H}_y - \frac{1}{s_y}\frac{\partial}{\partial y}\breve{H}_x \tag{10.26c}$$

which is identical to (10.23). This procedure is repeated for the split-field equations of (10.19) rewritten for the time-harmonic case, leading exactly to (10.24). Specifically, we see that the stretched-coordinate form of the PML is equivalent to the split-field PML; however, it re-poses it in a nonsplit form.

The principal advantage the complex stretched-coordinate formulation offers is the ease of mathematically manipulating the PML equations, thereby simplifying the understanding of the behavior of the PML. It also provides a pathway to mapping the PML into other coordinate systems such as cylindrical and spherical coordinates, as shown by TEIXEIRA and CHEW [1997], as well as utilizing the split-field PML in frequency-domain finite-element methods based on unstructured discretizations, as shown by RAPPAPORT [1995] and CHEW and JIN [1996].

10.5. An anisotropic PML absorbing medium

The split-field PML introduced by Berenger is a hypothetical, nonphysical medium based on a mathematical model. Due to the coordinate-dependence of the loss terms, if such a physical medium exists, it must be anisotropic.

Indeed, a physical model based on an anisotropic, perfectly matched medium can be formulated. This was first discussed by SACKS, KINGSLAND, LEE, and LEE [1995]. For a single interface, the anisotropic medium is uniaxial and is composed of both electric and magnetic constitutive tensors. The uniaxial material performs as well as Berenger's PML while avoiding the nonphysical field splitting. This section introduces the theoretical basis of the uniaxial PML and compares its formulation with Berenger's PML and the stretched-coordinate PML.

10.5.1. Perfectly matched uniaxial medium
We consider an arbitrarily polarized time-harmonic plane wave propagating in Region 1, the isotropic half-space $x < 0$. This wave is assumed to be incident on Region 2, the half-space $x > 0$ comprised of a uniaxial anisotropic medium having the permittivity and permeability tensors

$$\bar{\bar{\varepsilon}}_2 = \varepsilon_2 \begin{bmatrix} a & 0 & 0 \\ 0 & b & 0 \\ 0 & 0 & b \end{bmatrix}, \tag{10.27a}$$

$$\bar{\bar{\mu}}_2 = \mu_2 \begin{bmatrix} c & 0 & 0 \\ 0 & d & 0 \\ 0 & 0 & d \end{bmatrix}. \tag{10.27b}$$

Here, $\varepsilon_{yy} = \varepsilon_{zz}$ and $\mu_{yy} = \mu_{zz}$ since the medium is assumed to be rotationally symmetric about the x-axis.

The fields excited within Region 2 are also plane-wave in nature and satisfy Maxwell's curl equations. We obtain

$$\vec{\beta}_2 \times \breve{\vec{E}} = \omega \bar{\bar{\mu}}_2 \breve{\vec{H}}; \tag{10.28a}$$

$$\vec{\beta}_2 \times \breve{\vec{H}} = -\omega \bar{\bar{\varepsilon}}_2 \breve{\vec{E}} \tag{10.28b}$$

where $\vec{\beta}_2 = \hat{x}\beta_{2_x} + \hat{y}\beta_{2_y}$ is the wavevector in anisotropic Region 2. In turn, this permits derivation of the wave equation

$$\vec{\beta}_2 \times \left(\bar{\bar{\varepsilon}}_2^{-1}\vec{\beta}_2\right) \times \vec{\breve{H}} + \omega^2\bar{\bar{\mu}}_2\vec{\breve{H}} = 0. \tag{10.29}$$

Expressing the cross products as matrix operators, this wave equation can be expressed in matrix form as

$$\left[\begin{array}{ccc} k_2^2 c - (\beta_{2_y})^2 b^{-1} & \beta_{2_x}\beta_{2_y}b^{-1} & 0 \\ \beta_{2_x}\beta_{2_y}b^{-1} & k_2^2 d - (\beta_{2_x})^2 b^{-1} & 0 \\ 0 & 0 & k_2^2 d - (\beta_{2_x})^2 b^{-1} - (\beta_{2_y})^2 a^{-1} \end{array}\right]$$

$$\times \left[\begin{array}{c} \breve{H}_x \\ \breve{H}_y \\ \breve{H}_z \end{array}\right] = 0, \tag{10.30}$$

where $k_2^2 = \omega^2\mu_2\varepsilon_2$. The dispersion relation for the uniaxial medium in Region 2 is derived from the determinant of the matrix operator. Solving for β_{2_x}, we find that there are four eigenmode solutions. Conveniently, these solutions can be decoupled into forward and backward TE$_z$ and TM$_z$ modes, which satisfy the dispersion relations

$$k_2^2 - (\beta_{2_x})^2 b^{-1}d^{-1} - (\beta_{2_y})^2 a^{-1}d^{-1} = 0: \quad \text{TE}_z(\breve{H}_x, \breve{H}_y = 0), \tag{10.31}$$

$$k_2^2 - (\beta_{2_x})^2 b^{-1}d^{-1} - (\beta_{2_y})^2 b^{-1}c^{-1} = 0: \quad \text{TM}_z(\breve{H}_z = 0). \tag{10.32}$$

The reflection coefficient at the interface $x = 0$ of Regions 1 and 2 can now be derived. Let us assume a TE$_z$ incident wave in Region 1. Then, in isotropic Region 1, the fields are expressed as a superposition of the incident and reflected fields as

$$\vec{\breve{H}}_1 = \hat{z}H_0\left(1 + \Gamma e^{2j\beta_{1x}x}\right)e^{-j\beta_{1x}x-j\beta_{1y}y},$$

$$\vec{\breve{E}}_1 = \left[-\hat{x}\frac{\beta_{1y}}{\omega\varepsilon_1}\left(1 + \Gamma e^{2j\beta_{1x}x}\right) + \hat{y}\frac{\beta_{1x}}{\omega\varepsilon_1}\left(1 - \Gamma e^{2j\beta_{1x}x}\right)\right]H_0e^{-j\beta_{1x}x-j\beta_{1y}y}. \tag{10.33}$$

The wave transmitted into Region 2 is also TE$_z$ with propagation characteristics governed by (10.31). These fields are expressed as

$$\vec{\breve{H}}_2 = \hat{z}H_0\tau e^{-j\beta_{2x}x-j\beta_{2y}y},$$

$$\vec{\breve{E}}_2 = \left(-\hat{x}\frac{\beta_{2y}}{\omega\varepsilon_2 a} + \hat{y}\frac{\beta_{2x}}{\omega\varepsilon_2 b}\right)H_0\tau e^{-j\beta_{2x}x-j\beta_{2y}y}, \tag{10.34}$$

where Γ and τ are the H-field reflection and transmission coefficients, respectively. These are derived by enforcing continuity of the tangential E and H fields across $x = 0$, and are given by

$$\Gamma = \frac{\beta_{1x} - \beta_{2x}b^{-1}}{\beta_{1x} + \beta_{2x}b^{-1}}; \tag{10.35a}$$

$$\tau = 1 + \Gamma = \frac{2\beta_{1x}}{\beta_{1x} + \beta_{2x}b^{-1}}. \tag{10.35b}$$

Further, for all angles of wave incidence we have

$$\beta_{2_y} = \beta_{1_y} \tag{10.36}$$

due to phase-matching across the $x = 0$ interface. Substituting (10.36) into (10.31) and solving for β_{2_x} yields

$$\beta_{2_x} = \sqrt{k_2^2 bd - (\beta_{1_y})^2 a^{-1} b}. \tag{10.37}$$

Then, if we set $\varepsilon_1 = \varepsilon_2$, $\mu_1 = \mu_2$, $d = b$, and $a^{-1} = b$, we have $k_2 = k_1$ and

$$\beta_{2_x} = \sqrt{k_1^2 b^2 - (\beta_{1_y})^2 b^2} = b\sqrt{k_1^2 - (\beta_{1_y})^2} \equiv b\beta_{1_x}. \tag{10.38}$$

Substituting (10.38) into (10.35a) yields $\Gamma = 0$ for all β_{1_x}. Thus, the interface between Regions 1 and 2 is reflectionless for angles of wave incidence.

The above exercise can be repeated for TM$_z$ polarization. Here, the E-field reflection coefficient is the dual of (10.35a) and is found by replacing b with d (and vice versa), and a with c. For this case, the reflectionless condition holds if $b = d$ and $c^{-1} = d$.

Combining the results for the TE$_z$ and TM$_z$ cases, we see that reflectionless wave transmission into Region 2 occurs when it is composed of a uniaxial medium having the ε and μ tensors

$$\bar{\bar{\varepsilon}}_2 = \varepsilon_1 \bar{\bar{s}}; \tag{10.39a}$$

$$\bar{\bar{\mu}}_2 = \mu_1 \bar{\bar{s}}; \tag{10.39b}$$

$$\bar{\bar{s}} = \begin{bmatrix} s_x^{-1} & 0 & 0 \\ 0 & s_x & 0 \\ 0 & 0 & s_x \end{bmatrix}. \tag{10.39c}$$

This reflectionless property is completely independent of the angle of incidence, polarization, and frequency of the incident wave. Further, from (10.31) and (10.32), the propagation characteristics of the TE- and TM-polarized waves are identical. We call this medium a *uniaxial PML* (UPML) in recognition of its uniaxial anisotropy and perfect matching.

Similar to Berenger's PML, the reflectionless property of the UPML in Region 2 is valid for any s_x. For example, choose $s_x = 1 + \sigma_x/j\omega\varepsilon_1 = 1 - j\sigma_x/\omega\varepsilon_1$. Then, from (10.38) we have

$$\beta_{2_x} = (1 - j\sigma_x/\omega\varepsilon_1)\beta_{1_x}. \tag{10.40}$$

We note that the real part of β_{2_x} is identical to β_{1_x}. Combined with (10.36), this implies that the phase velocities of the impinging and transmitted waves are identical for all incident angles. The characteristic wave impedance in Region 2 is also identical to that in Region 1, a consequence of the fact that the media are perfectly matched.

Finally, substituting (10.36) and (10.40) into (10.34) and (10.35b) yields the fields transmitted into the Region-2 UPML for a TE$_z$ incident wave:

$$\begin{aligned} \breve{\vec{H}}_2 &= \hat{z} H_0 e^{-j\beta_{1_x} x - j\beta_{1_y} y} e^{-\sigma_x x \eta_1 \cos\theta}, \\ \breve{\vec{E}}_2 &= (-\hat{x} s_x \eta_1 \sin\theta + \hat{y}\eta_1 \cos\theta) H_0 e^{-j\beta_{1_x} x - j\beta_{1_y} y} e^{-\sigma_x x \eta_1 \cos\theta}. \end{aligned} \tag{10.41}$$

Here, $\eta_1 = \sqrt{\mu_1/\varepsilon_1}$ and θ is the angle of incidence relative to the x-axis. Thus, the transmitted wave in the UPML propagates with the same phase velocity as the incident wave, while simultaneously undergoing exponential decay along the x-axis normal to the interface between Regions 1 and 2. The attenuation factor is independent of frequency, although it is dependent on θ and the UPML conductivity σ_x.

10.5.2. Relationship to Berenger's split-field PML

Comparing the E- and H-fields transmitted into the UPML in (10.41) with the corresponding fields for Berenger's split-field PML in (10.14)–(10.16), we observe identical fields and identical propagation characteristics. Further examination of (10.8) and (10.29) reveals that the two methods result in the same wave equation. Consequently, the plane waves satisfy the same dispersion relation.

However, in the split-field formulation, E_x is continuous across the $x = 0$ boundary, whereas for UPML, E_x is discontinuous and $D_x = s_x^{-1} E_x$ is continuous. This implies that the two methods host different divergence theorems. Within the UPML, Gauss' Law for the E-field is explicitly written as

$$\nabla \cdot \vec{D} = \nabla \cdot (\varepsilon \bar{\bar{s}} \vec{E}) = \frac{\partial}{\partial x}\left(\varepsilon s_x^{-1} E_x\right) + \frac{\partial}{\partial y}(\varepsilon s_x E_y) + \frac{\partial}{\partial z}(\varepsilon s_x E_z) = 0. \qquad (10.42)$$

This implies that $D_x = \varepsilon s_x^{-1} E_x$ must be continuous across the $x = 0$ interface since no sources are assumed here. It was shown that ε must be continuous across the interface for a perfectly matched condition. Thus, D_x and $s_x^{-1} E_x$ must be continuous across $x = 0$. Comparing (10.41) with (10.33), this is indeed true for a TE$_z$-polarized wave.

Next, consider Gauss' Law for Berenger's split-field PML formulation. The ∇ operator in the stretched-coordinate space of interest is defined as

$$\nabla = \hat{x}\frac{\partial}{s_x \partial x} + \hat{y}\frac{\partial}{\partial y} + \hat{z}\frac{\partial}{\partial z}. \qquad (10.43)$$

Therefore, we can express the divergence of the electric flux density as

$$\frac{1}{s_x}\frac{\partial}{\partial x}(\varepsilon E_x) + \frac{\partial}{\partial y}(\varepsilon E_y) + \frac{\partial}{\partial z}(\varepsilon E_z) = 0. \qquad (10.44)$$

Since ε is continuous across the boundary and s_x^{-1} occurs outside the derivative, both E_x and D_x are continuous.

In summary, Berenger's split-field PML and the UPML have the same propagation characteristics since they both result in the same wave equation. However, the two formulations have different Gauss' Laws. Hence, the E- and H-field components that are normal to the PML interface are different.

10.5.3. A generalized three-dimensional formulation

We now show that properly defining a general constitutive tensor $\bar{\bar{s}}$ allows the UPML medium to be used throughout the entire FDTD space lattice. This tensor provides for both a lossless, isotropic medium in the primary computation zone, *and* individual UPML absorbers adjacent to the outer lattice boundary planes for mitigation of spurious wave reflections.

For a matched condition, the time-harmonic Maxwell's curl equations in the UPML can be written in their most general form as

$$\nabla \times \bar{\tilde{H}} = j\omega\varepsilon\bar{\bar{s}}\bar{\tilde{E}};$$ (10.45a)

$$\nabla \times \bar{\tilde{E}} = -j\omega\mu\bar{\bar{s}}\bar{\tilde{H}}$$ (10.45b)

where $\bar{\bar{s}}$ is the diagonal tensor defined by

$$
\bar{\bar{s}} = \begin{bmatrix} s_x^{-1} & 0 & 0 \\ 0 & s_x & 0 \\ 0 & 0 & s_x \end{bmatrix} \begin{bmatrix} s_y & 0 & 0 \\ 0 & s_y^{-1} & 0 \\ 0 & 0 & s_y \end{bmatrix} \begin{bmatrix} s_z & 0 & 0 \\ 0 & s_z & 0 \\ 0 & 0 & s_z^{-1} \end{bmatrix}
$$

$$
= \begin{bmatrix} s_y s_z s_x^{-1} & 0 & 0 \\ 0 & s_x s_z s_y^{-1} & 0 \\ 0 & 0 & s_x s_y s_z^{-1} \end{bmatrix}.
$$ (10.46)

Allowing for a nonunity real part κ, the multiplicative components of the diagonal elements of $\bar{\bar{s}}$ are given by

$$s_x = \kappa_x + \frac{\sigma_x}{j\omega\varepsilon};$$ (10.47a)

$$s_y = \kappa_y + \frac{\sigma_y}{j\omega\varepsilon};$$ (10.47b)

$$s_z = \kappa_z + \frac{\sigma_z}{j\omega\varepsilon}.$$ (10.47c)

Now, given the above definitions, the following lists all of the special cases involved in implementing the strategy of using $\bar{\bar{s}}$ throughout the entire FDTD lattice.

Lossless, isotropic interior zone
$\bar{\bar{s}}$ is the identity tensor realized by setting $s_x = s_y = s_z = 1$ in (10.46). This requires $\sigma_x = \sigma_y = \sigma_z = 0$ and $\kappa_x = \kappa_y = \kappa_z = 1$ in (10.47).

UPML absorbers at x_{min} and x_{max} outer-boundary planes
$\bar{\bar{s}}$ is the tensor given in (10.39), which is realized by setting $s_y = s_z = 1$ in (10.46). This requires $\sigma_y = \sigma_z = 0$ and $\kappa_y = \kappa_z = 1$ in (10.47).

UPML absorbers at y_{min} and y_{max} outer-boundary planes
We set $s_x = s_z = 1$ in (10.46). This requires $\sigma_x = \sigma_z = 0$ and $\kappa_x = \kappa_z = 1$ in (10.47).

UPML Absorbers at z_{min} and z_{max} outer-boundary planes
We set $s_x = s_y = 1$ in (10.46). This requires $\sigma_x = \sigma_y = 0$ and $\kappa_x = \kappa_y = 1$ in (10.47).

Overlapping UPML absorbers at x_{min}, x_{max} and y_{min}, y_{max} dihedral corners
We set $s_z = 1$ in (10.46). This requires $\sigma_z = 0$ and $\kappa_z = 1$ in (10.47).

Overlapping UPML absorbers at $x_{\min}$, $x_{\max}$ and $z_{\min}$, $z_{\max}$ dihedral corners
We set $s_y = 1$ in (10.46). This requires $\sigma_y = 0$ and $\kappa_y = 1$ in (10.47).

Overlapping UPML absorbers at $y_{\min}$, $y_{\max}$ and $z_{\min}$, $z_{\max}$ dihedral corners
We set $s_x = 1$ in (10.46). This requires $\sigma_x = 0$ and $\kappa_x = 1$ in (10.47).

Overlapping UPML absorbers at all trihedral corners
We use the complete general tensor in (10.46).

The generalized constitutive tensor defined in (10.46) is no longer uniaxial by strict definition, but rather is anisotropic. However, the anisotropic PML is still referenced as uniaxial since it is uniaxial in the nonoverlapping PML regions.

10.5.4. Inhomogeneous media
At times, we need to use the PML to terminate an inhomogeneous material region in the FDTD space lattice. An example is a printed circuit constructed on a dielectric substrate backed by a metal ground plane. A second example is a long optical fiber. In such cases, the inhomogeneous material region extends through the PML to the outer boundary of the FDTD lattice. GEDNEY [1998] has shown that the PML can be perfectly matched to such a medium. However, care must be taken to properly choose the PML parameters to maintain a stable and accurate formulation.

Consider an x-normal UPML boundary. Let an inhomogeneous dielectric $\varepsilon(y, z)$, assumed to be piecewise constant in the transverse y- and z-directions, extend into the UPML. From fundamental electromagnetic theory, $D_y = \varepsilon E_y$ must be continuous across any y-normal boundary, and $D_z = \varepsilon E_z$ must be continuous across any z-normal boundary. Then, from Gauss' Law for the UPML in (10.42), we see that s_x must be independent of y and z to avoid surface charge at the boundaries of the discontinuity.

In the previous discussions, the dielectric in the UPML was assumed to be homogeneous. For this case in (10.47a), $s_x = \kappa_x + \sigma_x/j\omega\varepsilon$ was chosen. However, if $\varepsilon = \varepsilon(y, z)$ and is piecewise constant, then s_x is also piecewise constant in the transverse directions. Thus, surface charge densities result at the material boundaries as predicted by Gauss' Law in (10.42) due to the derivative of a discontinuous function. This nonphysical charge leads to an ill-posed formulation. To avoid this, s_x must be independent of y and z. This holds only if σ_x/ε is maintained constant. This can be done in a brute-force manner by modifying σ_x in the transverse direction such that $\sigma_x(y, z)/\varepsilon(y, z)$ is a constant. A much simpler approach is to normalize σ_x by the relative permittivity, rewriting (10.47a) as

$$s_x = \kappa_x + \sigma_x'/j\omega\varepsilon_0, \tag{10.48}$$

where ε_0 is the free-space permittivity. In this case, σ_x' is simply a constant in the transverse y- and z-directions, although it is still scaled along the normal x-direction. Now, Gauss' Law is satisfied within the UPML, leading to a well-posed formulation. This also leads to a materially independent formulation of the UPML.

Next, consider Berenger's split-field PML. To understand the constraints of this technique in an inhomogeneous medium, it is simpler to work with its stretched-coordinate representation. In stretched coordinates, Gauss' Law is represented in (10.44). Here, it appears that there are no further constraints on s_x. However, conservation laws require that the charge continuity equation be derived from Ampere's Law and Gauss' Law. To this end, the divergence of Ampere's Law in (10.23) is performed in the stretched coordinates using (10.44). We see that the divergence of the curl of $\vec{H}$ is zero only if s_x is independent of the transverse coordinates y and z. This holds only if σ_x/ε is independent of y and z. Again, this can be easily managed by representing s_x by (10.48), thus leading to a material-independent PML.

In summary, an inhomogeneous medium that is infinite in extent can be terminated by either a split-field PML or UPML medium. Both are perfectly matched to arbitrary electromagnetic waves impinging upon the PML boundary. The method is accurate and stable provided that the PML parameters s_w (s_x, s_y, or s_z) are posed to be independent of the transverse directions. This can be readily accomplished by normalizing σ_w by the relative permittivity, and hence posing $s_w = \kappa_w + \sigma'_w/j\omega\varepsilon_0$, where σ'_w is constant in the transverse direction.

10.6. Theoretical performance of the PML

10.6.1. The continuous space
When used to truncate an FDTD lattice, the PML has a thickness d and is terminated by the outer boundary of the lattice. If the outer boundary is assumed to be a PEC wall, finite power reflects back into the primary computation zone. For a wave impinging upon the PML at angle θ relative to the w-directed surface normal, this reflection can be computed using transmission line analysis, yielding

$$R(\theta) = e^{-2\sigma_w \eta d \cos\theta}. \tag{10.49}$$

Here, η and σ_w are, respectively, the PML's characteristic wave impedance and its conductivity, referred to propagation in the w-direction. In the context of an FDTD simulation, $R(\theta)$ is referred to as the "reflection error" since it is a nonphysical reflection due to the PEC wall that backs the PML. We note that the reflection error is the same for both the split-field PML and the UPML, since both support the same wave equation. This error decreases exponentially with σ_w and d. However, the reflection error increases as $\exp(\cos\theta)$, reaching the worst case for $\theta = 90°$. At this grazing angle of incidence, $R = 1$ and the PML is completely ineffective. To be useful within an FDTD simulation, we want $R(\theta)$ to be as small as possible. Clearly, for a thin PML, we must have σ_w as large as possible to reduce $R(\theta)$ to acceptably small levels, especially for θ approaching $90°$.

10.6.2. The discrete space
Grading of the PML loss parameters. Theoretically, reflectionless wave transmission can take place across a PML interface regardless of the local step-discontinuity in σ and σ^* presented to the continuous impinging electromagnetic field. However, in FDTD or any discrete representation of Maxwell's equations, numerical artifacts arise due to the

finite spatial sampling. Consequently, implementing PML as a single step-discontinuity of σ and σ^* in the FDTD lattice leads to significant spurious wave reflection at the PML surface.

To reduce this reflection error, BERENGER [1994] proposed that the PML losses gradually rise from zero along the direction normal to the interface. Assuming such a grading, the PML remains matched, as seen from the stretched-coordinate theory in Section 10.4. Pursuing this idea, we consider as an example an x-directed plane wave impinging at angle θ upon a PEC-backed PML slab of thickness d, with the front planar interface located in the $x = 0$ plane. Assuming the graded PML conductivity profile $\sigma_x(x)$, we have from (10.20) and (10.14)–(10.16) or (10.41)

$$R(\theta) = e^{-2\eta \cos\theta \int_0^d \sigma_x(x)\,dx}. \tag{10.50}$$

Polynomial grading. Several profiles have been suggested for grading $\sigma_x(x)$ (and $\kappa_x(x)$ in the context of the UPML). The most successful use a polynomial or geometric variation of the PML loss with depth x. Polynomial grading is simply

$$\sigma_x(x) = (x/d)^m \sigma_{x,\max}; \tag{10.51a}$$

$$\kappa_x(x) = 1 + (\kappa_{x,\max} - 1) \cdot (x/d)^m. \tag{10.51b}$$

This increases the value of the PML σ_x from zero at $x = 0$, the surface of the PML, to $\sigma_{x,\max}$ at $x = d$, the PEC outer boundary. Similarly, for the UPML, κ_x increases from one at $x = 0$ to $\kappa_{x,\max}$ at $x = d$. Substituting (10.51a) into (10.50) yields

$$R(\theta) = e^{-2\eta \sigma_{x,\max} d \cos\theta / (m+1)}. \tag{10.52}$$

For a fixed d, polynomial grading provides two parameters: $\sigma_{x,\max}$ and m. A large m yields a $\sigma_x(x)$ distribution that is relatively flat near the PML surface. However, deeper within the PML, σ_x increases more rapidly than for small m. In this region, the field amplitudes are substantially decayed and reflections due to the discretization error contribute less. Typically, $3 \leqslant m \leqslant 4$ has been found to be nearly optimal for many FDTD simulations (see, for example, BERENGER [1996]).

For polynomial grading, the PML parameters can be readily determined for a given error estimate. For example, let m, d, and the desired reflection error $R(0)$ be known. Then, from (10.52), $\sigma_{x,\max}$ is computed as

$$\sigma_{x,\max} = -\frac{(m+1)\ln[R(0)]}{2\eta d}. \tag{10.53}$$

Geometric grading. The PML loss profile for this case was defined by BERENGER [1997] as

$$\sigma_x(x) = \left(g^{1/\Delta}\right)^x \sigma_{x,0}; \tag{10.54a}$$

$$\kappa_x(x) = \left(g^{1/\Delta}\right)^x, \tag{10.54b}$$

where $\sigma_{x,0}$ is the PML conductivity at its surface, g is the scaling factor, and Δ is the FDTD space increment. Here, the PML conductivity increases from $\sigma_{x,0}$ at its surface to $g^{d/\Delta}\sigma_{x,0}$ at the PEC outer boundary. Substituting (10.54a) into (10.50) results in

$$R(\theta) = e^{-2\eta\sigma_{x,0}\Delta(g^{d/\Delta}-1)\cos\theta/\ln g}. \qquad (10.55)$$

For a fixed d, geometric grading provides two parameters: g and $\sigma_{x,0}$. $\sigma_{x,0}$ must be small to minimize the initial discretization error. Large values of g flatten the conductivity profile near $x = 0$, and steepen it deeper into the PML. Usually, g, d, and $R(0)$ are predetermined. This yields

$$\sigma_{x,0} = -\frac{\ln[R(0)]\ln(g)}{2\eta\Delta(g^{d/\Delta}-1)}. \qquad (10.56)$$

Typically, $2 \leqslant g \leqslant 3$ has been found to be nearly optimal for many FDTD simulations.

Discretization error. The design of an effective PML requires balancing the theoretical reflection error $R(\theta)$ and the numerical discretization error. For example, (10.53) provides $\sigma_{x,\max}$ for a polynomial-graded conductivity given a predetermined $R(0)$ and m. If $\sigma_{x,\max}$ is small, the primary reflection from the PML is due to its PEC backing, and (10.50) provides a fairly accurate approximation of the reflection error. Now, we normally choose $\sigma_{x,\max}$ to be as large as possible to minimize $R(\theta)$. However, if $\sigma_{x,\max}$ is too large, the discretization error due to the FDTD approximation dominates, and the actual reflection error is potentially orders of magnitude higher than what (10.50) predicts. Consequently, there is an optimal choice for $\sigma_{x,\max}$ that balances reflection from the PEC outer boundary and discretization error.

BERENGER [1996], BERENGER [1997] postulated that the largest reflection error due to discretization occurs at $x = 0$, the PML surface. Any wave energy that penetrates further into the PML and then is reflected undergoes attenuation both before and after its point of reflection, and typically is not as large a contribution. Thus, it is desirable to minimize the discontinuity at $x = 0$. As discussed earlier, one way to achieve this is by flattening the PML loss profile near $x = 0$. However, if the subsequent rise of loss with depth is too rapid, reflections from deeper within the PML can dominate.

Through extensive numerical experimentation, GEDNEY [1996] and HE [1997] found that, for a broad range of applications, an optimal choice for a 10-cell-thick, polynomial-graded PML is $R(0) \approx e^{-16}$. For a 5-cell-thick PML, $R(0) \approx e^{-8}$ is optimal. From (10.53), this leads to an optimal $\sigma_{x,\max}$ for polynomial grading:

$$\sigma_{x,\mathrm{opt}} \approx -\frac{(m+1)\cdot(-16)}{(2\eta)\cdot(10\Delta)} = \frac{0.8(m+1)}{\eta\Delta}. \qquad (10.57)$$

This expression has proven to be quite robust for many applications. However, its value may be too large when the PML terminates highly elongated resonant structures or sources with a very long time duration, such as a unit step. For a detailed discussion, the reader is referred to GEDNEY [1998].

10.7. Efficient implementation of UPML in FDTD

This section discusses mapping the UPML presented in Section 10.5 into the discrete FDTD space. The FDTD approximation is derived from the time-harmonic Maxwell's curl equations within the generalized uniaxial medium as defined in (10.45)–(10.47).

10.7.1. Derivation of the finite-difference expressions

Starting with (10.45a) and (10.46), Ampere's Law in a matched UPML is expressed as

$$
\begin{bmatrix}
\frac{\partial \breve{H}_z}{\partial y} - \frac{\partial \breve{H}_y}{\partial z} \\[4pt]
\frac{\partial \breve{H}_x}{\partial z} - \frac{\partial \breve{H}_z}{\partial x} \\[4pt]
\frac{\partial \breve{H}_y}{\partial x} - \frac{\partial \breve{H}_x}{\partial y}
\end{bmatrix}
= j\omega\varepsilon
\begin{bmatrix}
\frac{s_y s_z}{s_x} & 0 & 0 \\[4pt]
0 & \frac{s_x s_z}{s_y} & 0 \\[4pt]
0 & 0 & \frac{s_x s_y}{s_z}
\end{bmatrix}
\begin{bmatrix}
\breve{E}_x \\[4pt]
\breve{E}_y \\[4pt]
\breve{E}_z
\end{bmatrix},
\tag{10.58}
$$

where s_x, s_y, and s_z are defined in (10.47). Directly inserting (10.47) into (10.58) and then transforming into the time domain would lead to a convolution between the tensor coefficients and the E-field. This is not advisable because implementing this convolution would be computationally intensive. As shown by GEDNEY [1995], GEDNEY [1996], a much more efficient approach is to define the proper constitutive relationship to decouple the frequency-dependent terms. Specifically, let

$$
\breve{D}_x = \varepsilon \frac{s_z}{s_x} \breve{E}_x;
\tag{10.59a}
$$

$$
\breve{D}_y = \varepsilon \frac{s_x}{s_y} \breve{E}_y;
\tag{10.59b}
$$

$$
\breve{D}_z = \varepsilon \frac{s_y}{s_z} \breve{E}_z.
\tag{10.59c}
$$

Then, (10.58) is rewritten as

$$
\begin{bmatrix}
\frac{\partial \breve{H}_z}{\partial y} - \frac{\partial \breve{H}_y}{\partial z} \\[4pt]
\frac{\partial \breve{H}_x}{\partial z} - \frac{\partial \breve{H}_z}{\partial x} \\[4pt]
\frac{\partial \breve{H}_y}{\partial x} - \frac{\partial \breve{H}_x}{\partial y}
\end{bmatrix}
= j\omega
\begin{bmatrix}
s_y & 0 & 0 \\
0 & s_z & 0 \\
0 & 0 & s_x
\end{bmatrix}
\begin{bmatrix}
\breve{D}_x \\[4pt]
\breve{D}_y \\[4pt]
\breve{D}_z
\end{bmatrix}.
\tag{10.60}
$$

Now, we substitute s_x, s_y, and s_z from (10.47) into (10.60), and then apply the inverse Fourier transform using the identity $j\omega f(\omega) \to (\partial/\partial t) f(t)$. This yields an equivalent system of time-domain differential equations for (10.60):

$$
\begin{bmatrix}
\frac{\partial H_z}{\partial y} - \frac{\partial H_y}{\partial z} \\[4pt]
\frac{\partial H_x}{\partial z} - \frac{\partial H_z}{\partial x} \\[4pt]
\frac{\partial H_y}{\partial x} - \frac{\partial H_x}{\partial y}
\end{bmatrix}
= \frac{\partial}{\partial t}
\begin{bmatrix}
\kappa_y & 0 & 0 \\
0 & \kappa_z & 0 \\
0 & 0 & \kappa_x
\end{bmatrix}
\begin{bmatrix}
D_x \\
D_y \\
D_z
\end{bmatrix}
+ \frac{1}{\varepsilon}
\begin{bmatrix}
\sigma_y & 0 & 0 \\
0 & \sigma_z & 0 \\
0 & 0 & \sigma_x
\end{bmatrix}
\begin{bmatrix}
D_x \\
D_y \\
D_z
\end{bmatrix}.
\tag{10.61}
$$

The system of equations in (10.61) can be discretized on the standard Yee lattice. It is suitable to use normal leapfrogging in time wherein the loss terms are time-averaged according to the semi-implicit scheme. This leads to explicit time-stepping expressions for D_x, D_y, and D_z. For example, the D_x update is given by

$$
D_x\big|_{i+1/2,j,k}^{n+1}
= \left(\frac{2\varepsilon\kappa_y - \sigma_y \Delta t}{2\varepsilon\kappa_y + \sigma_y \Delta t} \right) D_x\big|_{i+1/2,j,k}^{n} + \left(\frac{2\varepsilon \Delta t}{2\varepsilon\kappa_y + \sigma_y \Delta t} \right)
$$

$$\times \left(\frac{H_z|_{i+1/2,j+1/2,k}^{n+1/2} - H_z|_{i+1/2,j-1/2,k}^{n+1/2}}{\Delta y} \right.$$
$$\left. - \frac{H_y|_{i+1/2,j,k+1/2}^{n+1/2} - H_y|_{i+1/2,j,k-1/2}^{n+1/2}}{\Delta z} \right). \tag{10.62}$$

Next, we focus on (10.59a)–(10.59c). For example, we consider (10.59a). After multiplying both sides by s_x and substituting for s_x and s_z from (10.47a), (10.47c), we have

$$\left(\kappa_x + \frac{\sigma_x}{j\omega\varepsilon} \right) \breve{D}_x = \varepsilon \left(\kappa_z + \frac{\sigma_z}{j\omega\varepsilon} \right) \breve{E}_x. \tag{10.63}$$

Multiplying both sides by $j\omega$ and transforming into the time domain leads to

$$\frac{\partial}{\partial t}(\kappa_x D_x) + \frac{\sigma_x}{\varepsilon} D_x = \varepsilon \left[\frac{\partial}{\partial t}(\kappa_z E_x) + \frac{\sigma_z}{\varepsilon} E_x \right]. \tag{10.64a}$$

Similarly, from (10.59b) and (10.59c), we obtain

$$\frac{\partial}{\partial t}(\kappa_y D_y) + \frac{\sigma_y}{\varepsilon} D_y = \varepsilon \left[\frac{\partial}{\partial t}(\kappa_x E_y) + \frac{\sigma_x}{\varepsilon} E_y \right], \tag{10.64b}$$

$$\frac{\partial}{\partial t}(\kappa_z D_z) + \frac{\sigma_z}{\varepsilon} D_z = \varepsilon \left[\frac{\partial}{\partial t}(\kappa_y E_z) + \frac{\sigma_y}{\varepsilon} E_z \right]. \tag{10.64c}$$

The time derivatives in (10.64) are discretized using standard Yee leapfrogging and time-averaging the loss terms. This yields explicit time-stepping expressions for E_x, E_y, and E_z. For example, the E_x update is given by

$$E_x|_{i+1/2,j,k}^{n+1} = \left(\frac{2\varepsilon\kappa_z - \sigma_z\Delta t}{2\varepsilon\kappa_z + \sigma_z\Delta t} \right) E_x|_{i+1/2,j,k}^{n} + \left[\frac{1}{(2\varepsilon\kappa_z + \sigma_z\Delta t)\varepsilon} \right]$$
$$\times \left[(2\varepsilon\kappa_x + \sigma_x\Delta t)D_x|_{i+1/2,j,k}^{n+1} - (2\varepsilon\kappa_x - \sigma_x\Delta t)D_x|_{i+1/2,j,k}^{n} \right]. \tag{10.65}$$

Overall, updating the components of $\vec{E}$ in the UPML requires two steps in sequence: (1) obtaining the new values of the components of $\vec{D}$ according to (10.62), and (2) using these new $\vec{D}$ components to obtain new values of the $\vec{E}$-components according to (10.65).

A similar two-step procedure is required to update the components of $\vec{H}$ in the UPML. Starting with Faraday's Law in (10.45b) and (10.46), the first step involves developing the updates for the components of $\vec{B}$. A procedure analogous to that followed in obtaining (10.62) yields, for example, the following update for B_x:

$$B_x|_{i,j+1/2,k+1/2}^{n+3/2}$$
$$= \left(\frac{2\varepsilon\kappa_y - \sigma_y\Delta t}{2\varepsilon\kappa_y + \sigma_y\Delta t} \right) B_x|_{i,j+1/2,k+1/2}^{n+1/2} - \left(\frac{2\varepsilon\Delta t}{2\varepsilon\kappa_y + \sigma_y\Delta t} \right)$$
$$\times \left(\frac{E_z|_{i,j+1,k+1/2}^{n+1} - E_z|_{i,j,k+1/2}^{n+1}}{\Delta y} - \frac{E_y|_{i,j+1/2,k+1}^{n+1} - E_y|_{i,j+1/2,k}^{n+1}}{\Delta z} \right), \tag{10.66}$$

The second step involves updating the $\vec{H}$ components in the UPML using the values of the $\vec{B}$ components just obtained with (10.66) and similar expressions for B_y and B_z. For example, employing the dual constitutive relation $\check{B}_x = \mu(s_z/s_x)\check{H}_x$, a procedure analogous to that followed in obtaining (10.65) yields the following update for H_x:

$$
H_x\big|_{i,j+1/2,k+1/2}^{n+3/2}
$$
$$
= \left(\frac{2\varepsilon\kappa_z - \sigma_z\Delta t}{2\varepsilon\kappa_z + \sigma_z\Delta t}\right) H_x\big|_{i,j+1/2,k+1/2}^{n+1/2} + \left[\frac{1}{(2\varepsilon\kappa_z + \sigma_z\Delta t)\mu}\right]
$$
$$
\times \left[(2\varepsilon\kappa_x + \sigma_x\Delta t)B_x\big|_{i,j+1/2,k+1/2}^{n+3/2} - (2\varepsilon\kappa_x - \sigma_x\Delta t)B_x\big|_{i,j+1/2,k+1/2}^{n+1/2}\right].
$$
(10.67)

Similar expressions can be derived for H_y and H_z.

NEHRBASS, LEE, and LEE [1996] showed that such an algorithm is numerically stable within the Courant limit. Further, ABARBANEL and GOTTLIEB [1997] showed that the resulting discrete fields satisfy Gauss' Law, and the UPML is well posed.

10.7.2. Computer implementation of the UPML

Each $\vec{E}$ and $\vec{H}$ component within the UPML is computed using an explicit two-step time-marching scheme as illustrated in (10.62) and (10.65) for E_x, and in (10.66) and (10.67) for H_x. Based on these updates, the UPML is easily and efficiently implemented within the framework of existing FDTD codes. We now illustrate this in FORTRAN using the time-stepping expressions for E_x given in (10.62) and (10.65). First, we pre-compute six coefficient arrays to be used in the field updates:

$$
C1(j) = \frac{2\varepsilon\kappa_y(j) - \sigma_y(j)\Delta t}{2\varepsilon\kappa_y(j) + \sigma_y(j)\Delta t}, \tag{10.68a}
$$
$$
C2(j) = \frac{2\varepsilon\Delta t}{2\varepsilon\kappa_y(j) + \sigma_y(j)\Delta t}, \tag{10.68b}
$$
$$
C3(k) = \frac{2\varepsilon\kappa_z(k) - \sigma_z(k)\Delta t}{2\varepsilon\kappa_z(k) + \sigma_z(k)\Delta t}, \tag{10.68c}
$$
$$
C4(k) = \frac{1}{[2\varepsilon\kappa_z(k) + \sigma_z(k)\Delta t]\varepsilon}, \tag{10.68d}
$$
$$
C5(i) = 2\varepsilon\kappa_x(i) + \sigma_x(i)\Delta t, \tag{10.68e}
$$
$$
C6(i) = 2\varepsilon\kappa_x(i) - \sigma_x(i)\Delta t. \tag{10.68f}
$$

Defining the field-updating coefficients in this manner permits a unified treatment of both the lossless interior working volume and the UPML slabs. In effect, UPML is assumed to fill the entire FDTD space lattice. We set $\sigma_w = 0$ and $\kappa_w = 1$ in the working volume to model free space. However, in the UPML slabs, σ_w and κ_w are assumed to have the polynomial-graded profile given in (10.51), or the geometric-graded profile given in (10.54), along the normal axes of the UPML slabs. As a result, the coefficients in (10.68) vary in only one dimension.

When defining the coefficient arrays specified in (10.68), it is critical to assign the proper value to the UPML loss parameters. To this end, σ_w and κ_w are computed at a physical coordinate using (10.51) or (10.54). The appropriate choice of the physical coordinate is at the edge center of the discrete field $E_x(i, j, k)$, which is $[(i + 1/2)\Delta x, j\Delta y, k\Delta z]$. Thus, in (10.68e) and (10.68f), $\sigma_x(i)$ and $\kappa_x(i)$ are computed at the physical coordinate $(i + 1/2)\Delta x$. Similarly, in (10.68a) and (10.68b), $\sigma_y(j)$ and $\kappa_y(j)$ are computed at the physical coordinate $j\Delta y$; and in (10.68c) and (10.68d), $\sigma_z(k)$ and $\kappa_z(k)$ are computed at the physical coordinate $k\Delta z$. This is similarly done for the updates of E_y and E_z.

The UPML loss parameters for the H-fields are chosen at the lattice face centers. For example, the physical coordinate of the discrete field $H_x(i, j, k)$ is $[i\Delta x, (j + 1/2)\Delta y, (k + 1/2)\Delta z]$. Thus, for the update of $H_x(i, j, k)$, $\sigma_x(i)$ and $\kappa_x(i)$ are computed at the physical coordinate $i\Delta x$; $\sigma_y(j)$ and $\kappa_y(j)$ are computed at the physical coordinate $(j + 1/2)\Delta y$; and $\sigma_z(k)$ and $\kappa_z(k)$ are computed at the physical coordinate $(k + 1/2)\Delta z$. This is similarly done for the updates of H_y and H_z.

Given the above "all-UPML" strategy, and assuming that the infinite region extending out of the space lattice has homogeneous material properties, then the FORTRAN program segment that executes the time-stepping of E_x *everywhere* in the FDTD space lattice can be written as a simple triply-nested loop:

```
do 10 k=2,nz-1
  do 10 j=2,ny-1
    do 10 i=1,nx-1
      dstore = dx(i,j,k)
      dx(i,j,k) = C1(j)*dx(i,j,k)
        + C2(j)*( (hz(i,j,k) - hz(i,j-1,k)) / deltay -
        (hy(i,j,k) - hy(i,j,k-1)) / deltaz )
      ex(i,j,k) = C3(k)*ex(i,j,k)
        + C4(k)*( C5(i)*dx(i,j,k) - C6(i)*dstore )
  10 continue
```
$$(10.69)$$

Assuming UPML throughout the entire FDTD lattice in this manner has the limitation that the flux densities D_x and B_x must be stored everywhere in the lattice. However, this approach offers the significant advantage of simplifying the modification of existing FDTD codes. An alternative is to write a triply-nested loop for the interior fields and separate loops for the various UPML slabs (segregating the corner regions). In this case, the auxiliary variables need to be stored only in the UPML region, leading to memory savings. Further, in this circumstance, the UPML requires considerably less storage than Berenger's split-field PML since only the normal fields require dual storage, as opposed to the two tangential fields. The memory requirement (real numbers) for the UPML truncation on all six outer lattice boundaries totals

$$6N_x N_y N_z + 8N_{\text{UPML}}(N_x N_y + N_y N_z + N_z N_x)$$
$$- 16N_{\text{UPML}}(N_x + N_y + N_z) + (24N_{\text{UPML}})^2, \tag{10.70}$$

where N_{UPML} is the thickness (in space cells) of the UPML. In contrast, approximately $6N_x N_y N_z$ real numbers must be stored when using FDTD with a local ABC. With these measures, it is straightforward to calculate the percentage of additional memory required to implement the UPML ABC in a cubic lattice ($N_x = N_y = N_z$) relative to the primary field storage. For 4-cell UPML, the storage burden drops below 10% for

$N_x > 90$; and for 10-cell UPML, this burden drops below 10% for $N_x > 240$. Note also that, without the use of the reflection-cancellation techniques of RAMAHI [1998], local ABCs must be placed *much* further out than the UPML, requiring even larger lattices. Consequently, the UPML ABC can lead to an overall decrease in the required memory to achieve a given desired outer-boundary reflectivity.

If we assume UPML throughout the entire FDTD space lattice as in (10.68) and (10.69), the computer memory requirement is $12N_x N_y N_z$ real numbers. This option is suitable when coding simplicity is desired and memory is not a constraint.

10.8. Numerical experiments with Berenger's split-field PML

10.8.1. Outgoing cylindrical wave in a two-dimensional open-region grid

We first review the numerical experiment of KATZ, THIELE, and TAFLOVE [1994] which used Berenger's split-field PML in a two-dimensional square-cell FDTD grid to absorb an outgoing cylindrical wave generated by a hard source centered in the grid. Using the methodology of MOORE, BLASCHAK, TAFLOVE, and KRIEGSMANN [1988], the accuracy of the Berenger's PML ABC was compared with that of the previously standard, second-order accurate, analytical ABC of MUR [1981]. The PML loss was assumed to be quadratically graded with depth from the interface of the interior free-space computation region. This allowed a direct comparison with the computed results reported by BERENGER [1994].

Fig. 10.2 graphs the global error power within a 100×50-cell TE$_z$ test grid for both the Mur ABC and a 16-cell Berenger PML ABC. At $n = 100$ time steps, the global reflection error in the PML grid is about 10^{-7} times the error in the Mur grid, dropping to a microscopic 10^{-12} times the global error in the Mur grid at $n = 500$ time steps.

We next consider the performance of Berenger's PML ABC for this open-region radiation problem as a function of frequency. Here, the local PML reflection coefficient versus frequency is obtained by using the discrete Fourier transform to calculate the

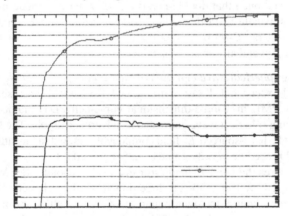

FIG. 10.2. Global error power within a 100×50-cell 2D TE$_z$ test grid for both the second-order Mur ABC and a 16-cell quadratically graded Berenger PML ABC, plotted as a function of time-step number on a logarithmic vertical scale. *Source*: D.S. Katz et al., *IEEE Microwave and Guided Wave Letters*, 1994, pp. 268–270, ©1994 IEEE.

204   *S.C. Hagness et al.*

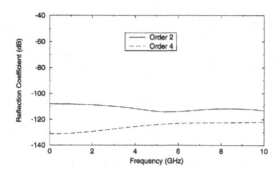

FIG. 10.3. PML reflection coefficient versus frequency for the order-2 (baseline quadratic grading) and order-4 grading cases. Two-dimensional grid with 16-cell-thick PML having $R(0) = 10^{-6}$ used for these calculations.

incident and reflected pulse spectra observed at the midpoint of the 100-cell air/PML interface, and dividing the reflected spectrum by the incident spectrum. The numerical procedure is otherwise similar to that used above, with the exception that the grading of the PML loss is either order 2 or order 4.

Fig. 10.3 graphs the results of this study, comparing the local PML reflection from 0–10 GHz. Here, the PML thickness is 16 cells with $R(0) = 10^{-6}$, and a uniform grid space increment of 1.5 mm (equivalent to $\lambda_0/20$ at 10 GHz) is used.

Fig. 10.3 shows that the local PML reflection coefficient is *virtually flat* from 0–10 GHz. Therefore, Berenger's PML is effective for absorbing ultrawideband pulses. We also observe that the order-4 PML loss grading has 10–24 dB less reflectivity than the baseline quadratic case. Additional studies of this type have shown similar results for a variety of FDTD models. These indicate that the optimum grading of the PML loss is generally not quadratic. It is apparent that a simple grading optimization provides a no-cost means of achieving the widest possible dynamic range of the PML ABC.

The reader is cautioned that double-precision computer arithmetic may be required to achieve the full benefit of PML grading. Simply shifting the test code from a Unix workstation to the Cray C-90 permitted the grading improvement of Fig. 10.3 to be observed. The improvement was not observed on the workstation.

10.8.2. *Outgoing spherical wave in a three-dimensional open-region lattice*
In this numerical experiment, KATZ, THIELE, and TAFLOVE [1994] used quadratically graded Berenger PML in a three-dimensional cubic-cell FDTD lattice to absorb an impulsive, outgoing spherical wave generated by a Hertzian dipole. The Hertzian dipole was simply a single, hard-sourced E_z field component centered in the lattice. Otherwise, the experimental procedure was the same as in Section 10.8.1.

Fig. 10.4 compares the local E-field error due to the second-order Mur and 16-cell PML ABCs for a $100 \times 100 \times 50$-cell three-dimensional test lattice. The observation was made along the x-axis at the outer boundary of the test lattice at time step $n = 100$, the time of maximum excitation of the PML by the outgoing wave. We see that the error due to the PML is on the order of 10^{-3} times that of the Mur ABC.

KATZ, THIELE, and TAFLOVE [1994] determined that, if one fixes the PML thickness, increasing the PML loss can reduce both the local and global reflection errors.

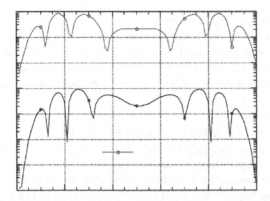

FIG. 10.4. Local E-field error at time-step $n = 100$ along the x-axis at the outer boundary of a $100 \times 100 \times 50$-cell three-dimensional test lattice for Mur's second-order ABC and 16-cell quadratically graded PML, plotted on a logarithmic vertical scale. *Source*: D.S. Katz et al., *IEEE Microwave and Guided Wave Letters*, 1994, pp. 268–270, © 1994 IEEE.

TABLE 10.1

Tradeoff of error reduction for quadratically graded PML relative to Mur's second-order abc versus computer resources for a three-dimensional test lattice of $100 \times 100 \times 50$ cells. *Source*: D.S. Katz et al., *IEEE Microwave and Guided Wave Letters*, 1994, pp. 268–270, © 1994 IEEE

ABC	Avg. local field error reduction relative to second-order Mur	Computer resources one CPU, Cray C-90	If free-space buffer is reduced by 10 cells
Mur	1 (0 dB)	10 Mwords, 6.5 s	–
4-cell PML	22 (27 dB)	16 Mwords, 12 s	7 Mwords, 10 s
8-cell PML	580 (55 dB)	23 Mwords, 37 s	12 Mwords, 27 s
16-cell PML	5800 (75 dB)	43 Mwords, 87 s	25 Mwords, 60 s

However, this benefit levels off when $R(0)$ drops to less than 10^{-5}. Similarly, the local and global error can drop as the PML thickness increases. Here, however, a tradeoff with the computer burden must be factored.

Table 10.1 compares for the test case of Fig. 10.4 the error reduction and computer resources of Mur's second-order ABC with a quadratically graded Berenger PML of varying thickness. Here, the arithmetic average of the absolute values of the E-field errors over a complete planar cut through the $100 \times 100 \times 50$-cell lattice at $y = 0$ and $n = 100$ is compared for the Mur and PML ABCs. The last column indicates the effect of reducing the free-space buffer between the interior working zone and the PML interface by 10 cells relative to that needed for Mur, taking advantage of the transparency of the PML ABC. From these results, a PML that is 4–8 cells thick appears to present a good balance between error reduction and computer burden. Relative to the outer-boundary reflection noise caused by Mur's ABC, PMLs in this thickness range improve the FDTD computational dynamic range by 27–55 dB.

In summary, these results show that Berenger's split-field PML achieves orders-of-magnitude less outer-boundary reflection than previous ABCs when used to model radiating sources in open regions. Depending upon the grading order of the PML loss, 16-cell PML is 60–80 dB less reflective than the second-order Mur ABC. Berenger's PML is also effective over ultrawideband frequency ranges. Unlike previous analytical ABCs (see TAFLOVE and HAGNESS [2000, Chapter 6]) used without the reflection cancellation techniques of RAMAHI [1998], the PML ABC can realize close to its theoretical potential.

10.8.3. Dispersive wave propagation in metal waveguides

FDTD is being used increasingly to model the electromagnetic behavior of not only open-region scattering problems, but also propagation of waves in microwave and optical circuits. An outstanding problem here is the accurate termination of guided-wave structures extending beyond the FDTD lattice boundaries. The key difficulty is that the propagation in a waveguide can be multimodal and dispersive, and the ABC used to terminate the waveguide must be able to absorb energy having widely varying transverse distributions and group velocities v_g.

REUTER, JOSEPH, THIELE, KATZ, and TAFLOVE [1994] used Berenger's split-field PML to obtain an ABC for an FDTD model of dispersive wave propagation in a two-dimensional, parallel-plate metal waveguide. This paper assumed a waveguide filled with air and having perfectly conducting walls separated by 40 mm (f_{cutoff} = 3.75 GHz). The waveguide was assumed to be excited by a Gaussian pulse of temporal width 83.3 ps (full width at half maximum – FWHM) modulating a 7.5-GHz carrier. This launched a $+x$-directed TM$_1$ mode having the field components E_x, E_y, and H_z towards an 8-cell or 32-cell PML absorber. In effect, the waveguide plunged into the PML, which provided an absorbing "plug." The two waveguide plates continued to the outer boundary of the FDTD grid, where they electrically contacted the perfectly conducting wall backing the PML medium. For the 8-cell PML trial, a quadratic loss grading was assumed with $R(0) = 10^{-6}$; cubic loss grading with $R(0) = 10^{-7}$ was used for the 32-cell PML trial.

Fig. 10.5(a) shows the spectrum of the input pulse used by REUTER, JOSEPH, THIELE, KATZ, and TAFLOVE [1994] superimposed upon the normalized group velocity for the TM$_1$ mode. The incident pulse contained significant spectral energy below cutoff, and the group velocity of the pulse's spectral components varied over an enormous range from zero at f_{cutoff} to about 0.98c well above f_{cutoff}. Because of this huge range, REUTER, JOSEPH, THIELE, KATZ, and TAFLOVE [1994] allowed the wave reflected from the PML to fully evolve over many thousands of time steps before completing the simulation. This properly modeled the very slowly propagating spectral components near f_{cutoff}, which generated an equally slowly decaying impulse response for the PML termination. Using a discrete Fourier transformation run concurrently with the FDTD time-stepping, this allowed calculation of the PML reflection coefficient versus frequency by dividing the reflected spectrum by the incident spectrum as observed at the air/PML interface.

Fig. 10.5(b) graphs the resulting reflection coefficient of the waveguide PML ABC versus frequency. For the 8-cell PML, reflections were between −60 dB and −100 dB in the frequency range 4–20 GHz. For the 32-cell PML, the reflection coefficient was below −100 dB in the frequency range 6–18 GHz. (Note that Cray word precision

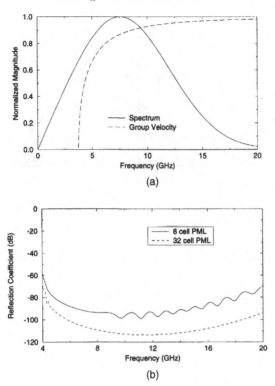

FIG. 10.5. Test of PML ABC for two-dimensional PEC parallel-plate waveguide propagating an ultrawide-band pulsed TM$_1$ mode. (a) Excitation spectrum superimposed upon the group velocity versus frequency (cutoff = 3.75 GHz). (b) PML reflection coefficient versus frequency. *Adapted from*: C.E. Reuter et al., *IEEE Microwave and Guided Wave Letters*, 1994, pp. 344–346, © 1994 IEEE.

was used for these studies.) This example demonstrates the ability of the PML ABC to absorb ultrawideband energy propagating in a waveguide having strong dispersion.

10.8.4. Dispersive and multimode wave propagation in dielectric waveguides

REUTER, JOSEPH, THIELE, KATZ, and TAFLOVE [1994] also reported numerical experiments using Berenger's split-field PML to terminate the FDTD model of a two-dimensional, asymmetric, dielectric-slab optical waveguide. This consisted of a 1.5-μm film of permittivity $\varepsilon_r = 10.63$ sandwiched between an infinite substrate of $\varepsilon_r = 9.61$ and an infinite region of air. The excitation introduced at the left edge of the three-layer system was a 17-fs FWHM Gaussian pulse modulating a 200-THz carrier. Fig. 10.6(a) shows the spectrum of this excitation superimposed upon the normalized propagation factors of the three modes supported by the optical waveguide in the frequency range 100–300 THz. We see that the incident pulse contained significant spectral energy over this entire range. Therefore, all three of the waveguide modes were launched. The model of the optical waveguiding system was terminated by extending each of its three dielectric layers into matching PML regions at the right side of the grid. The PML was 16 cells thick with (σ_x, σ_x^*) varying quadratically in the x-direction. Further, the PML loss

S.C. Hagness et al.

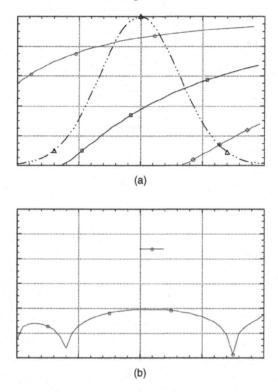

(a)

(b)

FIG. 10.6. Test of PML ABC for 2D asymmetric three-layer dielectric optical waveguide propagating a pulsed tri-modal wave: (a) excitation spectrum superimposed upon propagation factors for the three modes; (b) PML reflection coefficient versus frequency. *Source*: C.E. Reuter et al., *IEEE Microwave and Guided Wave Letters*, 1994, pp. 344–346, © 1994 IEEE.

parameter was chosen such that s_x was constant in the transverse direction, as described in Section 10.5.4.

Fig. 10.6(b) graphs the composite reflection coefficient representing the total retrodirected energy in all three regions, as computed at the PML interface. The PML ABC exhibited reflections below −80 dB across the entire spectrum of the incident field. This demonstrates the absorptive capability of the PML ABC for dispersive multimodal propagation.

In summary, these results show that Berenger's split-field PML can achieve highly accurate, ultrawideband terminations of PEC and dielectric waveguides in FDTD space lattices. The PML ABC can provide broadband reflection coefficients better than −80 dB, absorbing dispersive and multimodal energy. Relative to previous approaches for this purpose, the PML ABC has the advantages of being local in time and space, extremely accurate over a wide range of group velocities, and requiring no a priori knowledge of the modal distribution or dispersive nature of the propagating field. PML provides a combination of broadband effectiveness, robustness, and computational efficiency that is unmatched by previous ABCs for FDTD models.

10.9. Numerical experiments with UPML

In this section, the UPML termination of FDTD grids is presented for a number of sample applications. The goal is to provide an understanding of how the UPML parameters and grading functions impact its effectiveness as an ABC. In this manner, the reader can better understand how to properly choose these parameters.

10.9.1. Current source radiating in an unbounded two-dimensional region

Fig. 10.7 illustrates the first example, as reported by GEDNEY and TAFLOVE [2000]. This involves an electric current source $\vec{J}$ centered in a 40×40-cell FDTD grid. The source is vertically directed and invariant along the axial direction. Hence, it radiates two-dimensional TE waves. It has the time signature of a differentiated Gaussian pulse

$$J_y(x_0, y_0, t) = -2\big[(t - t_0)/t_w\big] \exp\big\{-\big[(t - t_0)/t_w\big]^2\big\}, \tag{10.71}$$

where $t_w = 26.53$ ps and $t_0 = 4t_w$.

The grid has 1-mm square cells and a time step of 0.98 times the Courant limit. The E-field is probed at two points, A and B, as shown in the figure. Point A is in the same plane as the source and two cells from the UPML, and point B is two cells from the bottom and side UPMLs. Time-stepping runs over 1000 iterations, well past the steady-state response. Both 5-cell and 10-cell UPML ABCs are used with polynomial grading $m = 4$.

For this case, the reference solution $E_{\text{ref}}|_{i,j}^n$ is obtained using a 1240×1240-cell grid. An identical current source is centered within this grid, and the field-observation point (i, j) is at the same position relative to the source as in the test grid. The reference grid is sufficiently large such that there are no reflections from its outer boundaries during the time stepping. This allows a relative error to be defined as

$$\text{Rel.error}|_{i,j}^n = \big|E|_{i,j}^n - E_{\text{ref}}|_{i,j}^n\big|/\big|E_{\text{ref max}}|_{i,j}\big|, \tag{10.72}$$

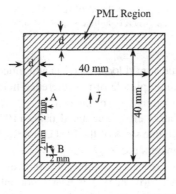

FIG. 10.7. Vertically directed electric current source centered in a 2D FDTD grid. The working volume of 40×40 mm is surrounded by UPML of thickness d. E-fields are probed at points A and B.

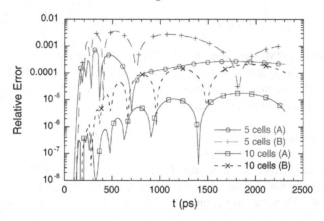

FIG. 10.8. Relative error at points A and B of Fig. 10.7 over 1000 time steps for 5-cell and 10-cell UPMLs with $\sigma_{max} = \sigma_{opt}$ and $\kappa = 1$.

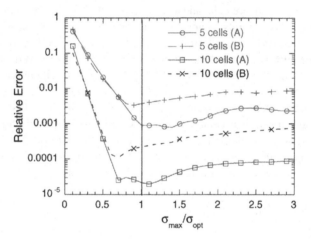

FIG. 10.9. Maximum relative error due to 5-cell and 10-cell UPML ABCs in the grid of Fig. 10.7 versus $\sigma_{max}/\sigma_{opt}$ over a 1000-time-step observation period.

where $E|_{i,j}^{n}$ is the field value at grid location (i, j) and time step n in the test grid, and $E_{\text{ref max}}|_{i,j}$ is the maximum amplitude of the reference field at grid location (i, j), as observed during the time-stepping span of interest.

Fig. 10.8 graphs the relative error calculated using (10.72) at points A and B of Fig. 10.7 over the first 1000 time steps of the FDTD run for 5-cell and 10-cell UPMLs. Here, the key UPML parameters are $\sigma_{max} = \sigma_{opt}$, where σ_{opt} is given by (10.57), and $\kappa_{max} = 1$. We note that the error at A is always less than that at B. This is because the wave impinging on the UPML near A is nearly normally incident and undergoes maximum absorption. At B, while the amplitude of the outgoing wave is smaller due to the radiation pattern of the source, the wave impinges on the UPML obliquely at $45°$.

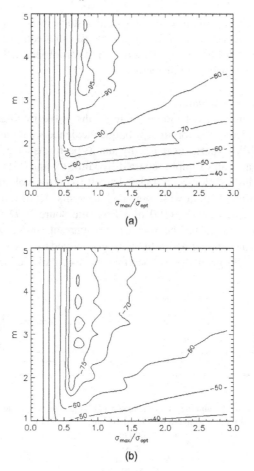

FIG. 10.10. Contour plots of the maximum relative error in dB in the grid of Fig. 10.7 versus $\sigma_{max}/\sigma_{opt}$ and polynomial order m for a 10-cell UPML: (a) at point A; (b) at point B.

Fig. 10.9 provides additional information by graphing as a function of $\sigma_{max}/\sigma_{opt}$ the maximum relative error at points A and B during the 1000-time-step simulation. As before, polynomial grading is used with $m = 4$, and the same σ_{max} is used for each of the four UPML absorbers at the outer boundary planes of the grid. We see that the optimal choice for σ_{max} is indeed close to σ_{opt}. Again, the maximum error at B is about an order of magnitude larger than that at A.

Figs. 10.10(a) and 10.10(b) are contour plots resulting from a comprehensive parametric study of the 10-cell UPML. These figures map the maximum relative error at A and B, respectively, during the 1000-time-step simulation. The horizontal axis of each plot provides a scale for $\sigma_{max}/\sigma_{opt}$, and the vertical axis provides a scale for the polynomial order m, not necessarily an integer. The minimum error is found for $3 < m < 4$ and $\sigma_{max} \cong 0.75\sigma_{opt}$, and is approximately -95 dB at A and -80 dB at B.

212   S.C. Hagness et al.

GEDNEY and TAFLOVE [2000] reported a similar study involving geometric grading
of the UPML parameters. For $g = 2.2$, about -85 dB of reflection error was realized at
both A and B for $\ln[R(0)]$ between -12 and -16. It was observed that the effectiveness
of the UPML is quite sensitive to the choice of g.

10.9.2. Highly elongated domains
In the previous example, the incident angle of the wave impinging on the UPML
never exceeded 45°. However, for highly oblique incidence angles approaching 90°, the
reflectivity of the UPML increases markedly and its performance as an ABC degrades.

Fig. 10.11 illustrates an example reported by GEDNEY and TAFLOVE [2000] of just
such a situation, a highly elongated 435×15-cell (working volume) TE$_z$ grid. Here,
the relative error in the E-field was computed at points A (10 mm from the source),
B (50 mm from the source), C (100 mm from the source), D (200 mm from the
source), and E (400 mm from the source). The current source was polarized such
that the radiated signal impinging on the long grid boundary was maximum in amplitude. From the problem geometry, we see that the specularly reflected wave from the

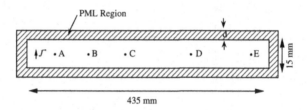

FIG. 10.11. Current element $\vec{J}$ radiating in an elongated FDTD grid (not to scale) terminated by UPML.
Distance of each observation point from the source: A, 10 mm; B, 50 mm; C, 100 mm; D, 200 mm; and E,
400 mm.

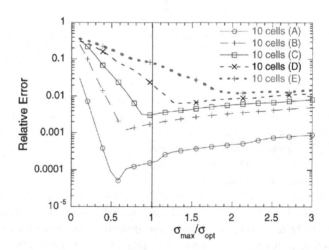

FIG. 10.12. Maximum relative error at points A, B, C, D, and E in the grid of Fig. 10.11 due to a polynomial-scaled 10-cell PML vs. $\sigma_{max}/\sigma_{opt}$. Parameters $m = 4$ and $\kappa = 1$ are fixed.

long-grid-boundary UPML arriving at E was incident on the UPML at 89°, implying that the reflection error at E should be degraded from that observed at A.

The model of Fig. 10.11 used 1-mm square grid cells and Δt set at 0.98 times the Courant limit. The source had the same differentiated Gaussian-pulse waveform used previously. A 10-cell UPML absorber with polynomial spatial scaling ($m = 4$ and $\kappa_{max} = 1$) was used on all sides.

Fig. 10.12 graphs the maximum relative error recorded at each of the observation points over the initial 1000 time-steps as a function of $\sigma_{max}/\sigma_{opt}$. While the error at

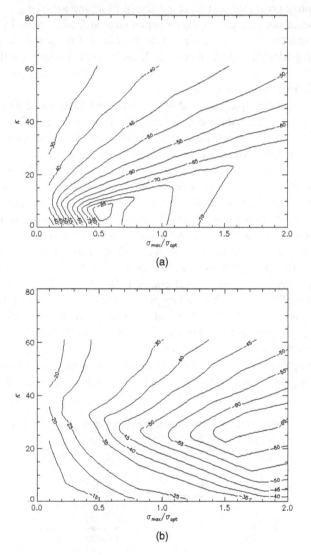

(a)

(b)

FIG. 10.13. Contour plots of the maximum relative error in dB in the grid of Fig. 10.11 vs. $\kappa_{y\text{-max}}$ and $\sigma_{y\text{-max}}/\sigma_{opt}$ for a polynomial-scaled 10-cell UPML: (a) at point A; (b) at point E.

A can be less than 0.0001, or -80 dB, the error progressively worsens at $B-E$. We expect that larger values of σ_{max} could reduce the reflection error. However, larger values of σ_{max} lead to larger step discontinuities in the UPML profile, and hence, a larger discretization error.

We note that the radiation due to the current source is characterized by a spectrum of waves containing evanescent as well as propagating modes. However, the evanescent modes are not absorbed by the UPML when $\kappa = 1$. Increasing κ should help this situation. To investigate this possibility, Figs. 10.13(a) and (b)10.13 plot contours of the maximum relative error at A and E, respectively, as a function of $\sigma_{max}/\sigma_{opt}$ and κ_{max}. We see that increasing κ_{max} causes the reflection error at E to decrease by two orders of magnitude to less than -65 dB in the vicinity of $\sigma_{max} \approx 1.6\sigma_{opt}$ and $\kappa_{max} \approx 25$. While this strategy degrades the reflection error at A, the error at A is still less than -65 dB.

10.9.3. Microstrip transmission line

Our final numerical example, reported by GEDNEY and TAFLOVE [2000], involves the use of UPML to terminate a three-dimensional FDTD model of a 50 Ω microstrip transmission line. This is a case wherein an inhomogeneous dielectric medium penetrates into the UPML, and ultralow levels of wave reflection are required.

Fig. 10.14 illustrates the cross-section of the microstrip line. The metal trace was assumed to be 254 μm wide with a negligible thickness compared to its width. This trace was assumed printed on a 254 μm thick alumina substrate having $\varepsilon_r = 9.8$, with the region above the substrate being air. The line was assumed to be excited at one end by a voltage source with a Gaussian profile and a 40-GHz bandwidth.

A uniform lattice discretization $\Delta x = \Delta y = 42.333$ μm was used in the transverse (x, y)-plane of the microstrip line, while $\Delta z = 120$ μm was used along the direction of wave propagation z. Polynomial-graded UPML backed by perfectly conducting walls terminated all lattice outer boundary planes. The metal trace extended completely through the z-normal UPML and electrically contacted the backing wall. In addition, the substrate, ground plane, and air media each continued into their respective adjacent UPMLs, maintaining their permittivities and geometries.

The wave impinging on the z-normal boundary was a quasi-TEM mode supported by the microstrip line. Even though the media were inhomogeneous, the UPML was

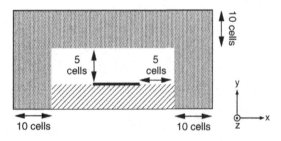

FIG. 10.14. Cross section of a 50 Ω microstrip line printed on a 254 μm thick alumina substrate terminating in UPML. The FDTD lattice sidewalls are also terminated by UPML.

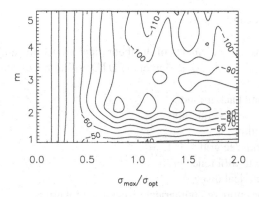

Fɪɢ. 10.15. Reflection error (in dB) of a 10-cell z-normal UPML in the microstrip line of Fig. 10.14 as a function of the polynomial grading order m and the normalized conductivity.

perfectly matched to this wave. The goal was to study the reflection behavior of the UPML as a function of its loss profile when used to terminate this line.

Because the conductivity within the UPML was polynomial graded, the optimal value of σ_{zmax} could be predicted by (10.57). However, an ambiguity existed because the dielectric penetrating into the z-normal UPML was inhomogeneous. Thus, as recommended in Section 10.5.4, the UPML conductivity was scaled by the relative permittivity as defined in (10.48). This yielded

$$\sigma_{z,\text{opt}} = \frac{0.8(m+1)}{\eta_0 \Delta \sqrt{\varepsilon_{\text{eff}}}}, \tag{10.73}$$

where ε_{eff} is the effective relative permittivity for the inhomogeneous media extending into the UPML. For the microstrip line, ε_{eff} could be estimated via quasistatic theory (see, for example, Poᴢᴀʀ [1998]), and in the case of Fig. 10.14 equals 6.62.

Fig. 10.15 illustrates the results of a parametric study of the reflection-error performance of a 10-cell UPML at the z-normal lattice outer boundary. The parameters investigated were $\sigma_{z\,\text{max}}$ and the polynomial grading order m. We see that the optimal value of σ_{zmax} was well predicted by (10.73), and further choosing m in the range of 3–5 was sufficient for minimizing the reflection error. Additional studies showed that the UPML reflection was better than -100 dB from 0–50 GHz for $m = 4$ and $\sigma_{z,\text{opt}}$.

10.10. UPML terminations for conductive and dispersive media

For certain applications, it is necessary to simulate electromagnetic wave interactions within conductive or frequency-dispersive media of significant spatial extent. Examples include wave propagation within microwave circuits printed on lossy dielectric substrates, and impulsive scattering by objects buried in the earth or embedded within biological tissues. In such cases, it is desirable to simulate the lossy or dispersive material extending to infinity through the use of a PML absorbing boundary. The reader is referred to the work of Gᴇᴅɴᴇʏ and Tᴀғʟᴏᴠᴇ [2000], which discusses in detail the extension of the UPML formulation presented previously in this section for purposes of terminating conductive and dispersive media.

11. Summary and conclusions

This chapter reviewed key elements of the theoretical foundation and numerical implementation of finite-difference time-domain (FDTD) solutions of Maxwell's equations. The chapter included:

- Introduction and background
- Review of Maxwell's equations
- The Yee algorithm
- The nonuniform Yee grid
- Alternative finite-difference grids
- Theory of numerical dispersion
- Algorithms for improved numerical dispersion properties
- Theory of numerical stability
- Alternating-direction implicit time-stepping algorithm for operation beyond the Courant limit
- Perfectly matched layer (PML) absorbing boundary conditions, including Berenger's split-field PML, the stretched-coordinate PML formulation, and the uniaxial anisotropic PML (UPML).

With literally hundreds of papers on FDTD methods and applications published each year, it is clear that FDTD is one of the most powerful and widely used numerical modeling approaches for electromagnetic wave interaction problems. With expanding developer and user communities within an increasing number of disciplines in science and engineering, FDTD technology is continually evolving in terms of its theoretical basis, numerical implementation, and technological applications. The latter now literally approach the proverbial spectral range from dc to light.

References

ABARBANEL, S., GOTTLIEB, D. (1997). A mathematical analysis of the PML method. *J. Comput. Phys.* **134**, 357–363.

BERENGER, J.P. (1994). A perfectly matched layer for the absorption of electromagnetic waves. *J. Comput. Phys.* **114**, 185–200.

BERENGER, J.P. (1996). Perfectly matched layer for the FDTD solution of wave-structure interaction problems. *IEEE Trans. Antennas Propagat.* **44**, 110–117.

BERENGER, J.P. (1997). Improved PML for the FDTD solution of wave-structure interaction problems. *IEEE Trans. Antennas Propagat.* **45**, 466–473.

CHEW, W.C., JIN, J.M. (1996). Perfectly matched layers in the discretized space: an analysis and optimization. *Electromagnetics* **16**, 325–340.

CHEW, W.C., WEEDON, W.H. (1994). A 3D perfectly matched medium from modified Maxwell's equations with stretched coordinates. *IEEE Microwave Guided Wave Lett.* **4**, 599–604.

CHURCHILL, R.V., BROWN, J.W., VERHEY, R.F. (1976). *Complex Variables and Applications* (McGraw-Hill, New York).

DEY, S., MITTRA, R. (1997). A locally conformal finite-difference time-domain algorithm for modeling three-dimensional perfectly conducting objects. *IEEE Microwave Guided Wave Lett.* **7**, 273–275.

ENGQUIST, B., MAJDA, A. (1977). Absorbing boundary conditions for the numerical simulation of waves. *Math. Comput.* **31**, 629–651.

FANG, J. (1989). Time-domain finite difference computations for Maxwell's equations, Ph.D. dissertation, Univ. of California, Berkeley, CA.

GEDNEY, S.D. (1995). An anisotropic perfectly matched layer absorbing medium for the truncation of FDTD lattices, Report EMG-95-006, University of Kentucky, Lexington, KY.

GEDNEY, S.D. (1996). An anisotropic perfectly matched layer absorbing medium for the truncation of FDTD lattices. *IEEE Trans. Antennas Propagat.* **44**, 1630–1639.

GEDNEY, S.D. (1998). The perfectly matched layer absorbing medium. In: Taflove, A. (ed.), *Advances in Computational Electrodynamics: The Finite-Difference Time-Domain Method* (Artech House, Norwood, MA), pp. 263–343.

GEDNEY, S.D., LANSING, F. (1995). Nonuniform orthogonal grids. In: Taflove, A. (ed.), *Computational Electrodynamics: The Finite-Difference Time-Domain Method* (Artech House, Norwood, MA), pp. 344–353.

GEDNEY, S.D., TAFLOVE, A. (2000). Perfectly matched layer absorbing boundary conditions. In: Taflove, A., Hagness, S.C. (eds.), *Computational Electrodynamics: The Finite-Difference Time-Domain Method*, second ed. (Artech House, Norwood, MA), pp. 285–348.

GONZALEZ GARCIA, S., LEE, T.W., HAGNESS, S.C. (2002). On the accuracy of the ADI-FDTD method. *IEEE Antennas Wireless Propagation Lett.* **1**, 31–34.

HARRINGTON, R.F., (1968). *Field Computation by Moment Methods* (MacMillan, New York).

HE, L. (1997). FDTD – Advances in sub-sampling methods, UPML, and higher-order boundary conditions, M.S. thesis, University of Kentucky, Lexington, KY.

HIGDON, R.L. (1986). Absorbing boundary conditions for difference approximations to the multi-dimensional wave equation. *Math. Comput.* **47**, 437–459.

HOLLAND, R. (1984). Implicit three-dimensional finite-differencing of Maxwell's equations. *IEEE Trans. Nucl. Sci.* **31**, 1322–1326.

HOLLAND, R., CHO, K.S. (1986). Alternating-direction implicit differencing of Maxwell's equations: 3D results. Computer Sciences Corp., Albuquerque, NM, Technical report to Harry Diamond Labs., Adelphi, MD, Contract DAAL02-85-C-0200

HOLLAND, R., WILLIAMS, J. (1983). Total-field versus scattered-field finite-difference. *IEEE Trans. Nucl. Sci.* **30**, 4583–4587.

JURGENS T.G., TAFLOVE, A., UMASHANKAR, K.R., MOOREA, T.G. (1992). Finite-difference time-domain modeling of curved surfaces. *IEEE Trans. Antennas Propagat.* **40**, 357–366.

KATZ, D.S., THIELE, E.T., TAFLOVE, A. (1994). Validation and extension to three-dimensions of the Berenger PML absorbing boundary condition for FDTD meshes. *IEEE Microwave Guided Wave Lett.* **4**, 268–270.

KELLER, J.B. (1962). Geometrical theory of diffraction. *J. Opt. Soc. Amer.* **52**, 116–130.

KOUYOUMJIAN, R.G., PATHAK, P.H. (1974). A uniform geometrical theory of diffraction for an edge in a perfectly conducting surface. *Proc. IEEE* **62**, 1448–1461.

KREISS, H., MANTEUFFEL, T., SCHWARTZ, B., WENDROFF, B., WHITE, J.A.B. (1986). Supraconvergent schemes on irregular meshes. *Math. Comput.* **47**, 537–554.

KRUMPHOLZ, M., KATEHI, L.P.B. (1996). MRTD: new time-domain schemes based on multiresolution analysis. *IEEE Trans. Microwave Theory Tech.* **44**, 555–572.

LIAO, Z.P., WONG, H.L., YANG, B.P., YUAN, Y.F. (1984). A transmitting boundary for transient wave analyses. *Scientia Sinica (series A)* **XXVII**, 1063–1076.

LIU, Q.H. (1996). The PSTD algorithm: a time-domain method requiring only two grids per wavelength, Report NMSU-ECE96–013, New Mexico State Univ., Las Cruces, NM

LIU, Q.H. (1997). The pseudospectral time-domain (PSTD) method: a new algorithm for solutions of Maxwell's equations. In: Proc. 1997 IEEE Antennas & Propagation Soc. Internat. Symp. **I** (IEEE, Piscataway, NJ), pp. 122–125 (catalog no. 97CH36122).

LIU, Y. (1996). Fourier analysis of numerical algorithms for the Maxwell's equations. *J. Comput. Phys.* **124**, 396–416.

MADSEN, N.K., ZIOLKOWSKI, R.W. (1990). A three-dimensional modified finite volume technique for Maxwell's equations. *Electromagnetics* **10**, 147–161.

MANTEUFFEL, T.A., WHITE, J.A. (1986). The numerical solution of second-order boundary value problems on nonuniform meshes. *Math. Comput.* **47**, 511–535.

MIN, M.S., TENG, C.H. (2001). The instability of the Yee scheme for the "magic time step." *J. Comput. Phys.* **166**, 418–424.

MONK, P. (1994). Error estimates for Yee's method on non-uniform grids. *IEEE Trans. Magnetics.* **30**, 3200–3203.

MONK, P., SULI, E. (1994). A convergence analysis of Yee's scheme on non-uniform grids. *SIAM J. Numer. Anal.* **31**, 393–412.

MOORE, T.G. BLASCHAK, J.G., TAFLOVE, A., KRIEGSMANN, G.A. (1988). Theory and application of radiation boundary operators. *IEEE Trans. Antennas Propagation* **36**, 1797–1812.

MUR, G. (1981). Absorbing boundary conditions for the finite-difference approximation of the time-domain electromagnetic field equations. *IEEE Trans. Electromagnetic Compatibility* **23**, 377–382.

NAMIKI, T. (2000). 3-D ADI-FDTD method – unconditionally stable time-domain algorithm for solving full vector Maxwell's equations. *IEEE Trans. Microwave Theory and Techniques* **48**, 1743–1748.

NEHRBASS, J.W., LEE, J.F., LEE, R. (1996). Stability analysis for perfectly matched layered absorbers. *Electromagnetics* **16**, 385–389.

POZAR, D.M. (1998). Microwave Engineering, second ed. (Wiley, New York)

RAMAHI, O.M. (1997). The complementary operators method in FDTD simulations. *IEEE Antennas Propagat. Magazine* **39/6**, 33–45.

RAMAHI, O.M. (1998). The concurrent complementary operators method for FDTD mesh truncation. *IEEE Trans. Antennas Propagat.* **46**, 1475–1482.

RAPPAPORT, C.M. (1995) Perfectly matched absorbing boundary conditions based on anisotropic lossy mapping of space. *IEEE Microwave Guided Wave Lett.* **5**, 90–92.

REUTER, C.E., JOSEPH, R.M., THIELE, E.T., KATZ, D.S., TAFLOVE, A. (1994). Ultrawideband absorbing boundary condition for termination of waveguiding structures in FDTD simulations. *IEEE Microwave Guided Wave Lett.* **4**, 344–346.

SACKS, Z.S., KINGSLAND, D.M., LEE, R., LEE, J.F. (1995). A perfectly matched anisotropic absorber for use as an absorbing boundary condition. *IEEE Trans. Antennas Propagat.* **43**, 1460–1463.

SCHNEIDER, J.B., WAGNER, C.L. (1999). FDTD dispersion revisited: Faster-than-light propagation. *IEEE Microwave Guided Wave Lett.* **9**, 54–56.

SHANKAR, V., MOHAMMADIAN, A.H., HALL, W.F. (1990). A time-domain finite-volume treatment for the Maxwell equations. *Electromagnetics*, **10**, 127–145.

SHEEN, D. (1991). Numerical modeling of microstrip circuits and antennas, Ph.D. thesis, Massachusetts Institute of Technology, Cambridge, MA

SHLAGER, K.L., SCHNEIDER, J.B. (1998). A survey of the finite-difference time-domain literature. In: Taflove, A. (ed.), *Advances in Computational Electrodynamics: The Finite-Difference Time-Domain Method* (Artech House, Norwood, MA), pp. 1–62.

SONG, J. CHEW, W.C. (1998). The fast Illinois solver code: requirements and scaling properties. *IEEE Comp. Sci. Engrg.* **5**.

TAFLOVE, A. (1995). *Computational Electrodynamics: The Finite-Difference Time-Domain Method*, first ed. (Artech House, Norwood, MA).

TAFLOVE, A., BRODWIN, M.E. (1975). Numerical solution of steady-state electromagnetic scattering problems using the time-dependent Maxwell's equations. *IEEE Trans. Microwave Theory and Techniques* **23**, 623–630.

TAFLOVE, A., HAGNESS, S.C. (2000). *Computational Electrodynamics: The Finite-Difference Time-Domain Method*, second ed. (Artech House, Norwood, MA).

TAFLOVE, A., UMASHANKAR, K.R., BEKER, B., HARFOUSH F.A., YEE, K.S. (1988). Detailed FDTD analysis of electromagnetic fields penetrating narrow slots and lapped joints in thick conducting screens. *IEEE Trans. Antennas Propagation* **36**, 247–257.

TEIXEIRA, F.L., CHEW, W.C. (1997). PML-FDTD in cylindrical and spherical coordinates. *IEEE Microwave Guided Wave Lett.* **7** 285–287.

TULINTSEFF, A. (1992). The finite-difference time-domain method and computer program description applied to multilayered microstrip antenna and circuit configurations, Technical Report AMT: 336.5-92-041, Jet Propulsion Laboratory, Pasadena, CA.

TURKEL, E. (1998). In: Taflove, A. (ed.), *Advances in Computational Electrodynamics: The Finite-Difference Time-Domain Method* (Artech House, Norwood, MA). Chapter 2.

YEE, K.S. (1966). Numerical solution of initial boundary value problems involving Maxwell's equations in isotropic media. *IEEE Trans. Antennas Propagat.* **14**, 302–307.

ZHENG, F., CHEN, Z., ZHANG, J. (2000). Toward the development of a three-dimensional unconditionally stable finite-difference time-domain method. *IEEE Trans. Microwave Theory and Techniques.* **48**, 1550–1558.

ZHENG, F., CHEN, Z. (2001). Numerical dispersion analysis of the unconditionally stable 3-D ADI-FDTD method. *IEEE Trans. Microwave Theory and Techniques* **49**, 1006–1009.

Simulation of EMC Behavior

A.J.H. Wachters

Philips Research Laboratories, Prof. Holstlaan 4,
5656 AA, Eindhoven, The Netherlands
E-mail address: wachters@natlab.research.philips.com

W.H.A. Schilders

Philips Research Laboratories, IC Design, Prof. Holstlaan 4,
5656 AA, Eindhoven, The Netherlands
E-mail address: wil.schilders@philips.com

1. Introduction

In this chapter we describe methods that have been used to perform simulations of the electromagnetic behavior of multilayer interconnection systems. Such a system consists of a number of planar conductors immersed in a configuration of homogeneous media of different permittivity bound by parallel planes. Examples of such systems are printed circuit boards, IC packages, filters, and passive IC's. The program Fasterix, developed within Philips Research, has been used to simulate the electromagnetic behavior of a variety of such systems (see DU CLOUX, MAAS, and WACHTERS [1994]). One of the reasons for developing this program were the strict regulations as far as electromagnetic compatibility are concerned. Electronic devices influence each other, but this influence should be kept to a minimum. This explains the large interest in simulations of EMC behavior in the past 10 years. The present chapter is devoted to this subject, and gives a very detailed impression of how these simulations are enabled in practice. The development of numerical algorithms to solve the EMC problems is rather involved. A strong interplay is required between analytical and numerical techniques in order to obtain an efficient way of simulating devices. Evaluating fourfold integrals with singularities is not a trivial task, and requires a lot of tedious work. The chapter also shows that, even

Essential Numerical Methods in Electromagnetics
Special Volume (W.H.A. Schilders and E.J.W. ter Maten, Guest Editors) of
HANDBOOK OF NUMERICAL ANALYSIS, VOL. XIII
P.G. Ciarlet (Editor)

ISSN 1570-8659
DOI 10.1016/B978-0-444-53756-0.00003-3

in rather elementary tasks like numerical integration, sophisticated algorithms must be used to handle the complexity of the problem. This is characteristic for the present chapter: several methods that are not very well known will be discussed, such as the Kronrod and Patterson quadrature rules, and Orden's method for solving indefinite linear systems.

The structure of this chapter is as follows. Section 2 contains a derivation of the *Kirchhoff equations* from the *Maxwell equations* (also see Chapter 1 in this book), whereas Sections 3–7 contain the numerical methods required for the evaluation of the fourfold integrals that form the matrix elements in these equations. These matrix elements represent the resistors, inductors, and capacitors in the equivalent circuit model for the interconnection system.

Section 8 presents an improvement of the method, treated in Section 5, for the analytical integration of the inner integrals for vector valued basis functions for a quadrilateral element. This improvement makes it possible to use nonplanar conducting structures, such as bond wires, screens, and boxes in a simulation. The analytical integration of the inner integrals over a triangular element for scalar and vector valued basis functions is discussed in Section 9.

Section 10 presents the solution methods used for the Kirchhoff equations. The linear algebra methods used for the solution of the linear system of equations and the generalized eigenvalue problems involved are discussed in Section 11. An efficient method for solving equations with large matrices is given in Section 12.

2. Derivation of Kirchhoff equations

In this section a derivation will be given of the Kirchhoff equations which describe the behavior of an equivalent circuit of a PCB. For this purpose an equivalent boundary value problem will be derived from the Maxwell's equations. Next, we present a variational formulation of this problem and the function spaces that contain its weak solutions. To be able to compute these solutions the problem domain is subdivided into quadrilateral elements and the solutions are approximated by linear combinations of basis functions in finite dimensional subspaces of the original function spaces. We obtain a linear set of equations, which correspond to the *Kirchhoff equations*, the solutions of which represent the currents, charges, and potentials in an electronic circuit. The matrix elements belonging to this set of equations are integrals. The fourfold integrals that represent the electromagnetic interaction between charges and currents in two elements of the discrete domain will be called *interaction integrals*.

Since in this chapter our attention is restricted to the evaluation methods for the interaction integrals, the derivation of the Kirchhoff equations will be given in a simplified form. For a more rigorous derivation see DU CLOUX, MAAS, and WACHTERS [1994], or Chapter 1 in this volume.

2.1. Introduction

A PCB consists of a set of thin metal layers separated by dielectric layers. The metal layers are the conductors (see Fig. 2.1) that connect the external electronic components

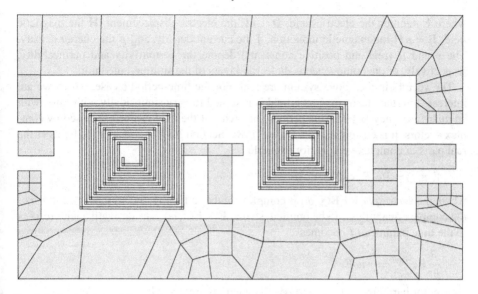

FIG. 2.1. A printed circuit board viewed from above. The patterns of the conductors are described by poly-gons, which are divided into quadrilaterals.

mounted on the PCB. By currents through the conductors electromagnetic fields will be generated that can cause crosstalk between the parts of an electronic system. This crosstalk often interferes with the desired signals and manifests itself as noise. For high frequencies this disturbance can sometimes be significantly large. A study of these elec-tromagnetic fields may provide understanding of this disturbing interference between parts of an electronic system.

As will be demonstrated later in this chapter, the electromagnetic properties of a PCB can be translated into an equivalent circuit model (see also Chapter 4 in this volume for more general techniques). By putting such model, together with those for the external components, in a circuit analysis program, the potentials at the circuit nodes can be obtained. From these potentials the currents through the conductors and the radiated electromagnetic fields can be calculated. In fact, this provides a way of performing a coupled analysis of circuit behavior and electromagnetics effects.

2.2. Maxwell's equations

The electromagnetic field in this electronic system can be described by the *Maxwell's equations*:

$$\nabla \times \mathbf{E} = -\frac{\partial \mathbf{B}}{\partial t},$$

$$\nabla \times \mathbf{H} = \mathbf{J} + \frac{\partial \mathbf{D}}{\partial t},$$

$$\nabla \cdot \mathbf{B} = 0,$$

$$\nabla \cdot \mathbf{D} = \rho,$$

where $\mathbf{E}$ denotes the electric field, $\mathbf{D} = \varepsilon\mathbf{E}$ the electric displacement, $\mathbf{H}$ the magnetic field, $\mathbf{B} = \mu\mathbf{H}$ the magnetic induction, $\mathbf{J}$ the current density, and ρ the charge density. The ε and μ (real and positive constants) denote the permittivity and permeability, respectively, of the homogeneous dielectric layers of the stratified medium.

The vectors in the above system are real. For the time-periodic case, which we are interested in, the electromagnetic field is assumed to vary sinusoidally with time, with angular frequency $\omega \geqslant 0$. This periodic behavior of the field can be expressed by complex vectors. It is customary (see RAMO [1984, Section 3.8]) to consider the fields as the real parts of complex vectors, for example

$$\mathbf{E}(\mathbf{x}, t) = \mathrm{Re}\big[\mathbf{E}(\mathbf{x}, \omega)e^{-i\omega t}\big],$$

where the *vector phasor* $\mathbf{E}(\mathbf{x}, \omega)$ is complex. In the following, the arguments $\mathbf{x}$ and ω of the complex quantities will be omitted, so that $\mathbf{E} = \mathbf{E}(\mathbf{x}, \omega)$. The derivative with respect to the time parameter t becomes

$$\frac{\partial}{\partial t}\mathbf{E}e^{-i\omega t} = -i\omega\mathbf{E}.$$

Hence, for harmonic fields the Maxwell equations are given by

$$\nabla \times \mathbf{E} = i\omega\mathbf{B}, \tag{2.1}$$

$$\nabla \times \mathbf{H} = \mathbf{J} - i\omega\mathbf{D}, \tag{2.2}$$

$$\nabla \cdot \mathbf{B} = 0, \tag{2.3}$$

$$\nabla \cdot \mathbf{D} = \rho. \tag{2.4}$$

Since $\nabla \cdot (\nabla \times \mathbf{H}) = 0$, from (2.2) and (2.4) the *current continuity equation* follows:

$$\nabla \cdot \mathbf{J} - i\omega\rho = 0. \tag{2.5}$$

Further the following relations hold:

$$\mathbf{J} = \sigma\mathbf{E} \quad \text{(Ohm's law)}, \tag{2.6}$$

$$\mathbf{D} = \varepsilon\mathbf{E}, \tag{2.7}$$

$$\mathbf{B} = \mu\mathbf{H}, \tag{2.8}$$

where the material properties σ and ε are assumed to be constant for each layer and μ is constant for the whole problem region.

2.3. Equations to be solved

The Maxwell equations form a set of coupled first order partial differential equations which give relations between electric and magnetic fields. In view of later applications it is convenient to introduce potentials (see JACKSON [1975, Section 6.4, pp. 219–220]) to obtain a smaller number of (second-order) equations, equivalent to the Maxwell equations.

The properties $\nabla \cdot (\nabla \times \mathbf{u}) = 0$ and $\nabla \times (\nabla u) = 0$ for any $\mathbf{u} \in \mathbb{R}^3$ lead to the following observations.

Since $\nabla \cdot \mathbf{B} = 0$, the magnetic induction $\mathbf{B}$ can be defined in terms of a magnetic vector potential that satisfies

$$\mathbf{B} = \nabla \times \mathbf{A}. \tag{2.9}$$

Then, Eq. (2.1) can be rewritten as

$$\nabla \times (\mathbf{E} - i\omega\mathbf{A}) = 0.$$

The argument of the rotation $(\nabla \times)$ can be written as the gradient of some scalar function, namely the electric potential φ, so that

$$\mathbf{E} - i\omega\mathbf{A} = -\nabla\varphi. \tag{2.10}$$

These definitions of the potentials are consistent with (2.1) and (2.3). By the other Maxwell's equations, (2.2) and (2.4), restrictions are imposed on these potentials. From (2.8) and (2.2) it follows that

$$\nabla \times \mathbf{B} = \mu(\nabla \times \mathbf{H}) = \mu\mathbf{J} - i\omega\mu\mathbf{D}.$$

After substitution of (2.9), (2.7), (2.10), and property

$$\nabla \times (\nabla \times \mathbf{A}) = \nabla(\nabla \cdot \mathbf{A}) - \Delta\mathbf{A},$$

in this expression and

$$\mathbf{D} = \varepsilon\mathbf{E} = \varepsilon(-\nabla\varphi + i\omega\mathbf{A}),$$

in (2.4) one obtains

$$\nabla(\nabla \cdot \mathbf{A}) - \Delta\mathbf{A} - i\omega\mu\varepsilon\nabla\varphi - \omega^2\mu\varepsilon\mathbf{A} = \mu\mathbf{J}, \tag{2.11}$$

$$-\nabla \cdot (\varepsilon\nabla\varphi) + i\omega\varepsilon\nabla \cdot \mathbf{A} = \rho. \tag{2.12}$$

Hence, the eight Maxwell equations have been reduced to four equations. However, the potentials $\mathbf{A}$ and φ are still not uniquely defined. If $(\mathbf{A}, \varphi)$ is a solution of (2.11) and (2.12), then $(\mathbf{A} + \nabla f, \varphi + i\omega f)$ for an arbitrary scalar function f is also a solution. To express $\mathbf{A}$ and φ uniquely, an extra condition must be added. One possibility is to use the *Lorentz gauge condition*

$$\nabla \cdot \mathbf{A} - i\omega\mu\varepsilon\varphi = 0. \tag{2.13}$$

After substitution of condition (2.13) in (2.11) and (2.12) one obtains the *Helmholtz equations*

$$(\Delta + k^2)\mathbf{A} = -\mu\mathbf{J}, \tag{2.14}$$

$$\nabla \cdot (\varepsilon\nabla\varphi) + \varepsilon k^2\varphi = -\rho, \tag{2.15}$$

where $k = \omega\sqrt{\varepsilon\mu}$. It can be shown that the solutions of Helmholtz equations are unique for appropriate Dirichlet and Neumann boundary conditions (see COLTON and KRESS [1992, Section 3]).

In summary, the total system of equations to be solved for $\mathbf{A}$, φ, $\mathbf{J}$, and ρ is as follows:

$$(\Delta + k^2)\mathbf{A} = -\mu\mathbf{J}, \tag{2.16}$$

$$\nabla \cdot (\varepsilon\nabla\varphi) + \varepsilon k^2\varphi = -\rho, \tag{2.17}$$

$$\mathbf{J} = \sigma\mathbf{E} = \sigma(-\nabla\varphi + i\omega\mathbf{A}), \tag{2.18}$$

$$\nabla \cdot \mathbf{J} - i\omega\rho = 0, \tag{2.19}$$

including appropriate boundary conditions.

2.4. The boundary value problem

In this subsection an equivalent boundary value problem will be posed. First, we will introduce the Green's functions, which are the solutions of an inhomogeneous Helmholtz equation for a homogeneous medium. Let the Green's function $G(\mathbf{x}', \mathbf{x}; k)$ be defined as the solution of the following equation

$$(\Delta + k^2)G(\mathbf{x}', \mathbf{x}; k) = -\delta(\mathbf{x}' - \mathbf{x}), \quad \mathbf{x} \in \Omega, \tag{2.20}$$

for a fixed point $\mathbf{x}'$ in a domain $\Omega \subset \mathbb{R}^3$ and the *Dirichlet* condition

$$G(\mathbf{x}', \mathbf{x}; k) = g(\mathbf{x}) \quad \text{on } \delta\Omega.$$

Here, $\delta(\mathbf{x}' - \mathbf{x})$ is the *Dirac delta function* with the properties:

$$\delta(\mathbf{x}' - \mathbf{x}) = 0, \quad \text{if } |\mathbf{x}' - \mathbf{x}| > 0,$$

$$\int_{B_R(\mathbf{x})} \delta(\mathbf{x}' - \mathbf{x})\,d\mathbf{x}' = 1,$$

$$\int_{B_R(\mathbf{x})} \delta(\mathbf{x}' - \mathbf{x})\xi(\mathbf{x}')\,d\mathbf{x}' = \xi(\mathbf{x}),$$

where the "sphere" $B_R(\mathbf{x}) = \{\mathbf{x}' \in \mathbb{R}^3; |\mathbf{x}' - \mathbf{x}| \leqslant R; \mathbf{x} \in \mathbb{R}^3\}$, and $\xi(\mathbf{x})$ is an arbitrary function over $B_R(\mathbf{x})$. The fundamental solution of Eq. (2.20) is

$$G_0(\mathbf{x}', \mathbf{x}; k) = \frac{e^{ik|\mathbf{x}'-\mathbf{x}|}}{4\pi|\mathbf{x}' - \mathbf{x}|}.$$

Thus, restricting the domain Ω to the conductors, the solutions of the Helmholtz equations (2.16) and (2.17) can be formulated as:

$$\mathbf{A}(\mathbf{x}, \omega) = \int_\Omega G_\mathbf{A}(\mathbf{x}', \mathbf{x}; k)\mu\mathbf{J}(\mathbf{x}', \omega)\,d\mathbf{x}' + \mathbf{A}_0(\mathbf{x}, \omega), \tag{2.21}$$

$$\varphi(\mathbf{x}, \omega) = \int_\Omega G_\varphi(\mathbf{x}', \mathbf{x}; k)\frac{\rho(\mathbf{x}', \omega)}{\varepsilon}\,d\mathbf{x}' + \varphi_0(\mathbf{x}, \omega), \tag{2.22}$$

where $\mathbf{A}_0$ and φ_0 are solutions of the homogeneous problem and

$$G_\varphi = \frac{e^{ik|\mathbf{x}'-\mathbf{x}|}}{4\pi|\mathbf{x}' - \mathbf{x}|}, \qquad G_\mathbf{A} = \begin{pmatrix} 1 & 0 & 0 \\ 0 & 1 & 0 \\ 0 & 0 & 1 \end{pmatrix}\frac{e^{ik|\mathbf{x}'-\mathbf{x}|}}{4\pi|\mathbf{x}' - \mathbf{x}|},$$

the fundamental solutions of the Helmholtz equations (2.16) and (2.17) for a homogeneous medium.

For a stratified inhomogeneous medium it is more difficult to obtain the Green's function. For $k|\mathbf{x}' - \mathbf{x}| \ll 1$, known as the *quasi-static* case, the method of images (see JACKSON [1975, Section 2.1, pp. 54–55]) can be used. Then the Green's function for the scalar potential becomes

$$G_\varphi = \sum_{j \in \text{Images}(\varphi)} c_j \frac{e^{ik|\mathbf{x}'_j - \mathbf{x}|}}{4\pi |\mathbf{x}'_j - \mathbf{x}|}, \quad \text{for } \mathbf{x}'_j = (x', y', z'_j).$$

The images for φ are due to reflections at the dielectric interfaces and the ground plane, all of which are perpendicular to the z-axis. The constants c_j only depend on the dielectric constants of the layers. If there are two or more of such reflection planes, the number of images is infinite. The Green's function for the vector potential has the form

$$G_\mathbf{A} = \sum_{j \in \text{Images}(\mathbf{A})} M_j \frac{e^{ik|\mathbf{x}'_j - \mathbf{x}|}}{4\pi |\mathbf{x}'_j - \mathbf{x}|}, \quad \text{for } \mathbf{x}'_j = (x', y', z'_j).$$

Since μ is constant, the images for $\mathbf{A}$ are only due to reflections at the groundplane, i.e., for each source point $\mathbf{x}'$ there is only one image point. If all metal layers are parallel to the ground layer and the ground plane is a perfect conductor, then

$$M_1 = -M_2 = \begin{pmatrix} 1 & 0 \\ 0 & 1 \end{pmatrix},$$

otherwise

$$M_1 = \begin{pmatrix} 1 & 0 & 0 \\ 0 & 1 & 0 \\ 0 & 0 & 1 \end{pmatrix}, \qquad M_2 = \begin{pmatrix} -1 & 0 & 0 \\ 0 & -1 & 0 \\ 0 & 0 & 1 \end{pmatrix}.$$

Now, we can formulate the following boundary value problem. Let Ω be the interior of the finite conductor regions, let $\Gamma = \delta\Omega$ be the boundary of these regions, and let Γ_V be that part of Γ that is restricted to the *connection ports*, where the external wires are connected to the conductors. For more details see DU CLOUX, MAAS, and WACHTERS [1994]. After substitution of (2.21) in (2.18) the following boundary value problem can be formulated:

$$\frac{\mathbf{J}}{\sigma} + \nabla\varphi - i\omega \int_\Omega G_\mathbf{A}\mu\mathbf{J}\,d\mathbf{x}' = \mathbf{E}_0, \tag{2.23}$$

$$\nabla \cdot \mathbf{J} - i\omega\rho = 0, \tag{2.24}$$

$$\varphi - \int_\Omega G_\varphi \frac{\rho}{\varepsilon}\,d\mathbf{x}' = \varphi_0, \tag{2.25}$$

under the boundary conditions

$$\varphi(\mathbf{x}) = V_{\text{fixed}}, \quad \mathbf{x} \in \Gamma_V,$$
$$\mathbf{J} \cdot \mathbf{n} = 0, \quad \mathbf{x} \in \Gamma,$$

where φ, $\mathbf{J}$, and ρ are elements of the function spaces

$$\rho \in L^2(\Omega),$$
$$\varphi \in H^1(\Omega) = \{u \in L^2(\Omega) \mid \nabla u \in L^2(\Omega)^3\},$$
$$\mathbf{J} \in H^{\mathrm{div}}(\Omega) = \{\mathbf{v} \in L^2(\Omega)^3 \mid \nabla \cdot \mathbf{v} \in L^2(\Omega)\},$$

where $L^2(\Omega) = \{u \mid \int_\Omega u^2 \, dx < \infty\}$. Further, $\mathbf{E}_0$ and φ_0 are due to irradiation from external sources, associated with the homogeneous solutions $\mathbf{A}_0$ and φ_0, respectively, of the Helmholtz equations, and $\mathbf{n}$ is the unit normal vector perpendicular to the boundary surface. The physical meaning of the boundary conditions is that no current will flow through the boundary, except through the connection ports.

2.5. Variational formulation

Assuming that the irradiation is zero, i.e., $\mathbf{E}_0 \equiv 0$ and $\varphi_0 \equiv 0$, the variational formulation of the boundary value problem is obtained by multiplying (2.23)–(2.25) with test functions (denoted by a tilde over the symbol) and integrating over the domain Ω of the conductors:

$$\int_\Omega \left\{ \frac{\mathbf{J}}{\sigma} + \nabla\varphi - i\omega \int_\Omega G_A \mu \mathbf{J} \, dx' \right\} \cdot \tilde{\mathbf{J}} \, dx = 0, \tag{2.26}$$

$$\int_\Omega \{\nabla \cdot \mathbf{J} - i\omega\rho\}\tilde{\varphi} \, dx = 0, \tag{2.27}$$

$$\int_\Omega \left\{ \varphi - \int_\Omega G_\varphi \frac{\rho}{\varepsilon} \, dx' \right\} \tilde{\rho} \, dx = 0, \tag{2.28}$$

where $\tilde{\mathbf{J}}$, $\tilde{\varphi}$, and $\tilde{\rho}$ are test functions in the infinite dimensional function spaces associated with $\mathbf{J}$, φ, and ρ, respectively:

$$\rho, \tilde{\rho} \in L^2(\Omega),$$
$$\varphi, \tilde{\varphi} \in H^1(\Omega) = \{u \in L^2(\Omega) \mid \nabla u \in L^2(\Omega)^3\},$$
$$\mathbf{J}, \tilde{\mathbf{J}} \in H_0^{\mathrm{div}}(\Omega) = \{\mathbf{v} \in L^2(\Omega)^3 \mid \nabla \cdot \mathbf{v} \in L^2(\Omega); \ \mathbf{v} \cdot \mathbf{n} = 0 \text{ on } \Gamma\}.$$

After integration by parts, which is allowed since $\tilde{\mathbf{J}} \in H_0^{\mathrm{div}}(\Omega)$, and substitution of the boundary condition $\tilde{\mathbf{J}} \cdot \mathbf{n} = 0$, the following relation holds:

$$\int_\Omega \tilde{\mathbf{J}} \cdot \nabla\varphi \, dx = -\int_\Omega \varphi \nabla \cdot \tilde{\mathbf{J}} \, dx.$$

Substituting this expression in (2.26) gives the following weak formulation of the boundary value problem:

$$\rho(\mathbf{x}), \varphi(\mathbf{x}) \in L^2(\Omega), \quad \text{and} \quad \mathbf{J}(\mathbf{x}) \in H_0^{\mathrm{div}}(\Omega),$$

$$\int_\Omega \left\{ \frac{\mathbf{J}}{\sigma} \cdot \tilde{\mathbf{J}} - \varphi \nabla \cdot \tilde{\mathbf{J}} - i\omega \int_\Omega G_A \mu \mathbf{J} \, dx' \cdot \tilde{\mathbf{J}} \right\} dx = 0 \quad \text{for all } \tilde{\mathbf{J}} \in H_0^{\mathrm{div}}(\Omega), \tag{2.29}$$

$$\int_{\Omega} \{\nabla \cdot \mathbf{J} - i\omega\rho\}\tilde{\varphi}\,d\mathbf{x} = 0 \quad \text{for all } \tilde{\varphi} \in L^2(\Omega), \tag{2.30}$$

$$\int_{\Omega} \left\{\varphi - \int_{\Omega} G_\varphi \frac{\rho}{\varepsilon}\,d\mathbf{x}'\right\}\tilde{\rho}\,d\mathbf{x} = 0 \quad \text{for all } \tilde{\rho} \in L^2(\Omega). \tag{2.31}$$

We assume that the conductors are planar and very thin so that, for the frequencies we are interested in, the quantities $\mathbf{J}$, φ, and ρ are constant in the direction perpendicular to the conductors. Therefore, the dependence of the above expressions on the coordinate direction perpendicular to the layers may be separated from the dependence in parallel direction. Hence, the 3D integrals over the volume of the conductors may be replaced by 2D integrals over the surfaces, that result when the thickness of the conductor layers becomes zero. In the following Ω will be considered as a 2D manifold embedded in $\mathbb{R}^3$.

The system of Eqs. (2.23)–(2.25) is called the *operational formulation* of the problem and (2.29)–(2.31) is the *variational formulation*. It is easily seen that if $(\mathbf{J}, \varphi, \rho)$ is a solution of (2.23)–(2.25), it is also a solution of (2.29)–(2.31). Conversely, it can be shown (see AUBIN [1972, Section 1.5, p. 27]) that if the material constants σ^{-1}, ω, μ, and ε^{-1} are bounded, and $(\mathbf{J}, \varphi, \rho)$ satisfy the variational formulation (2.29)–(2.31) for all $(\tilde{\mathbf{J}}, \tilde{\varphi}, \tilde{\rho})$ in the associated function spaces, the functions $(\mathbf{J}, \varphi, \rho)$ also satisfy the operational formulation of the boundary value problem.

2.6. Discretization

To find an approximating solution of Eqs. (2.29)–(2.31), the function spaces are approximated by finite dimensional subspaces. Let us assume that the planar regions to which the conductors reduce when their thickness becomes zero consist of polygons, and let the domain of these regions be denoted by Ω_h. Then, the domain can be subdivided into convex quadrilaterals Ω_j as illustrated in Fig. 2.1. Since the planar conductor regions often have quadrilateral shapes with large aspect ratios, we have chosen quadrilateral elements instead of triangles. The set of quadrilaterals is referred to as the set of elements Ω_j, $j = 1, \ldots, N_{\text{elem}}$. The edges of the quadrilaterals inside the domain Ω_h, i.e., excluding the element edges in the boundary, are referred to as the set of edges $\mathcal{E}_l$, $l = 1, \ldots, N_{\text{edge}}$. On the domain Ω_h finite dimensional subspaces U_h, W_h, and $H_{h,0}^{\text{div}}$ of the infinite dimensional function spaces L^2 and H_0^{div} are taken. The *discrete formulation* associated with the problem (2.29)–(2.31) is to find the functions $(\varphi_h, \mathbf{J}_h, \rho_h)$ for which

$$\int_{\Omega_h} \left\{\frac{\mathbf{J}_h}{\sigma} \cdot \tilde{\mathbf{J}}_h - \varphi_h \nabla \cdot \tilde{\mathbf{J}}_h - i\omega \int_{\Omega_h} G_A \mu \mathbf{J}_h\,d\mathbf{x}' \cdot \tilde{\mathbf{J}}_h\right\}d\mathbf{x} = 0 \quad \text{for all } \tilde{\mathbf{J}}_h \in H_{h,0}^{\text{div}}, \tag{2.32}$$

$$\int_{\Omega_h} \{\nabla \cdot \mathbf{J}_h - i\omega\rho_h\}\tilde{\varphi}_h\,d\mathbf{x} = 0 \quad \text{for all } \tilde{\varphi}_h \in U_h, \tag{2.33}$$

$$\int_{\Omega_h} \left\{\varphi_h - \int_{\Omega_h} G_\varphi \frac{\rho_h}{\varepsilon}\,d\mathbf{x}'\right\}\tilde{\rho}_h\,d\mathbf{x} = 0 \quad \text{for all } \tilde{\rho}_h \in W_h. \tag{2.34}$$

The functions φ_h, $\mathbf{J}_h$, and ρ_h are expanded in terms of basis functions, which span the finite dimensional subspaces defined above.

The scalar potential is expanded as

$$\varphi_h(\mathbf{x}) = \sum_{j=1}^{N_{\text{elem}}} V_j b_j(\mathbf{x}),$$

where V_j is the potential of element j, and $b_j(\mathbf{x})$ is defined by

$$b_j(\mathbf{x}) = \begin{cases} 1 & \text{for } \mathbf{x} \in \Omega_j, \\ 0 & \text{elsewhere.} \end{cases}$$

The surface charge density is expanded as

$$\rho_h(\mathbf{x}) = \sum_{j=1}^{N_{\text{elem}}} Q_j c_j(\mathbf{x}),$$

where Q_j is the charge of element j, while $c_j(\mathbf{x})$ are basis functions on the elements, adapted to include the singularity of the charge density (see Appendix A) near the conductor edge

$$c_j(\mathbf{x}) = f(\mathbf{x}) b_j(\mathbf{x}).$$

The function $f(\mathbf{x})$ is defined in Appendix A. It satisfies the following condition

$$\int_{\Omega_j} f(\mathbf{x}) \, d\mathbf{x} = |\Omega_j|, \tag{2.35}$$

where $|\Omega_j|$ is defined as the area of Ω_j.

Finally, the surface current density is expanded as

$$\mathbf{J}_h(\mathbf{x}) = \sum_{l=1}^{N_{\text{edge}}} I_l \tilde{\mathbf{w}}_l(\mathbf{x}),$$

where I_l is the current through edge l, and $\tilde{\mathbf{w}}_l(\mathbf{x})$ is defined by

$$\tilde{\mathbf{w}}_l(\mathbf{x}) = \begin{cases} f(\mathbf{x}) \mathbf{w}_l(\mathbf{x}) & \text{for } \mathbf{x} \in \Omega_i \cup \Omega_j \text{ and } \mathcal{E}_l = \Omega_i \cap \Omega_j, \\ 0 & \text{otherwise.} \end{cases} \tag{2.36}$$

The form and properties of the basis functions $\mathbf{w}_l$ are defined in Appendix B. Note that the component of $\mathbf{w}_l$ normal to the edge is continuous at $\mathbf{x} \in \mathcal{E}_l$ when passing from Ω_i to Ω_j.

After substitution of the expansions of φ_h, $\mathbf{J}_h$, and ρ_h in (2.29)–(2.31) we obtain the following linear system of equations:

$$\sum_{l=1}^{N_{\text{edge}}} (R_{kl} - i\omega L_{kl}) I_l - \sum_{j=1}^{N_{\text{elem}}} P_{kj} V_j = 0,$$

$$i\omega \sum_{j=1}^{N_{\text{elem}}} M_{ij} Q_j - \sum_{l=1}^{N_{\text{edge}}} P_{li} I_l = 0,$$

$$\sum_{j=1}^{N_{\text{elem}}} (M_{ij}V_j - D_{ij}Q_j) = 0.$$

The matrix elements of $\mathbf{R}$, $\mathbf{L}$, $\mathbf{P}$, $\mathbf{M}$, and $\mathbf{D}$ ($k, l = 1 \ldots N_{\text{edge}}$ and $i, j = 1 \ldots N_{\text{elem}}$) are given by

$$R_{kl} = \int_{\Omega_h} \frac{1}{\sigma} \tilde{\mathbf{w}}_l(\mathbf{x}) \cdot \tilde{\mathbf{w}}_k(\mathbf{x}) \, d\mathbf{x},$$

$$L_{kl} = \int_{\Omega_h} \tilde{\mathbf{w}}_l(\mathbf{x}) \cdot \left\{ \int_{\Omega_h} G_{\mathbf{A}}(\mathbf{x}, \mathbf{x}') \mu \tilde{\mathbf{w}}_k(\mathbf{x}') \, d\mathbf{x}' \right\} d\mathbf{x}, \qquad (2.37)$$

$$P_{kj} = \int_{\Omega_j} b_j(\mathbf{x}) \nabla \cdot \tilde{\mathbf{w}}_k(\mathbf{x}) \, d\mathbf{x},$$

$$M_{ij} = \int_{\Omega_j} c_j(\mathbf{x}) b_i(\mathbf{x}) \, d\mathbf{x},$$

$$D_{ij} = \int_{\Omega_j} c_j(\mathbf{x}) \left\{ \int_{\Omega_i} G_{\varphi}(\mathbf{x}, \mathbf{x}') \frac{c_i(\mathbf{x}')}{\varepsilon} \, d\mathbf{x}' \right\} d\mathbf{x}. \qquad (2.38)$$

$\mathbf{R}$ is a sparse matrix and $\mathbf{L}$ and $\mathbf{D}$ are symmetrical, full matrices. From the definition of the basis functions b_i and c_j and Eq. (2.35) it follows that

$$M_{ij} = \delta_{ij} |\Omega_j|, \quad \delta_{ij} = \begin{cases} 1 & \text{if } i = j, \\ 0 & \text{if } i \neq j \end{cases}$$

so that $\mathbf{M}$ is a diagonal matrix of which the elements are the areas of the Ω_j.

LEMMA 2.1. *Let J be the Jacobian defined by the transformation (B.1) for $\mathbf{x} \in \Omega_j$. For one of the following conditions*

(1) $f(\mathbf{x}) \equiv 1$,

(2) $\nabla f \cdot \mathbf{w}_k = 0$, *and* $J = |\Omega_j|$,

the matrix $\mathbf{P}$ has the form

$$P_{kj} = \begin{cases} \pm 1 & \text{if } \mathcal{E}_k \subset \Omega_j, \\ 0 & \text{otherwise.} \end{cases}$$

PROOF. Let $\mathcal{E}_k \subset \Omega_j$, then Lemma B.2 shows that $\nabla \cdot \mathbf{w}_k = \frac{\pm 1}{J}$.
    If condition (1) holds

$$P_{kj} = \int_{\Omega_j} b_j(\mathbf{x}) \nabla \cdot \tilde{\mathbf{w}}_k(\mathbf{x}) \, d\mathbf{x} = \int_{\Omega_j} \nabla \cdot \mathbf{w}_k(\mathbf{x}) \, d\mathbf{x} = \int_{\Omega_j} \frac{\pm 1}{J} \, d\mathbf{x} = \pm 1.$$

If $\nabla f \cdot \mathbf{w}_k = 0$ of condition (2) holds

$$P_{kj} = \int_{\Omega_j} b_j(\mathbf{x}) \nabla \cdot \big(f(\mathbf{x}) \mathbf{w}_k(\mathbf{x})\big) \, d\mathbf{x} = \int_{\Omega_j} f(\mathbf{x}) \nabla \cdot \mathbf{w}_k(\mathbf{x}) \, d\mathbf{x} = \pm 1 \int_{\Omega_j} \frac{f(\mathbf{x})}{J} \, d\mathbf{x},$$

thus if $J = |\Omega_j|$ it follows from (2.35) that

$$P_{kj} = \frac{\pm 1}{|\Omega_j|} \int_{\Omega_j} f(\mathbf{x})\, d\mathbf{x} = \pm 1.$$

If $\mathcal{E}_k \not\subset \Omega_j$, then the supports of b_j and $\nabla \cdot \tilde{\mathbf{w}}_k(\mathbf{x})$ are disjoint so that $P_{kj} = 0$.   □

Condition (2) is fulfilled if the element Ω_j is rectangular and has two opposite edges lying in the boundary.

2.7. The Kirchhoff's equations

If Lemma 2.1 holds matrix $\mathbf{P}$ is an incidence matrix. Therefore, the elements and edges may be associated with the nodes and branches of a directed graph, so that our quasi-static electromagnetic model of a PCB is equivalent to a circuit of which the behavior is described by the following set of $N_{\text{branches}} + 2N_{\text{nodes}}$ equations:

$$(\mathbf{R} - i\omega\mathbf{L})I - \mathbf{P}V = 0,$$
$$-\mathbf{P}^{\mathsf{T}}I + i\omega\mathbf{M}Q = 0,$$
$$\mathbf{M}^{\mathsf{T}}V - \mathbf{D}Q = 0,$$

and at particular nodes j, corresponding to elements $\Omega_j \subset \Gamma_V$, the excitation conditions

$$V_j = V_{\text{fixed},j}.$$

These equations are the *Kirchhoff's equations*, which are used in classical circuit theory. The meaning of the quantities in these equations is given below:

$V \sim$ Potentials at nodes,

$I \sim$ Currents over branches,

$\mathbf{P} \sim$ Incidence matrix between nodes and branches,

$\mathbf{R} \sim$ Resistance (of branches),

$\mathbf{L} \sim$ Inductance (of branches),

$\mathbf{M}\mathbf{D}^{-1}\mathbf{M}^{\mathsf{T}} \equiv \mathbf{C} \sim$ Capacitance (between nodes).

After elimination of Q we obtain a system of $N_{\text{branches}} + N_{\text{nodes}}$ equations for the unknown I and V:

$$(\mathbf{R} - i\omega\mathbf{L})I - \mathbf{P}V = 0 \quad \textit{(Kirchhoff's voltage law)},$$
$$-\mathbf{P}^{\mathsf{T}}I + i\omega\mathbf{C}V = 0 \qquad \textit{(Kirchhoff's current law)}.$$

If $N_{V,\text{fixed}}$ is the number of potentials for which

$$V_j = V_{\text{fixed},j},$$

and $N_V = N_{\text{nodes}} - N_{V,\text{fixed}}$ the final system of equations has $N_{\text{branches}} + N_V$ unknowns.

3. Interaction integrals

The Kirchhoff equations, which have been derived in Section 2, describe the behavior of an equivalent circuit of a PCB. This system of equations is linear and the coefficients of the matrices associated with this system are integrals. The fourfold integrals that represent an electromagnetic interaction between charges and currents in two elements are called *interaction integrals*. These are the subject of the following sections.

There are two types of interaction integrals: the scalar-type interaction integral (2.38), representing the *capacitive* coupling between charges on the elements, and the vector-type interaction integral (2.37), representing the *inductive* coupling between currents flowing through the element edges. Since edges lie in two adjacent quadrilaterals the vector-type interaction integral (2.37) is an assemblage of four integrals over quadrilaterals. The integrals, defined on the quadrilaterals Ω_j and Ω_i, are of the following form

$$I = \int_{\Omega_j} \tilde{\psi}_j(\mathbf{x}) \cdot \mathbf{I}_i(\mathbf{x}) \, d\mathbf{x}, \tag{3.1}$$

where the inner integral $\mathbf{I}_i$, called the *source integral*, is

$$\mathbf{I}_i(\mathbf{x}) = \int_{\Omega_i} G(\mathbf{x}', \mathbf{x}) \tilde{\psi}_i(\mathbf{x}') \, d\mathbf{x}'. \tag{3.2}$$

Here $\tilde{\psi}_i$ and $\tilde{\psi}_j$ are vector valued edge functions (cf. $\tilde{\mathbf{w}}_k$, $\tilde{\mathbf{w}}_l$ in Section (2.6) for vector-type integrals and scalar valued element functions (cf. c_i, c_j for scalar-type integrals. The $\tilde{\psi}_i$ and $\tilde{\psi}_j$ contain a factor $f(\mathbf{x}) = \frac{1}{\sqrt{d(s_1, s_2)}}$, defined in Appendix A. This function is singular on the boundary of the domain of the conductors. G is the Green's function, which is singular if $|\mathbf{x}' - \mathbf{x}| = 0$, i.e., in the case of self-interaction ($\Omega_i = \Omega_j$) or if the integration elements are neighbors. In the following sections it will be assumed that $k|\mathbf{x}' - \mathbf{x}| \ll 1$ so that the expressions of Green's functions given in Section 2.4 reduce to the *quasi-static* form

$$G(\mathbf{x}', \mathbf{x}) = \sum_{k=1}^{N} c_k |\mathbf{x}'_k - \mathbf{x}|^{-1}, \tag{3.3}$$

where c_k is a scalar or matrix depending on the type of the integral, and N the number of images. Further, the factor 4π in the denominator of $G(\mathbf{x}', \mathbf{x})$ is omitted.

In some special cases (a part of) the integral can be evaluated analytically. For constant $\tilde{\psi}_j$ and $\tilde{\psi}_i$ and rectangular quadrilaterals Ω_j and Ω_i with corresponding edges in parallel the integral I can be evaluated analytically. For constant $\tilde{\psi}_j$ and $\tilde{\psi}_i$ and arbitrary, convex, quadrilaterals Ω_j and Ω_i the inner integral $\mathbf{I}_i$ can be evaluated analytically. For the vector valued $\tilde{\psi}_j$ and $\tilde{\psi}_i$ with constant f only the inner integral of $\mathbf{I}_i$ can be evaluated analytically. The analytical approach for these integrals is discussed in Sections 5 (scalar case) and 8 (vector-valued case).

In all other cases the integral I has to be evaluated numerically. For the numerical integration *quadrature rules* are needed. Several quadrature rules are discussed in

Section 4, in particular the *Patterson's quadrature rules*, which have the fastest convergence rates. However, as will be shown in that section, quadrature rules only reach fast convergence if the integrand is smooth enough, i.e., the integrand must be n times differentiable over the whole integration interval, for n large enough. Singularity of the integrand leads to very slow convergence. Therefore, methods to regularize or eliminate these singularities have to be investigated.

In Section 6 some methods are discussed by which the inner integrand is regularized such that it is smooth enough for integration by a quadrature process. The sources of the singularities in the inner integral are the Green's function and the factor for the boundary singularity. Regularization of the Green's function is done by transformation to polar coordinates. For this purpose the element has to be divided into triangles, each of which has one edge of the element as base and the projection of the quadrature point of the outer integral I on the source element as vertex. The regularization of the factor for the boundary singularity has to be treated in a special way. In particular, the treatment of the "flat" triangles deserves special attention.

After numerical evaluation of the inner integral the outer integral only contains a factor for the boundary singularity. Regularization of this singularity is done by a simple substitution, which is discussed in the last subsection of Section 6.

Throughout this chapter the integration domain of the outer integral, Ω_j, is referred to as the *object element* and the integration domain of the inner integral, Ω_i, as the *source element*. If the "distance" between the source and object element is large enough, then the Green's function can be approximated satisfactorily with a *Taylor expansion*. In Section 7 an error estimate of the Taylor expansion dependent on the distance will be given. In the same section the *moment integrals* are introduced. These are *twofold* integrals which, possibly, still contain the boundary singularity. After regularization of this singularity, which can be done by an analogous substitution as discussed for the outer integral in the last subsection of Section 6, the moment integral can be evaluated by Patterson's quadrature process. The interaction integral I can be written as a linear combination of products of these moments.

This method has the advantage that, instead of a large number (quadratic with the number of elements or edges) of fourfold integrals, only a small number (linear with the number of elements or edges) of twofold moment integrals has to be evaluated. Moreover, these moment integrals can be evaluated in advance. Since it saves a lot of computer time, this method is preferred as an alternative for the numerical treatment of Section 6, if the distance between the source and object element is sufficiently large.

4. Numerical integration

Since most of the integrals, discussed in the previous chapters, cannot be evaluated analytically, we have to rely on numerical integration methods. For a detailed discussion of these methods see DAVIS and RABINOWITZ [1984].

In this chapter several numerical integration methods are discussed, in particular *Patterson's* quadrature formulae. Because special transformations are needed to regularize the singularity of the integrand, only one-dimensional Patterson's rules are used. First follows a short introduction to *quadrature formulae*.

4.1. Quadrature formulae

The essence of numerical quadrature is the approximation of an integral by a linear combination of the values of the integrand. Consider the integral

$$I(f) = \int_a^b f(x)\,dx,$$

which has to be numerically evaluated. A quadrature formula is given by

$$K(f) = \sum_{i=0}^{n-1} w_i f(x_i), \tag{4.1}$$

where $f(x)$ is an arbitrary (smooth) function, x_i are different abscissae in the integration interval and w_i are the corresponding weights.

Often, we choose the weights and abscissae such that the rule is exact for polynomials up to a certain degree. Given an arbitrary set of n distinct abscissae $x_i \in [a, b]$, the corresponding weights can be determined by solving the linear system

$$\sum_{i=0}^{n-1} w_i x_i{}^k = \int_a^b x^k\,dx, \quad \text{for } k = 0, \dots, n-1. \tag{4.2}$$

Note that the coefficient matrix (x_i^k) of the above system is a Vandermonde matrix, and therefore, nonsingular if $x_i \ne x_j$ for $i \ne j$. Hence, there is always a unique solution.

If the nodes $a = x_0 < \cdots < x_{n-1} = b$ are equidistant, these quadrature formulae are called *Newton–Cotes* formulae.

For the w_i's obtained by solving the system (4.2) the integration formula (4.1) is at least *exact* for polynomials of degree $n-1$. If f is sufficiently smooth, say $f \in C^n[a, b]$, then the error is given by

$$\left| I(f) - K(f) \right| = C f^{(n)}(\xi)(b-a)^{n+1}, \quad \xi \in (a, b),$$

where C is a constant and $f^{(n)}$ denotes the nth derivative of f.

When the behavior of the function to be integrated is very distinct on different parts of the integration interval, it is advantageous to subdivide the interval. A Newton–Cotes formula can be applied to each subinterval. If the subdivision of the interval is done automatically it is called an *adaptive* subdivision. A *nonadaptive* subdivision is characterized by a predetermined choice of subdivision points.

Some examples of quadrature rules. Some integration formulae of the Newton–Cotes type, that approximate the integral $\int_a^b f(x)\,dx$, and the corresponding error formulae are given below:

$$K_M = (b-a)f\left(\frac{b+a}{2}\right),$$

$$I - K_M = \frac{1}{24}(b-a)^3 f''(\xi_M) \quad \text{(Midpoint rule)},$$

$$K_T = \frac{b-a}{2}\{f(a) + f(b)\},$$

$$I - K_T = \frac{-1}{12}(b-a)^3 f''(\xi_T) \quad \text{(Trapezoidal rule)},$$

$$K_S = \frac{b-a}{6}\left\{f(a) + 4f\left(\frac{b+a}{2}\right) + f(b)\right\},$$

$$I - K_S = \frac{-1}{90}\left(\frac{b-a}{2}\right)^5 f^{(4)}(\xi_S) \quad \text{(Simpson's rule)},$$

where the ξ_M, ξ_T, ξ_S are points in the interval (a, b) depending on the function f to be integrated.

For n-point Newton–Cotes formulae ($n \leqslant 8$) the associated weights are positive. The next theorem shows that these formulae are stable.

THEOREM 4.1 (Stability). *Consider the n-point quadrature formula*

$$K(f) = \sum_{i=1}^{n} w_i f(x_i),$$

to approximate the integral $\int_a^b f(x)\,\mathrm{d}x$. If the weights are all nonnegative, i.e., $w_i \geqslant 0$, the quadrature formula is stable.

PROOF. Let $\varepsilon(x)$ be a perturbation of $f(x)$ and let ε_f be a constant such that $|\varepsilon(x)| \leqslant \varepsilon_f$ for all $x \in [a, b]$. Then, from the positivity of the weights it follows that

$$\int_a^b \mathrm{d}x = (b-a) = \sum_{i=1}^{n} w_i = \sum_{i=1}^{n} |w_i|,$$

hence,

$$\sum_{i=1}^{n} w_i \varepsilon(x_i) \leqslant (b-a)\varepsilon_f. \qquad \square$$

A reason not to use higher order Newton–Cotes formulae ($n > 8$) is the possible instability due to negative weights. Instead one could use lower order formulae on subintervals of $[a, b]$.

Repeated quadrature. The successive application of a quadrature formula on ever smaller subintervals of $[a, b]$ to obtain an increasingly better approximation of the integral is called *repeated quadrature*. Here, we give an example for the trapezoidal rule.

Divide the interval $[a, b]$ in n equal subintervals $[x_{i-1}, x_i]$ of length h, such that $a = x_0 < x_1 < \cdots < x_{n-1} < x_n = b$ and where $h = \frac{b-a}{n}$. Repeated application of the trapezoidal rule gives

$$T_n = h\left\{\tfrac{1}{2}f(x_0) + f(x_1) + \cdots + f(x_{n-1}) + \tfrac{1}{2}f(x_n)\right\}.$$

Assuming that f is sufficiently differentiable, the error is given by (for a proof see DAVIS and RABINOWITZ [1984, Section 2.9])

$$I - T_n = C_1 h^2 + C_2 h^4 + \cdots + C_n h^{2n} + \mathcal{O}(h^{2n+2}),$$

where the constants C_i depend on f, but are independent of h. The following application of repeated quadrature is based on this error formula.

Romberg integration. Suppose one has a function $f \in C^{2(n+2)}[a, b]$, where $[a, b]$ is an interval of length h_0 in $\mathbb{R}$, and the approximations $T_0^{(0)}, T_1^{(0)}, \ldots, T_n^{(0)}$ of the integral of f over $[a, b]$. These $T_i^{(0)}$ have been obtained by applying repeated trapezoidal rules on the 2^i subintervals of length h_i, where $h_i = 2^{-i} h_0$. Again, the errors for the approximations are given by

$$I - T_i^{(0)} = C_1 h_i^2 + C_2 h_i^4 + \cdots + \mathcal{O}(h_i^{2(n+2)})$$
$$= C_1 h_0^2 2^{-2i} + C_2 h_0^4 2^{-4i} + \cdots + \mathcal{O}(h_i^{2(n+2)}).$$

Since the first terms in $I - T_{i-1}^{(0)}$ and $I - T_i^{(0)}$ are $C_1 h_{i-1}^2$ and $C_1 (2^{-1} h_{i-1})^2 = \frac{1}{4} C_1 h_{i-1}^2$, respectively, one can eliminate these terms by applying *Richardson extrapolation* to the sequence $T_0^{(0)}, \ldots, T_n^{(0)}$:

$$T_i^{(1)} = \frac{4 T_i^{(0)} - T_{i-1}^{(0)}}{3}, \quad \text{for } i = 1, \ldots, n.$$

The errors for the newly obtained sequence of approximations are

$$I - T_i^{(1)} = D_2 h_i^4 + D_3 h_i^6 + \cdots + \mathcal{O}(h_i^{2(n+2)})$$
$$= D_2 h_0^4 2^{-4i} + D_3 h_0^6 2^{-6i} + \cdots + \mathcal{O}(h_i^{2(n+2)}),$$

where $D_i = \frac{2^2 - 2^{2i}}{2^2 - 1} C_i$. This process can be applied recursively on the sequences by

$$T_i^{(k)} = \frac{2^{2k} T_i^{(k-1)} - T_{i-1}^{(k-1)}}{2^{2k} - 1}, \quad \text{for } i = k, \ldots, n,$$

and the error for $T_i^{(n)}$ is

$$I - T_i^{(n)} = 2^{n(n+1)} C_{n+1} h_i^{2(n+1)} + \mathcal{O}(h_i^{2(n+2)}) = \mathcal{O}\left(\frac{2^{n(n+1)}}{4^{i(n+1)}} h_0^{2(n+1)}\right).$$

Thus, by *repeated* Richardson extrapolation, we get the *Romberg integration method*, so that the following theorem holds.

THEOREM 4.2. *Suppose $f \in C^\infty[a, b]$, then for $i \to \infty$ the sequences $T_i^{(k)}$, constructed as described above, converge towards $\int_a^b f(x)\,dx$ for every $k = 0, 1, \ldots$ with error $\mathcal{O}(h_i^{2k+2})$.*

Moreover, the diagonal $T_n^{(n)}$ converges with error $\mathcal{O}(h_n^{2n+2})$.

TABLE 4.1
Romberg integration of the exponential function

i	$T_i^{(0)}$	$T_i^{(1)}$	$T_i^{(2)}$	$T_i^{(3)}$	$T_i^{(4)}$
0	2.03663128				
1	1.28898621	1.03977118			
2	1.06215961	0.98655075	0.98300272		
3	1.00205141	0.98201534	0.98171298	0.98169251	
4	0.98679198	0.98170551	0.98168485	0.98168441	0.98168437

EXAMPLE. Table 4.1 shows the results of the Romberg method for the integral $\int_0^4 e^{-x}\,dx$ (≈ 0.98168436). We can see that $T_n^{(n)}$ for $n > 0$ gives a much more accurate approximation than $T_n^{(0)}$. For the calculation of $T_0^{(0)}, \ldots, T_n^{(0)}$ a total of $2^n + 1$ function evaluations are needed.

Note, that these function evaluations are only necessary for the calculation of $T_i^{(0)}$, so that by repeated Richardson extrapolation with little extra cost an even better approximation can be obtained. It can also be proven that the sequence $(T_i^{(k)})_{i=k}^\infty$ converges faster towards $\int_a^b f(x)\,dx$ than $(T_i^{(k-1)})_{i=k-1}^\infty$.

4.1.1. The Gaussian quadrature formulae

Gauss has proven that an n-point quadrature formula can be found which is exact for polynomials up to degree $2n - 1$ and that this is the highest possible degree. This will be shown after the following definitions.

DEFINITION 4.1. Let $w(x)$ be a continuous function on (a, b), with $w(x) \geqslant 0$ on $[a, b]$. We define the *inner product* with respect to the weight function $w(x)$ as

$$\langle f, g \rangle = \int_a^b w(x) f(x) g(x)\,dx, \quad f, g \in C[a, b].$$

DEFINITION 4.2. Let $\{F_i(x), i = 0, \ldots, n\}$ be a set of nonzero polynomials with F_i of degree i. The polynomials are said to be *orthogonal* on $[a, b]$ with respect to the inner product $\langle , \rangle$ if they satisfy

$$\langle F_i, F_j \rangle = 0 \quad \text{if } i \neq j.$$

Suppose $F_{2n-1}(x)$ is an arbitrary polynomial of degree $2n - 1$. This can be expressed as

$$F_{2n-1}(x) = P_n(x) Q_{n-1}(x) + R_{n-1}(x),$$

where P_n is an nth degree polynomial with n distinct roots in $[a, b]$. Let these roots be the abscissae of a new quadrature formula. Then, this formula will integrate $P_n(x)Q_{n-1}(x)$ to zero and, by construction of the corresponding weights (see (4.2)), it will integrate $R_{n-1}(x)$ exactly, since this is a polynomial of degree $n - 1$. Hence, if

$P_n(x)$ satisfies

$$\int_a^b P_n(x) Q_{n-1}(x) \, dx = 0 \tag{4.3}$$

for any polynomial $Q_{n-1}(x)$ of degree $n-1$, $F_{2n-1}(x)$ will be integrated exactly by this quadrature formula. Since $P_n(x)^2 \geq 0$ is not integrated exactly, this is also the highest possible degree.

We must find a $P_n(x)$ such that condition (4.3) will be satisfied. This is the same as

$$\int_a^b P_n(x) \sum_{i=0}^{n-1} a_i x^i \, dx = 0, \quad \text{for arbitrary } a_i,$$

so that for each individual term:

$$\int_a^b P_n(x) x^i \, dx = 0, \quad \text{for } i = 0, \ldots, n-1. \tag{4.4}$$

If the ith degree polynomials $P_i(x)$, $i = 0, \ldots, n$, form an orthogonal set these conditions are satisfied, because any $(n-1)$th degree polynomial can be expressed as a linear combination of $P_i(x)$, $i = 0, \ldots, n-1$.

Consider the inner product

$$\langle f, g \rangle = \int_a^b f(x) g(x) \, dx,$$

and take $a = -1$ and $b = 1$. The *Legendre*-polynomials (see Appendix C) are polynomials, that are mutually orthogonal with respect to this inner product, and therefore, they satisfy property (4.4). Hence, the quadrature formulae, of which the abscissae are the roots of the Legendre polynomials and the weights are constructed by solving Eq. (4.2), have the property of integrating $F_{2n-1}(x)$ exactly on $[-1, 1]$. The following theorems hold. Theorem 4.4 proves the stability of Gaussian quadrature formulae.

THEOREM 4.3. *The n-point quadrature formula $K_G^{(n)}$, with abscissae the roots of the Legendre polynomial $P_n(x)$ (the Gaussian quadrature formula), is exact for all polynomials in $\Pi_{2n-1}[-1, 1]$. If $f(x) \in C^{2n}[-1, 1]$, the error incurred in integrating $f(x)$ is given by*

$$I(f) - K_G(f) = \frac{2^{2n+1}(n!)^4}{(2n+1)((2n)!)^3} f^{(2n)}(\xi), \quad \xi \in (-1, 1).$$

PROOF. For the error estimate see DAVIS and RABINOWITZ [1961, pp. 428–437]. □

THEOREM 4.4. *The weights w_i for a Gaussian formula $K_G^{(n)}$ are positive.*

PROOF. Let

$$l_i(x) = \prod_{\substack{j=1 \\ j \neq i}}^n (x - x_j) \quad (\in \Pi_{n-1}).$$

Then, $l_i(x_j) \neq 0$ for $i = j$ and $l_i(x_j) = 0$ for $i \neq j$.

The Gaussian formula certainly is exact for $l_i(x)^2$. Hence, from

$$0 < \int_{-1}^{1} l_i(x)^2 \, dx = \sum_{j=1}^{n} w_j l_i(x_j)^2 = w_i l_i(x_i)^2$$

it follows that $w_i > 0$ for all $i = 1, \ldots, n$. $\square$

In a *quadrature process* quadrature rules of increasing order are applied successively, until the (estimated) relative error is smaller than a given tolerance. The absolute error can be estimated by taking the difference between the last two integral approximations, so that the relative error is estimated by the ratio of the absolute error to the last approximation. The following corollary holds.

COROLLARY 4.1. *The Gaussian quadrature process $K_G^{(n)}$ is convergent for every function $f(x)$ which is Riemann-integrable in $[-1, 1]$, i.e.,*

$$\lim_{n \to \infty} K_G^{(n)}(f) = \int_{-1}^{1} f(x) \, dx.$$

PROOF. See DAVIS and RABINOWITZ [1984, Section 2.7.8]. $\square$

Unfortunately, all the roots of the different Legendre polynomials, except zero, are different. Thus, the Gaussian quadrature process is rather inefficient, since in a step of the process no use is made of the integrands evaluated in the preceding steps. In the next subsection a more efficient method will be discussed.

4.2. Kronrod's extension of quadrature formulae

KRONROD [1965, p. 597] has suggested an extension of an n-point quadrature formula, by adding $n+1$ new abscissae to the original set, to yield a quadrature formula of degree $3n + 1$ (n even) or $3n + 2$ (n odd). This has the advantage that integrand evaluations needed for an n-point quadrature rule can be used again for the $(2n + 1)$-point quadrature rule. In the discussion of the Kronrod scheme we will restrict ourself to integrals with integration interval $[-1, 1]$. However, the results are applicable to integrals with an arbitrary finite interval $[a, b]$.

Let p be the number of points added to the original set of points and let F_{n+2p-1} be an arbitrary polynomial of degree $n + 2p - 1$. After division with remainder, this can be expressed as

$$F_{n+2p-1} = \tilde{P}_{n+p} Q_{p-1} + R_{n+p-1}.$$

Here $\tilde{P}_{n+p}$ is a polynomial whose roots are the $n + p$ abscissae of the new, extended quadrature formula. Since R_{n+p-1} is some polynomial of degree $n + p - 1$, it can always be exactly integrated by a $(n + p)$-point formula. Furthermore, Q_{p-1} can be expressed as

$$Q_{p-1} = \sum_{i=0}^{p-1} c_i x^i.$$

Therefore, if $\tilde{P}_{n+p}$ satisfies

$$\int_{-1}^{1} \tilde{P}_{n+p} x^i \, dx = 0, \quad \text{for every } i = 0, \ldots, p-1,$$

then

$$\int_{-1}^{1} \tilde{P}_{n+p} Q_{p-1} \, dx = 0.$$

Since $\int_{-1}^{1} F_{n+2p-1} \, dx = \int_{-1}^{1} \tilde{P}_{n+p} Q_{p-1} \, dx + \int_{-1}^{1} R_{n+p-1} \, dx$, the quadrature formula is exact for all polynomials in Π_{n+2p-1}.

4.2.1. Application to Gauss–Legendre

Take $p = n + 1$ and P_n the nth degree Legendre polynomial. This choice of p yields the number of points required to subdivide the intervals spanned by the n original Gauss points and the boundaries (see next subsection). Let

$$\tilde{P}_{n+p} = K_{n+1} P_n,$$

then K_{n+1} can be determined by expanding it as a polynomial,

$$K_{n+1}(x) = x^{n+1} + \sum_{i=0}^{n} a_i x^i.$$

The coefficients a_i are calculated by solving the linear system

$$\int_{-1}^{1} K_{n+1}(x) P_n(x) x^k \, dx = 0, \quad k = 0, \ldots, n.$$

As a result we can construct a quadrature formula, of which the abscissae are the n Gauss-points and the $n + 1$ roots of K_{n+1}. The corresponding weights are determined by solving the system (4.2), the method described earlier. The obtained formula is exact for F_{3n+1}. From an n'-point quadrature formula, with $n' = 2n + 1$, a new quadrature formula can be constructed by applying Kronrod's method to these n' points. The abscissae of this formula are the original n' points and the $n' + 1$ added points. The resulting quadrature formulae is exact for $F_{3n'+1}$. Since the formulae are symmetrical in the range interval $[-1, 1]$ (if x_i is a root, also $-x_i$ is a root) odd functions are always integrated exactly. Hence, the effective degree can be increased to $3n + 2$ when n is odd.

4.2.2. Patterson's quadrature formulae

PATTERSON [1968] has applied the Kronrod's method successively, starting with a 3-point Gaussian quadrature formula and developed a stable algorithm to calculate the nodes and corresponding weights of these quadrature formulae. Thus, he has derived a sequence of quadrature formulae of degrees $n = 7, 15, 31, 63, 127, 255$, and 511. The great advantage of these formulae compared to the Gaussian formulae is that all function evaluations of an n-point formula can be used in the extended $(2n + 1)$-point formula. The condition for Patterson's quadrature formulae to be stable is the positivity of the weights.

First we will give a justification for the choice of adding $n + 1$ points to the n original Gauss-points.

LEMMA 4.1. *Let $x_1^{(n)}, \ldots, x_n^{(n)}$ be the zeros of the Legendre polynomial $P_n(x)$ of degree n and let $y_1^{(n)}, \ldots, y_p^{(n)}$ be $p > \frac{n}{2}$ new points within the interval $(-1, 1)$. Let $K(f)$ be an extended quadrature formula*

$$K(f) = \sum_{j=1}^{n} w_j^{(n)} f(x_j^{(n)}) + \sum_{i=1}^{p} \tilde{w}_i^{(n)} f(y_i^{(n)}) \tag{4.5}$$

to approximate $I(f) = \int_{-1}^{1} f(x)\, dx$. If this formula is exact for $f \in \Pi_{n+2p-1}$, then $p > n$.

PROOF. For $k \in \{1, \ldots, p\}$ define $P_n^*(x) = \prod_{j=1}^{n}(x - x_j^{(n)})$, the nth degree Legendre polynomial with leading coefficient 1, and $s_k(x) = \prod_{\substack{i=1 \\ i \neq k}}^{p} (x - y_i^{(n)})$. Let

$$g_k(x) = P_n^*(x)s_k(x), \quad P_n^* \in \Pi_n, \ s_k \in \Pi_{p-1}.$$

If $p \leqslant n$

$$I(g_k) = \int_{-1}^{1} g_k(x)\, dx = \int_{-1}^{1} P_n^*(x)s_k(x)\, dx = 0,$$

since $P_n^* \perp s_k \in \Pi_{n-1}$. But since $g_k \in \Pi_{n+p-1}$, it follows that

$$0 = I(g_k) = K(g_k)$$
$$= \sum_{j=1}^{n} w_j^{(n)} g_k(x_j^{(n)}) + \sum_{i=1}^{p} \tilde{w}_i^{(n)} g_k(y_i^{(n)})$$
$$= 0 + \tilde{w}_k^{(n)} g_k(y_k^{(n)}).$$

Since $g_k(y_k^{(n)}) \neq 0$ we find that $\tilde{w}_k^{(n)} = 0$. This holds for all $k \leqslant p$, which means that we have $K(f) = \sum_{j=1}^{n} w_j^{(n)} f(x_j^{(n)})$, which is the original Gauss formula. Therefore, $P_n(x)^2 \in \Pi_{2n}$ would be integrated to zero. However, if $p > \frac{n}{2}$, the formula should be exact for $f \in \Pi_{2n+1}$. But since $P_n(x)^2 \geqslant 0$ we have a contradiction. Hence $p > n$. $\square$

From this lemma it follows that for exact integration of functions $f \in \Pi_{n+2p-1}$ the condition $p \geqslant n + 1$ is necessary. Let us consider $p = n + 1$. MONEGATO [1976a] has proven the following theorem.

THEOREM 4.5. *Let (4.5) be an extended Gauss–Legendre rule, then all $\tilde{w}_i^{(n)} > 0$ if and only if the nodes $x_j^{(n)}$ and $y_i^{(n)}$ interlace.*

SZEGÖ [1934] has proven that the nodes $x_j^{(n)}$ and $y_i^{(n)}$ interlace. Hence, the previous theorem shows that the weights $\tilde{w}_i^{(n)}$ are all positive. Furthermore, MONEGATO [1976b] has proven the next theorem.

THEOREM 4.6. *The weights $w_j^{(n)}$ of the extended Gauss–Legendre rules are positive.*

It should be noted that the weights $w_j^{(n)}$ are *not* the original Gauss weights. Although the weights of the extended Gauss–Legendre rules are all positive, the positivity of the weights of *all* successively obtained Patterson's rules has not been proven yet. However, from the tables it follows that the weights associated with the successive Patterson's quadrature rules (based on the 3-point Gauss–Legendre rule) until order 511 are all positive. Hence, the quadrature process by these quadrature rules is stable. Looking at Fig. 4.1 the conjecture might raise that the weights for higher order formulae are also positive.

Furthermore, tests have shown that the Patterson's rules converge *faster* to the true values of the integrals than the Gauss–Legendre rules for the same number of quadrature points. This was a reason for PATTERSON [1968] to state the following conjecture.

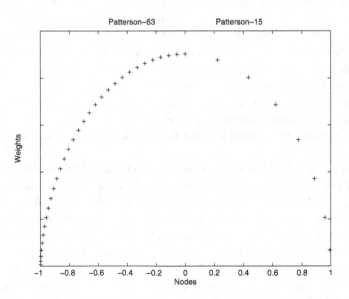

FIG. 4.1. The nodes x_i plotted against the weights w_i of the Patterson's 63-point formula in $(-1, 0]$ and the 15-point formulae in $[0, 1)$. The weights are scaled such that they can be compared.

CONJECTURE 4.1. *The quadrature process by Patterson's rules is stable and uniformly convergent for every function $f(x)$ which is Riemann-integrable in $[-1, 1]$.*

4.3. Comparison of integration methods

Next, we will compare the following four integration methods for some typical examples:

1. **Romberg's rule**

    Repeated trapezoidal rules and Richardson extrapolation until a certain accuracy has been reached. Error: $\mathcal{O}((\frac{b-a}{2^k})^{2(1+k)})$ for depth k, i.e., $2^k + 1$ function evaluations.

2. **Adaptive, recursive Simpson's rule**

    Repeated Simpson's rules on each subinterval recursively, until a certain tolerance level has been reached on each of the subintervals. Error: $\mathcal{O}((\frac{b-a}{2^k})^4)$ for depth k (the smallest subinterval has length $(b-a)2^{-k}$).

3. **Adaptive, recursive Newton–Cotes 8 panel rule**

    Repeated Newton–Cotes rules, where each interval is divided into 8 subintervals, recursively, until a certain accuracy has been reached on the subintervals. Error: $\mathcal{O}((\frac{b-a}{2^k})^9)$ for depth k.

4. **Patterson's quadrature rules**

    Successive extension of the Gauss–Legendre rule, up to degree 63.

The benefit of the Romberg's rule is its high theoretical convergence rate and the simplicity of the algorithm, and of an adaptive method its local refinement in the neighborhood of a singularity. The first three methods use a 2-, 3-, and 9-point Newton–Cotes formulae, and therefore, are only exact up to the corresponding degrees. For polynomial functions Patterson's rules of sufficiently high degree are exact. Therefore, we have chosen irregular integrands for comparison of the methods, so that none of these methods will be exact, but all converge to the exact values of integrals.

A measure for the convergence rate of the different methods is the number of necessary function evaluations to obtain a result with a given accuracy. In Table 4.2 the number of function evaluations is given for several tolerance levels for the relative error.

From the tables it is obvious that for the integrands chosen the Patterson's method is always the best choice. Compared to the other methods the number of function evaluations is very small. Since the Romberg's algorithm available to us makes use of global refinement for a singularity somewhere on the interval, the number of evaluation points becomes very large over the complete interval. Since Simpson's rule is a low order formula, it needs many subdivisions to obtain a sufficiently small error.

It appears that the available Romberg's algorithm and the adaptive Simpson's method are too expensive to obtain a satisfactory result. The adaptive Newton–Cotes 8 panel method is globally quite good, but the number of iterations, and therefore, the number of function evaluations is still quite large, because of its low order formula.

TABLE 4.2

Comparisons of the number of function evaluations needed for the different methods discussed, tested for several irregular functions. The tolerance is an upper bound for the absolute relative error

$I = \int_0^1 e^x \, dx$					$I = \int_0^1 \frac{1}{1+x^2} \, dx$				
Tol.	10^{-3}	10^{-6}	10^{-9}	10^{-12}	Tol.	10^{-3}	10^{-6}	10^{-9}	10^{-12}
Romb	5	9	17	33	Romb	9	33	65	129
AdSi	9	33	513	4097	AdSi	9	89	937	7885
ANC8	33	33	33	33	ANC8	33	33	49	113
Patt	7	7	15	15	Patt	7	15	31	31
$I = \int_0^1 x\sqrt{x} \, dx$					$I = \int_0^\pi \frac{1}{5+4\cos x} \, dx$				
Romb	9	129	2049	32769	Romb	17	129	129	513
AdSi	61	269	1965	6765	AdSi	37	321	3625	8193
ANC8	33	177	273	497	ANC8	33	65	129	289
Patt	7	31	63	63	Patt	15	31	63	63

4.3.1. Analysis of Patterson's quadrature rules

In this subsection two extra examples will be shown, that are of special interest for the study of the interaction integrals. The performance of the Patterson's quadrature rules for these integrals will be compared with two other quadrature rules.

The two integrals are:

$$I_1(\varepsilon) = \int_\varepsilon^1 \frac{1}{\sqrt{x}} \, dx,$$

$$I_2(\varepsilon) = \int_\varepsilon^{1-\varepsilon} \frac{1}{\sqrt{x(1-x)}} \, dx.$$

The smaller ε becomes, the better the integration interval approximates the interval $[0, 1]$, and the more irregular the integrands become. Note that the singular behavior near $x = 0$ for both integrals is about the same. Therefore, only the results for $I_2(\varepsilon)$ are shown. Those for $I_1(\varepsilon)$ are similar.

Table 4.3 shows the fast convergence rate of Patterson's quadrature rules compared to the other two methods. The results show that the number of function evaluations for the available Romberg's algorithm is too large to get a satisfactory result and that also the adaptive NC8 method needs too many evaluations. Although the number of evaluations for Patterson's rules is restricted to 63, the approximation of the integral is still quite accurate.

Next, we will compare the integrating power of Patterson's quadrature rules with that of the regular Gauss–Legendre rules. Fig. 4.2 shows the results obtained when these methods are applied to the integrand of $I_2(\varepsilon)$, which is not expected to be integrated exactly by these methods. The relative error is plotted against ε, which determines the integration bounds. For comparison this figure also shows the results for the Gauss–Legendre formula using the same number of points. Since an n-point Gauss rule is exact for polynomials of degree $2n - 1$ and an n-point Patterson rule is "only" exact for

TABLE 4.3

(Absolute) relative errors $(|I_2 - K|)/|I_2|$ for different ε and the number of function evaluations needed. The maximum number of function evaluations for the different methods has been set to respectively 63, 8193, and 1048577

ε	Patterson		Adaptive NC8		Romberg	
10^{-10}	$3.3 \cdot 10^{-3}$	(63)	$2.6 \cdot 10^{-0}$	(8189)	$1.8 \cdot 10^{-2}$	(1048577)
10^{-7}	$3.0 \cdot 10^{-3}$	(63)	$7.3 \cdot 10^{-2}$	(8181)	$1.4 \cdot 10^{-4}$	(1048577)
10^{-5}	$7.7 \cdot 10^{-4}$	(63)	$2.6 \cdot 10^{-3}$	(8185)	$1.3 \cdot 10^{-11}$	(1048577)
10^{-4}	$2.1 \cdot 10^{-5}$	(63)	$5.5 \cdot 10^{-5}$	(8189)	$1.0 \cdot 10^{-14}$	(524289)
10^{-3}	$1.6 \cdot 10^{-8}$	(63)	$4.7 \cdot 10^{-8}$	(8185)	$7.4 \cdot 10^{-16}$	(65537)
10^{-2}	exact*	(63)	$1.6 \cdot 10^{-11}$	(8189)	$4.4 \cdot 10^{-15}$	(8193)

*Exact indicates an accuracy in excess of 16 digits.

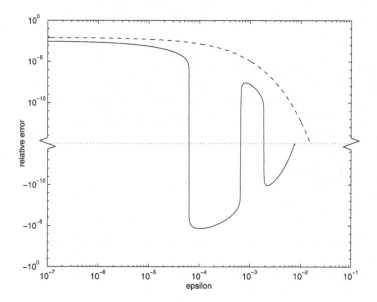

FIG. 4.2. Relative error in evaluating $I_2(\varepsilon)$ using Patterson-63. The dashed line shows the corresponding result for Gauss-63. The dotted line (rel. error $= 0$) indicates an accuracy in excess of 15 digits.

polynomials of $\frac{3}{2}n + 1$, one might expect that the Gauss–Legendre rules are better for higher degree polynomials. However, this is not the case. Tests have been performed, which show that integrals of high powers of x are better approximated by Patterson's rules than by Gauss–Legendre rules using the same number of points.

Apparently, the performance of Patterson's rules is superior to that of Gauss–Legendre rules. Tests on other almost singular integrands, show also that Patterson's rules are more accurate than the Gauss–Legendre rules for the same number of quadrature points. In summary, the advantages of the Patterson's rules above the Gauss–Legendre rules are

- All function evaluations of a quadrature rule can be used again in higher order quadrature rules,
- The relative error can be easily estimated, by taking the difference between two successive approximations.

5. Analytical integration

As we have already mentioned in Section 3, there are several cases where (a part of) the integral can be evaluated analytically. Basically, this is the case when the edges of the interaction domains are not a part of the boundary of the conductor region. A very special situation occurs when the quadrilateral elements are rectangular and parallel to the axes of the coordinate system. In that case the interaction integral with scalar valued basis functions can be evaluated completely analytically. For quadrilaterals which are not necessarily rectangular the partly analytical evaluation of the integrals is discussed in this section.

5.1. Analytical formula for scalar inner integral

In this subsection we consider the interaction integral with scalar valued basis functions. If none of the edges of the source element lie in the boundary of the conductor region, the factor for the boundary singularity is constant, therefore, the scalar basis functions are constant. In that case the inner part of the interaction integral can be evaluated completely analytically. According to expression (3.2) the inner integral has the form

$$I_i = \int_{\Omega_i} G(\mathbf{x}', \mathbf{x}) \, d\mathbf{x}',$$

where $\mathbf{x} = (x, y, z)$ is a fixed point of the interior of the object element. In this section only one term of the Green's function, $G(\mathbf{x}', \mathbf{x})$, of expression (3.3) will be taken into account.

The source quadrilateral Ω_i can be divided into triangles, of which the projection of $\mathbf{x}$ on the plane of the Ω_i is the common vertex (cf. Fig. 6.1). After transformation to polar coordinates, the inner integral I_i can be written as the sum of the four integrals over the triangles:

$$\int_{\Omega_i} \frac{1}{|\mathbf{x} - \mathbf{x}'|} \, d\mathbf{x}' = \sum_{j=1}^{4} \int_{\varphi_{1j}}^{\varphi_{2j}} \int_0^{h_j/\cos\varphi} \frac{r}{\sqrt{r^2 + z^2}} \, dr \, d\varphi,$$

where φ_{1j} and φ_{2j} are the polar angles corresponding to the jth edge of Ω_i, $r^2 = (x - x')^2 + (y - y')^2$ and z stands for $z - z'$ (see Fig. 5.1).

Integration over r gives us

$$I_i = \sum_{j=1}^{4} \left\{ \int_{\varphi_{1j}}^{\varphi_{2j}} \sqrt{\frac{h_j^2}{\cos^2\varphi} + z^2} \, d\varphi - |z|(\varphi_{2j} - \varphi_{1j}) \right\} \cdot \text{sign}(h_j).$$

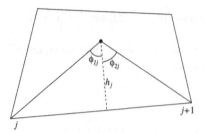

FIG. 5.1.

Finally, after the introduction of $p_j^2 = \frac{z^2}{h_j^2 + z^2}$, $t_j = \sqrt{1 - p_j^2 \sin^2 \varphi}$, and $q_j = \sqrt{1 - p_j^2}$, we obtain

$$
I_i = \sum_{j=1}^4 \left\{ \sqrt{h_j^2 + z^2} \int_{\varphi_{1j}}^{\varphi_{2j}} \frac{\sqrt{1 - p_j^2 \sin^2 \varphi}}{\cos \varphi} \, d\varphi - |z|(\varphi_{2j} - \varphi_{1j}) \right\} \cdot \text{sign}(h_j)
$$

$$
= \sum_{j=1}^4 \left\{ \sqrt{h_j^2 + z^2} \left[\frac{1}{2} q_j \ln \frac{t_j + q_j \sin \varphi}{t_j - q_j \sin \varphi} + p_j \arcsin(p_j \sin \varphi) \right]_{\varphi_{1j}}^{\varphi_{2j}} \right.
$$

$$
\left. - |z|(\varphi_{2j} - \varphi_{1j}) \right\} \cdot \text{sign}(h_j).
$$

Note that if $h_j = 0$, then $\cos \varphi_{1j} = \cos \varphi_{2j} = 0$, $p_j = 1$, $q_j = 0$, and $t_{1j} = t_{2j} = 0$, so that

$$
\left\{ |z| \arcsin(\sin \varphi) \big|_{\varphi_{1j}}^{\varphi_{2j}} - |z|(\varphi_{2j} - \varphi_{1j}) \right\} \cdot \text{sign}(h_j) = 0.
$$

If $z = 0$, then

$$
I_i = \sum_{j=1}^4 \left[\frac{1}{2} h_j \ln \frac{1 + \sin \varphi}{1 - \sin \varphi} \right]_{\varphi_{1j}}^{\varphi_{2j}}.
$$

5.2. Analytical formula for vector valued inner integral

Let us now consider the interaction integrals with vector valued basis functions, with a constant factor for the boundary singularity. In that case only the inner integral of $\mathbf{I}_i$ can be evaluated analytically. For this purpose we have to define some auxiliary quantities.

5.2.1. Definitions of some auxiliary quantities
Consider the quadrilateral $\mathbf{x}_1 \ldots \mathbf{x}_4$, with the vectors $\mathbf{v}_1 = \mathbf{x}_{12} - (\mathbf{x}_{12} + \mathbf{x}_{34})s_2$ and $\mathbf{v}_2 = -\mathbf{x}_{41} - (\mathbf{x}_{12} + \mathbf{x}_{34})s_1$ for $\mathbf{x}_{ij} = \mathbf{x}_j - \mathbf{x}_i$, and $\mathbf{w}_i$ as defined in Appendix B. After transformation to the isoparametric coordinates s_1 and s_2, the integrals over the edge functions $\mathbf{w}_i$ of the source element Ω_i, for a fixed object point $\mathbf{x}_m = (x_m, y_m, z_m)$, have

the form

$$\int_0^1 \int_0^1 \frac{\mathbf{w}_i(s_1, s_2)}{|\mathbf{x}_m - \mathbf{x}'|} J(s_1, s_2) \, ds_1 \, ds_2,$$

where J is the Jacobian $|\mathbf{v}_1 \times \mathbf{v}_2|$. This can be rewritten as

$$\int_0^1 \int_0^1 \frac{\mathbf{q}_i(s_1, s_2)}{|\mathbf{x}_m - \mathbf{x}'|} \, ds_1 \, ds_2, \tag{5.1}$$

where $\mathbf{q}_i$, for $j = 1, \ldots, 4$ are the quadratic functions

$$\mathbf{q}_1 = (1 - s_2)\mathbf{v}_2, \quad \mathbf{q}_2 = -s_1 \mathbf{v}_1, \quad \mathbf{q}_3 = -s_2 \mathbf{v}_2, \quad \mathbf{q}_4 = (1 - s_1)\mathbf{v}_1. \tag{5.2}$$

Further, we introduce the vector $\mathbf{v}_0 = (1 - s_2)\mathbf{x}_1 + s_2 \mathbf{x}_4 - \mathbf{x}_m$.

Since $\mathbf{x}'(s_1, s_2) = \mathbf{x}_1 + \mathbf{x}_{12}s_1 - \mathbf{x}_{41}s_2 - (\mathbf{x}_{12} + \mathbf{x}_{34})s_1 s_2$, the square of the denominator of the integrand of (5.1) can be rewritten as

$$|\mathbf{x}_m - \mathbf{x}'|^2 = Q(s_1, s_2) = a(s_2) + b(s_2)s_1 + c(s_2)s_1^2,$$

where a, b, c are quadratic functions of s_2:

$$
\begin{aligned}
a(s_2) &= \mathbf{v}_0 \cdot \mathbf{v}_0, \\
&= |\mathbf{x}_{41}|^2 s_2^2 - 2\mathbf{x}_{41} \cdot \mathbf{x}_{m1}s_2 + |\mathbf{x}_{m1}|^2, \\
b(s_2) &= 2\mathbf{v}_1 \cdot \mathbf{v}_0, \\
&= 2\{(\mathbf{x}_{12} + \mathbf{x}_{34}) \cdot \mathbf{x}_{41}s_2^2 - \{(\mathbf{x}_{12} + \mathbf{x}_{34}) \cdot \mathbf{x}_{m1} + \mathbf{x}_{12} \cdot \mathbf{x}_{41}\}s_2 + \mathbf{x}_{12} \cdot \mathbf{x}_{m1}\}, \\
c(s_2) &= \mathbf{v}_1 \cdot \mathbf{v}_1, \\
&= |\mathbf{x}_{12} + \mathbf{x}_{34}|^2 s_2^2 - 2(\mathbf{x}_{12} + \mathbf{x}_{34}) \cdot \mathbf{x}_{12}s_2 + |\mathbf{x}_{12}|^2.
\end{aligned}
$$

5.2.2. Form of the integrals

Since the functions $\mathbf{q}_i$ in the expressions (5.2) are linear with respect to s_1 and s_2, the integral (5.1) is a linear combination of two integrals of the form:

$$\int_0^1 \frac{1}{\sqrt{Q(s_1, s_2)}} \, ds_1 = \int_0^1 \frac{1}{\sqrt{a + bs_1 + cs_1^2}} \, ds_1,$$

$$\int_0^1 \frac{s_1}{\sqrt{Q(s_1, s_2)}} \, ds_1 = \int_0^1 \frac{s_1}{\sqrt{a + bs_1 + cs_1^2}} \, ds_1$$

where $Q(s_1, s_2)$ is quadratic with respect to s_1 and s_2 and $a(s_2)$, $b(s_2)$, and $c(s_2)$ are the quadratic functions as defined above. The integrals can be readily evaluated:

$$\frac{1}{\sqrt{c}} \ln\left(\frac{\sqrt{c}\sqrt{a+b+c} + c + \frac{1}{2}b}{\sqrt{c}\sqrt{a} + \frac{1}{2}b}\right), \quad \frac{\sqrt{a+b+c} - \sqrt{a}}{c} - \frac{b}{2c}I_0.$$

After substitution of these integrals in (5.1) we obtain

$$\int_0^1 \int_0^1 \frac{\mathbf{q}_i(s_1, s_2)}{|\mathbf{x}_m - \mathbf{x}'|} \, ds_1 \, ds_2 = \int_0^1 \mathcal{F}_i \ln\left(\frac{\sqrt{c}\sqrt{a+b+c} + c + \frac{1}{2}b}{\sqrt{c}\sqrt{a} + \frac{1}{2}b}\right) ds_2$$

$$+ \int_0^1 \mathcal{G}_i \left(\sqrt{a+b+c} - \sqrt{a}\right) ds_2,$$

where the vector valued functions $\overline{\mathcal{F}}_i$ and $\overline{\mathcal{G}}_i$ are defined by:

$$\overline{\mathcal{F}}_1(s_2) = (1 - s_2)\frac{1}{\sqrt{c}}\left\{-\mathbf{x}_{41} + \frac{b}{2c}(\mathbf{x}_{12} + \mathbf{x}_{34})\right\},$$

$$\overline{\mathcal{G}}_1(s_2) = -\frac{1}{c}(1 - s_2)(\mathbf{x}_{12} + \mathbf{x}_{34}),$$

$$\overline{\mathcal{F}}_2(s_2) = \frac{b}{2c\sqrt{c}}\left\{\mathbf{x}_{12} - s_2(\mathbf{x}_{12} + \mathbf{x}_{34})\right\},$$

$$\overline{\mathcal{G}}_2(s_2) = -\frac{1}{c}\left\{\mathbf{x}_{12} - s_2(\mathbf{x}_{12} + \mathbf{x}_{34})\right\},$$

$$\overline{\mathcal{F}}_3(s_2) = -s_2\frac{1}{\sqrt{c}}\left\{-\mathbf{x}_{41} + \frac{b}{2c}(\mathbf{x}_{12} + \mathbf{x}_{34})\right\},$$

$$\overline{\mathcal{G}}_3(s_2) = \frac{1}{c}s_2(\mathbf{x}_{12} + \mathbf{x}_{34}),$$

$$\overline{\mathcal{F}}_4(s_2) = \frac{1}{\sqrt{c}}\left(1 + \frac{b}{2c}\right)\left\{\mathbf{x}_{12} - s_2(\mathbf{x}_{12} + \mathbf{x}_{34})\right\},$$

$$\overline{\mathcal{G}}_4(s_2) = -\frac{1}{c}\left\{\mathbf{x}_{12} - s_2(\mathbf{x}_{12} + \mathbf{x}_{34})\right\}.$$

After substitution of the expressions for a, b, and c in the integrands, we obtain

$$\int_0^1\int_0^1 \frac{\mathbf{q}_i(s_1, s_2)}{|\mathbf{x}_m - \mathbf{x}'|}\,ds_1\,ds_2 = \int_0^1 \overline{\mathcal{F}}_i \ln\left(\frac{|\mathbf{v}_1||\mathbf{v}_1 + \mathbf{v}_0| + \mathbf{v}_1 \cdot (\mathbf{v}_1 + \mathbf{v}_0)}{|\mathbf{v}_1||\mathbf{v}_0| + \mathbf{v}_1 \cdot \mathbf{v}_0}\right)\,ds_2$$
$$+ \int_0^1 \overline{\mathcal{G}}_i\left(|\mathbf{v}_1 + \mathbf{v}_0| - |\mathbf{v}_0|\right)\,ds_2. \qquad (5.3)$$

5.2.3. Evaluation for nonsingular integrand

If the object point $\mathbf{x}_m$ is not in the source element, which means that either $z_m \neq 0$ or the projection point lies outside the element, the integrands of (5.3) are nonsingular. Therefore, the integral (5.3) can be evaluated numerically with a satisfactory accuracy. However, if the distance, h, of $\mathbf{x}_m$ to the line $s_2 = $ constant through the integration point (s_1, s_2), which intersects the edges $\mathbf{x}_{23}$ and $\mathbf{x}_{41}$ of the source element, is almost zero, the integrand of the first integral of (5.3) is irregular and has to be analyzed separately.

Behavior of the integrand for $h \to 0$. Let d be defined as illustrated in Fig. 5.2. For $\alpha < \frac{\pi}{2}$ it holds that $d < 0$, and for $\alpha > \frac{\pi}{2}$ that $d > 0$. Note that $|\mathbf{v}_1| > 0$, since $\mathbf{v}_1$ is a convex combination of the edges $\mathbf{x}_{12}$ and $\mathbf{x}_{34}$, which are both nonzero.

An analysis of the argument L of the logarithm in the first integral of (5.3) shows that

$$L = \frac{|\mathbf{v}_1||\mathbf{v}_1 + \mathbf{v}_0| + \mathbf{v}_1 \cdot (\mathbf{v}_1 + \mathbf{v}_0)}{|\mathbf{v}_1||\mathbf{v}_0| + \mathbf{v}_1 \cdot \mathbf{v}_0} = \frac{|\mathbf{v}_1 + \mathbf{v}_0| - d}{|\mathbf{v}_0| - |\mathbf{v}_1| - d}$$
$$= \frac{\sqrt{h^2 + d^2} - d}{\sqrt{h^2 + (|\mathbf{v}_1| + d)^2} - |\mathbf{v}_1| - d}. \qquad (5.4)$$

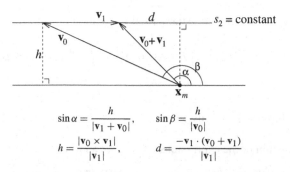

$$\sin\alpha = \frac{h}{|\mathbf{v}_1 + \mathbf{v}_0|}, \qquad \sin\beta = \frac{h}{|\mathbf{v}_0|}$$

$$h = \frac{|\mathbf{v}_0 \times \mathbf{v}_1|}{|\mathbf{v}_1|}, \qquad d = \frac{-\mathbf{v}_1 \cdot (\mathbf{v}_0 + \mathbf{v}_1)}{|\mathbf{v}_1|}$$

FIG. 5.2. For the analysis of the integrand of the first integral of (5.3) a plane has been drawn through $\mathbf{x}_m$ and the line $s_2 = $ constant. Note that this line is in the plane of the quadrilateral, but $\mathbf{x}_m$ is not, if $z_m \neq 0$.

For $d < 0$, L is singular if $\sqrt{h^2 + (|\mathbf{v}_1| - |d|)^2} = |\mathbf{v}_1| - |d|$, i.e., $h = 0$ and $|d| \leqslant |\mathbf{v}_1|$, so that for $h \to 0$ formula (5.4) can only be used for $d < -|\mathbf{v}_1|$. For $d = 0$, L is singular if $h = 0$.

If $d > 0$ and $h \to 0$, then $|\mathbf{v}_1 + \mathbf{v}_0| \to d$ and $|\mathbf{v}_0| - |\mathbf{v}_1| \to d$, so that the two terms in the numerator and the denominator of L have opposite signs and their value may become inaccurate due to cancellation of significant digits. This may affect the accuracy of L. However, L can be reformulated as follows:

$$L = \frac{|\mathbf{v}_1||\mathbf{v}_0| - \mathbf{v}_1 \cdot \mathbf{v}_0}{|\mathbf{v}_1||\mathbf{v}_1 + \mathbf{v}_0| - \mathbf{v}_1 \cdot (\mathbf{v}_1 + \mathbf{v}_0)} = \frac{|\mathbf{v}_0| + |\mathbf{v}_1| + d}{|\mathbf{v}_1 + \mathbf{v}_0| + d}$$
$$= \frac{\sqrt{h^2 + (|\mathbf{v}_1| + d)^2} + |\mathbf{v}_1| + d}{\sqrt{h^2 + d^2} + d}. \tag{5.5}$$

Since for $d > 0$ all terms in the numerator and in the denominator of the right-hand side are positive, both the numerator and denominator do not vanish, so that L can be evaluated accurately. If $d \leqslant 0$ and $h \to 0$, then $|\mathbf{v}_0| - |\mathbf{v}_1| \to -d$, so that L is singular.

In summary, for $d < -|\mathbf{v}_1|$ formula (5.4) must be used and for $d > 0$ formula (5.5). However, for $-|\mathbf{v}_1| \leqslant d \leqslant 0$, L is singular for $h = 0$. In this case the first integral of (5.3) has to be handled differently as will be shown in the next subsection.

5.2.4. Evaluation for singular integrand
If the integrand of the first integral of (5.3) is singular, some more provisions have to be made. Note that, if $z_m = 0$, $\mathbf{x}_m$ can also be written in isoparametric coordinates $s_i^{(m)}$.

The first integral of (5.3),

$$\int_0^1 \mathcal{F}_i \ln\left(\frac{\sqrt{c}\sqrt{a+b+c} + c + \frac{1}{2}b}{\sqrt{c}\sqrt{a} + \frac{1}{2}b}\right) ds_2,$$

can be rewritten as

$$\int_0^1 \mathcal{F}_i \ln\left(\left(\sqrt{c}\sqrt{a+b+c} + c + \frac{1}{2}b\right)\left(\sqrt{c}\sqrt{a} - \frac{1}{2}b\right)\right) ds_2$$
$$- \int_0^1 \mathcal{F}_i \ln\left(ca - \left(\frac{1}{2}b\right)^2\right) ds_2,$$

the last term of which is singular. After partial integration this integral becomes

$$\int_0^1 \overline{\mathcal{F}}_i \ln\!\left(ca - \left(\tfrac{1}{2}b\right)^2\right) ds_2 = \left[\overline{\mathcal{F}}_i(s_2)\mathcal{L}(s_2)\right]_0^1 - \int_0^1 \overline{\mathcal{F}}_i'(s_2)\mathcal{L}(s_2)\,ds_2,$$

where

$$\mathcal{L}(s_2) = \int \ln\!\left(ca - \left(\tfrac{1}{2}b\right)^2\right) ds_2. \tag{5.6}$$

After reformulation of the expressions for $\mathbf{x}_m$, $\mathbf{v}_0$, and $\mathbf{v}_1$:

$$\mathbf{x}_m = \mathbf{x}_1 + \mathbf{x}_{12}s_1^{(m)} - \mathbf{x}_{41}s_2^{(m)} - (\mathbf{x}_{12}+\mathbf{x}_{34})s_1^{(m)}s_2^{(m)},$$

$$\mathbf{v}_0 = (1-s_2)\mathbf{x}_1 + s_2\mathbf{x}_4 - \mathbf{x}_m = \mathbf{x}_1 - \mathbf{x}_{41}s_2 - \mathbf{x}_m$$

$$= -\mathbf{x}_{12}s_1^{(m)} + \mathbf{x}_{41}\!\left(s_2^{(m)} - s_2\right) + (\mathbf{x}_{12}+\mathbf{x}_{34})s_1^{(m)}s_2^{(m)}$$

$$= \mathbf{c}_0\!\left(s_2^{(m)} - s_2\right) - \mathbf{d}s_1^{(m)},$$

$$\mathbf{v}_1 = \mathbf{x}_{12}(1-s_2) - \mathbf{x}_{34}s_2$$

$$= \left\{\mathbf{x}_{12}\!\left(1 - s_2^{(m)}\right) - \mathbf{x}_{34}s_2^{(m)}\right\} + (\mathbf{x}_{12}+\mathbf{x}_{34})\!\left(s_2^{(m)} - s_2\right)$$

$$= \mathbf{d} + \mathbf{c}_1\!\left(s_2^{(m)} - s_2\right),$$

where $\mathbf{c}_0 = \mathbf{x}_{41}$, $\mathbf{c}_1 = \mathbf{x}_{12} + \mathbf{x}_{34}$, and $\mathbf{d} = \mathbf{x}_{12}(1 - s_2^{(m)}) - \mathbf{x}_{34}s_2^{(m)}$, we obtain for the argument of the logarithm of integral (5.6):

$$ca - \left(\tfrac{1}{2}b\right)^2 = |\mathbf{v}_0|^2|\mathbf{v}_1|^2 - (\mathbf{v}_0 \cdot \mathbf{v}_1)^2 = |\mathbf{v}_0|^2|\mathbf{v}_1|^2\!\left(1 - (\cos\beta)^2\right)$$

$$= |\mathbf{v}_0|^2|\mathbf{v}_1|^2(\sin\beta)^2 = |\mathbf{v}_0 \times \mathbf{v}_1|^2$$

$$= \left|(\mathbf{c}_0 \times \mathbf{c}_1)\!\left(s_2^{(m)} - s_2\right)^2 + \left(\mathbf{c}_0 + \mathbf{c}_1 s_1^{(m)}\right) \times \mathbf{d}\!\left(s_2^{(m)} - s_2\right)\right|^2$$

$$= \left(s_2^{(m)} - s_2\right)^2\!\left\{C^2\!\left(s_2^{(m)} - s_2\right)^2 + B\!\left(s_2^{(m)} - s_2\right) + A^2\right\},$$

where

$$C = |\mathbf{c}_0 \times \mathbf{c}_1|, \qquad B = 2(\mathbf{c}_0 \times \mathbf{c}_1) \cdot \left(\left(\mathbf{c}_0 + \mathbf{c}_1 s_1^{(m)}\right) \times \mathbf{d}\right),$$

$$A = \left|\left(\mathbf{c}_0 + \mathbf{c}_1 s_1^{(m)}\right) \times \mathbf{d}\right|.$$

Since the vectors $\mathbf{c}_0$, $\mathbf{c}_1$, and $\mathbf{d}$ all lie in the same plane, $\mathbf{c}_0 \times \mathbf{c}_1 \parallel (\mathbf{c}_0 + \mathbf{c}_1 s_1^{(m)}) \times \mathbf{d}$, so that

$$ca - \left(\tfrac{1}{2}b\right)^2 = y^2\!\left(Cy + \text{sign}(B)A\right)^2,$$

where $y = s_2^{(m)} - s_2$. Therefore, the integral (5.6) becomes

$$\int \ln y^2\,ds_2 + \int \ln\!\left(Cy + \text{sign}(B)A\right)^2 ds_2.$$

Since $dy = -ds_2$, the first integral becomes

$$\int \ln y^2\,ds_2 = -\int \ln y^2\,dy = -y(\ln y^2 - 2),$$

and the second integral:

$$\int \ln(Cy + \text{sign}(B)A)^2 \, ds_2 = -\int \ln(Cy + \text{sign}(B)A)^2 \, dy$$

$$= -\frac{1}{C}(Cy + \text{sign}(B)A)\left(\ln(Cy + \text{sign}(B)A)^2 - 2\right).$$

Hence,

$$\mathcal{L}(s_2) = -y\left\{\ln y^2 \left(Cy + \text{sign}(B)A\right)^2 - 4\right\}$$

$$- \text{sign}(B)\frac{A}{C}\left(\ln(Cy + \text{sign}(B)A)^2 - 2\right).$$

If $C = 0$ the integral becomes:

$$\mathcal{L}(s_2) = -y\left(\ln(yA)^2 - 2\right).$$

6. Regularizations

In the previous section, integrands were discussed that did not contain a factor for the boundary singularity. These could be integrated partly analytically. For integrals with a factor for boundary singularity numerical methods must be used. As shown in Section 4 a very important condition for the numerical integration to give a satisfactory result is the smoothness of the integrand. So if the integrand contains singularities quadrature rules can give very inaccurate results. Since the interaction integral contains the Green's singularity as well as the boundary singularity, straightforward numerical integration is rather unreliable. In this section we will treat the elimination of the singularities by regularization of the integrands. The Green's singularity and the boundary singularity will be treated separately.

6.1. The inner part of the interaction integral

As shown in Section 3 the interaction integral has the general form:

$$\int_{\Omega_j} \frac{\psi_j(\mathbf{x})}{\sqrt{d_j}} \left\{\int_{\Omega_i} G(\mathbf{x}', \mathbf{x})\frac{\psi_i(\mathbf{x}')}{\sqrt{d_i}} \, d\mathbf{x}'\right\} d\mathbf{x},$$

with d_j and d_i the distance functions $d_j(s_1, s_2)$ and $d_i(s_1', s_2')$, the specific form of which is given in Appendix A. Without any loss of generality we will assume that the z-coordinate of $\mathbf{x}'$ is always zero.

The first four subsections of this section deal with the *inner part* of the interaction integral

$$\mathbf{I}_i = \int_{\Omega_i} G(\mathbf{x}', \mathbf{x})\frac{\psi_i(\mathbf{x}')}{\sqrt{d_i}} \, d\mathbf{x}', \tag{6.1}$$

the last subsection with the outer part. In this section only one term of the Green's function, $G(\mathbf{x}', \mathbf{x})$, of expression (3.3) will be taken into account.

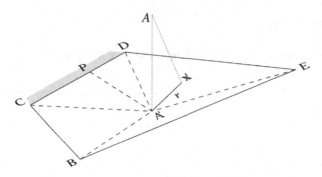

FIG. 6.1. Source quadrilateral divided into triangles, determined by the projection A' of the object point A.

Let A be the object point $\mathbf{x}$, and $A' = (x, y, 0)$, the projection of A on the source element plane (see Fig. 6.1). Let $r(x, y) = \sqrt{(x' - x)^2 + (y' - y)^2}$, then $G(\mathbf{x}', \mathbf{x})$ can be written as $|\mathbf{x}' - \mathbf{x}|^{-1} = 1/\sqrt{r^2 + z^2}$. Then (6.1) becomes

$$\iint_{\Omega_i} \frac{1}{\sqrt{r^2 + z^2}} \frac{\psi_i(\mathbf{x}')}{\sqrt{d_i}} \, d\mathbf{x}'.$$

This integral has two sources of singularities: the Green's function $(r^2 + z^2)^{-1/2}$ and the boundary singularity $d_i^{-1/2} = d_i(s_1, s_2)^{-1/2}$. First we will concentrate on the former.

6.2. The Green's singularity

The expression $(r^2 + z^2)^{-1/2}$ depends on the distance z between the planes of the interacting quadrilaterals; for $z = 0$ it has a singularity in $r = 0$.

To remove the source of singularity, the quadrilateral will be divided into four triangles, each of which has an edge of the quadrilateral as base and midpoint A' as vertex (see Fig. 6.1). After transformation to polar coordinates the integral for one of four triangles becomes:

$$\int_{\varphi_1}^{\varphi_2} \int_0^{R(\varphi)} \frac{r}{\sqrt{r^2 + z^2}} \frac{\psi_i}{\sqrt{d}} \, dr \, d\varphi. \qquad (6.2)$$

For example, for the triangle $\triangle A'CD$, φ_1 and φ_2 are the polar angles corresponding to $A'D$ and $A'C$, and $R(\varphi) = |A'P|$, with P the point on the boundary CD corresponding to the angle φ (see also Fig. 6.3). From the expression (6.2) one can see that if $z = 0$, i.e., $\mathbf{x}$ and $\mathbf{x}'$ are in the same plane, the Green's singularity has been completely eliminated. Even for $z \neq 0$ the function $r/\sqrt{r^2 + z^2}$ is regular.

6.3. The boundary singularity

To regularize the boundary singularity due to d in (6.2) is much more complicated. In Appendix A d is given as a function of the isoparametric coordinates (s_1, s_2). But the integrand of (6.2) must be integrated over the variables r and φ. Therefore, we need an expression of $\tilde{s}_i(r, \varphi)$ as function of r, φ, or an expression $s_i(r, \varphi)$, since the expressions $x'(r, \varphi)$ and $y'(r, \varphi)$ are well known. An unique expression $s_i(r, \varphi)$ only exists for points (x', y') inside and at the edges of the source quadrilateral, because an extra condition, $0 \leqslant s_i \leqslant 1$, must also be fulfilled.

Before deriving the expression $s_i(x', y')$ we will first define some auxiliary quantities. Having the quadrilateral $\mathbf{x}_1\mathbf{x}_2\mathbf{x}_3\mathbf{x}_4$ (in counterclockwise order), the edges can be defined as $\mathbf{x}_{ij} \equiv (\mathbf{x}_j - \mathbf{x}_i)$, for $j = i \bmod 4 + 1$.

The transformation can be given by

$$\mathbf{x}'(s_1, s_2) = \mathbf{x}_1 + \mathbf{x}_{12}s_1 - \mathbf{x}_{41}s_2 - (\mathbf{x}_{12} + \mathbf{x}_{34})s_1 s_2, \tag{6.3}$$

or

$$\mathbf{x}' - \mathbf{x}_1 - \mathbf{x}_{12}s_1 = -\{(\mathbf{x}_{12} + \mathbf{x}_{34})s_1 + \mathbf{x}_{41}\}s_2.$$

Taking the outer product of the left-hand side with the right-hand side, we obtain an implicit expression for s_1 in terms of $\mathbf{x}' = (x', y')$:

$$f(s_1)\mathbf{e}_z = (\mathbf{x}_{12} \times \mathbf{x}_{34})s_1^2 + \{(\mathbf{x}_{12} \times \mathbf{x}_{41}) - ((\mathbf{x}' - \mathbf{x}_1) \times (\mathbf{x}_{12} + \mathbf{x}_{34}))\}s_1$$
$$- ((\mathbf{x}' - \mathbf{x}_1) \times \mathbf{x}_{41}) = 0,$$

where $\mathbf{e}_z$ is the unit vector in the z-direction and $(\mathbf{a} \times \mathbf{b}) = (a_x b_y - a_y b_x)\mathbf{e}_z$, since the z-coordinates were assumed to be zero.

Note that $f(0)\mathbf{e}_z = -((\mathbf{x}' - \mathbf{x}_1) \times \mathbf{x}_{41})$ and $f(1)\mathbf{e}_z = ((\mathbf{x}' - \mathbf{x}_2) \times \mathbf{x}_{23})$. For a point $\mathbf{x}'$ inside or at the edges of the source quadrilateral Fig. 6.2 shows that $f(0) \leqslant 0$ and $f(1) \geqslant 0$, and that if $f(0) = 0$, then $f(1) \neq 0$, and if $f(1) = 0$, then $f(0) \neq 0$. Thus, f has exactly one root for $0 \leqslant s_1 \leqslant 1$. For a point $\mathbf{x}'$ outside the source quadrilateral there are generally two roots! From (6.3) it follows now

$$s_2 = -\frac{x' - x_1 - x_{12}s_1}{x_{41} + (x_{12} + x_{34})s_1} = -\frac{y' - y_1 - y_{12}s_1}{y_{41} + (y_{12} + y_{34})s_1}. \tag{6.4}$$

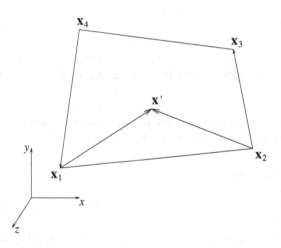

FIG. 6.2. $\mathbf{x}'$ inside the quadrilateral $\mathbf{x}_1\mathbf{x}_2\mathbf{x}_3\mathbf{x}_4$: f has one root for $0 \leqslant s_1 \leqslant 1$.

Hence we have an expression $s_i(x', y')$ for both isoparametric coordinates in terms of x' and y'. As a function of polar coordinates we have

$$\tilde{s}_i(r, \varphi) = s_i\big(x'(r, \varphi), y'(r, \varphi)\big),$$

so that the boundary singularity, in polar coordinates, has the form

$$d\big(\tilde{s}_1(r, \varphi), \tilde{s}_2(r, \varphi)\big)^{-1/2}.$$

The expression for d is too complicated for an analysis of the behavior of the boundary singularity. Although d is a linear function of s_1 or s_2, generally, it is not the geometric distance from point $\mathbf{x}'$ to the boundary. Let $s' = |\mathbf{x}'(s_1, s_2) - \mathbf{x}'(0, \tilde{s}_2)|$ represent the distance from $\mathbf{x}'$ to the boundary of the conductor region. The following lemma holds:

LEMMA 6.1. *Let s' be defined as the distance $|\mathbf{x}'(s_1, s_2) - \mathbf{x}'(0, \tilde{s}_2)|$ to the boundary, where $\tilde{s}_2$ corresponds to the projection of $\mathbf{x}'$ to the boundary, and let d be the isoparametric distance to the boundary in the unit square. Then, if $\mathbf{x}'$ approaches the boundary, s' is approximately proportional to d, i.e., $s' \approx Cd$, where C is a constant.*

PROOF. Since the source for the boundary singularity is not constant, it may be assumed that the quadrilateral is rectangular. Suppose $d = s_2$. Let then $\mathbf{x}(s_1, s_2) = a(s_1)s_2 + b(s_1)$, for linear functions a and b. Since the line "$s_1 = \text{constant}$" is perpendicular to the boundary, the point on the boundary with the shortest distance to $\mathbf{x}(s_1, s_2)$ is $\mathbf{x}(s_1, 0)$. So $s' = |\mathbf{x}(s_1, s_2) - \mathbf{x}(s_1, 0)| = |\{a(s_1)s_2 + b(s_1)\} - b(s_1)| = |a(s_1)s_2| = Cs_2$, since $a(s_1)$ is independent of s_2, hence constant for fixed s_1. The proof is analogous for other boundary singularities.   □

The lemma says that, except for a constant, s' and d are equivalent functions. Therefore, in the following d will be represented as s'.

6.3.1. Transformations for one boundary

In order to evaluate the integrals over the triangles, we have to determine the integration bounds explicitly, in the particular cases. In Fig. 6.1 it can be seen that if one or more of the edges of the quadrilateral is a boundary of the domain, we have different situations for the integration triangles. We will first consider the case where the quadrilateral has only one boundary.

Triangle tangent to the boundary. The integrand with the boundary singularity has, after transformation to polar coordinates, the general form

$$\int_{\varphi_1}^{\varphi_2} \int_0^{R(\varphi)} \frac{\psi_i}{\sqrt{s'}} \, dr \, d\varphi,$$

where $R(\varphi) = |A'P|$, as can be seen in Fig. 6.3. Here we have the situation that one edge of the triangle is a boundary.

We want to regularize the boundary singularity by a certain substitution. So we have to find an expression for s' in terms of the integration variables. In the picture one

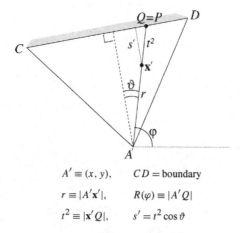

$$A' \equiv (x, y), \qquad CD = \text{boundary}$$
$$r \equiv |A'\mathbf{x}'|, \qquad R(\varphi) \equiv |A'Q|$$
$$t^2 \equiv |\mathbf{x}'Q|, \qquad s' = t^2 \cos \vartheta$$

FIG. 6.3. $\triangle A'CD$ with polar coordinates, where the edge CD is a boundary of the source domain.

can see that we have chosen t^2 such that $r = R(\varphi) - t^2$. To eliminate the $\sqrt{s'}$ singularity, which is now dependent on t and φ, we may substitute $s' = t^2 \cos \vartheta$, where $\vartheta = \frac{\pi}{2} - \angle A'QC$ is dependent on φ, since the position of Q is. Thus finally, with these substitutions the integral becomes

$$\int_{\varphi_1}^{\varphi_2} \int_0^{\sqrt{R(\varphi)}} \frac{\psi_i}{\sqrt{\cos \vartheta}} 2t \, dt \, d\varphi.$$

Since $\vartheta < \frac{\pi}{2}$, this is a regular integral, which could be evaluated rather accurate by Patterson, unless the triangle is very flat, i.e., $|\frac{\pi}{2} - \vartheta|$ small, then we will approach this integrand in a special way (see Section 6.4).

Triangle without tangent to the boundary. The latter case was a situation where one edge of the triangle is a boundary. The treatment changes a little if none of the edges is a boundary. Take, for instance, $\triangle A'DE$, where P, which is the intersubsection point of $A'\mathbf{x}'$ with the edge DE, is not a boundary point and Q is the intersubsection point of $A'P$ with the boundary CD. We will use the same kind of substitutions, but for that purpose some adaptions have to be made: let $q(\varphi) = |A'Q|$ and $R(\varphi) = |A'P|$, so we may substitute $r = q(\varphi) - t^2$. Then the inner integral becomes

$$\int_{\sqrt{q(\varphi)-R(\varphi)}}^{\sqrt{q(\varphi)}} \frac{\psi_i}{\sqrt{\cos \vartheta}} 2t \, dt.$$

At first sight this seems satisfactory. However, if $A'P \| CD$ then ϑ becomes zero, or at least small if approximately $A'P \| CD$. This would introduce another singularity, namely $(\cos \vartheta)^{-1/2}$. Yet, if ϑ might become small then substitution won't be necessary anymore, for in that case $\sqrt{s'}$ changes hardly along AP, which means that it behaves very much like a constant. So here we would simply have the original integral

$$\int_0^{R(\varphi)} \frac{\psi_i}{\sqrt{s'}} \, dr \, d\varphi.$$

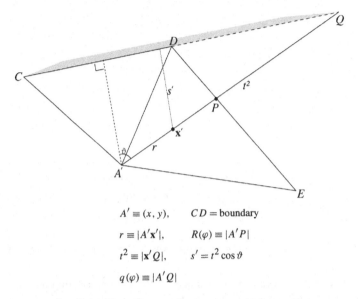

$$A' \equiv (x, y), \qquad CD = \text{boundary}$$

$$r \equiv |A'\mathbf{x}'|, \qquad R(\varphi) \equiv |A'P|$$

$$t^2 \equiv |\mathbf{x}'Q|, \qquad s' = t^2 \cos \vartheta$$

$$q(\varphi) \equiv |A'Q|$$

FIG. 6.4. Integration over $\triangle A'DE$, with boundary edge CD.

Suppose (see Fig. 6.4) P comes in the neighborhood of E, then Q lies at the other side of D so that A' lies in between P and Q. This means that the substitution becomes $r = t^2 - q(\varphi)$. So the integration bounds will be influenced differently as in the former case. The actual integral becomes then

$$\int_{\sqrt{q(\varphi)}}^{\sqrt{q(\varphi)+R(\varphi)}} \frac{\psi_i}{\sqrt{\cos \vartheta}} 2t \, dt.$$

To distinguish the different cases we could split up the triangle in wedges, as illustrated in Fig. 6.5, such that we may use the former (or latter) substitution if the angle between $A'P$ and CD becomes larger than, say ε, or if $|\cos \vartheta| > \varepsilon$, and otherwise we will use no substitution. Hence for $\triangle A'DE$ we finally have the sum over the integrals.

For the remaining triangles we can use analogous substitutions to these.

Still leaves us the case where the projection A' of the object point lies outside the quadrilateral.

A' outside the quadrilateral. Suppose now that A is situated in such a way that A' lies outside the quadrilateral. The main difference from the previous cases is that, if we divide the quadrilateral into triangles, these triangles will overlap and we only want to integrate over the parts of the triangles that are inside the quadrilateral. If we consider the integral over $\triangle A'BC$, then even the whole triangle lies outside the quadrilateral, and so does $\triangle A'CD$. So we only have to integrate over the triangles $\triangle A'ED$ and $\triangle A'BE$. Let us consider the latter. In Fig. 6.6 we can see that a part of the triangle lies outside the quadrilateral. We only want to integrate over the part that intersects with the quadrilateral, so instead of the bounds 0 and $R(\varphi)$ for integration over r, we will use

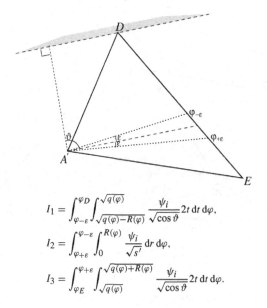

$$I_1 = \int_{\varphi-\varepsilon}^{\varphi_D} \int_{\sqrt{q(\varphi)-R(\varphi)}}^{\sqrt{q(\varphi)}} \frac{\psi_i}{\sqrt{\cos\vartheta}} 2t \, dt \, d\varphi,$$

$$I_2 = \int_{\varphi+\varepsilon}^{\varphi-\varepsilon} \int_0^{R(\varphi)} \frac{\psi_i}{\sqrt{s'}} \, dr \, d\varphi,$$

$$I_3 = \int_{\varphi_E}^{\varphi+\varepsilon} \int_{\sqrt{q(\varphi)}}^{\sqrt{q(\varphi)+R(\varphi)}} \frac{\psi_i}{\sqrt{\cos\vartheta}} 2t \, dt \, d\varphi.$$

Fɪɢ. 6.5.

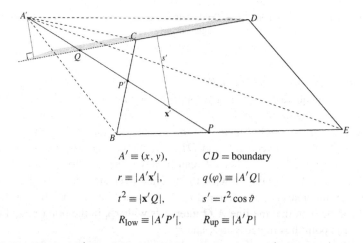

$A' \equiv (x, y)$,	$CD =$ boundary
$r \equiv \lvert A'\mathbf{x}'\rvert$,	$q(\varphi) \equiv \lvert A'Q\rvert$
$t^2 \equiv \lvert \mathbf{x}'Q\rvert$,	$s' = t^2 \cos\vartheta$
$R_{\text{low}} \equiv \lvert A'P'\rvert$,	$R_{\text{up}} \equiv \lvert A'P\rvert$

Fɪɢ. 6.6. A situation where A' lies outside the quadrilateral, where both $\triangle A'BC$ and $\triangle A'CD$ lie completely outside the quadrilateral, so only integration over $\triangle A'ED$ and $\triangle A'BE$, therefore the lower integration bound for r has to be calculated.

the lower bound R_{low} and the upper bound R_{up} (see the legend with Fig. 6.6), so that after substitution of $r = t^2 + q(\varphi)$ we obtain the integral

$$\int_{\varphi_B}^{\varphi_E} \int_{\sqrt{R_{\text{low}}-q(\varphi)}}^{\sqrt{R_{\text{up}}-q(\varphi)}} \frac{\psi_i}{\sqrt{\cos\vartheta}} 2t \, dt \, d\varphi.$$

The further approach is the same as in the previous subsections.

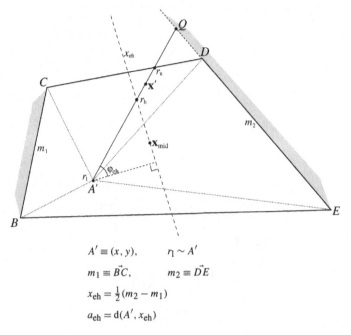

$$A' \equiv (x, y), \qquad r_1 \sim A'$$

$$m_1 \equiv \vec{BC}, \qquad m_2 \equiv \vec{DE}$$

$$x_{\text{eh}} = \tfrac{1}{2}(m_2 - m_1)$$

$$a_{\text{eh}} = \text{d}(A', x_{\text{eh}})$$

FIG. 6.7. Quadrilateral with two boundaries. On the left side of x_{eh} the singularity in m_1 dominates, on the right side m_2.

6.3.2. Transformations for two boundaries

In the case that the quadrilateral has two opposite boundaries m_1 and m_2, the approach for the boundary singularity $\{s_i(1 - s_i)\}^{-1/2}$ is almost the same as for one boundary, except that we distinct the left-hand and the right-hand side of x_{eh}, where x_{eh} is the isoparametric midline between BC and DE. The picture shows a particular case, but we consider the general case, where A' can be anywhere, so also outside the quadrilateral. Consider $\triangle A'CD$ and we want to integrate over the line $A'Q$. Then we have the integration boundaries r_1 ($\sim A'$) and r_u (this means that $\mathbf{x}'$ goes from 0 to r_u). In Fig. 6.7 it can be seen that the line $A'Q$ intersects with x_{eh} in the point r_h and that this intersubsection point lies in between r_1 and r_u.

Over the line segment $r_1 r_h$ the m_1-singularity dominates and over the line segment $r_h r_u$ the m_2-singularity dominates. So we can split the integral into a lower and upper part. For the lower part we use a substitution for $m = m_1$, which means: approach the integrand as if m_1 were the only boundary. For the upper part we will use $m = m_2$, which means: use the same treatment as with one boundary m_2. That is, if we let $s_1' \equiv \text{dist}(\mathbf{x}', m_2)$ and $t_2^2 \equiv |\mathbf{x}'Q|$, then we use the substitution $s_1' = t_2^2 \cos \vartheta$. The substitution for t in r is completely analogue to the previous cases.

If a_{eh}, defined as the distance of A' to the midline x_{eh}, might become zero (A' lies on the midline) and the angle $\varphi_{\text{eh}} \approx \frac{\pi}{2}$, then we will not use a substitution, since the contribution of the singularity is nearly constant.

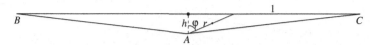

FIG. 6.8. Very flat triangle of height h.

6.4. Flat triangles

Sometimes we are dealing with flat triangles, which means that they have a very wide (half) top angle, or, which is the same, a small height compared to the base. Numerical integration over these kind of triangles (see e.g. Fig. 6.8) to the polar coordinates is not very accurate, since the inner integral of

$$\int_{-\varphi_t}^{\varphi_t} \int_0^{h/\cos\varphi} f(r,\varphi)\, dr\, d\varphi$$

varies very rapidly for $\varphi \approx \pm\frac{\pi}{2}$. First let us examine this problem in general.

It may be clear that for flat triangles the inner integral of

$$I = \int_{\varphi_1}^{\varphi_2} \int_{R_l}^{R_u} f(r,\varphi)\, dr\, d\varphi,$$

as function of φ, varies very rapidly. The reason for this is that the range for r, or the length of the radius over which is integrated, changes fast. A remedy for this would be adapting the inner integral, such that it becomes a function of φ expressing the average along the radius. Such an integral $F(\varphi)$ could be the following:

$$F(\varphi) = \left(\int_{R_l}^{R_u} f(r,\varphi)\, dr \right) \Big/ (R_u - R_l),$$

so that we can rewrite

$$I = \int_{\varphi_1}^{\varphi_2} F(\varphi)(R_u - R_l)\, d\varphi.$$

For the polar radii we have $R_l = \frac{k'}{\cos\varphi'}$ and $R_u = \frac{k}{\cos\varphi}$, where k' and k are the distances of the origin to the intersecting edges, φ' and φ are the angles with the normal of the corresponding edge (see Fig. 6.9). So the integral becomes

$$I = \int_{\varphi_1}^{\varphi_2} F(\varphi)\left(\frac{k}{\cos\varphi} - \frac{k'}{\cos\varphi} \right) d\varphi.$$

Since $\int \frac{1}{\cos\varphi} = \frac{1}{2}\ln\frac{1+\sin\varphi}{1-\sin\varphi}$, we can eliminate the factors $\frac{1}{\cos\varphi}$ by the substitution

$$u = \frac{1}{2}k\ln\left(\frac{1+\sin\varphi}{1-\sin\varphi} \right) - \frac{1}{2}k'\ln\left(\frac{1+\sin\varphi'}{1-\sin\varphi'} \right), \tag{6.5}$$

for which $\frac{du}{d\varphi} = \frac{k}{\cos\varphi} - \frac{k'}{\cos\varphi'}$. After this substitution the integral has the form

$$\int_{u(\varphi_1)}^{u(\varphi_2)} F(\varphi)\, du.$$

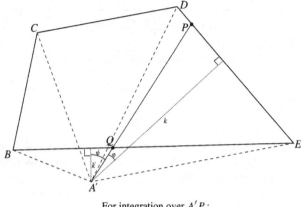

For integration over $A'P$:

$$R_l \sim A'Q \qquad R_u \sim A'P$$

$$\varphi_1 \sim A'E \qquad \varphi_2 \sim A'D$$

FIG. 6.9. Integration over $\triangle A'DE$ in a situation that A' near to the edge BE: if the polar angle φ goes from φ_1 to φ_2, the intersubsection point Q goes in the neighborhood of E very rapidly along EB, so that the lower bound for the polar radius R_l also changes very rapidly.

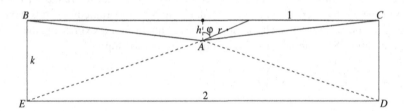

FIG. 6.10. Flat triangle of height h in a rectangular element of height k: for small enough h the integral over the triangle can be neglected.

In general this substitution can be used for any triangle, if the range of the polar radius changes rapidly.

However, in the case that one of the integration bounds for the polar angle is absolute approximately $\frac{\pi}{2}$, we have another tedious situation. When integrating over φ, the polar angle goes along the edge opposite to the origin. The intersubsection point with this edge goes very rapidly in the neighborhood of the absolute angles $\frac{\pi}{2}$ (cf. Figs. 6.8 and 6.11). This will result in a function F, that does not depend very nicely on φ. We will discuss this situation in the following subsections.

6.4.1. Triangles inside the quadrilateral

If the origin (the integration point of the object domain) of the polar coordinate system lies inside the quadrilateral, we have the following situation: suppose the triangle ABC has base BC, a boundary of the quadrilateral $BCDE$ which is a rectangle of height k (cf. Fig. 6.10). Assuming that the base of the triangle has length 2, the height h depends on the size of the half top angle φ_t, for which holds: $|1 - \sin\varphi_t| = \varepsilon$. Then we can express $h(\varepsilon) = \tan(\arcsin(1 - \varepsilon))^{-1}$ and for very small ε holds $h(\varepsilon) \ll 1$.

In particular we want to examine the singularity $s^{-1/2}$, where s is the normalized distance to the boundary BC. For the rectangle (size $2 \times k$) the integral becomes then

$$I_r = \int_0^k \int_{-1}^1 \frac{\sqrt{k}}{\sqrt{k-y}}\, dx\, dy = \int_0^k 2\frac{\sqrt{k}}{\sqrt{k-y}}\, dy = \left[-4\sqrt{k}\sqrt{k-y}\right]_0^k = 4k,$$

and for the triangle (Fig. 6.8) the integral becomes, if r goes from A to a point on BC,

$$I_t = \int_{-\varphi_t}^{\varphi_t} \int_0^{h/\cos\varphi} \frac{r\sqrt{k}}{\sqrt{h - r\cos\varphi}}\, dr\, d\varphi = \int_{-\varphi_t}^{\varphi_t} \frac{4}{3}\frac{h\sqrt{hk}}{\cos(\varphi)^2}\, d\varphi = \frac{8}{3}h\sqrt{hk}\tan\varphi_t,$$

where φ_t is the half top angle, so $\tan\varphi_t = \frac{1}{h}$.

Hence we can give the ratio of the two integrals, $\frac{I_t}{I_r}$, for different h, to get an idea of the contribution of the integral over the triangle to the integral over the complete quadrilateral:

$$\frac{I_t}{I_r} = \frac{\frac{8}{3}\sqrt{hk}}{4k} = \frac{2}{3}\sqrt{\frac{h}{k}}.$$

What we have here is a formula for the rate of contribution of the integral over the triangle to the integral over the rectangle. For h small enough the contribution will be significantly small and can therefore be neglected.

6.4.2. Triangle partly outside the quadrilateral

If the origin of the polar coordinate system lies outside the quadrilateral, but very near to the edge, we would have a situation that leads to a very unsmooth function $F(\varphi)$, to be considered as the average for a certain φ, as defined above. So here the approach will be different. Suppose we have a situation as illustrated in Fig. 6.11, where ε is small. For the new integration variable we will choose a point u that goes along $\overline{VM'}$. In order to determine such an u a few relations have to be looked at. Assume first that we have an u as described. Then we have

$$\overline{MS} = \overline{MV} + \overline{VS} = \overline{MV} + u\overline{VM'},$$

and we can give an expression for $\tan\varphi$ (for readability the vectors will now be denoted without an overline):

$$\tan\varphi = \frac{y_{MS}}{x_{MS}} = \frac{y_{MV} + uy_{VM'}}{x_{MV} + ux_{VM'}} = \frac{y_{MV} + uy_{VM'}}{x_{MV}(1 - u)} = \frac{\tan\varphi_l - u\tan\beta}{1 - u},$$

since $x_{VM'} = -x_{MV}$. From this we can determine u:

$$u = 1 - \frac{\tan\varphi_l - \tan(\varphi_l - \varepsilon)}{\tan\varphi - \tan(\varphi_l - \varepsilon)},$$

so that

$$\frac{du}{d\varphi} = \frac{(1 - u)^2 + (\tan\varphi_l - u\tan(\varphi_l - \varepsilon))^2}{\tan\varphi_l - \tan(\varphi_l - \varepsilon)}.$$

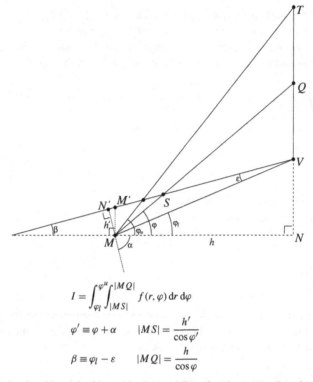

$$I = \int_{\varphi_l}^{\varphi^u} \int_{|MS|}^{|MQ|} f(r, \varphi) \, dr \, d\varphi$$

$$\varphi' \equiv \varphi + \alpha \qquad |MS| = \frac{h'}{\cos \varphi'}$$

$$\beta \equiv \varphi_l - \varepsilon \qquad |MQ| = \frac{h}{\cos \varphi}$$

FIG. 6.11. Flat triangle with origin M outside the quadrilateral and very small ε: for integration over $\triangle MVT$, the polar angle φ is substituted such that u becomes the new integration variable, that goes along the edge VM'.

If u goes along VM', and comes in the neighborhood of M', we have the same difficulty again. To prevent this we can split up the triangle in wedges such that on the lower wedge (say $u \in [0, 0.9]$) we use the latter method and on the other wedge, the method discussed in Section 6.4, the substitution (6.5). Unless the lower integration bound of the polar angle of the upper wedge is still too flat; in that case we do again an analogue splitting up of the upper wedge, and so on.

6.5. The outer integral

Now that we have a method to evaluate the inner integral for an arbitrary object point $\mathbf{x}$, we are able to evaluate the outer part of interaction integral. This has the following form:

$$I = \int_{\Omega_j} \tilde{\psi}_j(\mathbf{x}) \cdot \mathbf{I}_i(\mathbf{x}) \, d\mathbf{x},$$

where the inner integral $\mathbf{I}_i$ is

$$\mathbf{I}_i(\mathbf{x}) = \int_{\Omega_i} G(\mathbf{x}', \mathbf{x}) \tilde{\psi}_i(\mathbf{x}') \, d\mathbf{x}'.$$

After transformation to isoparametric coordinates the outer integral has the form

$$I = \int_0^1 \int_0^1 \tilde{\psi}_j(s_1, s_2) J(s_1, s_2) \cdot \mathbf{I}_i(s_1, s_2) \, ds_1 \, ds_2$$

$$= \int_0^1 \int_0^1 \frac{\psi(s_1, s_2)}{\sqrt{d(s_1, s_2)}} \, ds_2 \, ds_1.$$

The boundary singularities can be divided in three cases and are for each case regularized as follows:

If none of the edges of the corresponding element is a part of the boundary, then there is no boundary singularity. In that case we simply have:

$$I = \int_0^1 \int_0^1 \psi(s_1, s_2) \, ds_2 \, ds_1,$$

so no substitution is used here.

The approach for an integrand with a factor for the boundary singularity is quite simple and divided into two cases:

With one edge in the boundary there are four possibilities, where $t = \sqrt{d}$ is substituted:

$$d = s_2 \quad \Longrightarrow \quad I = 2 \int_0^1 \int_0^1 \psi\big(s_1, s_2(t)\big) \, dt \, ds_1, \quad \text{where } s_2 = t^2,$$

$$d = 1 - s_1 \quad \Longrightarrow \quad I = 2 \int_0^1 \int_0^1 \psi\big(s_1(t), s_2\big) \, ds_2 \, dt, \quad \text{where } s_1 = 1 - t^2,$$

$$d = 1 - s_2 \quad \Longrightarrow \quad I = 2 \int_0^1 \int_0^1 \psi\big(s_1, s_2(t)\big) \, dt \, ds_1, \quad \text{where } s_2 = 1 - t^2,$$

$$d = s_1 \quad \Longrightarrow \quad I = 2 \int_0^1 \int_0^1 \psi\big(s_1(t), s_2\big) \, ds_2 \, dt, \quad \text{where } s_1 = t^2.$$

With two boundary edges, d has the form $s_i(1 - s_i)$, since the boundaries are always opposite to each other. Here we substitute $t = \arcsin(2s_i - 1)$:

$$d = s_2(1 - s_2) \quad \Longrightarrow \quad I = \int_0^1 \int_{-1/2\pi}^{1/2\pi} \psi\big(s_1, s_2(t)\big) \, dt \, ds_1,$$

where $s_2 = \frac{1}{2}(1 + \sin t)$,

$$d = s_1(1 - s_1) \quad \Longrightarrow \quad I = \int_{-1/2\pi}^{1/2\pi} \int_0^1 \psi\big(s_1(t), s_2\big) \, ds_2 \, dt,$$

where $s_1 = \frac{1}{2}(1 + \sin t)$.

After these substitutions we have obtained nonsingular integrals, which can be evaluated numerically by the method described in Section 4.

It is clear that the complete numerical integration takes a lot of function evaluations, namely $\mathcal{O}(k^4)$, if k is the average number of function evaluations needed for the quadrature with respect to each of the four integration variables. Totally, there are N^2 of such interaction integrals, where $N = n + m$ is the number of elements and edges, so for the evaluation of these integrals a total of $\mathcal{O}(N^2 k^4)$ function evaluations would be needed. This number grows very rapidly if n is of order 10^3 or 10^4, and k might be 15 or 31 for irregular integrands. For a great part of the integrals an alternative method can be used. This is discussed in the next section.

7. Taylor expansion

In the previous section we have seen that the evaluation of the fourfold interaction integrals by numerical quadrature can take very much computer time. If the "distance" between two elements is larger than a given tolerance, the Green's function in the integrand of the interaction integral can be approximated by *Taylor expansion*. This method will be discussed in this section. In Section 7.7 a relation will be derived between the distance and the relative error made in the evaluation of the integral by this method.

7.1. Interaction integral in general form

The interaction integrals (2.37) and (2.38) belonging to the matrices **L** and **D** have the general form

$$I = \int_{\Omega_i} \tilde{\psi}_i(\mathbf{x}) \cdot \int_{\Omega_j} \tilde{\psi}_j(\mathbf{x}') G(\mathbf{x}' - \mathbf{x}) \, d\mathbf{x}' \, d\mathbf{x}. \tag{7.1}$$

Here $\tilde{\psi}_i(\mathbf{x})$, $\tilde{\psi}_j(\mathbf{x})$ are vector valued basis functions (cf. $\tilde{\mathbf{w}}_k$, $\tilde{\mathbf{w}}_l$ in (2.37)) or scalar valued basis functions (cf. c_i, c_j in (2.38)) belonging to Ω_i and Ω_j, respectively, possibly containing a factor for singularity. Further, $\mathbf{x}$ and $\mathbf{x}'$ represent points in the object domain Ω_i and the source domain Ω_j, respectively.

The *distance* between two disjoint elements Ω_i and Ω_j (see Fig. 7.1) will be defined by

$$r_{\min}(\Omega_i, \Omega_j) = \min\{|\mathbf{x}' - \mathbf{x}|; \ \mathbf{x} \in \Omega_i, \ \mathbf{x}' \in \Omega_j\}.$$

If the distance is large enough we can apply Taylor expansion to the Green's function G with respect to $(\mathbf{x}' - \mathbf{x}'_m)$ and $(\mathbf{x} - \mathbf{x}_m)$, where $\mathbf{x}'_m$ and $\mathbf{x}_m$ are the midpoints of the source and object element, respectively.

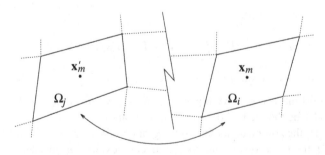

FIG. 7.1. Interaction between quadrilateral elements, $\Omega_i \ni \mathbf{x}$ and $\Omega_j \ni \mathbf{x}'$. These domains must be disjunct. $\mathbf{x}'_m$ and $\mathbf{x}_m$ are the midpoints of the elements.

7.2. Taylor expansion of a one-term Green's function

Consider the one-term Green's function

$$G(\mathbf{x}' - \mathbf{x}) = \frac{1}{|\mathbf{x}' - \mathbf{x}|}.$$

After the substitution $\mathbf{y} = \mathbf{x}' - \mathbf{x}$ and the second order Taylor expansion of $G(\mathbf{y})$ with respect to $\mathbf{y}_m = \mathbf{x}'_m - \mathbf{x}_m$:

$$G(\mathbf{y}) = G(\mathbf{y}_m) + (\nabla G)(\mathbf{y}_m)^{\mathrm{T}}[\mathbf{y} - \mathbf{y}_m]$$
$$+ \tfrac{1}{2}[\mathbf{y} - \mathbf{y}_m]^{\mathrm{T}}(\nabla(\nabla G))(\mathbf{y}_m)[\mathbf{y} - \mathbf{y}_m] + \mathcal{O}(|\mathbf{y} - \mathbf{y}_m|^3), \tag{7.2}$$

where the expressions for the gradient and the Hessian are given by

$$(\nabla G)(\mathbf{y}_m) = -G^3(\mathbf{y}_m)\mathbf{y}_m,$$
$$(\nabla(\nabla G))(\mathbf{y}_m) = -G^3(\mathbf{y}_m)\mathbf{I} + 3G^5(\mathbf{y}_m)(\mathbf{y}_m \otimes \mathbf{y}_m).$$

$\mathbf{I}$ is the identity matrix and $\mathbf{u} \otimes \mathbf{v}$ stands for $\mathbf{u}\mathbf{v}^{\mathrm{T}}$. Using the following shorthand notation

$$g_m = G(\mathbf{y}_m), \qquad \mathbf{g}_m = (\nabla G)(\mathbf{y}_m), \qquad \mathbf{G}_m = (\nabla(\nabla G))(\mathbf{y}_m),$$

the Taylor approximation for the one-term Green's function becomes

$$G(\mathbf{y}) \approx g_m + \mathbf{g}_m^{\mathrm{T}}(\mathbf{y} - \mathbf{y}_m) + \tfrac{1}{2}(\mathbf{y} - \mathbf{y}_m)^{\mathrm{T}}\mathbf{G}_m(\mathbf{y} - \mathbf{y}_m). \tag{7.3}$$

7.3. Taylor expansion of a multiple-term Green's function

In this subsection we consider the multiterm Green's function

$$G(\mathbf{x}' - \mathbf{x}) = \sum_{i=0}^{N} \frac{c_i}{|\mathbf{x}'_i - \mathbf{x}|},$$

where N is the number of images. Analogously to Section 7.2 we substitute $\mathbf{y}_i = \mathbf{x}'_i - \mathbf{x}$ and $\mathbf{y}_{im} = \mathbf{x}'_{im} - \mathbf{x}_m$. If $\mathbf{y}_i - \mathbf{y}_{im} = \mathbf{y} - \mathbf{y}_m$, the Taylor expansion for images becomes

$$G(\mathbf{y}) \approx g_m + \mathbf{g}_m^{\mathrm{T}}(\mathbf{y} - \mathbf{y}_m) + \tfrac{1}{2}(\mathbf{y} - \mathbf{y}_m)^{\mathrm{T}}\mathbf{G}_m(\mathbf{y} - \mathbf{y}_m), \tag{7.4}$$

where for $r_i = |\mathbf{y}_{im}|$

$$g_m = \sum_{i=0}^{N} G_i(\mathbf{y}_{im}) = \sum_{i=0}^{N} \frac{c_i}{r_i}, \tag{7.5}$$

$$\mathbf{g}_m = \sum_{i=0}^{N} (\nabla G_i)(\mathbf{y}_{im}) = \sum_{i=0}^{N} \frac{-c_i}{r_i^3}\mathbf{y}_{im}, \tag{7.6}$$

$$\mathbf{G}_m = \sum_{i=0}^{N} (\nabla(\nabla G_i))(\mathbf{y}_{im}) = \sum_{i=0}^{N} \frac{-c_i}{r_i^3}\mathbf{I} + 3\frac{c_i}{r_i^5}(\mathbf{y}_{im} \otimes \mathbf{y}_{im}). \tag{7.7}$$

Note, that the sum of images only contributes to the expansion coefficients.

Special attention has to be paid to the evaluation of the sums. If the value of a sum is relatively small compared to the individual terms, straightforward evaluation might result in a cancellation of significant digits. This occurs, for example, when $\sum c_i \approx 0$ and the r_i's are large, but their range is very small, i.e., $(r_i - r_0) \ll r_0$. By using the following identity for the sum in the expression of g_m

$$\sum \frac{c_i}{r_i} \quad \longrightarrow \quad \sum \frac{c_i(r_0 - r_i)}{r_i r_0} + \frac{1}{r_0} \sum c_i,$$

the terms in the first sum are much smaller than the terms in the original sum, so that less cancellation will occur. A similar treatment can be used for the derivatives.

7.4. Substitution of the Taylor expansion in the interaction integral

After substitution of the Taylor approximation (7.3) of $G(\mathbf{x}' - \mathbf{x})$ in (7.1) one obtains

$$\begin{aligned}
\tilde{I} &= \int_{\Omega_i} \tilde{\psi}_i(\mathbf{x}) \cdot \int_{\Omega_j} \tilde{\psi}_j(\mathbf{x}') g_m \, d\mathbf{x}' \, d\mathbf{x} \\
&+ \int_{\Omega_i} \tilde{\psi}_i(\mathbf{x}) \cdot \int_{\Omega_j} \tilde{\psi}_j(\mathbf{x}') \mathbf{g}_m^{\mathrm{T}} (\mathbf{y} - \mathbf{y}_m) \, d\mathbf{x}' \, d\mathbf{x} \\
&+ \frac{1}{2} \int_{\Omega_i} \tilde{\psi}_i(\mathbf{x}) \cdot \int_{\Omega_j} \tilde{\psi}_j(\mathbf{x}') (\mathbf{y} - \mathbf{y}_m)^{\mathrm{T}} \mathbf{G}_m (\mathbf{y} - \mathbf{y}_m) \, d\mathbf{x}' \, d\mathbf{x},
\end{aligned}$$

where $\mathbf{y} = \mathbf{x}' - \mathbf{x}$ and $\mathbf{y}_m = \mathbf{x}'_m - \mathbf{x}_m$. Since g_m, $\mathbf{g}_m$, and $\mathbf{G}_m$ are independent of $\mathbf{x}$ and $\mathbf{x}'$, they appear as constant terms in the integral, so that it may be rewritten as

$$\begin{aligned}
\tilde{I} &= g_m \int_{\Omega_i} \tilde{\psi}_i(\mathbf{x}) \cdot \int_{\Omega_j} \tilde{\psi}_j(\mathbf{x}') \, d\mathbf{x}' \, d\mathbf{x} \\
&- g_m^3 \int_{\Omega_i} \tilde{\psi}_i(\mathbf{x}) \cdot \int_{\Omega_j} \tilde{\psi}_j(\mathbf{x}') \{ \mathbf{y}_m^{\mathrm{T}} (\mathbf{y} - \mathbf{y}_m) \} \, d\mathbf{x}' \, d\mathbf{x} \\
&- \frac{1}{2} g_m^3 \int_{\Omega_i} \tilde{\psi}_i(\mathbf{x}) \cdot \int_{\Omega_j} \tilde{\psi}_j(\mathbf{x}') \{ \mathbf{I} \cdot [(\mathbf{y} - \mathbf{y}_m) \otimes (\mathbf{y} - \mathbf{y}_m)] \} \, d\mathbf{x}' \, d\mathbf{x} \\
&+ \frac{3}{2} g_m^5 \int_{\Omega_i} \tilde{\psi}_i(\mathbf{x}) \cdot \int_{\Omega_j} \tilde{\psi}_j(\mathbf{x}') \{ [\mathbf{y}_m \otimes \mathbf{y}_m] \cdot [(\mathbf{y} - \mathbf{y}_m) \otimes (\mathbf{y} - \mathbf{y}_m)] \} \, d\mathbf{x}' \, d\mathbf{x}.
\end{aligned}$$

The expressions between brackets {...} are scalar. For the $\otimes$-notation and the definition of inner products for matrices see Appendix D. The integral $\tilde{I}$ can be written in terms of moment integrals $M_{\alpha\beta}$.

$$\begin{aligned}
\tilde{I} &= g_m M_{00} M_{00}' - g_m^3 \left\{ M_{00} \cdot \sum_{\alpha \geqslant 1} \{\mathbf{y}_m\}_\alpha M_{0\alpha}' - M_{00}' \cdot \sum_{\alpha \geqslant 1} \{\mathbf{y}_m\}_\alpha M_{0\alpha} \right\} \\
&- \frac{1}{2} g_m^3 \left\{ M_{00} \cdot \sum_{\alpha \geqslant 1} M_{\alpha\alpha}' + M_{00}' \cdot \sum_{\alpha \geqslant 1} M_{\alpha\alpha} - 2 \sum_{\alpha \geqslant 1} M_{0\alpha} \cdot M_{0\alpha}' \right\} \\
&+ \frac{3}{2} g_m^5 \left\{ M_{00} \cdot \sum_{\alpha,\beta \geqslant 1} \{\mathbf{y}_m\}_\alpha \{\mathbf{y}_m\}_\beta M_{\alpha\beta}' + M_{00}' \cdot \sum_{\alpha,\beta \geqslant 1} \{\mathbf{y}_m\}_\alpha \{\mathbf{y}_m\}_\beta M_{\alpha\beta} \right. \\
&\left. - 2 \sum_{\alpha \geqslant 1} \{\mathbf{y}_m\}_\alpha M_{0\alpha} \cdot \sum_{\beta \geqslant 1} \{\mathbf{y}_m\}_\beta M_{0\beta}' \right\}. \tag{7.8}
\end{aligned}$$

The moment integrals are defined by

$$M_{\alpha\beta} = \int_{\Omega_i} \tilde{\psi}_i(\mathbf{x})\{\mathbf{x} - \mathbf{x}_m\}_\alpha \{\mathbf{x} - \mathbf{x}_m\}_\beta \, d\mathbf{x},$$

$$M'_{\alpha\beta} = \int_{\Omega_j} \tilde{\psi}_j(\mathbf{x}')\{\mathbf{x}' - \mathbf{x}'_m\}_\alpha \{\mathbf{x}' - \mathbf{x}'_m\}_\beta \, d\mathbf{x}',$$

where $\{\mathbf{x} - \mathbf{x}_m\}_\alpha = 1$, $(x - x_m)$, $(y - y_m)$, or $(z - z_m)$ for $\alpha = 0$, 1, 2, or 3. They are scalars for scalar valued functions $\tilde{\psi}_i$, and vectors for vector valued functions $\tilde{\psi}_i$. Let n_x be the upper bound of α. In general, $n_x = 3$. However, if all elements are parallel to the x, y-plane then $z - z_m = 0$, so that $n_x = 2$. After transformation to isoparametric coordinates the moment integrals have the form

$$M = \int_0^1 \int_0^1 \tilde{\mu}(s_1, s_2) \, ds_2 \, ds_1.$$

If one or more of the edges of the integration domain are a part of a boundary, the function $\tilde{\mu}$ contains a factor for boundary singularity $d^{-1/2}$, and the general form of the moment integrals becomes

$$M = \int_0^1 \int_0^1 \frac{\mu(s_1, s_2)}{\sqrt{d(s_1, s_2)}} \, ds_2 \, ds_1,$$

where μ is a smooth function. For the regularization of this boundary singularity a similar method can used as for that of the outer integral described in Section 6.5. The integrals obtained can be evaluated numerically by the Patterson's method described in Section 4.

If N is the number of elements and k is the average number of function evaluations for the quadrature with respect to each of the isoparametric parameters s_1 and s_2, the total number of function evaluations for calculating the moment integrals is of the order of Nk^2. Since the moment integrals can be evaluated in advance, the computer time for the evaluation of the interaction integrals is reduced considerably.

7.5. Efficiency improvement of the algorithm

In this subsection we concentrate on the efficiency of the algorithm to evaluate the interaction integral by expression (7.8) for a given set of moment integrals. Assuming that two identical terms are computed only once, the number of operations to evaluate the expression (7.8) with scalar moment integrals is

$$N_{\text{scalar}} = n_s n_o \left[\{1\} + \{2(n_x + 1)\} + \{2(n_x + 1)\} + \{\tfrac{3}{2} n_x (n_x + 1) + 3\} + 4 \right]$$
$$= n_s n_o \left[(n_x + 1)(\tfrac{3}{2} n_x + 4) + 4 + 4 \right],$$

where n_s and n_o are the number of different basis functions on the source and object quadrilaterals, i.e., $n_s = n_o = 1$. If $n_x = 2$, then $N_{\text{scalar}} = 29$.

For the expression (7.8) with vector moment integrals the number of operations is

$$N_{\text{vector}} = n_x n_s n_o \left\{ (n_x + 1)(\tfrac{3}{2} n_x + 4) + 5 \right\} + 4 n_s n_o,$$

where the first factor n_x is due to taking the inner product. The maximum value of n_s and n_o is 2. This is because for a vector-type interaction integral the source or object domain usually consists of two adjacent elements with an edge in common. If $n_x = 2$, then $N_{\text{vector}} = 56 n_s n_o$.

These numbers of operations can be reduced by calculating the inner products between the vectors and matrices which occur in expression (7.8) beforehand. Using the following shorthand notation for the moments due to the source element:

$$
S_{00} = M'_{00}, \qquad
s_0 = \begin{bmatrix} M'_{01} \\ \vdots \\ M'_{0n_x} \end{bmatrix}, \qquad
S = \begin{bmatrix} M'_{11} & \cdots & M'_{1n_x} \\ \vdots & \ddots & \vdots \\ M'_{n_x 1} & \cdots & M'_{n_x n_x} \end{bmatrix},
$$

and an analogous notation $\mathbf{m}_0$ and $\mathbf{M}$ for the object element, we obtain a compact expression for the integral of (7.8)

$$
\begin{aligned}
\tilde{I} &= M_{00}\left(g_m S_{00} + \mathbf{g}_m \cdot \mathbf{s}_0 + \tfrac{1}{2}\mathbf{G}_m \cdot \mathbf{S}\right) \\
&\quad + S_{00}\left(-\mathbf{g}_m \cdot \mathbf{m}_0 + \tfrac{1}{2}\mathbf{G}_m \cdot \mathbf{M}\right) - \mathbf{G}_m \cdot (\mathbf{m}_0 \otimes \mathbf{s}_0) \\
&= M_{00}\left(g_m S_{00} + \mathbf{g}_m \cdot \mathbf{s}_0 + \tfrac{1}{2}\mathbf{G}_m \cdot \mathbf{S}\right) \\
&\quad + \mathbf{m}_0 \cdot \left(-\mathbf{g}_m S_{00} - \mathbf{G}_m \mathbf{s}_0\right) + \mathbf{M} \cdot \tfrac{1}{2}\mathbf{G}_m S_{00},
\end{aligned} \tag{7.9}
$$

where g_m, $\mathbf{g}_m$, and $\mathbf{G}_m$ are defined in Sections 7.2 and 7.3. For the definition of $\otimes$ see Appendix D.

The numbers of operations for the evaluation of expression (7.9) are:

$$
\begin{aligned}
N_{\text{scalar}} &= n_o n_s + n_s \left\{1 + n_x + \tfrac{1}{2}n_x(n_x + 1)\right\} \\
&\quad + n_o n_s n_x + n_s\left(n_x + n_x^2\right) + \left(\tfrac{1}{2}n_x(n_x + 1) + 1\right)n_o n_s \\
&= n_s\left\{(n_x + 1)\left(\tfrac{3}{2}n_x + 1\right) + n_o(n_m + 1)\right\}, \\
N_{\text{vector}} &= n_x n_s\left\{(n_x + 1)\left(\tfrac{3}{2}n_x + 1\right) + n_o(n_m + 1)\right\},
\end{aligned}
$$

where $n_m = 1 + n_x + \tfrac{1}{2}n_x(n_x + 1)$. Thus, for $n_x = 2, 3$

$$
\begin{aligned}
N_{\text{scalar}} &= 19, 33 \qquad (n_s = n_o = 1) \\
N_{\text{vector}} &\leqslant 104, 264 \quad (n_s = n_o = 2).
\end{aligned}
$$

7.6. Moment integrals in local coordinate system

So far, we have applied Taylor expansion with respect to midpoints of the elements in a global coordinate system. If all elements are on the same layer, or on parallel layers, the evaluation of the moment integrals is restricted to $n_x = 2$, otherwise $n_x = 3$. However, if the moment integral of each element is calculated in a local coordinate system for which the z-axis is perpendicular to the plane of that element, again $n_x = 2$. But for the evaluation of an interaction integral between an object and a source element, the moment integrals of the object element have to be transformed from the local coordinate system of the object element to the local coordinate system of the source element.

7.6.1. Transformation matrix

Let $\mathbf{Q}$ and $\mathbf{R}$ be the transformation matrices from a local to a global coordinate system of the source and object element, respectively, then

$$\mathbf{x} = \mathbf{x}_s + \mathbf{Q}\bar{\eta}, \qquad \mathbf{Q}\mathbf{Q}^{\mathrm{T}} = \mathbf{I},$$
$$\mathbf{x} = \mathbf{x}_o + \mathbf{R}\bar{\xi}, \qquad \mathbf{R}\mathbf{R}^{\mathrm{T}} = \mathbf{I},$$

so that

$$\bar{\eta} = \mathbf{Q}^{\mathrm{T}}(\mathbf{x}_o - \mathbf{x}_s + \mathbf{R}\bar{\xi}),$$

where $\bar{\eta} = (\eta_x, \eta_y, 0)$ and $\bar{\xi} = (\xi_x, \xi_y, 0)$ are the local coordinates in the source and object coordinate system, respectively, of point $\mathbf{x}$ with coordinates in the global coordinate system. Let $\bar{\xi}_m = (\xi_{m,x}, \xi_{m,y}, 0)$ and $\bar{\xi} = (\xi_x, \xi_y, 0)$ be points in the local object coordinate system, then

$$\bar{\eta}_m = \mathbf{Q}^{\mathrm{T}}(\mathbf{x}_o - \mathbf{x}_s + \mathbf{R}\bar{\xi}_m),$$
$$\bar{\eta} = \mathbf{Q}^{\mathrm{T}}(\mathbf{x}_o - \mathbf{x}_s + \mathbf{R}\bar{\xi}) = \bar{\eta}_m + \mathbf{Q}^{\mathrm{T}}\mathbf{R}(\bar{\xi} - \bar{\xi}_m)$$
$$= \bar{\eta}_m + \mathbf{T}(\bar{\xi} - \bar{\xi}_m)$$

are the coordinates of these points in the local source coordinate system, where $\mathbf{T} = \mathbf{Q}^{\mathrm{T}}\mathbf{R}$ is a rotation matrix.

7.6.2. Taylor expansion with transformation

Let $\bar{\eta}' = (\eta'_x, \eta'_y, 0)$ and $\bar{\eta} = \bar{\eta}_m + \mathbf{T}(\bar{\xi} - \bar{\xi}_m)$ be arbitrary points inside the source and object element, respectively, with coordinates in the local source coordinate system. Let $\Delta\bar{\eta}' = \bar{\eta}' - \bar{\eta}'_m$ and $\Delta\bar{\xi} = \bar{\xi} - \bar{\xi}_m$, then $\Delta\bar{\eta} = \mathbf{T}\Delta\bar{\xi}$, and

$$G(\bar{\eta}' - \bar{\eta}) = \sum_i c_i R_i^{-1},$$

$$R_i = \left| \bar{\eta}'_i - \bar{\eta}_m - \mathbf{T}\Delta\bar{\xi} \right|$$
$$= \left\{ (\eta'_x - \eta_{m,x} - T_{11}\Delta\xi_x - T_{12}\Delta\xi_y)^2 + (\eta'_y - \eta_{m,y} - T_{21}\Delta\xi_x - T_{22}\Delta\xi_y)^2 \right.$$
$$\left. + (\eta'_{i,z} - \eta'_{m,z} - T_{31}\Delta\xi_x - T_{32}\Delta\xi_y)^2 \right\}^{-1/2}.$$

Let $\mathbf{y} = \bar{\eta}' - \bar{\eta}$ and $\mathbf{y}_m = \bar{\eta}'_m - \bar{\eta}_m$. The Taylor expansion of $G(\mathbf{y})$ is similar to that of expression (7.2)

$$G(\mathbf{y}) = G(\mathbf{y}_m) + (\nabla G)(\mathbf{y}_m)^{\mathrm{T}}[\mathbf{y} - \mathbf{y}_m]$$
$$+ \tfrac{1}{2}[\mathbf{y} - \mathbf{y}_m]^{\mathrm{T}}\big(\nabla(\nabla G)\big)(\mathbf{y}_m)[\mathbf{y} - \mathbf{y}_m] + \mathcal{O}\big(|\mathbf{y} - \mathbf{y}_m|^3\big),$$
$$= g_m + \mathbf{g}_m^{\mathrm{T}}(\Delta\bar{\eta}' - \mathbf{T}\Delta\bar{\xi}) + \tfrac{1}{2}(\Delta\bar{\eta}' - \mathbf{T}\Delta\bar{\xi})^{\mathrm{T}}\mathbf{G}_m(\Delta\bar{\eta}' - \mathbf{T}\Delta\bar{\xi}),$$

where

$$g_m = G(\mathbf{y}_m), \qquad \mathbf{g}_m = (\nabla G)(\mathbf{y}_m), \qquad \mathbf{G}_m = \big(\nabla(\nabla G)\big)(\mathbf{y}_m).$$

∇ refers to derivatives with respect to the local source coordinates. Note that, in contrast to the expansion in Sections 7.2 and 7.3 for $n_x = 2$, all three elements of $\mathbf{g}_m$ and all nine elements of $\mathbf{G}_m$ are generally nonzero.

Let

$$\mathbf{s}_0 = (S_{01}, S_{02}, 0)^{\mathrm{T}}, \qquad \mathbf{m}_0 = (M_{01}, M_{02}, 0)^{\mathrm{T}},$$

$$\mathbf{S} = \begin{pmatrix} S_{11} & S_{12} & 0 \\ S_{21} & S_{22} & 0 \\ 0 & 0 & 0 \end{pmatrix}, \qquad \mathbf{M} = \begin{pmatrix} M_{11} & M_{12} & 0 \\ M_{21} & M_{22} & 0 \\ 0 & 0 & 0 \end{pmatrix}.$$

Then, similar to expression (7.9)

$$\tilde{I} = M_{00}\left(g_m S_{00} + \mathbf{g}_m \cdot \mathbf{s}_0 + \tfrac{1}{2}\mathbf{G}_m \cdot \mathbf{S}\right)$$
$$+ (-\mathbf{g}_m S_{00} - \mathbf{G}_m \mathbf{s}_0) \cdot \mathbf{T}\mathbf{m}_0 + (\mathbf{T}\mathbf{M}\mathbf{T}^{\mathrm{T}}) \cdot \tfrac{1}{2}\mathbf{G}_m S_{00}.$$

For the evaluation of $\tilde{I}$, the quantities $\mathbf{m}_0$, $\mathbf{M}$, and $\mathbf{T}$ may be redefined as follows:

$$\mathbf{m}_0 = (M_{01}, M_{02})^{\mathrm{T}}, \qquad \mathbf{M} = \begin{pmatrix} M_{11} & M_{12} \\ M_{21} & M_{22} \end{pmatrix}, \qquad \mathbf{T} = \begin{pmatrix} T_{11} & T_{12} \\ T_{21} & T_{22} \\ T_{31} & T_{32} \end{pmatrix}.$$

Since $\mathbf{T}$ is a 3×2-matrix and $\mathbf{G}_m$ is a full 3×3-matrix, a complete reduction to two dimensions as in the previous subsection for $n_x = 2$ is not possible. An operation count for this transformed integral gives us:

$$N_{\text{scalar}} = n_o n_s + n_s \left(1 + 2 + \tfrac{1}{2} \cdot 2 \cdot 3\right) + 3n_o n_s + n_s (3 + 3 \cdot 2)$$
$$+ n_o (3 \cdot 2) + n_o n_s \left(\tfrac{1}{2} \cdot 3 \cdot 4 + 1\right) + n_o (12 + 18)$$
$$= 15 n_s + 36 n_o + 10 n_o n_s = 61,$$
$$N_{\text{vector}} = 2(35 n_s + 72) = 284,$$

which is larger than the number of operations for the untransformed case with $n_x = 3$ of Section 7.5. This is due to the large number of operations needed for the transformation of $\mathbf{m}_0$ and $\mathbf{M}$.

7.7. Error estimates

To estimate the error in the second order Taylor expansion of (7.2) we have derived the following expression for the third order terms:

$$\tfrac{1}{6}\overline{\mathbf{G}}_m \cdot [\mathbf{y} - \mathbf{y}_m]^3,$$

where $\mathbf{t}^{\alpha} = \underbrace{\mathbf{t} \otimes \cdots \otimes \mathbf{t}}_{\alpha}$, and

$$\overline{\mathbf{G}}_m = 3 g_m^5 \left[\mathbf{I} \otimes \mathbf{y}_m + \nabla \mathbf{H}\right] - 15 g_m^7 \overline{\mathbf{T}},$$

with the Hessian matrix $\mathbf{H} = \mathbf{y}_m \otimes \mathbf{y}_m$ and tensor $\overline{\overline{\mathbf{T}}} = \mathbf{H} \otimes \mathbf{y}_m$. For the definition of $\otimes$ see Appendix D.

For any $\mathbf{y}$ there exists a point $\bar{\xi} = \mathbf{y}_m + \vartheta[\mathbf{y} - \mathbf{y}_m]$ for $0 < \vartheta < 1$ such that

$$G(\mathbf{y}) = g_m + \mathbf{g}_m \cdot [\mathbf{y} - \mathbf{y}_m] + \tfrac{1}{2}\mathbf{G}_m \cdot [\mathbf{y} - \mathbf{y}_m]^2 + \tfrac{1}{6}\overline{\mathbf{G}}(\bar{\xi}) \cdot [\mathbf{y} - \mathbf{y}_m]^3,$$

where

$$\overline{\mathbf{G}}(\bar{\xi}) = 3G^5(\bar{\xi})[\mathbf{I} \otimes \bar{\xi} + \nabla\mathbf{H}(\bar{\xi})] + 15G^7(\bar{\xi})\overline{\mathbf{T}}(\bar{\xi}),$$
$$\mathbf{H}(\bar{\xi}) = \bar{\xi} \otimes \bar{\xi},$$
$$\overline{\mathbf{T}}(\bar{\xi}) = \mathbf{H}(\bar{\xi}) \otimes \bar{\xi}.$$

Let

$$r_{\min} = \min_{\substack{\mathbf{x} \in \Omega_i \\ \mathbf{x}' \in \Omega_j}} \|\mathbf{x}' - \mathbf{x}\|$$

be the smallest distance between two elements and

$$d = \max_{\mathbf{y}} \|\mathbf{y} - \mathbf{y}_m\|$$

the upper bound for $\|\mathbf{y} - \mathbf{y}_m\|$, where $\mathbf{y} \in \{\mathbf{x}' - \mathbf{x} \mid \mathbf{x}' \in \Omega_j, \mathbf{x} \in \Omega_i\}$. Note, that d depends on the choice of the midpoints, $\mathbf{x}_m$ and $\mathbf{x}'_m$. The following theorem states the relation that holds between d, $r_{\min}$, and the upper bound for the relative error in the second order Taylor expansion.

THEOREM 7.1. *Let $\tilde{G}$ be an approximation of G with the second order Taylor expansion (7.3), then for every $\varepsilon > 0$ we have*

$$\frac{d}{r_{\min}} \leqslant \sqrt[3]{\frac{\varepsilon}{2}} \implies \frac{\|\tilde{G}(\mathbf{x}' - \mathbf{x}) - G(\mathbf{x}' - \mathbf{x})\|_\infty}{|G(\mathbf{x}' - \mathbf{x})|} \leqslant \varepsilon,$$

for every $\mathbf{x}' \in \Omega_j, \mathbf{x} \in \Omega_i$.

PROOF. Let

$$\delta_G = |G - \tilde{G}| = \left|\tfrac{1}{6}\overline{\mathbf{G}}(\bar{\xi}) \cdot [\mathbf{y} - \mathbf{y}_m]^3\right|$$

be the absolute error in the Taylor expansion and let further $\mathbf{y} = (\mathbf{x}' - \mathbf{x})$ and $\Delta\mathbf{y} = (\mathbf{y} - \mathbf{y}_m)$. We have

$$\mathbf{I} \otimes \bar{\xi} + \nabla\mathbf{H}(\bar{\xi}) = \begin{bmatrix} (3\xi_x, \xi_y, \xi_z) & (\xi_y, \xi_x, 0) & (\xi_z, 0, \xi_x) \\ (\xi_y, \xi_x, 0) & (\xi_x, 3\xi_y, \xi_z) & (0, \xi_z, \xi_y) \\ (\xi_z, 0, \xi_x) & (0, \xi_z, \xi_y) & (\xi_x, \xi_y, 3\xi_z) \end{bmatrix}.$$

If we multiply this by $\Delta\mathbf{y}^3 = \Delta\mathbf{y} \otimes \Delta\mathbf{y} \otimes \Delta\mathbf{y}$ and let $\mathbf{y} = (x, y, z)$, we obtain (see Appendix D)

$$\left(\mathbf{I} \otimes \bar{\xi} + \nabla\mathbf{H}(\bar{\xi})\right) \cdot \Delta\mathbf{y}^3 = 3(\Delta x^2 + \Delta y^2 + \Delta z^2)(\xi_x \Delta x + \xi_y \Delta y + \xi_z \Delta z)$$
$$= 3\|\Delta\mathbf{y}\|^2 \langle \bar{\xi}, \Delta\mathbf{y} \rangle.$$

Further, we have

$$\overline{\mathbf{T}}(\bar{\xi}) \cdot \Delta\mathbf{y}^3 = (\xi_x \Delta x + \xi_y \Delta y + \xi_z \Delta z)^3 = \langle \bar{\xi}, \Delta\mathbf{y} \rangle^3.$$

An upper boundary for the error can be given by

$$\delta_G = \left| \tfrac{1}{6} \{ 3G^5(\bar{\xi})[\mathbf{I} \otimes \bar{\xi} + \nabla \mathbf{H}(\bar{\xi})] - 15G^7(\bar{\xi})\overline{\mathbf{T}}(\bar{\xi}) \} \cdot \Delta \mathbf{y}^3 \right|$$

$$\leqslant \max \left| \frac{9\|\Delta \mathbf{y}\|^2 \langle \bar{\xi}, \Delta \mathbf{y} \rangle}{6\|\bar{\xi}\|^5} - \frac{15\langle \bar{\xi}, \Delta \mathbf{y} \rangle^3}{6\|\bar{\xi}\|^7} \right|$$

$$= \max \left| \langle \bar{\xi}, \Delta \mathbf{y} \rangle \left(\frac{9\|\Delta \mathbf{y}\|^2}{6\|\bar{\xi}\|^5} - \frac{15\langle \bar{\xi}, \Delta \mathbf{y} \rangle^2}{6\|\bar{\xi}\|^7} \right) \right|$$

$$= \max \left| \langle \bar{\xi}, \Delta \mathbf{y} \rangle \frac{9\|\bar{\xi}\|^2 \|\Delta \mathbf{y}\|^2 - 15\langle \bar{\xi}, \Delta \mathbf{y} \rangle^2}{6\|\bar{\xi}\|^7} \right|$$

$$= \max \left| \frac{\|\bar{\xi}\|^3 \|\Delta \mathbf{y}\|^3}{\|\bar{\xi}\|^7} \cdot \cos \varphi \frac{9 - 15 \cos^2 \varphi}{6} \right|$$

$$\leqslant \max \left| \frac{\|\Delta \mathbf{y}\|^3}{\|\bar{\xi}\|^4} \right|.$$

To compute the latter maximum we need to know the range of $\bar{\xi}$, which is the set $\varXi = \{ \mathbf{x}' - \mathbf{x} \mid \mathbf{x}' \in \varOmega_j, \mathbf{x} \in \varOmega_i \}$. This region can be obtained as follows. Translate $\varOmega_j$ over $-\mathbf{x}_m$, such that its midpoint is mapped onto $\mathbf{y}_m$. Move $-\varOmega_i$ with its midpoint $\mathbf{x}_m$ along the edges of the translated $\varOmega_j$, as illustrated in Fig. 7.2.

From the definitions of $r_{\min}$ and d it follows that $r_{\min}$ is the lower bound of $\|\bar{\xi}\|$ and d the upper bound of $\|\Delta \mathbf{y}\|$, for $\bar{\xi}, \mathbf{y} \in \varXi$, so that

$$\delta_G \leqslant \frac{d^3}{r_{\min}^4}.$$

Let $\mathbf{y}_{\min}$ be the point for which $\|\mathbf{y}_{\min}\| = r_{\min}$, then

$$\|\mathbf{y}\| = \|\mathbf{y}_{\min} + \mathbf{y} - \mathbf{y}_{\min}\| \leqslant \|\mathbf{y}_{\min}\| + \|\mathbf{y} - \mathbf{y}_{\min}\| \leqslant r_{\min} + d.$$

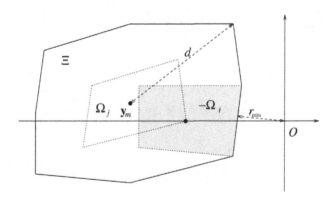

FIG. 7.2. Allowed region ($\varXi = \varOmega_j - \varOmega_i$) for $\bar{\xi}$, obtained by shifting the midpoint of $-\varOmega_i$ along the edge of $\varOmega_j$. Here $d = \max_{\mathbf{y} \in \varXi} \|\mathbf{y} - \mathbf{y}_m\|$ and $r_{\min} = \min_{\bar{\xi} \in \varXi} \|\bar{\xi}\|$.

Since $G(\mathbf{y}) = \|\mathbf{y}\|^{-1}$, we have for the relative error $\rho_G = \frac{\delta_G}{G}$:

$$\rho_G \leqslant \alpha^3 (1 + \alpha) \quad \left(\alpha = \frac{d}{r_{\min}} \right).$$

Therefore, $\rho_G \leqslant \varepsilon$ if

$$\alpha^3 (1 + \alpha) \leqslant \varepsilon, \quad \text{or} \quad \alpha^3 \leqslant \frac{\varepsilon}{2}, \quad \text{or} \quad \frac{d}{r_{\min}} \leqslant \sqrt[3]{\frac{\varepsilon}{2}}, \quad \text{for } \varepsilon < 1. \qquad \square$$

With somewhat more effort it can be shown that for a third order Taylor expansion the condition $\alpha^4 \leqslant \frac{\varepsilon}{2}$ should be satisfied in order to keep the error small enough. Hence, we might state the following conjecture.

CONJECTURE 7.1. *Let $\tilde{G}$ be an approximation of G with the nth order Taylor expansion, then for every $\varepsilon > 0$ holds*

$$\alpha \leqslant \sqrt[n+1]{\frac{\varepsilon}{2}} \quad \Longrightarrow \quad \frac{\|\tilde{G} - G\|_\infty}{G} \leqslant \varepsilon.$$

8. Analytical integration of the inner integrals for vector valued basis functions

This section presents an improvement of the method, treated in Section 5 for the derivation of analytical expressions for the inner integral over a quadrilateral source element, of which the integrand is irregular and contains vector valued basis functions.

The integral is the sum of the integrals

$$\int_0^1 \mathcal{F}_i(s_2) \ln\left(\frac{\sqrt{c}\sqrt{a+b+c} + c + \frac{1}{2}b}{\sqrt{c}\sqrt{a} + \frac{1}{2}b} \right) ds_2, \tag{8.1}$$

and

$$\int_0^1 \mathcal{G}_i(s_2)\left(\sqrt{a+b+c} - \sqrt{a} \right) ds_2, \tag{8.2}$$

where a, b, and c are quadratic functions of s_2. If the integration point of the outer integral, over the object element, lies in the interior or on the boundary of the source element, the integrand of the first integral is irregular. The integrand of the second integral has no singularity. In Section 5.2.4 the integral (8.1) is rewritten as

$$\int_0^1 \mathcal{F}_i(s_2) \ln\left(\sqrt{c}\sqrt{a+b+c} + c + \frac{1}{2}b \right)\left(\sqrt{c}\sqrt{a} - \frac{1}{2}b \right) ds_2$$

$$- \int_0^1 \mathcal{F}_i \ln\left(ca - \left(\tfrac{1}{2}b\right)^2 \right) ds_2. \tag{8.3}$$

Only the last term has a singular integrand. After partial integration this integral becomes

$$\int_0^1 \mathcal{F}_i(s_2) \ln\left(ca - \left(\tfrac{1}{2}b\right)^2 \right) ds_2 = \left[\mathcal{F}_i(s_2)\mathcal{L}(s_2) \right]_0^1 - \int_0^1 \mathcal{F}_i'(s_2)\mathcal{L}(s_2) ds_2, \tag{8.4}$$

where

$$\mathcal{L}(s_2) = \int \ln\left(ca - \left(\tfrac{1}{2}b\right)^2\right) ds_2. \tag{8.5}$$

In Section 5.2.4 an analytical expression for the integral (8.5) is derived. However, the integral in the right-hand side of (8.4) is evaluated numerically. Since the derivative of the integrand of (8.4) is singular, the numerical evaluation of this integral, by Patterson's rules, can become slowly convergent. Besides, the integral (8.5) can only be evaluated analytically if the integration point over the object element lies in the plane of the source element (see Section 5.2.4).

This section introduces another method, that allows for arbitrary orientation of object and source element. In this method the integration domain of the integrals (8.1) and (8.2) is split into at most three parts, for only one of which the integrand of (8.1) is singular. The domain of this singular part is chosen so that the approximation of (8.1) over it by an analytical expression is sufficiently accurate, and the integrals over the other parts can be evaluated by Patterson's rules. Let $v_1 = x_{12}(1 - s_2) - x_{34}s_2$ for $x_{ij} = x_j - x_i$, where $x_1 \ldots x_4$ are the vertices of the source element, and where s_2 is one of the iso-parametric coordinates of the integration point. It can be shown that only if the vector v_1 is independent of s_2, the integral (8.1) can be calculated analytically over the full integration domain. This is the case for a source element with $x_{12} + x_{34} = 0$. In all other cases the integral (8.1) must be approximated. The approximation method makes use of the Taylor expansion of the vector v_1 around the vector $d = x_{12}(1 - s_2^{(m)}) - x_{34}s_2^{(m)}$, where $s_2^{(m)}$ is the iso-parametric coordinate, s_2, of the projection of the object integration point in the plane of the source element. If $s_2^{(m)} - s_2 = 0$, the integrand of (8.1) is singular.

8.1. Definition of auxiliary quantities

In this subsection the quantities are defined that will be used in the evaluation of integrals (8.1) and (8.2). Let us define the following quantities:

$$c_0 = x_{41}, \qquad\qquad \hat{c}_0 = -x_{23}, \qquad\qquad c_1 = x_{12} + x_{34},$$
$$d = x_{12} - c_1 s_2^{(m)}, \qquad\qquad y = s_2^{(m)} - s_2, \qquad\qquad x = c_1 y,$$
$$v_0(y) = \tilde{v}_0(y) + x_{OM}, \qquad\qquad \tilde{v}_0(y) = c_0 y - ds_1^{(m)},$$
$$(\tilde{v}_0 + v_1)(y) = \hat{c}_0 y + d\left(1 - s_1^{(m)}\right), \qquad v_1(x) = d + x,$$
$$a(y) = \left|v_0(y)\right|^2, \qquad\qquad \tfrac{1}{2}b(y, x) = v_1(x)^T \tilde{v}_0(y), \qquad c(x) = \left|v_1(x)\right|^2.$$

Note that for a element with parallel opposite edges $x = 0$. The vector valued functions $\overline{\mathcal{F}}_i(y, x)$ and $\overline{\mathcal{G}}_i(y, x)$ can be written in terms of the above quantities as follows:

$$\overline{\mathcal{F}}_1(y, x) = -\frac{c_0}{|d|}\left\{f_{000}(y, x) + f_{010}(y, x)\right\} + \frac{c_1}{|d|}\left\{f_{001}(y, x) + f_{011}(y, x)\right\},$$

$$\overline{\mathcal{F}}_2(y, x) = \frac{d}{|d|}f_{001}(y, x) + \frac{c_1}{|d|}f_{101}(y, x),$$

$$\overline{\mathcal{F}}_3(y,\mathbf{x}) = -\frac{c_0}{|\mathbf{d}|}f_{010}(y,\mathbf{x}) + \frac{c_1}{|\mathbf{d}|}f_{011}(y,\mathbf{x}),$$

$$\overline{\mathcal{F}}_4(y,\mathbf{x}) = \frac{d}{|\mathbf{d}|}\{f_{000}(y,\mathbf{x}) + f_{001}(y,\mathbf{x})\} + \frac{c_1}{|\mathbf{d}|}\{f_{100}(y,\mathbf{x}) + f_{101}(y,\mathbf{x})\},$$

$$\overline{\mathcal{G}}_1(y,\mathbf{x}) = -\frac{c_1}{|\mathbf{d}|}\{g_{00}(y,\mathbf{x}) + g_{01}(y,\mathbf{x})\},$$

$$\overline{\mathcal{G}}_2(y,\mathbf{x}) = -\frac{d}{|\mathbf{d}|}g_{00}(y,\mathbf{x}) - \frac{c_1}{|\mathbf{d}|}g_{10}(y,\mathbf{x}),$$

$$\overline{\mathcal{G}}_3(y,\mathbf{x}) = -\frac{c_1}{|\mathbf{d}|}g_{01}(y,\mathbf{x}),$$

$$\overline{\mathcal{G}}_4(y,\mathbf{x}) = \overline{\mathcal{G}}_2(y,\mathbf{x}),$$

where

$$f_{jkl}(y,\mathbf{x}) = |\mathbf{d}|y^j\left(y - s_2^{(m)}\right)^k \tilde{f}_l(y,\mathbf{x}), \tag{8.6}$$

$$\tilde{f}_l(y,\mathbf{x}) = \frac{1}{\sqrt{c(\mathbf{x})}}\left(\frac{b(y,\mathbf{x})}{2c(\mathbf{x})}\right)^l = \frac{1}{|\mathbf{v}_1(\mathbf{x})|}\left(\frac{\mathbf{v}_1(\mathbf{x})^T\tilde{\mathbf{v}}_0(y)}{|\mathbf{v}_1(\mathbf{x})|^2}\right)^l, \tag{8.7}$$

$$g_{jk}(y,\mathbf{x}) = |\mathbf{d}|y^j\left(y - s_2^{(m)}\right)^k \tilde{g}(y,\mathbf{x}), \tag{8.8}$$

$$\tilde{g}(y,\mathbf{x}) = \frac{1}{|\mathbf{v}_1(\mathbf{x})|^2}. \tag{8.9}$$

Further, the arguments of the logarithm in the integrand of the integral (8.1) and the arguments in the integrand of the integral (8.2) are

$$\left(\sqrt{c}\sqrt{a} + \tfrac{1}{2}b\right)(y,\mathbf{x}) = |\mathbf{v}_1(\mathbf{x})||\mathbf{d}|\mathcal{R}(y) + \mathbf{v}_1(\mathbf{x})^T\tilde{\mathbf{v}}_0(y), \tag{8.10}$$

$$\left(\sqrt{c}\sqrt{a+b+c}+c+\tfrac{1}{2}b\right)(y,\mathbf{x}) = |\mathbf{v}_1(\mathbf{x})||\mathbf{d}|\hat{\mathcal{R}}(y) + \mathbf{v}_1(\mathbf{x})^T(\tilde{\mathbf{v}}_0 + \mathbf{v}_1)(y), \tag{8.11}$$

$$\left(\sqrt{a}\right)(y) = |\mathbf{d}|\mathcal{R}(y), \tag{8.12}$$

$$\left(\sqrt{a+b+c}\right)(y) = |\mathbf{d}|\hat{\mathcal{R}}(y), \tag{8.13}$$

where

$$\mathcal{R}(y) = \sqrt{\frac{|\tilde{\mathbf{v}}_0(y)|^2}{|\mathbf{d}|^2} + \frac{|\mathbf{x}_{OM}|^2}{|\mathbf{d}|^2}},$$

$$\hat{\mathcal{R}}(y) = \sqrt{\frac{|(\tilde{\mathbf{v}}_0 + \mathbf{v}_1)(y)|^2}{|\mathbf{d}|^2} + \frac{|\mathbf{x}_{OM}|^2}{|\mathbf{d}|^2}}.$$

The derivatives with respect to $\mathbf{x}$ of the quantities depending on $\mathbf{x}$ are

$$\nabla\mathbf{v}_1(\mathbf{x}) = I,$$
$$\nabla|\mathbf{v}_1(\mathbf{x})|^n = n|\mathbf{v}_1(\mathbf{x})|^{n-2}\mathbf{v}_1(\mathbf{x}),$$
$$\nabla\left(\mathbf{v}_1(\mathbf{x})^T\tilde{\mathbf{v}}_0(y)\right) = \tilde{\mathbf{v}}_0(y),$$
$$\nabla|\mathbf{v}_1(\mathbf{x})^T\tilde{\mathbf{v}}_0(y)|^2 = 2\mathbf{v}_1(\mathbf{x})^T\tilde{\mathbf{v}}_0(y)\tilde{\mathbf{v}}_0(y),$$

$$\nabla\big(\mathbf{v}_1(\mathbf{x})^{\mathrm{T}}(\tilde{\mathbf{v}}_0+\mathbf{v}_1)(y)\big) = (\tilde{\mathbf{v}}_0+\mathbf{v}_1)(y),$$

$$\nabla\big|\mathbf{v}_1(\mathbf{x})^{\mathrm{T}}(\tilde{\mathbf{v}}_0+\mathbf{v}_1)(y)\big|^2 = 2\mathbf{v}_1(\mathbf{x})^{\mathrm{T}}(\tilde{\mathbf{v}}_0+\mathbf{v}_1)(y)(\tilde{\mathbf{v}}_0+\mathbf{v}_1)(y).$$

The inner products of $\mathbf{x}$ with the first order derivatives with respect to $\mathbf{x}$ of the expressions in (8.7) for $l=0,1$ and (8.9) are

$$\mathbf{x}^{\mathrm{T}}(\nabla\tilde{f}_0)(y,\mathbf{x}) = -\frac{\mathbf{x}^{\mathrm{T}}\mathbf{v}_1(\mathbf{x})}{|\mathbf{v}_1(\mathbf{x})|^3},$$

$$\mathbf{x}^{\mathrm{T}}(\nabla\tilde{f}_1)(y,\mathbf{x}) = \frac{\mathbf{x}^{\mathrm{T}}\tilde{\mathbf{v}}_0(y)}{|\mathbf{v}_1(\mathbf{x})|^3} - 3\frac{\mathbf{v}_1(\mathbf{x})^{\mathrm{T}}\tilde{\mathbf{v}}_0(y)}{|\mathbf{v}_1(\mathbf{x})|^2}\frac{\mathbf{x}^{\mathrm{T}}\mathbf{v}_1(\mathbf{x})}{|\mathbf{v}_1(\mathbf{x})|^3},$$

$$\mathbf{x}^{\mathrm{T}}(\nabla\tilde{g})(y,\mathbf{x}) = -2\frac{\mathbf{x}^{\mathrm{T}}\mathbf{v}_1(\mathbf{x})}{|\mathbf{v}_1(\mathbf{x})|^4},$$

and with the second order derivatives with respect to $\mathbf{x}$ of the expressions in (8.7) for $l=0,1$ and (8.9) are

$$\mathbf{x}^{\mathrm{T}}\big(\nabla(\nabla\tilde{f}_0)\big)(y,\mathbf{x})\mathbf{x} = -\frac{1}{|\mathbf{v}_1(\mathbf{x})|^3}\left\{\mathbf{x}^{\mathrm{T}}\mathbf{x} - 3\frac{(\mathbf{x}^{\mathrm{T}}\mathbf{v}_1(\mathbf{x}))^2}{|\mathbf{v}_1(\mathbf{x})|^2}\right\},$$

$$\mathbf{x}^{\mathrm{T}}\big(\nabla(\nabla\tilde{f}_1)\big)(y,\mathbf{x})\mathbf{x} = -6\frac{(\mathbf{x}^{\mathrm{T}}\tilde{\mathbf{v}}_0(y))(\mathbf{x}^{\mathrm{T}}\mathbf{v}_1(\mathbf{x}))}{|\mathbf{v}_1(\mathbf{x})|^5}$$
$$-3\frac{\mathbf{v}_1(\mathbf{x})^{\mathrm{T}}\tilde{\mathbf{v}}_0(y)}{|\mathbf{v}_1(\mathbf{x})|^5}\left\{\mathbf{x}^{\mathrm{T}}\mathbf{x} - 5\frac{(\mathbf{x}^{\mathrm{T}}\mathbf{v}_1(\mathbf{x}))^2}{|\mathbf{v}_1(\mathbf{x})|^2}\right\},$$

$$\mathbf{x}^{\mathrm{T}}\big(\nabla(\nabla\tilde{g})\big)(y,\mathbf{x})\mathbf{x} = -\frac{2}{|\mathbf{v}_1(\mathbf{x})|^4}\left\{\mathbf{x}^{\mathrm{T}}\mathbf{x} - 4\frac{(\mathbf{x}^{\mathrm{T}}\mathbf{v}_1(\mathbf{x}))^2}{|\mathbf{v}_1(\mathbf{x})|^2}\right\}.$$

Those with the first order derivatives with respect to $\mathbf{x}$ of the expressions in (8.10) and (8.11) are

$$\mathbf{x}^{\mathrm{T}}\big(\nabla\sqrt{c}\sqrt{a}+\tfrac{1}{2}b\big)(y,\mathbf{x}) = |\mathbf{d}|\mathcal{R}(y)\frac{\mathbf{x}^{\mathrm{T}}\mathbf{v}_1(\mathbf{x})}{|\mathbf{v}_1(\mathbf{x})|} + \mathbf{x}^{\mathrm{T}}\tilde{\mathbf{v}}_0(y),$$

$$\mathbf{x}^{\mathrm{T}}\big(\nabla\sqrt{c}\sqrt{a+b+c}+c+\tfrac{1}{2}b\big)(y,\mathbf{x}) = |\mathbf{d}|\hat{\mathcal{R}}(y)\frac{\mathbf{x}^{\mathrm{T}}\mathbf{v}_1(\mathbf{x})}{|\mathbf{v}_1(\mathbf{x})|} + \mathbf{x}^{\mathrm{T}}(\tilde{\mathbf{v}}_0+\mathbf{v}_1)(y),$$

and with the second order derivatives with respect to $\mathbf{x}$ of the expressions in (8.10) and (8.11) are

$$\mathbf{x}^{\mathrm{T}}\big(\nabla(\nabla\sqrt{c}\sqrt{a}+\tfrac{1}{2}b)\big)(y,\mathbf{x})\mathbf{x} = \frac{|\mathbf{d}|\mathcal{R}(y)}{|\mathbf{v}_1(\mathbf{x})|}\left\{\mathbf{x}^{\mathrm{T}}\mathbf{x} - \frac{(\mathbf{x}^{\mathrm{T}}\mathbf{v}_1(\mathbf{x}))^2}{|\mathbf{v}_1(\mathbf{x})|^2}\right\},$$

$$\mathbf{x}^{\mathrm{T}}\big(\nabla(\nabla\sqrt{c}\sqrt{a+b+c}+c+\tfrac{1}{2}b)\big)(y,\mathbf{x})\mathbf{x} = \frac{|\mathbf{d}|\hat{\mathcal{R}}(y)}{|\mathbf{v}_1(\mathbf{x})|}\left\{\mathbf{x}^{\mathrm{T}}\mathbf{x} - \frac{(\mathbf{x}^{\mathrm{T}}\mathbf{v}_1(\mathbf{x}))^2}{|\mathbf{v}_1(\mathbf{x})|^2}\right\}.$$

When $\mathbf{v}_1(\mathbf{x})$ approaches $\mathbf{d}$, the expressions in (8.6) and (8.8), and the inner products of $\mathbf{x}$ with their derivatives with respect to $\mathbf{x}$ become:

$$f_{jkl}(y) = y^j\big(y-s_2^{(m)}\big)^k\left(\frac{\mathbf{d}^{\mathrm{T}}\tilde{\mathbf{v}}_0(y)}{|\mathbf{d}|^2}\right)^l = y^j\big(y-s_2^{(m)}\big)^k\big(\alpha y-s_1^{(m)}\big)^l,$$

$$g_{jk}(y) = \frac{1}{|\mathbf{d}|}y^j\left(y - s_2^{(m)}\right)^k,$$

$$\mathbf{x}^T(\nabla f_{jk0})(y) = -y^j\left(y - s_2^{(m)}\right)^k\frac{\mathbf{x}^T\mathbf{d}}{|\mathbf{d}|^2} = -y^{j+1}\left(y - s_2^{(m)}\right)^k\varepsilon,$$

$$\mathbf{x}^T(\nabla f_{jk1})(y) = y^j\left(y - s_2^{(m)}\right)^k\left\{\frac{\mathbf{x}^T\tilde{\mathbf{v}}_0(y)}{|\mathbf{d}|^2} - 3\frac{\mathbf{d}^T\tilde{\mathbf{v}}_0(y)}{|\mathbf{d}|^2}\frac{\mathbf{x}^T\mathbf{d}}{|\mathbf{d}|^2}\right\}$$

$$= y^{j+1}\left(y - s_2^{(m)}\right)^k\left\{\left(\gamma y - \varepsilon s_1^{(m)}\right) - 3\left(\alpha y - s_1^{(m)}\right)\varepsilon\right\},$$

$$\mathbf{x}^T(\nabla g_{jk})(y) = \frac{-2}{|\mathbf{d}|}y^j\left(y - s_2^{(m)}\right)^k\frac{\mathbf{x}^T\mathbf{d}}{|\mathbf{d}|^2} = \frac{-2}{|\mathbf{d}|}y^{j+1}\left(y - s_2^{(m)}\right)^k\varepsilon,$$

$$\mathbf{x}^T\left(\nabla(\nabla f_{jk0})\right)(y)\mathbf{x} = -y^j\left(y - s_2^{(m)}\right)^k\left(\frac{\mathbf{x}^T\mathbf{x}}{|\mathbf{d}|^2} - 3\frac{(\mathbf{x}^T\mathbf{d})^2}{|\mathbf{d}|^4}\right)$$

$$= -y^{j+2}\left(y - s_2^{(m)}\right)^k(\delta^2 - 3\varepsilon^2),$$

$$\mathbf{x}^T\left(\nabla(\nabla f_{jk1})\right)(y)\mathbf{x}$$
$$= y^j\left(y - s_2^{(m)}\right)^k\left\{-6\frac{\mathbf{x}^T\tilde{\mathbf{v}}_0(y)}{|\mathbf{d}|^2}\frac{\mathbf{x}^T\mathbf{d}}{|\mathbf{d}|^2} - 3\frac{\mathbf{d}^T\tilde{\mathbf{v}}_0(y)}{|\mathbf{d}|^2}\left(\frac{\mathbf{x}^T\mathbf{x}}{|\mathbf{d}|^2} - 5\frac{(\mathbf{x}^T\mathbf{d})^2}{|\mathbf{d}|^4}\right)\right\}$$
$$= y^{j+2}\left(y - s_2^{(m)}\right)^k\left\{-6\left(\gamma y - \varepsilon s_1^{(m)}\right)\varepsilon - 3\left(\alpha y - s_1^{(m)}\right)(\delta^2 - 5\varepsilon^2)\right\},$$

$$\mathbf{x}^T\left(\nabla(\nabla g_{jk})\right)(y)\mathbf{x} = \frac{-2}{|\mathbf{d}|}y^j\left(y - s_2^{(m)}\right)^k\left(\frac{\mathbf{x}^T\mathbf{x}}{|\mathbf{d}|^2} - 4\frac{(\mathbf{x}^T\mathbf{d})^2}{|\mathbf{d}|^4}\right)$$

$$= \frac{-2}{|\mathbf{d}|}y^{j+2}\left(y - s_2^{(m)}\right)^k(\delta^2 - 4\varepsilon^2),$$

the expressions in (8.10) and (8.11), and the inner products of $\mathbf{x}$ with their derivatives with respect to $\mathbf{x}$ become:

$$\left(\sqrt{c}\sqrt{a} + \tfrac{1}{2}b\right)(y) = |\mathbf{d}|^2\left\{\mathcal{R}(y) + \frac{\mathbf{d}^T\tilde{\mathbf{v}}_0(y)}{|\mathbf{d}|^2}\right\}$$

$$= |\mathbf{d}|^2\left\{\mathcal{R}(y) + \alpha y - s_1^{(m)}\right\},$$

$$\left(\sqrt{c}\sqrt{a+b+c} + c + \tfrac{1}{2}b\right)(y) = |\mathbf{d}|^2\left\{\hat{\mathcal{R}}(y) + \frac{\mathbf{d}^T(\tilde{\mathbf{v}}_0 + \mathbf{v}_1)(y)}{|\mathbf{d}|^2}\right\}$$

$$= |\mathbf{d}|^2\left\{\hat{\mathcal{R}}(y) + \hat{\alpha}y + \left(1 - s_1^{(m)}\right)\right\},$$

$$\mathbf{x}^T\left(\nabla\sqrt{c}\sqrt{a} + \tfrac{1}{2}b\right)(y) = |\mathbf{d}|^2\left\{\frac{\mathbf{x}^T\mathbf{d}}{|\mathbf{d}|^2}\mathcal{R}(y) + \frac{\mathbf{x}^T\tilde{\mathbf{v}}_0(y)}{|\mathbf{d}|^2}\right\}$$

$$= |\mathbf{d}|^2 y\left\{\varepsilon\mathcal{R}(y) + \gamma y - \varepsilon s_1^{(m)}\right\},$$

$$\mathbf{x}^T\left(\nabla\sqrt{c}\sqrt{a+b+c} + c + \tfrac{1}{2}b\right)(y) = |\mathbf{d}|^2\left\{\frac{\mathbf{x}^T\mathbf{d}}{|\mathbf{d}|^2}\hat{\mathcal{R}}(y) + \frac{\mathbf{x}^T(\tilde{\mathbf{v}}_0 + \mathbf{v}_1)(y)}{|\mathbf{d}|^2}\right\}$$

$$= |\mathbf{d}|^2 y\left\{\varepsilon\hat{\mathcal{R}}(y) + \hat{\gamma}y + \varepsilon\left(1 - s_1^{(m)}\right)\right\},$$

$$\mathbf{x}^{\mathrm{T}}(\nabla(\nabla\sqrt{c}\,\sqrt{a}+\tfrac{1}{2}b))(y)\mathbf{x} = |\mathbf{d}|^2\mathcal{R}(y)\left\{\frac{\mathbf{x}^{\mathrm{T}}\mathbf{x}}{|\mathbf{d}|^2}-\frac{(\mathbf{x}^{\mathrm{T}}\mathbf{d})^2}{|\mathbf{d}|^4}\right\}$$
$$= |\mathbf{d}|^2 y^2 \mathcal{R}(y)(\delta^2-\varepsilon^2),$$

$$\mathbf{x}^{\mathrm{T}}(\nabla(\nabla\sqrt{c}\,\sqrt{a+b+c}+c+\tfrac{1}{2}b))(y)\mathbf{x} = |\mathbf{d}|^2\hat{\mathcal{R}}(y)\left\{\frac{\mathbf{x}^{\mathrm{T}}\mathbf{x}}{|\mathbf{d}|^2}-\frac{(\mathbf{x}^{\mathrm{T}}\mathbf{d})^2}{|\mathbf{d}|^4}\right\}$$
$$= |\mathbf{d}|^2 y^2 \hat{\mathcal{R}}(y)(\delta^2-\varepsilon^2),$$

where

$$\mathcal{R}(y) = \sqrt{\beta^2 y^2 - 2\alpha s_1^{(m)} y + (s_1^{(m)})^2 + \zeta^2}, \qquad (8.14)$$
$$\hat{\mathcal{R}}(y) = \sqrt{\hat{\beta}^2 y^2 + 2\hat{\alpha}(1 - s_1^{(m)})y + (1 - s_1^{(m)})^2 + \zeta^2},$$

and

$$\alpha = \frac{\mathbf{d}^{\mathrm{T}}\mathbf{c}_0}{|\mathbf{d}|^2}, \qquad \beta = \frac{|\mathbf{c}_0|}{|\mathbf{d}|}, \qquad \gamma = \frac{\mathbf{c}_1^{\mathrm{T}}\mathbf{c}_0}{|\mathbf{d}|^2},$$

$$\hat{\alpha} = \frac{\mathbf{d}^{\mathrm{T}}\hat{\mathbf{c}}_0}{|\mathbf{d}|^2}, \qquad \hat{\beta} = \frac{|\hat{\mathbf{c}}_0|}{|\mathbf{d}|}, \qquad \hat{\gamma} = \frac{\mathbf{c}_1^{\mathrm{T}}\hat{\mathbf{c}}_0}{|\mathbf{d}|^2},$$

$$\varepsilon = \frac{\mathbf{d}^{\mathrm{T}}\mathbf{c}_1}{|\mathbf{d}|^2}, \qquad \delta = \frac{|\mathbf{c}_1|}{|\mathbf{d}|}, \qquad \zeta = \frac{|\mathbf{x}_{OM}|}{|\mathbf{d}|}.$$

8.2. Taylor expansion of the integrals for the functions f_{jkl}, g_{jk}

The integrals for scalar functions f_{kl} and g_k are as follows:

$$I_{jkl}(y,\mathbf{x}) = \int f_{jkl}(y,\mathbf{x})\ln F(y,\mathbf{x})\,dy, \qquad (8.15)$$

$$I_{jk}(y,\mathbf{x}) = \int g_{jk}(y,\mathbf{x})G(y)\,dy, \qquad (8.16)$$

where

$$F(y,\mathbf{x}) = \frac{(\sqrt{c}\,\sqrt{a}+\tfrac{1}{2}b)(y,\mathbf{x})}{(\sqrt{c}\,\sqrt{a+b+c}+c+\tfrac{1}{2}b)(y,\mathbf{x})},$$
$$G(y) = (\sqrt{a})(y) - (\sqrt{a+b+c})(y) = |\mathbf{d}|(\mathcal{R}(y) - \hat{\mathcal{R}}(y)).$$

The second order Taylor expansions of (8.15) and (8.16) with respect to $\mathbf{x}$ gives

$$I_{jkl}(y,\mathbf{x}) = I_{jkl}(y) + \mathbf{x}^{\mathrm{T}}(\nabla I_{jkl})(y) + \tfrac{1}{2}\mathbf{x}^{\mathrm{T}}(\nabla(\nabla I_{jkl}))(y)\mathbf{x} + \mathcal{O}(|\mathbf{x}|^3),$$
$$I_{jk}(y,\mathbf{x}) = I_{jk}(y) + \mathbf{x}^{\mathrm{T}}(\nabla I_{jk})(y) + \tfrac{1}{2}\mathbf{x}^{\mathrm{T}}(\nabla(\nabla I_{jk}))(y)\mathbf{x} + \mathcal{O}(|\mathbf{x}|^3),$$

where

$$I_{jkl}(y) = \int f_{jkl}(y)\ln F(y)\,dy,$$

$$\mathbf{x}^T(\nabla I_{jkl})(y) = \int \mathbf{x}^T(\nabla f_{jkl})(y)\ln F(y)\,dy + \int f_{jkl}(y)\mathbf{x}^T(\nabla \ln F)(y)\,dy,$$

$$\mathbf{x}^T\big(\nabla(\nabla I_{jkl})\big)(y)\mathbf{x} = \int \mathbf{x}^T\big(\nabla(\nabla f_{jkl})\big)(y)\mathbf{x}\ln F(y)\,dy$$

$$+ 2\int \mathbf{x}^T(\nabla f_{jkl})(y)\mathbf{x}^T(\nabla \ln F)(y)\,dy$$

$$+ \int f_{jkl}(y)\mathbf{x}^T\big(\nabla(\nabla \ln F)\big)(y)\mathbf{x}\,dy,$$

$$I_{jk}(y) = \int g_{jk}(y)G(y)\,dy,$$

$$\mathbf{x}^T(\nabla I_{jk})(y) = \int \mathbf{x}^T(\nabla g_{jk})(y)G(y)\,dy,$$

$$\mathbf{x}^T\big(\nabla(\nabla I_{jk})\big)(y)\mathbf{x} = \int \mathbf{x}^T\big(\nabla(\nabla g_{jk})\big)(y)\mathbf{x}G(y)\,dy,$$

and

$$\ln F(y) = \ln\big(\mathcal{R}(y) + \alpha y - s_1^{(m)}\big) - \ln\big(\hat{\mathcal{R}}(y) + \hat{\alpha} y + (1 - s_1^{(m)})\big),$$

$$\mathbf{x}^T(\nabla \ln F)(y) = y\left\{ \frac{\varepsilon\mathcal{R}(y) + \gamma y - \varepsilon s_1^{(m)}}{\mathcal{R}(y) + \alpha y - s_1^{(m)}} - \frac{\varepsilon\hat{\mathcal{R}}(y) + \hat{\gamma} y + \varepsilon(1 - s_1^{(m)})}{\hat{\mathcal{R}}(y) + \hat{\alpha} y + (1 - s_1^{(m)})} \right\},$$

$$\mathbf{x}^T\big(\nabla(\nabla \ln F)\big)(y)\mathbf{x}$$

$$= y^2\left\{ -\left(\frac{\varepsilon\mathcal{R}(y) + \gamma y - \varepsilon s_1^{(m)}}{\mathcal{R}(y) + \alpha y - s_1^{(m)}} \right)^2 + \left(\frac{\varepsilon\hat{\mathcal{R}}(y) + \hat{\gamma} y + \varepsilon(1 - s_1^{(m)})}{\hat{\mathcal{R}}(y) + \hat{\alpha} y + (1 - s_1^{(m)})} \right)^2 \right.$$

$$\left. + \frac{\mathcal{R}(y)(\delta^2 - \varepsilon^2)}{\mathcal{R}(y) + \alpha y - s_1^{(m)}} - \frac{\hat{\mathcal{R}}(y)(\delta^2 - \varepsilon^2)}{\hat{\mathcal{R}}(y) + \hat{\alpha} y + (1 - s_1^{(m)})} \right\}.$$

Therefore, the integrals to be evaluated are

$$I_{j00}(y) = K_j(y),$$

$$I_{j10}(y) = I_{j+1,00}(y) - s_2^{(m)} I_{j00}(y),$$

$$I_{j01}(y) = \alpha K_{j+1}(y) - s_1^{(m)} K_j(y),$$

$$I_{j11}(y) = I_{j+1,01}(y) - s_2^{(m)} I_{j01}(y),$$

$$\mathbf{x}^T(\nabla I_{j00})(y) = -\varepsilon\big(K_{j+1}(y) - L_{j+1}^{(1)}(y)\big),$$

$$\mathbf{x}^T(\nabla I_{j10})(y) = \mathbf{x}^T(\nabla I_{j+1,00})(y) - s_2^{(m)}\mathbf{x}^T(\nabla I_{j00})(y),$$

$$\mathbf{x}^T(\nabla I_{j01})(y) = (\gamma - 3\alpha\varepsilon)K_{j+2}(y) + \alpha\varepsilon L_{j+2}^{(1)}(y) + s_1^{(m)}\varepsilon\big(2K_{j+1} - L_{j+1}^{(1)}(y)\big),$$

$$\mathbf{x}^T(\nabla I_{j11})(y) = \mathbf{x}^T(\nabla I_{j+1,01})(y) - s_2^{(m)}\mathbf{x}^T(\nabla I_{j01})(y),$$

$$\mathbf{x}^{\mathrm{T}}\big(\nabla(\nabla I_{j00})\big)(y)\mathbf{x}$$
$$= -(\delta^2 - 3\varepsilon^2)K_{j+2}(y) - 2\varepsilon^2 L^{(1)}_{j+2}(y) - \varepsilon^2 L^{(2)}_{j+2}(y) + (\delta^2 - \varepsilon^2)\Lambda_{j+2}(y),$$

$$\mathbf{x}^{\mathrm{T}}\big(\nabla(\nabla I_{j10})\big)(y)\mathbf{x} = \mathbf{x}^{\mathrm{T}}\big(\nabla(\nabla I_{j+1,00})\big)(y)\mathbf{x} - s_2^{(m)}\mathbf{x}^{\mathrm{T}}\big(\nabla(\nabla I_{j00})\big)(y)\mathbf{x},$$

$$\mathbf{x}^{\mathrm{T}}\big(\nabla(\nabla I_{j01})\big)(y)\mathbf{x} = -3\big\{2\gamma\varepsilon + \alpha(\delta^2 - 5\varepsilon^2)\big\}K_{j+3}(y) + 2(\gamma - 3\alpha\varepsilon)\varepsilon L^{(1)}_{j+3}(y)$$
$$+ \alpha\big\{-\varepsilon^2 L^{(2)}_{j+3}(y) + (\delta^2 - \varepsilon^2)\Lambda_{j+3}(y)\big\}$$
$$+ s_1^{(m)}\big\{3(\delta^2 - 3\varepsilon^2)K_{j+2}(y) + 4\varepsilon^2 L^{(1)}_{j+2}(y) + \varepsilon^2 L^{(2)}_{j+2}(y)$$
$$- (\delta^2 - \varepsilon^2)\Lambda_{j+2}(y)\big\},$$

$$\mathbf{x}^{\mathrm{T}}\big(\nabla(\nabla I_{j11})\big)(y)\mathbf{x} = \mathbf{x}^{\mathrm{T}}\big(\nabla(\nabla I_{j+1,01})\big)(y)\mathbf{x} - s_2^{(m)}\mathbf{x}^{\mathrm{T}}\big(\nabla(\nabla I_{j01})\big)(y)\mathbf{x},$$

$$I_{j0}(y) = M_j(y),$$

$$I_{j1}(y) = I_{j+1,0}(y) - s_2^{(m)}I_{j0}(y),$$

$$\mathbf{x}^{\mathrm{T}}(\nabla I_{j0})(y) = -2\varepsilon M_{j+1}(y),$$

$$\mathbf{x}^{\mathrm{T}}(\nabla I_{j1})(y) = \mathbf{x}^{\mathrm{T}}(\nabla I_{j+1,0})(y) - s_2^{(m)}\mathbf{x}^{\mathrm{T}}(\nabla I_{j0})(y),$$

$$\mathbf{x}^{\mathrm{T}}\big(\nabla(\nabla I_{j0})\big)(y)\mathbf{x} = -2(\delta^2 - 4\varepsilon^2)M_{j+2}(y),$$

$$\mathbf{x}^{\mathrm{T}}\big(\nabla(\nabla I_{j1})\big)(y)\mathbf{x} = \mathbf{x}^{\mathrm{T}}\big(\nabla(\nabla I_{j+1,0})\big)(y)\mathbf{x} - s_2^{(m)}\mathbf{x}^{\mathrm{T}}\big(\nabla(\nabla I_{j0})\big)(y)\mathbf{x},$$

where the auxiliary integrals $K_i(y)$, $L_i^{(n)}(y)$, $\Lambda_i(y)$, and $M_i(y)$ are

$$K_i(y) = \mathcal{K}_i(y) - \hat{\mathcal{K}}_i(y),$$

$$L_i^{(n)}(y) = \mathcal{L}_i^{(n)}\Big(y; \frac{\gamma}{\varepsilon}, 1\Big) - \hat{\mathcal{L}}_i^{(n)}\Big(y; \frac{\hat{\gamma}}{\varepsilon}, 1\Big),$$

$$\Lambda_i(y) = \mathcal{L}_i^{(1)}(y; 0, 0) - \hat{\mathcal{L}}_i^{(1)}(y; 0, 0),$$

$$M_i(y) = \mathcal{M}_i(y) - \hat{\mathcal{M}}_i(y),$$

$$\mathcal{K}_i(y) = \int y^i \ln\big(\mathcal{R}(y) + \alpha y - s_1^{(m)}\big)\, dy, \tag{8.17}$$

$$\hat{\mathcal{K}}_i(y) = \int y^i \ln\big(\hat{\mathcal{R}}(y) + \hat{\alpha} y + (1 - s_1^{(m)})\big)\, dy, \tag{8.18}$$

$$\mathcal{L}_i^{(n)}(y; \rho, \sigma) = \int y^i \left(\frac{\mathcal{R}(y) + \rho y - \sigma s_1^{(m)}}{\mathcal{R}(y) + \alpha y - s_1^{(m)}}\right)^n dy, \tag{8.19}$$

$$\hat{\mathcal{L}}_i^{(n)}(y; \rho, \sigma) = \int y^i \left(\frac{\hat{\mathcal{R}}(y) + \rho y + \sigma(1 - s_1^{(m)})}{\hat{\mathcal{R}}(y) + \hat{\alpha} y + (1 - s_1^{(m)})}\right)^n dy, \tag{8.20}$$

$$\mathcal{M}_i(y) = \int y^i \mathcal{R}(y)\, dy, \tag{8.21}$$

$$\hat{\mathcal{M}}_i(y) = \int y^i \hat{\mathcal{R}}(y)\, dy. \tag{8.22}$$

Note that the integral $L_i^{(n)}(y)$ need only to be calculated if $\varepsilon \neq 0$.

8.3. Analytical evaluation of the auxiliary integrals

In this subsection the analytical expressions are obtained for the auxiliary integrals defined by (8.17), (8.19), and (8.21). Those for the integrals defined by (8.18), (8.20), and (8.22) are similar.

The expression (8.14) can be rewritten as follows:

$$R(y) = \sqrt{\beta^2 y^2 - 2\alpha s_1^{(m)} y + (s_1^{(m)})^2 + \zeta^2}$$
$$= \sqrt{\tilde{z}^2 + f^2},$$

where

$$\tilde{z} = \beta y - \frac{\alpha}{\beta} s_1^{(m)},$$

$$f = \sqrt{\left(1 - \frac{\alpha^2}{\beta^2}\right)(s_1^{(m)})^2 + \zeta^2}.$$

Since $|\frac{\alpha}{\beta}| < 1$, and $s_1^{(m)}$ and ζ are not equal to zero simultaneously, $f > 0$, so that

$$y = \frac{f}{\beta}(z + h),$$

$$R(y) + \alpha y - s_1^{(m)} = f(\sqrt{z^2 + 1} + pz + q),$$

where

$$z = \frac{\tilde{z}}{f}, \quad h = p\frac{s_1^{(m)}}{f}, \quad p = \frac{\alpha}{\beta}, \quad q = (p^2 - 1)\frac{s_1^{(m)}}{f},$$

and $|p| < 1$ and $|q| \leqslant 1$.

Similarly, one can write:

$$R(y) + \rho y - \sigma s_1^{(m)} = f(\sqrt{z^2 + 1} + rz + s),$$

where

$$r = \frac{\rho}{\beta}, \quad s = (pr - \sigma)\frac{s_1^{(m)}}{f}.$$

Therefore, the auxiliary integrals defined by (8.17), (8.19), and (8.21) can be written as follows:

$$\mathcal{K}_i(y) = \left(\frac{f}{\beta}\right)^{i+1}\left(\frac{\ln(f)}{i+1}(z+h)^{i+1} + \sum_{k=0}^{i}\binom{i}{k}h^{i-k}\mathcal{K}_k(z)\right), \tag{8.23}$$

$$\mathcal{L}_i^{(n)}(y; \rho, \sigma) = \left(\frac{f}{\beta}\right)^{i+1}\sum_{k=0}^{i}\binom{i}{k}h^{i-k}\mathcal{L}_k^{(n)}(z; r, s), \tag{8.24}$$

$$\mathcal{M}_i(y) = f\left(\frac{f}{\beta}\right)^{i+1}\sum_{k=0}^{i}\binom{i}{k}h^{i-k}\mathcal{M}_k(z), \tag{8.25}$$

where

$$\mathcal{K}_k(z) = \int z^k \ln\left(pz + q + \sqrt{z^2 + 1}\right) dz, \tag{8.26}$$

$$\mathcal{L}_k^{(n)}(z; r, s) = \int z^k \left(\frac{rz + s + \sqrt{z^2 + 1}}{pz + q + \sqrt{z^2 + 1}}\right)^n dz, \tag{8.27}$$

$$\mathcal{M}_k(z) = \int z^k \sqrt{z^2 + 1}\, dz. \tag{8.28}$$

The expression for the integrals (8.26), (8.27), and (8.28) for $k = 0$ and the recursion formulae $k > 0$ are given in the following subsection.

8.4. Analytical expressions and recursion formulae

For $p = q = 0$ the formulae for the integrals (8.26) are

$$\mathcal{K}_0(z) = -z + \arctan z + \tfrac{1}{2} z \ln(z^2 + 1),$$

$$\mathcal{K}_1(z) = \tfrac{1}{4}(z^2 + 1)\left(\ln(z^2 + 1) - 1\right),$$

$$\mathcal{K}_k(z) = \frac{1}{k+1}\left\{-\frac{z^{k+1}}{k+1} + \tfrac{1}{2}(z^{k+1} + z^{k-1})\ln(z^2 + 1) - (k-1)\mathcal{K}_{k-2}(z)\right\}.$$

For $p \neq 0$ or $q \neq 0$ the formulae for the integrals (8.26) are

$$\mathcal{K}_k(z) = \frac{1}{(k+1)(p^2 + q^2)}\left\{-\frac{q^2 z^{k+1}}{k+1} + (p^2 + q^2)z^{k+1}\ln\left(pz + q + \sqrt{z^2 + 1}\right)\right.$$

$$\left. - p\mathcal{I}_{k+1}(z) + q\mathcal{I}_k(z) - p(p^2 - 1)\mathcal{J}_{k+1}(z) + q(q^2 - 1)\mathcal{J}_k(z)\right\},$$

where

$$\mathcal{I}_0(z) = \int \frac{1}{\sqrt{z^2 + 1}}\, dz = \ln\left(z + \sqrt{z^2 + 1}\right),$$

$$\mathcal{I}_1(z) = \int \frac{z}{\sqrt{z^2 + 1}}\, dz = \sqrt{z^2 + 1}, \tag{8.29}$$

$$\mathcal{I}_k(z) = \int \frac{z^k}{\sqrt{z^2 + 1}}\, dz = \frac{1}{k}\left\{z^{k-1}\sqrt{z^2 + 1} - (k-1)\mathcal{I}_{k-2}(z)\right\},$$

and

$$\mathcal{J}_j(z) = \int \frac{z^j}{pz + q + \sqrt{z^2 + 1}}\, dz. \tag{8.30}$$

After the substitution $z = \tfrac{1}{2}(t - t^{-1})$, so that $\sqrt{z^2 + 1} = \tfrac{1}{2}(t + t^{-1})$, $dz = \tfrac{1}{2}(t + t^{-1}) \cdot t^{-1}\, dt$, and $t = z + \sqrt{z^2 + 1}$ the integral (8.30) can be written as follows:

$$\mathcal{J}_k(t) = \left(\frac{1}{2}\right)^k \int \frac{(t + t^{-1})(t - t^{-1})^k}{at^2 + bt + c}\, dt, = \left(\frac{1}{2}\right)^k \sum_{j=0}^{k} (-1)^j \binom{k}{j} T_{k-2j}(t),$$

$$T_i(t) = \tilde{J}_{i+1}^{(1)}(t) + \tilde{J}_{i-1}^{(1)}(t),$$

where

$$a = 1 + p, \quad b = 2q, \quad c = 1 - p, \quad -k \leqslant i \leqslant k,$$

and

$$\tilde{J}_j^{(n)}(t) = \int \frac{t^j}{(at^2 + bt + c)^n} \, dt. \tag{8.31}$$

Let constant $\tau \ll 1$, then the formulae for this integral are

$$\tilde{J}_0^{(1)}(t) = \begin{cases} \frac{2}{\sqrt{4ac-b^2}} \left(\arctan \frac{2at+b}{\sqrt{4ac-b^2}} - \frac{\pi}{2} \right), & \text{for } b^2 - 4ac < -\tau(2at+b)^2, \\[2mm] \frac{1}{\sqrt{b^2-4ac}} \ln\left(\frac{2at+b-\sqrt{b^2-4ac}}{2at+b+\sqrt{b^2-4ac}} \right), & \text{for } b^2 - 4ac > \tau(2at+b)^2, \\[2mm] \frac{-2}{2at+b} \sum_{k=0}^{\infty} \frac{1}{2k+1} \left(\frac{b^2-4ac}{(2at+b)^2} \right)^k, & \text{for } |b^2 - 4ac| < \tau(2at+b)^2, \\[2mm] & \text{i.e., } p^2 + q^2 \approx 1, \end{cases}$$

$$\tilde{J}_0^{(2)}(t) = \begin{cases} \frac{-1}{b^2-4ac} \left(\frac{2at+b}{at^2+bt+c} + 2a\tilde{J}_0^{(1)}(t) \right), & \text{for } |b^2 - 4ac| > \tau(2at+b)^2, \\[2mm] \frac{-8a}{3(2at+b)^3} \sum_{k=0}^{\infty} \frac{k+1}{2k+3} \left(\frac{b^2-4ac}{(2at+b)^2} \right)^k, & \text{for } |b^2 - 4ac| < \tau(2at+b)^2, \end{cases}$$

$$\tilde{J}_0^{(n)}(t) = \begin{cases} \frac{-1}{(n-1)(b^2-4ac)} \left(\frac{2at+b}{(at^2+bt+c)^{n-1}} + 2(2n-3)a\tilde{J}_0^{(n-1)}(t) \right), & \text{for } b^2 \neq 4ac, \\[2mm] \frac{-2^{2n-1}a^{n-1}}{(2n-1)(2at+b)^{2n-1}}, & \text{for } b^2 = 4ac, \end{cases}$$

$$\tilde{J}_{+1}^{(1)}(t) = \frac{1}{2a} \left(\ln(at^2 + bt + c) - b\tilde{J}_0^{(1)}(t) \right),$$

$$\tilde{J}_{-1}^{(1)}(t) = \frac{1}{2c} \left(\ln \frac{t^2}{at^2 + bt + c} - b\tilde{J}_0^{(1)}(t) \right),$$

$$\tilde{J}_{-1}^{(n)}(t) = \frac{1}{c} \left(\frac{1}{2(n-1)(at^2+bt+c)^{n-1}} - \frac{b}{2}\tilde{J}_0^{(n)}(t) + \tilde{J}_{-1}^{(n-1)}(t) \right),$$

$$\tilde{J}_{2n-1}^{(n)}(t) = \frac{1}{a} \left(\tilde{J}_{2n-3}^{(n-1)}(t) - b\tilde{J}_{2n-2}^{(n)}(t) - c\tilde{J}_{2n-3}^{(n)}(t) \right),$$

$$\tilde{J}_j^{(n)}(t) = \begin{cases} \frac{1}{(j-2n+1)a} \left(\frac{t^{j-1}}{(at^2+bt+c)^{n-1}} - (j-n)b\tilde{J}_{j-1}^{(n)}(t) - (j-1)c\tilde{J}_{j-2}^{(n)}(t) \right), \\[1mm] \quad \text{for } 0 < j \neq 2n-1, \\[3mm] \frac{1}{(j+1)c} \left(\frac{t^{j+1}}{(at^2+bt+c)^{n-1}} - (j-n+2)b\tilde{J}_{j+1}^{(n)}(t) - (j-2n+3)a\tilde{J}_{j+2}^{(n)}(t) \right), \\[1mm] \quad \text{for } j < -1, \end{cases}$$

where $n > 1$.

After the substitution $z = \frac{1}{2}(t - t^{-1})$, so that $\sqrt{z^2 + 1} = \frac{1}{2}(t + t^{-1})$, $dz = \frac{1}{2}(t + t^{-1}) \cdot t^{-1} \, dt$, and $t = z + \sqrt{z^2 + 1}$ the integral (8.27) can be written as follows:

$$\mathcal{L}_k^{(n)}(z; r, s)$$

$$= \left(\frac{1}{2}\right)^{k+1} \int (t + t^{-1}) t^{-1} (t - t^{-1})^k \left(\frac{r(t - t^{-1}) + 2s + (t + t^{-1})}{p(t - t^{-1}) + 2q + (t + t^{-1})}\right)^n dt$$

$$= \left(\frac{1}{2}\right)^{k+1} \sum_{j=0}^{k} (-1)^j \binom{k}{j} T_{k-2j}^{(n)}(t; r, s),$$

$$T_i^{(n)}(t; r, s) = \int (t + t^{-1}) t^{i-1} \left(\frac{\check{a}t^2 + \check{b}t + \check{c}}{at^2 + bt + c}\right)^n dt, \tag{8.32}$$

where

$$a = 1 + p, \qquad b = 2q, \qquad c = 1 - p,$$
$$\check{a} = 1 + r, \qquad \check{b} = 2s, \qquad \check{c} = 1 - r, \qquad -k \leqslant i \leqslant k.$$

The integral (8.32) can be rewritten as follows:

$$T_i^{(n)}(t; r, s) = \int (t + t^{-1}) t^{i-1} \left(A + \frac{Bt + C}{at^2 + bt + c}\right)^n dt,$$

where

$$A = \frac{\check{a}}{a}, \qquad B = \check{b} - bA, \qquad C = \check{c} - cA.$$

For $n = 1, 2$ the integral (8.32) can be expressed in terms of the integrals (8.31) as follows:

$$T_i^{(1)}(t; 0, 0) = \tilde{\mathcal{J}}_{i+2}^{(1)}(t) + \tilde{\mathcal{J}}_{i-2}^{(1)}(t) + 2\tilde{\mathcal{J}}_i^{(1)}(t),$$

$$T_i^{(1)}(t; r, s) = A\big(U_i(t) + U_{i-2}(t)\big) + B\big(\tilde{\mathcal{J}}_{i+1}^{(1)}(t) + \tilde{\mathcal{J}}_{i-1}^{(1)}(t)\big)$$
$$+ C\big(\tilde{\mathcal{J}}_i^{(1)}(t) + \tilde{\mathcal{J}}_{i-2}^{(1)}(t)\big),$$

$$T_i^{(2)}(t; r, s) = A^2\big(U_i(t) + U_{i-2}(t)\big) + 2AB\big(\tilde{\mathcal{J}}_{i+1}^{(1)}(t) + \tilde{\mathcal{J}}_{i-1}^{(1)}(t)\big)$$
$$+ 2AC\big(\tilde{\mathcal{J}}_i^{(1)}(t) + \tilde{\mathcal{J}}_{i-2}^{(1)}(t)\big) + B^2\big(\tilde{\mathcal{J}}_{i+2}^{(2)}(t) + \tilde{\mathcal{J}}_i^{(2)}(t)\big)$$
$$+ 2BC\big(\tilde{\mathcal{J}}_{i+1}^{(2)}(t) + \tilde{\mathcal{J}}_{i-1}^{(2)}(t)\big) + C^2\big(\tilde{\mathcal{J}}_i^{(2)}(t) + \tilde{\mathcal{J}}_{i-2}^{(2)}(t)\big),$$

where

$$U_i(t) = \int t^i \, dt = \begin{cases} \ln(t), & \text{for } i = -1, \\ \frac{1}{i+1} t^{i+1}, & \text{for } i \neq -1. \end{cases}$$

The formulae for the integrals (8.28) are as follows:

$$\mathcal{M}_k(z) = \mathcal{I}_{k+2}(z) + \mathcal{I}_k(z),$$

where $\mathcal{I}_k(z)$ are the integrals (8.29).

9. Analytical integration of integrals over a triangle for scalar and vector valued basis functions

This section presents the analytical evaluation of the inner and moment integrals for a quadrilateral source element with scalar valued basis functions, and for a triangular source element with scalar or vector valued basis functions. For scalar valued basis functions the analytical evaluation of the inner integrals are already described in Section 5.2.4. For vector valued basis functions on a triangular element the integrals are decomposed into a sum of integrals, some of which are of the same type as those for the scalar valued basis functions. Therefore, in this section all the integrals over triangles will be treated.

Let $\mathbf{x}$ be the coordinates of the object point, O, the integration point of the outer integral. Let the source element with n edges be divided into triangles $\triangle_k$ for $k = 1, \ldots, n$, and let each $\triangle_k$ be the triangle with as top the point M and as base, $\mathbf{e}_k$, the kth edge of the element. For the inner integral the point M is the projection of point, O, in the plane of the element, for the moment integral the point M is the midpoint of the element.

The inner and moment integrals are, respectively:

$$I_j(\mathbf{x}) = \sum_k \int_{\triangle_k} \frac{\psi_j(\mathbf{x}')}{|\mathbf{x}' - \mathbf{x}|} \, d\mathbf{x}', \tag{9.1}$$

$$M_{j,\alpha\beta} = \sum_k \int_{\triangle_k} \psi_j(\mathbf{x}')\{\mathbf{x}' - \mathbf{x}_M\}_\alpha \{\mathbf{x}' - \mathbf{x}_M\}_\beta \, d\mathbf{x}'. \tag{9.2}$$

For scalar valued basis functions $\psi_j(\mathbf{x}') = 1$, and vector valued basis functions $\psi_j(\mathbf{x}') = (\mathbf{x}'_j - \mathbf{x}')/2J$, where $\mathbf{x}'_j$ is the jth vertex of the element, and J is the area of the element. The expression $\{\mathbf{x}' - \mathbf{x}_M\}_\alpha = 1$, $(x' - x_M)$, $(y' - y_M)$, or $(z' - z_M)$ for $\alpha = 0, 1, 2$, or 3.

Let $\mathbf{x}_{i,k}$ for $(i = 1, 2)$ be the vertices of $\mathbf{e}_k$, $\mathbf{n}$ be the normal to the plane of the element, $\mathbf{h}_k$ the perpendicular from M to $\mathbf{e}_k$, φ the angle between $\mathbf{x}' - \mathbf{x}_M$ and $\mathbf{h}_k$, and $\varphi_{i,k}$ the angle between $\mathbf{x}_{i,k} - \mathbf{x}_M$ and $\mathbf{h}_k$. Let P_k be the intersubsection of $\mathbf{x}' - \mathbf{x}_M$ and $\mathbf{e}_k$, and M_k the projection of M on $\mathbf{e}_k$. Let $\hat{\mathbf{a}}$ be the unit vector along vector $\mathbf{a}$. Then

$$d = |OM|, \qquad h_k = |\mathbf{h}_k|,$$

$$\mathbf{h}_k = MM_k = (\mathbf{x}_{1,k} - \mathbf{x}_M) - \left((\mathbf{x}_{1,k} - \mathbf{x}_M) \cdot \hat{\mathbf{e}}_k\right)\hat{\mathbf{e}}_k,$$

$$\mathbf{x}' - \mathbf{x}_M = \left((\mathbf{x}' - \mathbf{x}_M) \cdot \hat{\mathbf{h}}_k\right)\hat{\mathbf{h}}_k + \left((\mathbf{x}' - \mathbf{x}_M) \cdot \hat{\mathbf{e}}_k\right)\hat{\mathbf{e}}_k.$$

9.1. Analytical formulae for the inner integrals

After transformation to the polar coordinates $r = |\mathbf{x}' - \mathbf{x}_M|$ and φ, the inner integrals for scalar and vector valued basis functions become, respectively:

$$I(\mathbf{x}) = \sum_k I_k(\mathbf{x}), \tag{9.3}$$

$$\mathbf{I}_j(\mathbf{x}) = \frac{1}{2J} \left\{ I(\mathbf{x}) f x (\mathbf{x}'_j - \mathbf{x}_M) - \sum_k \left(\mathbf{I}_k^{(c)}(\mathbf{x}) + \mathbf{I}_k^{(s)}(\mathbf{x})\right) \right\}, \tag{9.4}$$

where

$$I_k(\mathbf{x}) = \mathcal{I}(\varphi_{2,k}, h_k) - \mathcal{I}(\varphi_{1,k}, h_k), \tag{9.5}$$

$$\mathbf{I}_k^{(c)}(\mathbf{x}) = \{\mathcal{I}_c(\varphi_{2,k}, h_k) - \mathcal{I}_c(\varphi_{1,k}, h_k)\}\hat{\mathbf{h}}_k, \tag{9.6}$$

$$\mathbf{I}_k^{(s)}(\mathbf{x}) = \{\mathcal{I}_s(\varphi_{2,k}, h_k) - \mathcal{I}_s(\varphi_{1,k}, h_k)\}\hat{\mathbf{e}}_k, \tag{9.7}$$

and $\varphi_{i,k} = \arctan\{\frac{1}{h_k}(\mathbf{x}_{i,k} - \mathbf{x}_M) \cdot \hat{\mathbf{e}}_k\}$.

Dropping the indices k the integral $I(\varphi, h)$ is defined by:

$$\begin{aligned}
\mathcal{I}(\varphi, h) &= \int\int_0^{h/\cos\varphi} \frac{r}{\sqrt{r^2 + d^2}}\, dr\, d\varphi \\
&= \int [\sqrt{r^2 + d^2}]_0^{h/\cos\varphi}\, d\varphi = \frac{h}{q}\int \frac{\sqrt{1 + q^2 x^2}}{1 + x^2}\, dx - d\varphi \\
&= h\log\left(qx + \sqrt{1 + q^2 x^2}\right) + d\arctan\left(\frac{dq}{h\sqrt{1 + q^2 x^2}}\right) - d\varphi \\
&= h\log\left(\frac{h\tan\varphi + s(\varphi)}{\sqrt{d^2 + h^2}}\right) + d\arctan\left(\frac{d\tan\varphi}{s(\varphi)}\right) - d\varphi \\
&= h\log\left(\frac{h\tan\varphi + s(\varphi)}{\sqrt{d^2 + h^2}}\right) + d\arctan\left(\frac{(d - s(\varphi))\tan\varphi}{s(\varphi) + d\tan^2\varphi}\right), \tag{9.8}
\end{aligned}$$

where $x = \tan\varphi$, $q = \frac{h}{\sqrt{d^2 + h^2}}$, and $s(\varphi) = |OP| = \sqrt{\left(\frac{h}{\cos\varphi}\right)^2 + d^2} = \frac{h}{q}\sqrt{1 + q^2 x^2}$.

The integrals $I_c(\varphi, h)$ and $I_s(\varphi, h)$ are defined by:

$$\begin{aligned}
\mathcal{I}_c(\varphi, h) &= \int\int_0^{h/\cos\varphi} \frac{r^2\cos\varphi}{\sqrt{r^2 + d^2}}\, dr\, d\varphi \\
&= \frac{1}{2}\int \cos\varphi[r\sqrt{r^2 + d^2} - d^2\log(\sqrt{r^2 + d^2} + r)]_0^{h/\cos\varphi}\, d\varphi \\
&= -\frac{1}{2}d^2\sin\varphi\log\left(\frac{s(\varphi) + h/\cos\varphi}{d}\right) + \frac{h^2}{2q}\int \frac{1}{\sqrt{1 + q^2 x^2}}\, dx \\
&= -\frac{1}{2}d^2\sin\varphi\log\left(\frac{s(\varphi) + h/\cos\varphi}{d}\right) + \frac{1}{2}(h^2 + d^2)\log\left(\frac{h\tan\varphi + s(\varphi)}{\sqrt{h^2 + d^2}}\right), \tag{9.9}
\end{aligned}$$

$$\begin{aligned}
\mathcal{I}_s(\varphi, h) &= \int\int_0^{h/\cos\varphi} \frac{r^2\sin\varphi}{\sqrt{r^2 + d^2}}\, dr\, d\varphi \\
&= \frac{1}{2}\int \sin\varphi[r\sqrt{r^2 + d^2} - d^2\log(\sqrt{r^2 + d^2} + r)]_0^{h/\cos\varphi}\, d\varphi \\
&= \frac{1}{2}d^2\cos\varphi\log\left(\frac{s(\varphi) + h/\cos\varphi}{d}\right) + \frac{1}{2}h^2 q\int \frac{x}{\sqrt{1 + q^2 x^2}}\, dx \\
&= \frac{1}{2}d^2\cos\varphi\log\left(\frac{s(\varphi) + h/\cos\varphi}{d}\right) + \frac{1}{2}hs(\varphi), \tag{9.10}
\end{aligned}$$

where $x = \tan\varphi$, $q = \frac{h}{\sqrt{d^2 + h^2}}$, and $s(\varphi) = |OP| = \sqrt{(h/\cos\varphi)^2 + d^2} = \frac{h}{q}\sqrt{1 + q^2 x^2}$.

9.2. Analytical formulae for the moment integrals

After transformation to the polar coordinates $r = |\mathbf{x}' - \mathbf{x}_M|$ and φ, the moment integrals for scalar and vector valued basis functions become, respectively:

$$M_{\alpha\beta} = \sum_k M_{k,\alpha\beta}, \tag{9.11}$$

$$\mathbf{M}_{j,\alpha\beta} = \frac{1}{2J}\left\{ M_{\alpha\beta}\,(\mathbf{x}'_j - \mathbf{x}_M) - \sum_k (\mathbf{M}^{(c)}_{k,\alpha\beta} + \mathbf{M}^{(s)}_{k,\alpha\beta}) \right\}, \tag{9.12}$$

where

$$M_{k,\alpha\beta} = \mathcal{M}_{\alpha\beta}(\varphi_{2,k}, h_k) - \mathcal{M}_{\alpha\beta}(\varphi_{1,k}, h_k), \tag{9.13}$$

$$\mathbf{M}^{(c)}_{k,\alpha\beta} = \{\mathcal{M}^{(c)}_{\alpha\beta}(\varphi_{2,k}, h_k) - \mathcal{M}^{(c)}_{\alpha\beta}(\varphi_{1,k}, h_k)\}\hat{\mathbf{h}}_k, \tag{9.14}$$

$$\mathbf{M}^{(s)}_{k,\alpha\beta} = \{\mathcal{M}^{(s)}_{\alpha\beta}(\varphi_{2,k}, h_k) - \mathcal{M}^{(s)}_{\alpha\beta}(\varphi_{1,k}, h_k)\}\hat{\mathbf{e}}_k, \tag{9.15}$$

and $\varphi_{i,k} = \arctan\{\frac{1}{h_k}(\mathbf{x}_{i,k} - \mathbf{x}_M) \cdot \hat{\mathbf{e}}_k\}$.

Dropping the indices k the integrals $\mathcal{M}_{\alpha\beta}(\varphi, h)$, $\mathcal{M}^{(c)}_{\alpha\beta}(\varphi, h)$, and $\mathcal{M}^{(s)}_{\alpha\beta}(\varphi, h)$ are defined by:

$$\mathcal{M}_{00}(\varphi, h) = \iint_0^{h/\cos\varphi} r\,dr\,d\varphi = \tfrac{1}{2}h^2 \int \frac{1}{\cos^2\varphi}\,d\varphi = \tfrac{1}{2}h^2 \tan\varphi, \tag{9.16}$$

$$\begin{aligned}
\mathcal{M}_{0\alpha}(\varphi, h) &= \iint_0^{h/\cos\varphi} (\hat{h}_\alpha \cos\varphi + \hat{e}_\alpha \sin\varphi)r^2\,dr\,d\varphi \\
&= \tfrac{1}{3}h^3 \int \frac{1}{\cos^3\varphi}(\hat{h}_\alpha \cos\varphi + \hat{e}_\alpha \sin\varphi)\,d\varphi \\
&= \tfrac{1}{3}h^3 (\hat{h}_\alpha \tan\varphi + \tfrac{1}{2}\hat{e}_\alpha \tan^2\varphi),
\end{aligned} \tag{9.17}$$

$$\begin{aligned}
\mathcal{M}_{\alpha\beta}(\varphi, h) &= \iint_0^{h/\cos\varphi} (\hat{h}_\alpha \cos\varphi + \hat{e}_\alpha \sin\varphi)(\hat{h}_\beta \cos\varphi + \hat{e}_\beta \sin\varphi)r^3\,dr\,d\varphi \\
&= \tfrac{1}{4}h^4 \int \frac{1}{\cos^4\varphi}(\hat{h}_\alpha \cos\varphi + \hat{e}_\alpha \sin\varphi)(\hat{h}_\beta \cos\varphi + \hat{e}_\beta \sin\varphi)\,d\varphi \\
&= \tfrac{1}{4}h^4 \{\hat{h}_\beta(\hat{h}_\alpha \tan\varphi + \tfrac{1}{2}\hat{e}_\alpha \tan^2\varphi) + \hat{e}_\beta(\tfrac{1}{2}\hat{h}_\alpha \tan^2\varphi + \tfrac{1}{3}\hat{e}_\alpha \tan^3\varphi)\},
\end{aligned} \tag{9.18}$$

$$\mathcal{M}^{(c)}_{00}(\varphi, h) = \iint_0^{h/\cos\varphi} \cos\varphi\, r^2\,dr\,d\varphi = \tfrac{1}{3}h^3 \int \frac{1}{\cos^2\varphi}\,d\varphi = \tfrac{1}{3}h^3 \tan\varphi, \tag{9.19}$$

$$\begin{aligned}
\mathcal{M}^{(c)}_{0\alpha}(\varphi, h) &= \iint_0^{h/\cos\varphi} \cos\varphi(\hat{h}_\alpha \cos\varphi + \hat{e}_\alpha \sin\varphi)r^3\,dr\,d\varphi \\
&= \tfrac{1}{4}h^4 \int \frac{1}{\cos^3\varphi}(\hat{h}_\alpha \cos\varphi + \hat{e}_\alpha \sin\varphi)\,d\varphi \\
&= \tfrac{1}{4}h^4 (\hat{h}_\alpha \tan\varphi + \tfrac{1}{2}\hat{e}_\alpha \tan^2\varphi),
\end{aligned} \tag{9.20}$$

$$\mathcal{M}_{\alpha\beta}^{(c)}(\varphi, h) = \iint_0^{h/\cos\varphi} \cos\varphi (\hat{h}_\alpha \cos\varphi + \hat{e}_\alpha \sin\varphi)(\hat{h}_\beta \cos\varphi + \hat{e}_\beta \sin\varphi) r^4 \, dr \, d\varphi$$

$$= \tfrac{1}{5} h^5 \int \frac{1}{\cos^4\varphi} (\hat{h}_\alpha \cos\varphi + \hat{e}_\alpha \sin\varphi)(\hat{h}_\beta \cos\varphi + \hat{e}_\beta \sin\varphi) \, d\varphi$$

$$= \tfrac{1}{5} h^5 \{\hat{h}_\beta (\hat{h}_\alpha \tan\varphi + \tfrac{1}{2}\hat{e}_\alpha \tan^2\varphi) + \hat{e}_\beta (\tfrac{1}{2}\hat{h}_\alpha \tan^2\varphi + \tfrac{1}{3}\hat{e}_\alpha \tan^3\varphi)\},$$

$$\text{(9.21)}$$

$$\mathcal{M}_{00}^{(s)}(\varphi, h) = \iint_0^{h/\cos\varphi} \sin\varphi r^2 \, dr \, d\varphi = \tfrac{1}{3} h^3 \int \frac{\sin\varphi}{\cos^3\varphi} \, d\varphi = \tfrac{1}{6} h^3 \tan^2\varphi, \quad \text{(9.22)}$$

$$\mathcal{M}_{0\alpha}^{(s)}(\varphi, h) = \iint_0^{h/\cos\varphi} \sin\varphi (\hat{h}_\alpha \cos\varphi + \hat{e}_\alpha \sin\varphi) r^3 \, dr \, d\varphi$$

$$= \tfrac{1}{4} h^4 \int \frac{\sin\varphi}{\cos^4\varphi} (\hat{h}_\alpha \cos\varphi + \hat{e}_\alpha \sin\varphi) \, d\varphi$$

$$= \tfrac{1}{4} h^4 (\tfrac{1}{2}\hat{h}_\alpha \tan^2\varphi + \tfrac{1}{3}\hat{e}_\alpha \tan^3\varphi),$$

$$\text{(9.23)}$$

$$\mathcal{M}_{\alpha\beta}^{(s)}(\varphi, h) = \iint_0^{h/\cos\varphi} \sin\varphi (\hat{h}_\alpha \cos\varphi + \hat{e}_\alpha \sin\varphi)(\hat{h}_\beta \cos\varphi + \hat{e}_\beta \sin\varphi) r^4 \, dr \, d\varphi$$

$$= \tfrac{1}{5} h^5 \int \frac{\sin\varphi}{\cos^5\varphi} (\hat{h}_\alpha \cos\varphi + \hat{e}_\alpha \sin\varphi)(\hat{h}_\beta \cos\varphi + \hat{e}_\beta \sin\varphi) \, d\varphi$$

$$= \tfrac{1}{5} h^5 \{\hat{h}_\beta (\tfrac{1}{2}\hat{h}_\alpha \tan^2\varphi + \tfrac{1}{3}\hat{e}_\alpha \tan^3\varphi) + \hat{e}_\beta (\tfrac{1}{3}\hat{h}_\alpha \tan^3\varphi + \tfrac{1}{4}\hat{e}_\alpha \tan^4\varphi)\},$$

$$\text{(9.24)}$$

where $\hat{h}_\alpha = \hat{\mathbf{h}}_x, \hat{\mathbf{h}}_y$, or $\hat{\mathbf{h}}_z$, and $\hat{e}_\alpha = \hat{\mathbf{e}}_x, \hat{\mathbf{e}}_y$, or $\hat{\mathbf{e}}_z$, for $\alpha = 1, 2$, or 3.

10. Solution of Kirchhoff's equations

10.1. Kirchhoff's equations

This section presents the solution methods for solving the *Kirchhoff's equations* describing the behavior of a circuit which forms the electronic equivalent of an interconnection system consisting of a number of planar conductors immersed in a stratified medium. A derivation of Kirchhoff's equations from *Maxwell's equations* can be found in DU CLOUX, MAAS, and WACHTERS [1994], and in Chapter 1 of the present volume. In these references, a weak formulation and discretization of a mixed potential, boundary value problem is presented. Care is taken that, in the quasi-static approximation, the discretised equations admit an electronic circuit interpretation. The conductor surfaces are subdivided into a number of sufficiently small elements. The topology of the surfaces is described by the set of elements, the index set of which is denoted by $\mathcal{N}$, and a set of edges between adjacent elements, the index set of which is denoted by $\mathcal{E}$. The electric surface current, surface charge, and scalar potential, defined on the conductor surfaces, are expanded in a number of basis functions, defined on the elements.

The Kirchhoff equations are:

$$(\mathbf{R} + s\mathbf{L})I - \mathbf{P}V = 0, \tag{10.1}$$

$$\mathbf{P}^T I + sQ = J, \tag{10.2}$$

$$\mathbf{D}Q = V, \tag{10.3}$$

where I collects the edge currents, Q the element charges, V the element potentials, and J the external currents flowing into the interconnection system. Further, $\mathbf{R}$ denotes the resistance matrix, $\mathbf{L}$ the inductance matrix, and $\mathbf{D}$ the elastance matrix. The matrix $\mathbf{P}$ denotes the incidence matrix. It consists of entries 0 and ± 1, and represents the topology. Finally, s denotes the complex frequency. Its imaginary part is $-\omega$. It is assumed that $|\omega| \leqslant \Omega$, where Ω is the maximum frequency for which the generated equivalent circuit should be valid. The matrices $\mathbf{R}$, $\mathbf{L}$, $\mathbf{D}$, and $\mathbf{P}$ are independent of s.

Elimination of the charges from (10.1)–(10.3) gives

$$(\mathbf{R} + s\mathbf{L})I - \mathbf{P}V = 0, \tag{10.4}$$

$$\mathbf{P}^T I + s\mathbf{C}V = J. \tag{10.5}$$

The charges are obtained from the potentials according to

$$Q = \mathbf{C}V, \tag{10.6}$$

where $\mathbf{C} = \mathbf{D}^{-1}$ denotes the capacitance matrix.

The set of circuit nodes is defined to be a nonempty subset of the set of elements. Let N denote the index set of the set of circuit nodes, and N' the index set of its complement in $\mathcal{N}$. Introduction of this partitioning of the set of elements leads to the following equations

$$(\mathbf{R} + s\mathbf{L})\mathbf{I} - \mathbf{P}_{N'}\mathbf{V}_{N'} = \mathbf{P}_N\mathbf{V}_N, \tag{10.7}$$

$$-\mathbf{P}_{N'}^T\mathbf{I} - s\mathbf{C}_{N'N'}\mathbf{V}_{N'} = s\mathbf{C}_{N'N}\mathbf{V}_N, \tag{10.8}$$

and

$$\mathbf{J}_N = \mathbf{P}_N^T\mathbf{I} + s\mathbf{C}_{NN'}\mathbf{V}_{N'} + s\mathbf{C}_{NN}\mathbf{V}_N, \tag{10.9}$$

where $\mathbf{V}_N$ is the collection of prescribed vectors of circuit node voltages.

Let $\mathbb{R}$ and $\mathbb{C}$ be the sets of real and complex numbers, respectively, and $|.|$ denote the length of a set. From the discretization it follows that the matrices

$$\mathbf{R} \in \mathbb{R}^{|\mathcal{E}|\times|\mathcal{E}|}, \qquad \mathbf{L} \in \mathbb{R}^{|\mathcal{E}|\times|\mathcal{E}|}, \qquad \mathbf{P}_{N'} \in \mathbb{R}^{\mathcal{E}|\times|N'|}, \qquad \mathbf{P}_N \in \mathbb{R}^{\mathcal{E}|\times|N|},$$
$$\mathbf{C}_{N'N'} \in \mathbb{R}^{|N'|\times|N'|}, \quad \mathbf{C}_{N'N} \in \mathbb{R}^{|N'|\times|N|}, \quad \mathbf{C}_{NN'} \in \mathbb{R}^{|N|\times|N'|}, \quad \mathbf{C}_{NN} \in \mathbb{R}^{|N|\times|N|},$$
$$\mathbf{I} \in \mathbb{C}^{|\mathcal{E}|\times|N|}, \qquad \mathbf{V}_{N'} \in \mathbb{C}^{|N'|\times|N|}, \qquad \mathbf{V}_N \in \mathbb{R}^{|N|\times|N|}, \qquad \mathbf{J}_N \in \mathbb{C}^{|N|\times|N|}.$$

The matrices $\mathbf{R}$, $\mathbf{L}$, and $\mathbf{C}$ are symmetric and positive definite. The matrix $\mathbf{P}_{N'}$ has full column rank. From Eqs. (10.7)–(10.9) it follows that $\mathbf{J}_N$ is linearly related to $\mathbf{V}_N$, i.e.,

$$\mathbf{J}_N = \mathbf{Y}\mathbf{V}_N, \tag{10.10}$$

where $\mathbf{Y}$ is the admittance matrix of the interconnection system when observed from its circuit nodes.

10.2. Construction of the admittance matrix

Elimination of $\mathbf{I}$ from (10.7) and (10.8) gives

$$-\left(\mathbf{P}_{N'}^{\mathrm{T}}(\mathbf{R}+s\mathbf{L})^{-1}\mathbf{P}_{N'}+s\mathbf{C}_{N'N'}\right)\mathbf{V}_{N'} = \left(\mathbf{P}_{N'}^{\mathrm{T}}(\mathbf{R}+s\mathbf{L})^{-1}\mathbf{P}_{N}+s\mathbf{C}_{N'N}\right)\mathbf{V}_{N},$$
$$\tag{10.11}$$

$$\mathbf{I} = (\mathbf{R}+s\mathbf{L})^{-1}(\mathbf{P}_{N'}\mathbf{V}_{N'}+\mathbf{P}_{N}\mathbf{V}_{N}). \tag{10.12}$$

Let h be the mesh size, and $k_0 = \omega\sqrt{\varepsilon_0\mu_0}$ the free-space wavenumber, where ε_0 and μ_0 denote the free-space permittivity and permeability, respectively. When $k_0h \ll 1$, it follows from the expressions for the matrix elements of $\mathbf{C}$, $\mathbf{L}$, and $\mathbf{R}$ (see DU CLOUX, MAAS, and WACHTERS [1994]), that the orders of the matrix elements are

$$i\omega\mathbf{C}_{ij} = Z_0^{-1}\mathcal{O}(ik_0h), \qquad i\omega\mathbf{L}_{kl} = Z_0\mathcal{O}(ik_0h), \qquad \mathbf{R}_{kl} = Z_s\mathcal{O}(1),$$

where $i, j \in \mathcal{N}$ and $k, l \in \mathcal{E}$, $Z_0 = \sqrt{\mu_0/\varepsilon_0}$, and Z_s denotes the surface impedance of the conductors.

Therefore, the ratio between a matrix element of the second term and a corresponding matrix element of the first term in the left-hand side of (10.11) is $\mathcal{O}((ik_0h)^2)$, if $Z = 0$, and $\mathcal{O}(ik_0h)$, if $Z \neq 0$. Returning to (10.11), $\mathbf{V}_{N'}$ may be expanded in powers of ik_0h, which is then substituted into (10.12) to obtain $\mathbf{I}$. Neglecting higher order term in ik_0h, it follows that

$$\mathbf{V}_{N'} = \mathbf{V}_0 + \mathbf{V}_1, \qquad \mathbf{I} = \mathbf{I}_0 + \mathbf{I}_1, \tag{10.13}$$

where $(\mathbf{V}_0, \mathbf{I}_0)$ and $(\mathbf{V}_1, \mathbf{V}_1)$ may be obtained from two sets of equations,

$$(\mathbf{R}+s\mathbf{L})\mathbf{I}_0 - \mathbf{P}_{N'}\mathbf{V}_0 = \mathbf{P}_N\mathbf{V}_N, \tag{10.14}$$

$$-\mathbf{P}_{N'}^{\mathrm{T}}\mathbf{I}_0 = 0, \tag{10.15}$$

$$(\mathbf{R}+s\mathbf{L})\mathbf{I}_1 - \mathbf{P}_{N'}\mathbf{V}_1 = 0, \tag{10.16}$$

$$-\mathbf{P}_{N'}^{\mathrm{T}}\mathbf{I}_1 = s(\mathbf{C}_{N'N'}\mathbf{V}_0 + \mathbf{C}_{N'N}\mathbf{V}_N). \tag{10.17}$$

Let $\mathbf{V}_N$ be a unit matrix, then it follows from (10.10) that the admittance matrix $\mathbf{Y} = \mathbf{J}_N$. Substitution of (10.13) into (10.9) leads to

$$\mathbf{Y} = \mathbf{P}_N^{\mathrm{T}}(\mathbf{I}_0 + \mathbf{I}_1) + s\mathbf{C}_{NN'}\mathbf{V}_0 + s\mathbf{C}_{NN} + \mathcal{O}((ik_0h)^2). \tag{10.18}$$

From (10.18) it follows that treating capacitive effects as a perturbation is consistent with the quasi-static modeling of the interconnection system (see DU CLOUX, MAAS, and WACHTERS [1994]).

Depending on the frequency range of interest, four different methods can be distinguished to obtain a solution for these sets of equations. If one is only interested in the solution for a high (low) frequency range, it can be obtained by an expansion of $(\mathbf{V}_0, \mathbf{I}_0)$ and $(\mathbf{V}_1, \mathbf{I}_1)$ in s for relatively high (low) values of $|s|$.

However, if one is interested in the solution for the full frequency range there are two options. The first option is to solve Eqs. (10.14)–(10.17) for an appropriately chosen set of s values. The second option is to combine the solutions obtained for the high and low frequency ranges.

10.2.1. Admittance matrix for the high frequency range

In the high frequency range the $\mathbf{R}$ term in Eqs. (10.14) and (10.16) is considered as a perturbation of the $s\mathbf{L}$ term. Introducing the following expansions of $(\mathbf{V}_0, \mathbf{I}_0)$ and $(\mathbf{V}_1, \mathbf{V}_1)$

$$\mathbf{V}_0 = \mathbf{V}_{0,0} + s^{-1}\mathbf{V}_{0,1}, \qquad \mathbf{I}_0 = s^{-1}\mathbf{I}_{0,0} + s^{-2}\mathbf{I}_{0,1},$$
$$\mathbf{V}_1 = s^2\mathbf{V}_{1,0} + s\mathbf{V}_{1,1}, \qquad \mathbf{I}_1 = s\mathbf{I}_{1,0} + \mathbf{I}_{1,1},$$

and collecting the coefficients of s^0, s^{-1}, s^2, and s, respectively, the pairs $(\mathbf{V}_{i,j}, \mathbf{I}_{i,j})$ for $(i, j = 0, 1)$ may be obtained from the following four sets of equations

$$\mathbf{LI}_{0,0} - \mathbf{P}_{N'}\mathbf{V}_{0,0} = \mathbf{P}_N\mathbf{V}_N, \tag{10.19}$$

$$-\mathbf{P}_{N'}^T\mathbf{I}_{0,0} = 0, \tag{10.20}$$

$$\mathbf{LI}_{0,1} - \mathbf{P}_{N'}\mathbf{V}_{0,1} = -\mathbf{RI}_{0,0}, \tag{10.21}$$

$$-\mathbf{P}_{N'}^T\mathbf{I}_{0,1} = 0, \tag{10.22}$$

$$\mathbf{LI}_{1,0} - \mathbf{P}_{N'}\mathbf{V}_{1,0} = 0, \tag{10.23}$$

$$-\mathbf{P}_{N'}^T\mathbf{I}_{1,0} = \mathbf{C}_{N'N'}\mathbf{V}_{0,0} + \mathbf{C}_{N'N}\mathbf{V}_N, \tag{10.24}$$

$$\mathbf{LI}_{1,1} - \mathbf{P}_{N'}\mathbf{V}_{1,1} = -\mathbf{RI}_{1,0}, \tag{10.25}$$

$$-\mathbf{P}_{N'}^T\mathbf{I}_{1,1} = \mathbf{C}_{N'N'}\mathbf{V}_{0,1}. \tag{10.26}$$

The expansions of $(\mathbf{V}_0, \mathbf{I}_0)$ and $(\mathbf{V}_1, \mathbf{V}_1)$ are introduced to extend the validity of the high frequency range to lower frequencies. After substitution into the expression (10.18) and collection of the coefficients of powers of s one obtains

$$\mathbf{Y} = s^{-2}\mathbf{Y}_R + s^{-1}\mathbf{Y}_L + \mathbf{Y}_G + s\mathbf{Y}_C + \cdots, \tag{10.27}$$

where

$$\mathbf{Y}_L = \mathbf{P}_N^T\mathbf{I}_{0,0}, \qquad \mathbf{Y}_C = \mathbf{C}_{NN'}\mathbf{V}_{0,0} + \mathbf{C}_{NN} + \mathbf{P}_N^T\mathbf{I}_{1,0},$$
$$\mathbf{Y}_R = \mathbf{P}_N^T\mathbf{I}_{0,1}, \qquad \mathbf{Y}_G = \mathbf{C}_{NN'}\mathbf{V}_{0,1} + \mathbf{P}_N^T\mathbf{I}_{1,1}. \tag{10.28}$$

An equivalent circuit that represents the admittance matrix consists of branches between every pair of circuit nodes. For a circuit with frequency independent components each branch can be approximated by a series resistor R and inductor L, in parallel with a capacitor C and a resistor of conductance G, so that for the branch between the circuit nodes i and j

$$R = -\mathbf{y}_{R,ij}\mathbf{y}_{L,ij}^{-2}, \quad L = \mathbf{y}_{L,ij}^{-1}, \quad C = \mathbf{y}_{C,ij}, \quad G = \mathbf{y}_{G,ij}, \tag{10.29}$$

where the branch admittance matrix element, $\mathbf{y}_{ij}$, is related to the admittance matrix elements $\mathbf{Y}_{ij}$ through $(i, j \in N)$

$$\mathbf{y}_{ij} = -\mathbf{Y}_{ij} \quad (i \neq j), \tag{10.30}$$

$$\mathbf{y}_{ii} = \sum_{j \in N} \mathbf{Y}_{ij}. \tag{10.31}$$

294 A.J.H. Wachters and W.H.A. Schilders

The diagonal element, y_{ii}, represents the branch between the circuit node i and the ground plane or some reference at infinity. From Eqs. (10.19)–(10.22) it follows that $y_{R,ii} = y_{L,ii} = 0$.

If frequency dependent resistors are allowed each branch consists of a resistor $R = s^2 y_R^{-1}$ in parallel with an inductor L, a capacitor C, and a resistor of conductance G given in (10.29). For passive IC's it can be shown that this is a good approximation for the frequency range of interest.

10.2.2. Admittance matrix for the low frequency range

In the low frequency range the $s\mathbf{L}$ term in Eqs. (10.14) and (10.16) is considered as a perturbation of the term $\mathbf{R}$. Introducing the following expansions of $(\mathbf{V}_0, \mathbf{I}_0)$ and $(\mathbf{V}_1, \mathbf{V}_1)$

$$\mathbf{V}_0 = \mathbf{V}_{0,0} + s\mathbf{V}_{0,1}, \qquad \mathbf{I}_0 = \mathbf{I}_{0,0} + s\mathbf{I}_{0,1},$$

$$\mathbf{V}_1 = s\mathbf{V}_{1,0}, \qquad\qquad \mathbf{I}_1 = s\mathbf{I}_{1,0},$$

and collecting the coefficients of powers of s, the pairs $(\mathbf{V}_{i,j}, \mathbf{I}_{i,j})$ for $(i, j = 0, 1)$ may be obtained from the following three sets of equations

$$\mathbf{R}\mathbf{I}_{0,0} - \mathbf{P}_{N'}\mathbf{V}_{0,0} = \mathbf{P}_N\mathbf{V}_N, \tag{10.32}$$

$$-\mathbf{P}_{N'}^T\mathbf{I}_{0,0} = 0, \tag{10.33}$$

$$\mathbf{R}\mathbf{I}_{0,1} - \mathbf{P}_{N'}\mathbf{V}_{0,1} = -\mathbf{L}\mathbf{I}_{0,0}, \tag{10.34}$$

$$-\mathbf{P}_{N'}^T\mathbf{I}_{0,1} = 0, \tag{10.35}$$

$$\mathbf{R}\mathbf{I}_{1,0} - \mathbf{P}_{N'}\mathbf{V}_{1,0} = 0, \tag{10.36}$$

$$-\mathbf{P}_{N'}^T\mathbf{I}_{1,0} = \mathbf{C}_{N'N'}\mathbf{V}_{0,0} + \mathbf{C}_{N'N}\mathbf{V}_N, \tag{10.37}$$

The expansions of $(\mathbf{V}_0, \mathbf{I}_0)$ and $(\mathbf{V}_1, \mathbf{V}_1)$ are introduced to extend the validity of the low frequency range to higher frequencies. After substitution of them into the expression (10.18) and collection of the coefficients of powers of s one obtains

$$\mathbf{Y} = \mathbf{Y}_R + s\mathbf{Y}_C + \cdots, \tag{10.38}$$

where

$$\mathbf{Y}_R = \mathbf{P}_N^T\mathbf{I}_{0,0}, \qquad \mathbf{Y}_C = \mathbf{C}_{NN'}\mathbf{V}_{0,0} + \mathbf{C}_{NN} + \mathbf{P}_N^T(\mathbf{I}_{1,0} + \mathbf{I}_{1,0}).$$

An equivalent circuit that represents the admittance matrix consists of branches between every pair of circuit nodes. Each branch consists of a resistor R, in parallel with a capacitor C, so that for the branch between the circuit nodes i and j

$$R = y_{R,ij}^{-1}, \qquad C = y_{C,ij},$$

where the branch admittance matrix element y_{ij} is defined by the expressions (10.30) and (10.31) of Section 10.2.1. From Eqs. (10.32) and (10.33) it follows that $y_{R,ii} = 0$.

10.2.3. Approximate admittance matrix for the full frequency range

Returning to (10.14) and (10.15), an expression for $\mathbf{I}_0$ can be obtained by introducing the null space of $\mathbf{P}_{N'}^{\mathrm{T}}$. Let $\mathcal{C} \in \mathbb{R}^{|\mathcal{E}| \times (|\mathcal{E}| - |N'|)}$ and

$$\mathbf{P}_{N'}^{\mathrm{T}}\mathcal{C} = 0, \qquad \mathcal{C}^{\mathrm{T}}\mathbf{P}_{N'} = 0, \tag{10.39}$$

then it follows from (10.14) that

$$\mathbf{I}_0 = \mathcal{C}\big(\mathcal{C}^{\mathrm{T}}(\mathbf{R} + s\mathbf{L})\mathcal{C}\big)^{-1}\mathcal{C}^{\mathrm{T}}\mathbf{P}_N\mathbf{V}_N. \tag{10.40}$$

Let $\mathbf{A} = \mathcal{C}^{\mathrm{T}}\mathbf{R}\mathcal{C}$, $\mathbf{B} = \mathcal{C}^{\mathrm{T}}\mathbf{L}\mathcal{C}$, and $\mathbf{A}, \mathbf{B} \in \mathbb{R}^{n \times n}$, where $n = |\mathcal{E}| - |N'|$. Consider the generalized eigenvalue problem $\mathbf{A}\mathbf{x} = \lambda\mathbf{B}\mathbf{x}$. Since $\mathbf{R}$ and $\mathbf{L}$ are symmetric and positive definite, and $\mathcal{C}$ has full column rank, $\mathbf{A}$ and $\mathbf{B}$ are symmetric and positive definite. Pencils $\mathbf{A} - \lambda\mathbf{B}$ of this variety are referred to as symmetric-definite pencils. For such pencils there exists a nonsingular matrix $\mathbf{X} = [\mathbf{x}_1, \dots, \mathbf{x}_n]$ such that

$$\mathbf{X}^{\mathrm{T}}\mathbf{A}\mathbf{X} = \mathrm{diag}(a_1, \dots, a_n), \qquad \mathbf{X}^{\mathrm{T}}\mathbf{B}\mathbf{X} = \mathrm{diag}(b_1, \dots, b_n). \tag{10.41}$$

Moreover, $\mathbf{A}\mathbf{x}_i = \lambda_i\mathbf{x}_i$, for $i = 1, \dots, n$, where $\lambda_i = \frac{a_i}{b_i} > 0$ (see GOLUB and VAN LOAN [1986], Section 8.6).

From this it follows that

$$(\mathbf{A} + s\mathbf{B})^{-1} = \sum_{i=1}^{n} \frac{\mathbf{x}_i\mathbf{x}_i^{\mathrm{T}}}{a_i + sb_i}. \tag{10.42}$$

Let $\mathbf{Y}_{RL} = \mathbf{P}_N^{\mathrm{T}}\mathbf{I}_0$ be the contribution of $\mathbf{I}_0$ to the admittance matrix $\mathbf{Y}$ of (10.18) then after substitution of Section (10.42) into (10.40) it follows that

$$\mathbf{Y}_{RL} = \sum_{i=1}^{n} \frac{\mathbf{H}_i}{\lambda_i + s}, \tag{10.43}$$

where

$$\mathbf{H}_i = b_i^{-1}\mathbf{P}_N^{\mathrm{T}}\mathcal{C}\mathbf{x}_i\mathbf{x}_i^{\mathrm{T}}\mathcal{C}^{\mathrm{T}}\mathbf{P}_N. \tag{10.44}$$

Let for the contributions of $\mathbf{V}_0$ and $\mathbf{I}_1$ to $\mathbf{Y}$ of (10.18) the high frequency approximation of (10.27) be taken, then

$$\mathbf{Y} = \mathbf{Y}_{RL} + \mathbf{Y}_G + s\mathbf{Y}_C + \cdots. \tag{10.45}$$

The numerical computation of all eigenvalues and eigenvectors of the generalized eigenvalue problem becomes prohibitively expensive as soon as n becomes larger than a few hundred. Therefore, the only practical way to obtain an expression for the admittance matrix $\mathbf{Y}$ is through approximation. In view of the expression (10.45), it is natural to look for an approximation of $\mathbf{Y}_{RL}$ with a number of terms, $m \ll n$.

In a computer program this is accomplished by calculating m, low and high, eigenvalues, λ_i, of the generalized eigenvalue problem, and some admittance matrices, Y_k, for an appropriately chosen set of $m + 2$, negative, real values of s. The set of these match frequencies, s_k, consists of some large negative values between $-\Omega$ and

$- \max(\lambda_1, \ldots, \lambda_m)$, and some small negative values between $- \min(\lambda_1, \ldots, \lambda_m)$ and 0. They are chosen to be real, so that the components of the equivalent circuit will be real.

There are two options to obtain the Y_k's. The first option, the sampling method, is to solve Eqs. (10.14)–(10.17) for each s_k. The second option, the perturbation method, is to calculate the Y_k's for the large negative s_k values by the high frequency approximation (10.27), and those for the small negative s_k values by the low frequency approximation (10.38).

An element of the branch admittance matrix, defined by the (10.30) and (10.31) of Section 10.2.1, is approximated by

$$y_{ij}(s) = \mathbf{y}_{G,ij} + s\,\mathbf{y}_{C,ij} + \sum_{l=1}^{m} \frac{\mathbf{H}_{l,ij}}{\lambda_l + s}, \tag{10.46}$$

where the coefficients $\mathbf{y}_{G,ij}$, $\mathbf{y}_{C,ij}$, and $\mathbf{H}_{l,ij}$ are obtained by solving the following set of $m+2$ equations

$$\mathbf{y}_{G,ij} + s\mathbf{y}_{C,ij} + \sum_{l=1}^{m} \frac{\mathbf{H}_{l,ij}}{\lambda_l + s_k} = \mathbf{y}_{k,ij}, \quad \text{for } k = 1, \ldots, m+2. \tag{10.47}$$

An equivalent circuit which represents the admittance matrix consists of branches between every pair of circuit nodes. Each branch consists of m parallel connections of a series resistor R and inductor L, in parallel with a capacitor C, and a resistor of conductance G, so that for the branch between the circuit nodes i and j

$$R_l = \lambda_l \mathbf{H}_{l,ij}^{-1}, \qquad L_l = \mathbf{H}_{l,ij}^{-1}, \qquad C = \mathbf{y}_{C,ij}, \qquad G = \mathbf{y}_{G,ij}.$$

Since the components G are very small, and often introduce instabilities in the transient analysis of the equivalent circuit, in practice they are left out, so that instead of $m + 2$, only $m + 1$ match frequencies are needed.

10.3. Solution

The equivalent circuit for the interconnection system is submitted to a circuit analysis program, together with a description of the external components connected by the system, the bias conditions, and the frequency range or time domain for AC or transient analysis, respectively.

For AC analysis the output of the circuit analysis program is a list of nodal voltages for a number of frequencies. From these data the program can calculate the current density in the interconnection system by using the calculated transfer matrix. Next, the electromagnetic radiation can be calculated in the space around the system.

11. Linear algebra

This section presents linear algebra methods that can be used to solve the linear systems of equations obtained after discretization. Particularly, the methods used for the solution of the linear system of equations and of the generalized eigenvalue problem will be discussed. The discussion is relatively brief.

Complex geometries of the interconnection system imply that a relatively large number of elements have to be used for a proper discretization. As a result, the dimension of the coefficient matrices encountered in the linear system of equations is also large. This has a dramatic impact on the performance of the direct solution techniques: both the amount of storage needed and the time required to solve the linear system become prohibitively large. The former is often decisive, since it limits the size of problems which can be solved given the amount of memory space.

11.1. Solution of the linear systems of equations

In the solution of the Kirchhoff equations, discussed in Section 10, there are two types of linear system of equations.

In the first type of equation, e.g., $\mathbf{D}Q = V$ (see (10.3)), the matrix $\mathbf{D}$ is symmetric and positive definite. The solution of these systems can be performed by using the well known incomplete Cholesky conjugate gradient method, ICCG (see, e.g., BARRETT [1994]). For systems of a large dimension, n, it is often possible to perform the matrix vector multiplication in each iteration step in an efficient way. Therefore, the matrix $\mathbf{D}$ is approximated by the sum of a sparse matrix, $\mathbf{S}$, and a remainder matrix, $\mathbf{R} = \mathbf{V}\tilde{\mathbf{R}}\mathbf{V}$, where $\tilde{\mathbf{R}}$ is of dimension $m \ll n$, and $\mathbf{V}$ is an $n \times m$ prolongation matrix. This *matrix condensation* method will be discussed in Section 12. When there is a perfect ground plane present, as in the case of high frequency filter design, $\tilde{\mathbf{R}}$ is approximated by zero.

The second type of equations, e.g., (10.14) and (10.15), or (10.16) and (10.17), are of the form

$$\begin{pmatrix} \mathbf{A} & \mathbf{P} \\ \mathbf{P}^{\mathrm{T}} & 0 \end{pmatrix} \begin{pmatrix} I \\ V \end{pmatrix} = \begin{pmatrix} B_I \\ B_V \end{pmatrix}, \tag{11.1}$$

where $\mathbf{A} \in \mathbb{R}^{n \times n}$ is symmetric positive definite and $\mathbf{P} \in \mathbb{R}^{n \times m}$ such that

$$\mathbf{P}_{ij} \in \{-1, 0, 1\}, \quad \forall_{1 \leqslant i \leqslant n, 1 \leqslant j \leqslant m}.$$

Each row of $\mathbf{P}$ contains at most two nonzero elements, which are of opposite sign:

$$\sum_{j=1}^{m} |\mathbf{P}_{ij}| \leqslant 2, \quad -1 \leqslant \sum_{j=1}^{m} \mathbf{P}_{ij} \leqslant 1.$$

Finally, rank$(\mathbf{P}) = m$.

The coefficient matrix in (11.1) can be decomposed into the following form:

$$\begin{pmatrix} \mathbf{A} & \mathbf{P} \\ \mathbf{P}^{\mathrm{T}} & 0 \end{pmatrix} = \begin{pmatrix} \mathbf{A} & 0 \\ \mathbf{P}^{\mathrm{T}} & \mathbf{I} \end{pmatrix} \begin{pmatrix} \mathbf{A}^{-1} & 0 \\ 0 & -\mathbf{P}^{\mathrm{T}}\mathbf{A}^{-1}\mathbf{P} \end{pmatrix} \begin{pmatrix} \mathbf{A} & \mathbf{P} \\ 0 & \mathbf{I} \end{pmatrix},$$

which shows that there are n positive and m negative eigenvalues. Unfortunately, most iterative techniques can only be applied to the solution of positive definite systems. For systems with both positive and negative eigenvalues, such methods may break down. Furthermore, convergence will often be extremely slow since the polynomials generated in the methods have to locate both positive and negative eigenvalues. Because of these problems, direct solution techniques are often preferred.

A solution to the problems sketched in the above is to transform (11.1) into a number of linear systems which can be solved using standard iterative techniques. The latter is often more attractive than using direct solution techniques, for a number of reasons. Firstly, the amount of memory space needed is much smaller. Secondly, approximations to the coefficient matrix can be used in the matrix vector products occurring in iterative methods rather than the entire matrix (see Section 12). Hence, there is no need to fully assemble the matrix.

As noted in the above, if iterative methods are to be used it is desirable that the coefficient matrices involved are positive definite. There are essentially two ways of achieving this, namely either using the *range space method* or the *null space*. Often, the use of the latter method is ruled out because of the need to construct a basis for the null space of a large matrix. However, for interconnection systems, the null space method appears to be extremely useful. In the following, the method will be described and advantages will be listed.

11.1.1. Null space method
The basis for the null space method is the observation that the solution of the second set of equations in (11.1), i.e., $\mathbf{P}^{\mathrm{T}} I = B_V$, can be cast into the form

$$I = \mathbf{P}\tilde{V} + \mathcal{C}X, \tag{11.2}$$

where $\tilde{V} \in \mathbb{R}^m$, $X \in \mathbb{R}^{n-m}$, $\mathcal{C} \in \mathbb{R}^{n \times (n-m)}$, and $\mathbf{P}\tilde{V}$ is a special solution of the second set of equations, satisfying

$$\mathbf{P}^{\mathrm{T}}\mathbf{P}\tilde{V} = B_V,$$

and $\mathcal{C}$ is a matrix whose columns form a basis for the null space of $\mathbf{P}^{\mathrm{T}}$. Note that (11.2) is the most general form of solutions of the second set of equations in (11.1).

After substitution of (11.2) into the first set of equations, one obtains

$$A\mathcal{C}X + PV = B_I - \mathbf{A}\mathbf{P}X,$$

which, on multiplying by $\mathcal{C}^{\mathrm{T}}$ yields

$$\mathcal{C}^{\mathrm{T}}A\mathcal{C}X + \mathcal{C}^{\mathrm{T}}PV = \mathcal{C}^{\mathrm{T}}(B_I - \mathbf{A}\mathbf{P}X).$$

Since the columns of $\mathcal{C}$ constitute a basis of the null space of $\mathbf{P}^{\mathrm{T}}$, it holds that $\mathbf{P}^{\mathrm{T}}\mathcal{C} = 0$ or, equivalently, $\mathcal{C}^{\mathrm{T}}\mathbf{P} = 0$. Hence, the equation just derived is actually equal to

$$\mathcal{C}^{\mathrm{T}}A\mathcal{C}X = \mathcal{C}^{\mathrm{T}}(B_I - \mathbf{A}\mathbf{P}\tilde{V}). \tag{11.3}$$

The conclusion is that there are three steps involved in solving the original system:
1. First solve the system $\mathbf{P}^{\mathrm{T}}\mathbf{P}\tilde{V} = B_V$ to obtain $\tilde{V}$ and, subsequently calculate $\mathbf{P}\tilde{V}$, which is a special solution of the second set of equations.
2. Next determine the unknown vector X by solving the system (11.3). Combining the result with the special solution obtained in step 1 leads to the vector of unknown currents I.

3. Having found the current vector I, determine the vector of unknown potentials, V, by solving $\mathbf{P}^T\mathbf{P}V = \mathbf{P}^T(B_I - \mathbf{A}I)$.

The first and third step involve solving systems with a coefficient matrix $\mathbf{P}^T\mathbf{P}$ which, by the special structure of $\mathbf{P}$, is a positive definite symmetric matrix. In fact, it is an M-matrix, meaning that the diagonal entries are positive, the off-diagonal entries are nonpositive and the inverse is positive. The solution of these systems can be performed by using the ICCG method. Note that $\mathbf{P}^T\mathbf{P}$ is a sparse $m \times m$ matrix.

Crucial is the solution of the system in step 2 of the above procedure. Observe that, since $\mathbf{A}$ is a positive definite matrix, $\mathcal{C}^T\mathbf{A}\mathcal{C}$, is also positive definite. Hence, standard numerical solution techniques can be applied to the system (11.3). Thus, the problem of indefiniteness is avoided by using this approach. Note that $\mathcal{C}^T\mathbf{A}\mathcal{C}$ is an $(n-m) \times (n-m)$ matrix.

11.1.2. Construction of the null space matrix $\mathcal{C}$

The only problem is the construction of the null space matrix, $\mathcal{C}$. Fortunately, the matrix $\mathcal{C}$ only depends on the topology of the problem. Hence, the construction of the matrix $\mathcal{C}$ only has to be done once. A fortunate fact is that the elements of the null space can actually be interpreted physically. They are combinations of currents through branches constituting closed loops, the exterior of the problem area being considered as one node. This means that a considerable number of basis vectors can be constructed easily, by just finding all closed loops in the topology of the problem. Note that, in this way, a sparse basis is obtained. Most basis elements will consist of only a few nonzero entries. This is of importance when constructing the coefficient matrix $\mathcal{C}^T\mathbf{A}\mathcal{C}$, because it saves computer time. Since, for most topologies, this procedure does not lead to all elements of the null space, the set of basis functions found needs to be completed. This can be done in a fairly simple way. Suppose the matrix $\mathbf{P}^T$ is of the form

$$\mathbf{P}^T = \begin{pmatrix} F G \end{pmatrix},$$

where F is an $m \times m$ matrix and G an $m \times (n-m)$ matrix. Since rank$(\mathbf{P}) = m$, it is possible to choose a suitable permutation of unknown currents and voltages, so that the matrix F is nonsingular. It is even possible to have an upper triangular F. Since every column of F contains at most two nonzeroes, a simple elimination process leads to the situation where F is the identity matrix.

Now assume that the matrix $\mathcal{C}$ is of the form

$$\mathcal{C} = \begin{pmatrix} M \\ N \end{pmatrix},$$

with M an $m \times (n-m)$ matrix and N an $(n-m) \times (n-m)$ matrix. Then the requirement $\mathbf{P}^T\mathcal{C} = 0$ implies that

$$FM + GN = 0,$$

so that, if N is given, M follows from

$$M = -F^{-1}GN.$$

Orden (see ORDEN [1964]) already used this technique in 1964 to determine the null space. He chose N the identity matrix. Since a large number of basis elements are known, Orden's choice is not very efficient. Instead, we write

$$C = \begin{pmatrix} M_1 & M_2 \\ N_1 & N_2 \end{pmatrix},$$

such that the first set of columns

$$\begin{pmatrix} M_1 \\ N_1 \end{pmatrix}$$

correspond to the basis elements already generated. Suppose that this submatrix contains k columns. The k columns of the matrix N_1 constitute a subspace of $\mathbb{R}^{n-m}$, which is at most k-dimensional. This means that there are at least $n - m - k$ unit vectors from $\mathbb{R}^{n-m}$ that are not in this subspace. In other words, it is possible to fill the matrix N_2 with unit vectors that are not in the span of the columns of N_1. Having constructed the matrix N_2, the matrix M_2 can be produced simply by

$$M_2 = -F^{-1}GN_2.$$

If F is the identity matrix, the entries of M_2 are all in the set $-1, 0, 1$ (since this also holds for the elements of G). In this way, a complete set of basis vectors for the null space can be found.

11.2. The calculation of a subset of the eigenvalues of a generalized eigenvalue problem

The smallest and largest eigenvalues, λ, of the generalized eigenvalue problem, $\mathbf{Ax} = \lambda\mathbf{Bx}$, mentioned in Section 10.2.3, can be calculated by dedicated routines available in many software libraries. An effective procedure is to use a generalization of an algorithm, developed by PARLETT and REID [1981] for large symmetric eigenvalue problems. This algorithm is a reliable and efficient method for finding all or part of the spectrum of a large symmetric matrix $\mathbf{M}$, based on the Lanczos algorithm, by tracking the progress of the eigenvalues of the Lanczos tridiagonal matrices towards the eigenvalues of $\mathbf{M}$.

VAN DER VORST [1982] has generalized this algorithm for the computations of eigenvalues of the product matrix, $\mathbf{M} = \mathbf{AB}^{-1}$, where $\mathbf{A}$ is symmetric, and $\mathbf{B}$ is symmetric positive definite. The method allows for the computation of the eigenvalues of $\mathbf{B}^{-1}\mathbf{A}$ which are equal to those $\mathbf{AB}^{-1}$, without the explicit need for an LL^{T}-factorization of the matrix $\mathbf{B}$. This makes the generalized scheme very attractive, especially if $\mathbf{B}$ has a sparse structure. The method is attractive if fast solvers are available for the solution of linear systems of the form $\mathbf{B}y = z$.

Since the small eigenvalues of the partially solved eigenvalue problem, $\mathbf{AB}^{-1}\mathbf{x} = \lambda\mathbf{x}$, obtained by the above method, are often not accurate enough, they are obtained by partially solving the inverse eigenvalue problem, $\mathbf{A}^{-1}\mathbf{Bx} = \mu\mathbf{x}$, where $\mu = \lambda^{-1}$.

In recent years, the computation of eigenvalues and the solution of generalized eigenvalue problems has received much attention. A very effective method has been described in the work by Van der Vorst and his coworkers: the Jacobi–Davidson method. For more details, we refer the reader to the recent literature on this subject.

12. Matrix condensation

This section presents an efficient method for solving equations with large matrices, A, of the type discussed in the previous sections. In Section 7 it was shown that in the integrand of an interaction integral between an object and a source element that are sufficiently apart, the Green function can be approximated by Taylor expansion. In that case, the integral can be expressed as a sum of the products of a moment integral, $M_{\alpha\beta}(\mathbf{x}_m)$, corresponding to the object element, and a moment integral, $M_{\alpha\beta}(\mathbf{x}'_m)$, corresponding to the source element, and a factor resulting from the Green function, $G(\mathbf{x}'_m - \mathbf{x}_m)$, where $\mathbf{x}_m$ and $\mathbf{x}'_m$ are the midpoints of the object and source element. The indices α and β refer to the terms in the Taylor expansion. For more details see also Section 12.1.

In the *matrix condensation* method treated in this section the elements are clustered into cells. Let n be the total number of elements and m be the number of cells ($m \ll n$). Let the interaction integrals between elements belonging to adjacent cells form a sparse $n \times n$ matrix S with N_s nonzero elements ($N_s \ll n^2$). Let the interaction between the nonadjacent cells form an $m \times m$ matrix, $\tilde{R}$, the nonzero coefficients of which describe some kind of averaged interaction between the elements of one cell with the elements of another cell.

The matrix, $\tilde{R}$, may be considered as the product matrix $W M^{\mathrm{T}} : G : M W^{\mathrm{T}}$. The matrix W is an $m \times n$ restriction matrix, the ith row of which has only nonzero coefficients for the elements belonging to the ith cell. The matrix $M^{\mathrm{T}} : G : M$ is an $n \times n$ matrix, where M is an $n_T \times n$ matrix (n_T is the number of Taylor terms), and G is an $n_T \times n_T$ matrix. Matrix G results from the Green function, $G(\mathbf{x}'_m - \mathbf{x}_m)$, and is different for each combination of a column of M^{T} and a row of M, which represent the moments of an object element with midpoint $\mathbf{x}_m$ and the moments of a source element with midpoint $\mathbf{x}'_m$, respectively. This special matrix product is denoted by ":".

In the matrix condensation method the matrix, A, is approximated by matrix $\tilde{A} = S + V \tilde{R} V^{\mathrm{T}}$, where V is an $n \times m$ prolongation matrix the jth column of which has only nonzero coefficients for the elements belonging to the jth cell. The nonzero coefficients of V are set equal to 1. Hence, the total number of matrix coefficients of matrix, A, to be calculated is reduced from n^2 to $N_s + m^2$.

For the solution of the equations with these large matrices an iterative method is used. In each iteration step a matrix vector multiplication $v = Au$ is performed. In the matrix condensation method the vector v after multiplication is $v1 + v2$, where $v1 = Su$ and $v2 = V \tilde{R} V^{\mathrm{T}} u$. Hence, the total number of operations in a matrix vector multiplication is reduced from $\mathcal{O}(n^2)$ to $\mathcal{O}(n)$.

There are two different methods to obtain the vector $v2$. One method is to construct the matrix, $\tilde{R}$, in advance. Therefore, the $n_T \times m$ cell moments $M W^{\mathrm{T}}$ are calculated from the $n_T \times n$ element moments M assuming that the nonzero elements in a row of W are all equal, and their sums equal to 1. Physically, this meaning that the elements belonging to a particular cell will all have the same charge. Since it is not always possible to compose the cells so that this is a good approximation, an alternative method is to use in each iteration step the vector u to construct an $m \times n$ restriction matrix U and to calculate the $n_T \times m$ source cell moments MU^T. Each row of matrix U contains the elements of u belonging to the corresponding cell completed with zeroes. Next, the matrix MU^T is multiplied by the $m \times m$ matrix $WM^T : G : E$, where E is an $n_T \times m$ matrix with elements equal to 1. The matrix $WM^T : G : E$ can be calculated in advance. This method allows the elements in a cell to have different charges. Note that the matrix, $\tilde{R}$, is no longer symmetric, so that instead of the ICCG method an iterative solution method must taken that can handle nonsymmetric matrices, e.g., BICGSTAB (see BARRETT [1994]).

In the next subsection the coefficients of matrix $\tilde{R}$ will be derived.

12.1. Calculation of coefficients of matrix $\tilde{R}$

It has been shown that the interaction integrals belonging to the matrix $\mathbf{D}$ have the form:

$$I = \int_{\Omega_i} \tilde{\psi}_i(\mathbf{x}) \cdot \int_{\Omega_j} \tilde{\psi}_j(\mathbf{x}')G(\mathbf{x}' - \mathbf{x}) \, d\mathbf{x}' \, d\mathbf{x}. \tag{12.1}$$

Here, $\tilde{\psi}_i(\mathbf{x})$ and $\tilde{\psi}_j(\mathbf{x})$ are basis functions belonging to the object domain Ω_i and the source domain Ω_j, respectively. Further, $\mathbf{x}$ and $\mathbf{x}'$ represent points in Ω_i and Ω_j, respectively.

Let the Green function be of the form:

$$G(\mathbf{x}' - \mathbf{x}) = \sum_{i=0}^{N} \frac{c_i}{|\mathbf{x}'_i - \mathbf{x}|},$$

where N is the number of images.

If the distance between two disjoint elements Ω_i and Ω_j, defined by

$$\min\{|\mathbf{x}' - \mathbf{x}|; \ \mathbf{x} \in \Omega_i, \ \mathbf{x}' \in \Omega_j\},$$

is large enough one can apply Taylor expansion to the Green function G with respect to $(\mathbf{x}' - \mathbf{x}'_m)$ and $(\mathbf{x} - \mathbf{x}_m)$, where $\mathbf{x}'_m$ and $\mathbf{x}_m$ are the midpoints of the source and object element, respectively. After the substitutions $\mathbf{y}_i = \mathbf{x}'_i - \mathbf{x}$, $\mathbf{y}_{im} = \mathbf{x}'_{im} - \mathbf{x}_m$, and $\mathbf{y}_i - \mathbf{y}_{im} = \mathbf{y} - \mathbf{y}_m$, the second order Taylor expansion becomes:

$$G(\mathbf{y}) = g_m + \mathbf{g}_m^T(\mathbf{y} - \mathbf{y}_m) + \tfrac{1}{2}(\mathbf{y} - \mathbf{y}_m)^T\mathbf{G}_m(\mathbf{y} - \mathbf{y}_m) + \mathcal{O}(|\mathbf{y} - \mathbf{y}_m|^3), \tag{12.2}$$

where for $r_i = |\mathbf{y}_{im}|$

$$g_m = \sum_{i=0}^{N} G_i(\mathbf{y}_{im}) = \sum_{i=0}^{N} \frac{c_i}{r_i}, \tag{12.3}$$

$$\mathbf{g}_m = \sum_{i=0}^{N} (\nabla G_i)(\mathbf{y}_{im}) = \sum_{i=0}^{N} \frac{-c_i}{r_i^3}\mathbf{y}_{im}, \tag{12.4}$$

$$\mathbf{G}_m = \sum_{i=0}^{N} (\nabla(\nabla G_i))(\mathbf{y}_{im}) = \sum_{i=0}^{N} \frac{-c_i}{r_i^3}\mathbf{I} + 3\frac{c_i}{r_i^5}(\mathbf{y}_{im} \otimes \mathbf{y}_{im}). \tag{12.5}$$

For the definition of $\otimes$ see Appendix D. After substitution of the Taylor approximation (12.2) of $G(\mathbf{x}' - \mathbf{x})$ into the expression (12.1) one obtains

$$\tilde{I}(\mathbf{y}_m) = \int_{\Omega_i} \tilde{\psi}_i(\mathbf{x}) \cdot \int_{\Omega_j} \tilde{\psi}_j(\mathbf{x}')g_m \, d\mathbf{x}' \, d\mathbf{x} \tag{12.6}$$

$$+ \int_{\Omega_i} \tilde{\psi}_i(\mathbf{x}) \cdot \int_{\Omega_j} \tilde{\psi}_j(\mathbf{x}')\mathbf{g}_m^{\mathrm{T}}(\mathbf{y} - \mathbf{y}_m) \, d\mathbf{x}' \, d\mathbf{x} \tag{12.7}$$

$$+ \frac{1}{2} \int_{\Omega_i} \tilde{\psi}_i(\mathbf{x}) \cdot \int_{\Omega_j} \tilde{\psi}_j(\mathbf{x}')(\mathbf{y} - \mathbf{y}_m)^{\mathrm{T}}\mathbf{G}_m(\mathbf{y} - \mathbf{y}_m) \, d\mathbf{x}' \, d\mathbf{x}, \tag{12.8}$$

where $\mathbf{y} = \mathbf{x}' - \mathbf{x}$ and $\mathbf{y}_m = \mathbf{x}'_m - \mathbf{x}_m$.

Since g_m, $\mathbf{g}_m$, and $\mathbf{G}_m$ are independent of $\mathbf{x}$ and $\mathbf{x}'$, they appear as constant terms in
the integral. Let the moment integrals be defined by

$$M_{\alpha\beta}(\mathbf{x}_m) = \int_{\Omega_i} \tilde{\psi}_i(\mathbf{x})\{\mathbf{x} - \mathbf{x}_m\}_\alpha \{\mathbf{x} - \mathbf{x}_m\}_\beta \, d\mathbf{x},$$

where $\{\mathbf{x} - \mathbf{x}_m\}_\alpha = 1$, $(x - x_m)$, $(y - y_m)$, or $(z - z_m)$ for $\alpha = 0, 1, 2,$ or 3.

After substitution into (12.8)

$$\tilde{I}(\mathbf{y}_m) = M^{\mathrm{T}}GM, \tag{12.9}$$

where the n_T-dimensional vector M contains the elements $M_{\alpha\beta}$ in the row-wise order
$\alpha\beta = \{00, 01, \ldots, 33\}$. The $n_T \times n_T$ matrix G is of the form

$$G = \begin{bmatrix} g_m & \mathbf{g}_m^{\mathrm{T}} & \frac{1}{2}\tilde{\mathbf{G}}_m^{\mathrm{T}} \\ -\mathbf{g}_m & -\mathbf{G}_m & 0 \\ \frac{1}{2}\tilde{\mathbf{G}}_m & 0 & 0 \end{bmatrix},$$

where the scalar g_m, the 3-dimensional vector $\mathbf{g}_m$, and the 3×3 matrix $\mathbf{G}_m$ are defined
by (12.3)–(12.5). $\tilde{\mathbf{G}}_m$ is a 9-dimensional vector that contains the matrix elements of $\mathbf{G}_m$
in row-wise order $\alpha\beta = \{11, 12, \ldots, 33\}$.

The moment integrals are calculated in advance. Next, the cell moments are calculated, defined by:

$$\mathcal{M}_{\alpha\beta}(\mathbf{x}_{m_{\text{cell}}}) = \sum_{i \in \text{cell}} w_i M_{\alpha\beta}(\mathbf{x}_{m_i}),$$

where $\mathbf{x}_{m_i}$ are the midpoints of the elements belonging to the cell with midpoint $\mathbf{x}_{m_{\text{cell}}}$,
and w_i are weight factors.

An expression for the coefficients of matrix $\tilde{R}$ is obtained after substituting in (12.9)
the moments M by $\mathcal{M}$, and the matrix G by a similar matrix with the Green functions
$g_m, \mathbf{g}_m, \mathbf{G}_m$ of $\mathbf{y}_{i m_{\text{cell}}}$, where $\mathbf{y}_{m_{\text{cell}}} = \mathbf{x}'_{m_{\text{cell}}} - \mathbf{x}_{m_{\text{cell}}}$, and $\mathbf{x}_{m_{\text{cell}}}$ and $\mathbf{x}'_{m_{\text{cell}}}$ are the midpoints
of the object and source cell, respectively.

Appendix A. Boundary singularities

In this appendix we study the behavior of the potential, the fields, and the surface charge densities in the neighborhood of sharp "corners" or edges. We shall assume that they are infinitely sharp so that we can look at them closely enough that the behavior of the fields is determined in functional form solely by the properties of the "corner" being considered and not by the details of the overall configuration.

The general situation in two dimensions is shown in Fig. A.1. The two conducting planes intersect at an angle β. The planes are assumed to be held at potential V. Since we are interested in the functional behavior of the fields near the origin, we leave the "far away" behavior unspecified as much as possible.

The geometry of Fig. A.1 suggests the use of polar coordinates. In terms of the polar coordinates (r, φ), the Laplace equation for the potential Φ, in two dimensions is

$$\frac{1}{r}\mathbf{d}r\left(r\frac{\mathrm{d}\Phi}{\mathrm{d}r}\right) + \frac{1}{r^2}\frac{\mathrm{d}^2\Phi}{\mathrm{d}\varphi^2} = 0.$$

Using the separation of variables approach, we substitute

$$\Phi(r, \varphi) = R(r)F(\varphi).$$

This leads, upon multiplication by r^2/Φ, to

$$\frac{r}{R}\mathbf{d}r\left(r\frac{\mathrm{d}R}{\mathrm{d}r}\right) + \frac{1}{F}\frac{\mathrm{d}^2F}{\mathrm{d}\varphi^2} = 0.$$

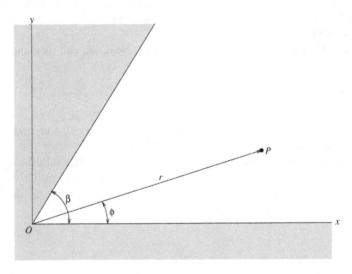

FIG. A.1. Intersubsection of two conducting planes, with potential V, defining a corner with opening angle β.

Since the two terms are separately function of r and φ, respectively, they each must be constant:

$$\frac{r}{R}\mathrm{d}r\left(r\frac{\mathrm{d}R}{\mathrm{d}r}\right) = v^2, \qquad \frac{1}{F}\frac{\mathrm{d}^2 F}{\mathrm{d}\varphi^2} = -v^2.$$

The solutions to these equations are

$$\left.\begin{array}{l} R(r) = ar^v + br^{-v}, \\ F(\varphi) = A\cos(v\varphi) + B\sin(v\varphi) \end{array}\right\} \tag{A.1}$$

and for the special circumstance of $v = 0$, the solutions are

$$\left.\begin{array}{l} R(r) = a_0 + b_0\ln r, \\ F(\varphi) = A_0 + B_0\varphi. \end{array}\right\} \tag{A.2}$$

These are the building blocks with which we construct the potential by linear superposition. For our situation the azimuthal angle is restricted to the range $0 \leqslant \varphi \leqslant \beta$. The boundary conditions are that $\Phi = V$ for all $r \geqslant 0$ when $\varphi = 0$ and $\varphi = \beta$. This requires that $b_0 = B_0 = 0$ in (A.2) and $b = A = 0$ in (A.1). Furthermore, it requires that v be chosen to make $\sin(v\beta) = 0$. Hence

$$v = \frac{m\pi}{\beta}, \quad m = 1, 2, \ldots$$

and the general solution becomes

$$\Phi(r, \varphi) = V + \sum_{m=1}^{\infty} a_m r^{m\pi/\beta} \sin(m\pi\varphi/\beta).$$

Since the series involves positive powers of $r^{\pi/\beta}$, for small enough r only the first term in the series will be important. Thus, near $r = 0$, the potential is approximately

$$\Phi(r, \varphi) \simeq V + a_1 r^{\pi/\beta} \sin(\pi\varphi/\beta).$$

The electric field components are

$$E_r(r, \varphi) = -\frac{\mathrm{d}\Phi}{\mathrm{d}r} \simeq -\frac{\pi a_1}{\beta} r^{(\pi/\beta)-1} \sin(\pi\varphi/\beta),$$

$$E_\varphi(r, \varphi) = -\frac{1}{r}\frac{\mathrm{d}\Phi}{\mathrm{d}\varphi} \simeq -\frac{\pi a_1}{\beta} r^{(\pi/\beta)-1} \cos(\pi\varphi/\beta).$$

The surface charge densities at $\varphi = 0$ and $\varphi = \beta$ are equal and are approximately

$$\rho(r) = \frac{E_\varphi(r, 0)}{4\pi} \simeq -\frac{a_1}{4\beta} r^{(\pi/\beta)-1}.$$

The components of the field and the surface charge density near $r = 0$ all vary with distance as $r^{(\pi/\beta)-1}$. This dependence on r gives us for $\beta = 2\pi$ (the edge of a thin sheet) the singularity as $r^{-1/2}$. This is still integrable so that the charge within a finite distance from the edge is finite, but it implies that field strengths become very large at the edges of conducting sheets.

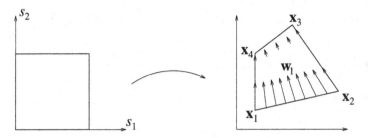

FIG. A.2. Transformation of the unit square to the quadrilateral in terms of the isoparametric coordinates s_1 and s_2, with the edge function $\mathbf{w}_1$.

To account for this boundary singularity the basis functions of the charge and the currents for the elements, of which one or more edges lie in the boundary of the 2D conductor region, will be adapted by the function $f(\mathbf{x})$ introduced in Section 2.6. Therefore, this function has the form $d^{-1/2}$. If one of the edges of the element lies in the boundary

$$d = s_i \quad \text{or} \quad d = 1 - s_i,$$

where s_i is one of the isoparametric coordinates (see Fig. A.2). If two opposite edges of the element lie in the boundary

$$d = s_i(1 - s_i).$$

Appendix B. Basis functions

In this appendix the vector valued basis functions for the current $\mathbf{J}$ on the edges of a quadrilateral element are defined. They span the function space H_h^{div}. For these basis functions some lemmas will be proven.

The mapping from a unit square with isoparametric coordinates s_1 and s_2 to the quadrilateral $\mathbf{x}_1 \ldots \mathbf{x}_4$, shown in Fig. A.2, is given by

$$\mathbf{x}(s_1, s_2) = (1 - s_2)\big[(1 - s_1)\mathbf{x}_1 + s_1\mathbf{x}_2\big] + s_2\big[(1 - s_1)\mathbf{x}_4 + s_1\mathbf{x}_3\big]. \tag{B.1}$$

For a particular element Ω_i the edge functions $\mathbf{w}_k$ are (see VAN WELIJ [1986, p. 371])

$$
\begin{aligned}
&\mathbf{w}_1 = \frac{(1 - s_2)\mathbf{v}_2}{|\mathbf{v}_1 \times \mathbf{v}_2|}, \qquad \mathbf{w}_2 = \frac{-s_1\mathbf{v}_1}{|\mathbf{v}_1 \times \mathbf{v}_2|}, \\
&\mathbf{w}_4 = \frac{(1 - s_1)\mathbf{v}_1}{|\mathbf{v}_1 \times \mathbf{v}_2|}, \qquad \mathbf{w}_3 = \frac{-s_2\mathbf{v}_2}{|\mathbf{v}_1 \times \mathbf{v}_2|}, \\
&\mathbf{v}_1 = (\mathbf{x}_2 - \mathbf{x}_1) + s_2(\mathbf{x}_1 - \mathbf{x}_2 + \mathbf{x}_3 - \mathbf{x}_4), \\
&\mathbf{v}_2 = (\mathbf{x}_4 - \mathbf{x}_1) + s_1(\mathbf{x}_1 - \mathbf{x}_2 + \mathbf{x}_3 - \mathbf{x}_4).
\end{aligned}
\tag{B.2}
$$

For these basis functions the following lemmas hold:

LEMMA B.1. *Let* $\mathbf{w}_k$, $k = 1, \ldots, 4$, *be defined as in* (B.2) *and let* $\mathbf{J}_0(s_1, s_2) = \sum_{k=1}^{4} I_k \mathbf{w}_k(s_1, s_2)$ *for certain* $I_1, \ldots, I_4$. *Then, for* $(s_1, s_2) \in [0, 1] \times [0, 1]$:

$$\sum_{k=1}^{4} I_k = 0 \implies \nabla \cdot \mathbf{J}_0 = 0.$$

PROOF. Consider the mapping (B.1) from the unit square to the quadrilateral. Let the vectors $\mathbf{v}_j$ and $\mathbf{v}_{12}$ be defined as $\mathbf{v}_j = \frac{\partial \mathbf{x}}{\partial s_j}$, and $\mathbf{v}_{12} = \frac{\partial^2 \mathbf{x}}{\partial s_1 \partial s_2}$. Then, we get the following relations:

$$\mathbf{v}_1 = (\mathbf{x}_2 - \mathbf{x}_1) + s_2 \mathbf{v}_{12},$$
$$\mathbf{v}_2 = (\mathbf{x}_4 - \mathbf{x}_1) + s_1 \mathbf{v}_{12},$$
$$\mathbf{v}_{12} = (\mathbf{x}_1 - \mathbf{x}_2 + \mathbf{x}_3 - \mathbf{x}_4).$$

In the following we will need the gradients ∇s_i ($i = 1, 2$). These gradients are the rows of the inverse of the Jacobian matrix $\{\frac{\partial \mathbf{x}}{\partial s_i}\}$, which has $\mathbf{v}_j$ ($j = 1, 2$) as columns. Hence $\mathbf{v}_j \cdot \nabla s_i = \delta_{ij}$. Let $J = (\mathbf{v}_1 \times \mathbf{v}_2) \cdot \mathbf{v}_3$, where $\mathbf{v}_3 = \frac{\mathbf{v}_1 \times \mathbf{v}_2}{|\mathbf{v}_1 \times \mathbf{v}_2|}$ is the unit normal vector to the quadrilateral, then the expressions for the gradients are:

$$\nabla s_1 = \frac{\mathbf{v}_2 \times \mathbf{v}_3}{J}, \qquad \nabla s_2 = \frac{\mathbf{v}_3 \times \mathbf{v}_1}{J}.$$

Now, let $f = I_4(1 - s_1) - I_2 s_1$ and $g = I_1(1 - s_2) - I_3 s_2$, then

$$\mathbf{J}_0 = (f\mathbf{v}_1 + g\mathbf{v}_2)/J,$$

and hence

$$\nabla \cdot \mathbf{J}_0 = \frac{1}{J}(\nabla f \cdot \mathbf{v}_1 + \nabla g \cdot \mathbf{v}_2) + \frac{1}{J^2}\{f(J\nabla \cdot \mathbf{v}_1 - \mathbf{v}_1 \cdot \nabla J)$$
$$+ g(J\nabla \cdot \mathbf{v}_2 - \mathbf{v}_2 \cdot \nabla J)\}.$$

Since $\nabla s_i \cdot \mathbf{v}_j = \delta_{ij}$,

$$\nabla f \cdot \mathbf{v}_1 = \frac{df}{ds_1}(\nabla s_1 \cdot \mathbf{v}_1) = -I_2 - I_4,$$

$$\nabla g \cdot \mathbf{v}_2 = \frac{dg}{ds_2}(\nabla s_2 \cdot \mathbf{v}_2) = -I_1 - I_3,$$

$$J\nabla \cdot \mathbf{v}_1 = J(\nabla s_2 \cdot \mathbf{v}_{12}) = (\mathbf{v}_3 \times \mathbf{v}_1) \cdot \mathbf{v}_{12} = \mathbf{v}_3 \cdot (\mathbf{v}_1 \times \mathbf{v}_{12})$$
$$= \mathbf{v}_3 \cdot ((\mathbf{x}_2 - \mathbf{x}_1) \times \mathbf{v}_{12}),$$

$$J\nabla \cdot \mathbf{v}_2 = J(\nabla s_1 \cdot \mathbf{v}_{12}) = (\mathbf{v}_2 \times \mathbf{v}_3) \cdot \mathbf{v}_{12} = \mathbf{v}_3 \cdot (\mathbf{v}_{12} \times \mathbf{v}_2)$$
$$= \mathbf{v}_3 \cdot (\mathbf{v}_{12} \times (\mathbf{x}_4 - \mathbf{x}_1)),$$

$$\mathbf{v}_1 \cdot \nabla J = (\mathbf{v}_1 \cdot \nabla s_1)((\mathbf{x}_2 - \mathbf{x}_1) \times \mathbf{v}_{12}) \cdot \mathbf{v}_3 = \mathbf{v}_3 \cdot ((\mathbf{x}_2 - \mathbf{x}_1) \times \mathbf{v}_{12}) = J\nabla \cdot \mathbf{v}_1,$$

$$\mathbf{v}_2 \cdot \nabla J = (\mathbf{v}_2 \cdot \nabla s_2)(\mathbf{v}_{12} \times (\mathbf{x}_4 - \mathbf{x}_1)) \cdot \mathbf{v}_3 = \mathbf{v}_3 \cdot (\mathbf{v}_{12} \times (\mathbf{x}_4 - \mathbf{x}_1)) = J\nabla \cdot \mathbf{v}_2,$$

so that $\nabla \cdot \mathbf{J}_0 = -(I_1 + I_2 + I_3 + I_4)/J$. $\qquad \square$

LEMMA B.2. *Let* $\mathbf{w}_1$ *for a particular element analogously be defined as in (B.2):*

$$\mathbf{w}_1 = \mp \frac{(1 - s_2)\mathbf{v}_2}{J},$$

then

$$\nabla \cdot \mathbf{w}_1 = \frac{\pm 1}{J} = \frac{\pm 1}{|\mathbf{v}_1 \times \mathbf{v}_2|}.$$

Further

$$\mathbf{w}_1 \cdot \mathbf{n} = \frac{\pm 1}{|\mathbf{x}_2 - \mathbf{x}_1|}.$$

PROOF.

$$\nabla \cdot \mathbf{w}_1 = \frac{\pm \nabla s_2 \cdot \mathbf{v}_2}{J} \mp \frac{(1 - s_2)}{J^2}(J\nabla \cdot \mathbf{v}_2 - \mathbf{v}_2 \cdot \nabla J) = \frac{\pm 1}{J},$$

since in the proof of Lemma B.1 we have shown that $\nabla s_i \cdot \mathbf{v}_j = \delta_{ij}$ and $\mathbf{v}_l \cdot \nabla J = J\nabla \cdot \mathbf{v}_l$.

Let $\mathbf{n}$ be the outward normal on edge 1 in the plane of the quadrilateral, then for $s_2 = 0$

$$\mathbf{w}_1 \cdot \mathbf{n} = \frac{\mp \mathbf{v}_2 \cdot \mathbf{n}}{((\mathbf{x}_2 - \mathbf{x}_1) \times \mathbf{v}_2) \cdot \mathbf{v}_3} = \frac{\mp \mathbf{v}_2 \cdot \mathbf{n}}{(\mathbf{v}_3 \times (\mathbf{x}_2 - \mathbf{x}_1)) \cdot \mathbf{v}_2}$$

$$= \frac{\mp \mathbf{v}_2 \cdot \mathbf{n}}{-|\mathbf{x}_2 - \mathbf{x}_1|\mathbf{v}_2 \cdot \mathbf{n}} = \frac{\pm 1}{|\mathbf{x}_2 - \mathbf{x}_1|},$$

i.e., $\mathbf{w}_1 \cdot \mathbf{n}$ depends only on the length of edge 1.   □

By similar reasoning it can be shown that the same holds for $\mathbf{w}_2$, $\mathbf{w}_3$, and $\mathbf{w}_4$.

In the following lemma it will be shown that, if $\mathbf{J}_h(\mathbf{x}) = \sum_{k=1}^{4} I_k \tilde{\mathbf{w}}_k(\mathbf{x})$, where $\tilde{\mathbf{w}}_k(\mathbf{x})$ is defined by Eq. (2.36) and $\sum_{k=1}^{4} I_k = 0$, which means that all the currents that enter an element Ω_i will leave the element, then $\nabla \cdot \mathbf{J}_h(\mathbf{x}) = 0$ will also hold for all $\mathbf{x} \in \Omega_i$.

LEMMA B.3. *If the following conditions hold:* $\nabla f(\mathbf{x}) = a\mathbf{n}$ *for some* $a \in \mathbb{R}$, $f(\mathbf{x}) \neq 0$, *the Jacobian* $J(\mathbf{x})$ *is bounded for all* $\mathbf{x} \in \Omega_i$, *and* $\mathbf{J}_h \cdot \mathbf{n} = 0$, *then*

$$\sum_{k=1}^{4} I_k = 0 \quad \Longleftrightarrow \quad \nabla \cdot \mathbf{J}_h = 0.$$

PROOF. Define $\mathbf{J}_0(\mathbf{x}) = \sum_{k=1}^{4} I_k \mathbf{w}_k(\mathbf{x})$. From Lemma B.1 it follows that $\mathbf{J}_0$ is divergence free, $\nabla \cdot \mathbf{J}_0 = 0$. Moreover, by the product rule

$$\nabla \cdot \mathbf{J}_h = \nabla \cdot (f\mathbf{J}_0) = \nabla f \cdot \mathbf{J}_0 + f\nabla \cdot \mathbf{J}_0$$

$$= \nabla f \cdot \mathbf{J}_0 + 0.$$

From $\mathbf{J}_0 \cdot \mathbf{n} = 0$ and $\nabla f(\mathbf{x}) = a\mathbf{n}$ for some $a \in \mathbb{R}$, if follows that $\nabla f \cdot \mathbf{J}_0 = 0$. Hence $\nabla \cdot \mathbf{J}_h = 0$.

Conversely, if $\nabla \cdot \mathbf{J}_h = 0$, then

$$\nabla f \cdot \mathbf{J}_0 + f \sum_{k=1}^{4} I_k \nabla \mathbf{w}_k = 0.$$

From $\mathbf{J}_0 \cdot \mathbf{n} = 0$ and $\nabla f(\mathbf{x}) = a\mathbf{n}$, it follows that $\nabla f \cdot \mathbf{J}_0 = 0$. Using Lemma B.2, it follows that

$$\frac{-f(\mathbf{x})}{J(\mathbf{x})} \sum_{k=1}^{4} I_k = 0.$$

Since $\frac{f(\mathbf{x})}{J(\mathbf{x})} \neq 0$ for all $\mathbf{x} \in \Omega_i$, it follows that $\sum_{k=1}^{4} I_k = 0$. $\qquad\square$

Appendix C. Legendre polynomials

DEFINITION C.1. By $\Pi_n[a, b]$ we denote the linear space of polynomials on $[a, b]$ of degree $\leqslant n$.

Consider the inner product $\langle \, , \rangle$ on $[a, b]$ with respect to the continuous weight function $w(x) > 0$ on (a, b):

$$\langle f, g \rangle = \int_{-1}^{1} w(x) f(x) g(x) \, \mathrm{d}x.$$

Then, one can build up a system of orthogonal polynomials by the Gram–Schmidt process:

THEOREM C.1. *Suppose one has orthogonal polynomials* $P_0, P_1, \ldots, P_{n-1}$ *of degree* $0, 1, \ldots, n-1$ *respectively, then* P_n, *constructed by (Gram–Schmidt)*

$$P_n = x^n - \frac{\langle P_0, x^n \rangle}{\langle P_0, P_0 \rangle} P_0 - \cdots - \frac{\langle P_{n-1}, x^n \rangle}{\langle P_{n-1}, P_{n-1} \rangle} P_{n-1},$$

is also orthogonal to $P_0, \ldots, P_{n-1}$. *Moreover, the orthogonal polynomials* P_n *are unique apart from a multiplicative constant.*

PROOF. From the orthogonality of $P_0, \ldots, P_{n-1}$ and Gram–Schmidt follows:

$$\langle P_k, P_n \rangle = \langle P_k, x^n \rangle - \frac{\langle P_k, x^n \rangle}{\langle P_k, P_k \rangle} \langle P_k, P_k \rangle = 0, \quad \forall k, 0 \leqslant k \leqslant n - 1,$$

so that $P_0, \ldots, P_n$ form a system of orthogonal polynomials.

Uniqueness: Let $\tilde{P}_n, P_n \in \Pi_n$, and let $P_0, \ldots, P_{n-1}$ be an orthogonal system. Suppose furthermore that $\tilde{P}_n, P_n \perp P_0, \ldots, P_{n-1}$. Then, $\tilde{P}_n = \sum_{j=0}^{n} \beta_j P_j$ implies that $\beta_j = 0$, for $j \neq n$. $\qquad\square$

THEOREM C.2. *All zeros of* P_n *are real, simple, and contained in* (a, b).

PROOF. Let $P_n \neq 0$. P_n changes sign at the distinct points $x_1, \ldots, x_k$ only, while $k < n$ and $x_1, \ldots, x_k \in (a, b)$. Then, $s(x)$, defined as

$$s(x) = w(x) P_n(x) \prod_{i=1}^{k} (x - x_i),$$

does not change sign on (a, b). Hence $\int_a^b s(x)\,dx \neq 0$.

Since $\prod_{i=1}^{k}(x - x_i) = \sum_{i=0}^{k} \beta_i P_i$ for some $\{\beta_i\}$, we also have, however,

$$\int_a^b s(x)\,dx = \sum_{i=0}^{k} \beta_i \langle P_n, P_i \rangle = 0,$$

which is a contradiction. Since $P_n \in \Pi_n$ we have $k = n$. $\qquad\square$

For the weight function $w(x) \equiv 1$ and $[a, b] = [-1, 1]$ the polynomials are called *Legendre polynomials*. These polynomials are uniquely determined by a multiplicative constant such that the leading coefficient is 1. All zeros of P_i ($i = 1, 2, \ldots$) are simple, real, and contained within $(-1, 1)$. Moreover, if $\tilde{x}$ is a zero of P_i, then $-\tilde{x}$ is also a zero.

THEOREM C.3. *A set of orthonormal polynomials $P_n^*(x)$, i.e., with leading coefficient 1, satisfy a three-term recurrence relationship*

$$P_n^*(x) = (a_n x + b_n) P_{n-1}^*(x) - c_n P_{n-2}^*(x), \quad n = 2, 3, \ldots.$$

PROOF. See DAVIS and RABINOWITZ [1961, pp. 167–168, 234–255]. $\qquad\square$

The Legendre polynomials (see Fig. C.1), P_n, can be defined by the three-term recursion

$$P_0(x) = 1,$$

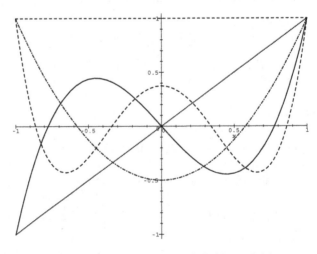

FIG. C.1. The Legendre polynomials $P_0(x), \ldots, P_4(x)$.

$$P_1(x) = x,$$
$$n P_n(x) = (2n - 1)x P_{n-1}(x) - (n - 1)P_{n-2}(x).$$

Appendix D. Inner products

We define the following inner products:

$$\mathbf{A} \cdot \mathbf{B} = \underbrace{\sum_{i_1=1}^{n} \cdots \sum_{i_d=1}^{n}}_{d} A_{i_1 \ldots i_d} B_{i_1 \ldots i_d}, \quad \text{where} \quad \begin{cases} d = 1 & \text{for vectors,} \\ d = 2 & \text{for matrices,} \\ d = 3 & \text{for tensors,} \\ \vdots \end{cases}$$

Let $\mathbf{x}_1 \otimes \cdots \otimes \mathbf{x}_\alpha$ be defined as

$$\mathbf{x}_1 \otimes \cdots \otimes \mathbf{x}_\alpha = \left\{ (x_1)_{i_1} (x_2)_{i_2} \cdots (x_\alpha)_{i_\alpha} \right\}_{i_1, i_2, \ldots, i_\alpha},$$

so that, for example, in the two-dimensional case for $\mathbf{x} = \mathbf{x}_1$, $\mathbf{y} = \mathbf{x}_2$, and $i_1, i_2 = 1, \ldots, n$

$$\mathbf{x} \otimes \mathbf{y} = \begin{pmatrix} x_1 \\ \vdots \\ x_n \end{pmatrix} (y_1 \ldots y_n) = \begin{pmatrix} x_1 y_1 & \cdots & x_1 y_n \\ \vdots & \ddots & \vdots \\ x_n y_1 & \cdots & x_n y_n \end{pmatrix}.$$

LEMMA D.1. *If* $\mathbf{x}$ *and* $\mathbf{y}$ *are vectors of dimension* n *and* $\mathbf{A}$ *is a* $(n \times n)$-*matrix, then*

$$(\mathbf{A}\mathbf{y})^T \mathbf{x} = \mathbf{A} \cdot (\mathbf{x} \otimes \mathbf{y}).$$

PROOF.

$$\mathbf{A}\mathbf{y} = \begin{pmatrix} A_{11} y_1 + \cdots + A_{1n} y_n \\ A_{21} y_1 + \cdots + A_{2n} y_n \\ \vdots \\ A_{n1} y_1 + \cdots + A_{nn} y_n \end{pmatrix},$$

$$(\mathbf{A}\mathbf{y})^T \mathbf{x} = \mathbf{x}^T (\mathbf{A}\mathbf{y}) = (x_1 A_{11} y_1 + \cdots + x_1 A_{1n} y_n + x_2 A_{21} y_1 + \cdots + x_n A_{nn} y_n)$$

$$= \sum_{i=1}^{n} \sum_{j=1}^{n} A_{ij} x_i y_j$$

$$= \mathbf{A} \cdot (\mathbf{x} \otimes \mathbf{y}). \qquad \square$$

We have, analogously to the lemma, for tensors $\overline{\mathbf{T}}$

$$\overline{\mathbf{T}} \cdot (\mathbf{x} \otimes \mathbf{y} \otimes \mathbf{z}) = \left((\overline{\mathbf{T}}\mathbf{z})\mathbf{y} \right)^T \mathbf{x}$$

where $\overline{\mathbf{T}}\mathbf{z} = \{ T_{ij1} z_1 + T_{ij2} z_2 + \cdots + T_{ijn} z_n \}_{i \times j}$. This follows directly from the lemma.

In general, the following theorem holds.

THEOREM D.1. *Let* $\mathbf{M}_\alpha$ *be a* $\underbrace{n \times \cdots \times n}_{\alpha}$ *-Tensor and let* $\mathbf{x}_1, \ldots, \mathbf{x}_\alpha$ *be vectors of dimension n. Then,*

$$\left(\ldots (\mathbf{M}_\alpha \mathbf{x}_\alpha) \ldots \mathbf{x}_3) \mathbf{x}_2\right)^T \mathbf{x}_1 = \mathbf{M}_\alpha \cdot (\mathbf{x}_1 \otimes \cdots \otimes \mathbf{x}_\alpha).$$

PROOF. Analogously to the lemma.   □

References

AUBIN, J.P. (1972). *Approximation of Elliptic Boundary Value Problems* (Wiley Interscience, New York).

BARRETT, R., et al. (1994). *Templates for the Solution of Linear Systems: Building Blocks for Iterative Methods*, second ed. (SIAM, Philadelphia, PA).

COLTON, D., KRESS, R. (1992). *Integral Equation Methods in Scattering Theory* (Krieger Publishing Company, Malabar, FL).

DAVIS, P.J., RABINOWITZ, P. (1961). Some geometrical theorems for abcissas and weights of Gauss type. *J. Math. Anal. Appl.* **2**, 167–168, 234–255, 428–437.

DAVIS, P.J., RABINOWITZ, P. (1984). *Methods of Numerical Integration*, second ed. (Academic Press, New York).

DU CLOUX, R., MAAS, G.P.J.F.M., WACHTERS, A.J.H. (1994). Quasi-static boundary element method for electromagnetic simulation of PCB's. *Philips J. Res.* **48**, 117–144.

GOLUB, G.H., VAN LOAN, C.F. (1986). *Matrix Computations* (North Oxford Academic Publishers Ltd, London).

JACKSON, J.D. (1975). *Classical Electrodynamics, vol. I* (John Wiley & Sons, New York).

KRONROD, A.S. (1965). *Nodes and Weights of Quadrature Formulas* (Consultants Bureau, New York), English transl. from Russian. MR 32:598.

MONEGATO, G. (1976a). A note on extended Gaussian quadrature rules. *Math. Comp.* **30**, 812–817.

MONEGATO, G. (1976b). Positivity of the weights of extended Gauss–Legendre quadrature rules. *Math. Comp.* **32**, 847–856.

ORDEN, A. (1964). Stationary points of quadratic functions under linear constraints. *Comput. J.* **7**, 238–242.

PARLETT, B.N., REID, J.K. (1981). Tracking the progress of the Lanczos algorithm for large symmetric eigenproblems. *IMA J. Numer. Anal.* **1**, 135–155.

PATTERSON, T.N.L. (1968). The optimum addition of points to quadrature formulae. *Math. Comp.* **22**, 847–856.

RAMO, S., WHINNERY, J.R., VAN DUZER, T. (1984). *Fields and Waves in Communication Electronics* (John Wiley and Sons, New York).

SZEGÖ, G. (1934). Über gewisse orthogonale Polynome, die zu einer oszilierenden Belegungsfunktion gehören. *Math. Ann.* **110**, 501–513.

VAN DER VORST, H.A. (1982). A generalized Lanczos scheme. *Math. Comp.* **39**, 559–561.

VAN WELIJ, J.S. (1986). Basis functions matching tangential components on element edges. In: *Proc. SISDEP-2 Conf., Swansea, UK*.

Reduced-Order Modeling

Zhaojun Bai

University of California, Department of Computer Science, One Shields Avenue,
3023 Engineering II, Davis, CA, USA
E-mail address: bai@cs.ucdavis.edu

Patrick M. Dewilde

TU Delft, Fac. Eletrotechniek Vakgroep CAS (Room L2.500), Mekelweg 4,
2628 CD, Delft, The Netherlands
E-mail address: p.dewile@dimes.tudelft.nl

Roland W. Freund

University of California, Davis, Department of Mathematics, One Shields Avenue,
Davis, CA 95616, USA
E-mail address: freund@math.ucdavis.edu

Abstract

In recent years, reduced-order modeling techniques have proven to be powerful tools for various problems in circuit simulation. For example, today, reduction techniques are routinely used to replace the large RCL subcircuits that model the interconnect or the pin package of VLSI circuits by models of much smaller dimension. In this chapter, we review the reduced-order modeling techniques that are most widely employed in VLSI circuit simulation.

1. Introduction to the problem of model reduction

Roughly speaking, the problem of model reduction is to replace a given mathematical model of a system or process by a model that is much "smaller" than the original model, but still describes – at least approximately – certain aspects of the system or process. Clearly, model reduction involves a number of interesting issues. First and foremost

Essential Numerical Methods in Electromagnetics
Special Volume (W.H.A. Schilders and E.J.W. ter Maten, Guest Editors) of
HANDBOOK OF NUMERICAL ANALYSIS, VOL. XIII
P.G. Ciarlet (Editor)
ISSN 1570-8659
DOI 10.1016/B978-0-444-53756-0.00004-5

is the issue of selecting appropriate approximation schemes that allow the definition of suitable reduced-order models. In addition, it is often important that the reduced-order model preserves certain crucial properties of the original system, such as stability or passivity. Other issues include the characterization of the quality of the models, the extraction of the data from the original model that is needed to actually generate the reduced-order models, and the efficient and numerically stable computation of the models.

In this paper, we discuss reduced-order modeling techniques for large-scale linear dynamical systems, especially those that arise in the simulation of electronic circuits and of microelectromechanical systems.

We begin with a brief description of reduced-order modeling problems in circuit simulation. Electronic circuits are usually modeled as networks whose branches correspond to the circuit elements and whose nodes correspond to the interconnections of the circuit elements. Such networks are characterized by three types of equations. The *Kirchhoff's current law* (KCL) states that, for each node of the network, the currents flowing in and out of that node sum up to zero. The *Kirchhoff's voltage law* (KVL) states that, for each closed loop of the network, the voltage drops along that loop sum up to zero. The *branch constitutive relations* (BCRs) are equations that characterize the actual circuit elements. For example, the BCR of a linear resistor is Ohm's law. The BCRs are linear equations for simple devices, such as linear resistors, capacitors, and inductors, and they are nonlinear equations for more complex devices, such as diodes and transistors. Furthermore, in general, the BCRs involve time-derivatives of the unknowns, and thus they are ordinary differential equations. On the other hand, the KCLs and KVLs are linear algebraic equations that only depend on the topology of the circuit.

The KCLs, KVLs, and BCRs can be summarized as a system of first-order, in general nonlinear, *differential-algebraic equations* (DAEs) of the form

$$\frac{\mathrm{d}}{\mathrm{d}t}q(\hat{x},t) + f(\hat{x},t) = 0, \tag{1.1}$$

together with suitable initial conditions. Here, $\hat{x} = \hat{x}(t)$ is the unknown vector of circuit variables at time t, the vector-valued function $f(\hat{x},t)$ represents the contributions of nonreactive elements such as resistors, sources, etc., and the vector-valued function $\frac{\mathrm{d}}{\mathrm{d}t}q(\hat{x},t)$ represents the contributions of reactive elements such as capacitors and inductors. There are a number of established methods, such as sparse tableau, nodal formulation, modified nodal analysis, etc. (see VLACH and SINGHAL [1994]), for generating a system of equations of the form (1.1) from a so-called *netlist* description of a given circuit. The vector functions $\hat{x}$, f, q, as well as their dimension, depend on the chosen formulation method. The most general method is sparse tableau, which consists of just listing all the KCLs, KVLs, and BCRs. The other formulation methods can be interpreted as starting from sparse tableau and eliminating some of the unknowns by using some of the KCL or KVL equations.

For all the standard formulation methods, the dimension of the system (1.1) is of the order of the number of elements in the circuit. Since today's VLSI circuits can have up to hundreds of millions of circuit elements, systems (1.1) describing such circuits can be of extremely large dimension. Reduced-order modeling allows to first replace large systems of the form (1.1) by systems of smaller dimension and then tackle these smaller systems by suitable DAE solvers. Ideally, one would like to apply nonlinear reduced-order modeling directly to the nonlinear system (1.1). However, since nonlinear reduction techniques are a lot less developed and less well-understood than linear ones, today, almost always linear reduced-order modeling is employed. To this end, one either linearizes the system (1.1) or decouples (1.1) into nonlinear and linear subsystems; see, e.g., FREUND [1999b] and the references given there.

For example, the first case arises in *small-signal analysis*; see, e.g., FREUND and FELDMANN [1996b]. Given a *DC operating point*, say $\hat{x}_0$, of the circuit described by (1.1), one linearizes the system (1.1) around $\hat{x}_0$. The resulting linearized version of (1.1) is of the following form:

$$E\frac{dx}{dt} = Ax + Bu(t), \tag{1.2}$$

$$y(t) = C^T x(t). \tag{1.3}$$

Here, $A = D_x f$ and $E = D_x q$ are the Jacobian matrices of f and q, respectively, at the DC operating point $\hat{x}_0$, $x(t) = \hat{x}(t) - \hat{x}_0$, $u(t)$ is the vector of excitations applied to the sources of the circuit, and $y(t)$ is the vector of circuit variables of interest. Eqs. (1.2) and (1.3) represent a *time-invariant linear dynamical system*. Its *state-space dimension*, N, is the length of the vector x of circuit variables. For a circuit with many elements, the system (1.2) and (1.3) is thus of very high dimension. The idea of reduced-order modeling is then to replace the original system (1.2) and (1.3) by one the same form,

$$E_n\frac{dz}{dt} = A_n z + B_n u(t),$$

$$y(t) = C_n^T z(t),$$

but of much smaller state-space dimension $n \ll N$.

Time-invariant linear dynamical systems of the form (1.2) and (1.3) also arise when equations describing linear subcircuits of a given circuit are decoupled from the system (1.1) that characterizes the whole circuit; see, e.g., [FREUND, 1999b]. For example, the interconnect or the pin package of VLSI circuits are often modeled as large linear RCL networks. Such linear subcircuits are described by systems of the form (1.2) and (1.3), where $x(t)$ is the vector of circuit variables associated with the subcircuit, and the vectors $u(t)$ and $y(t)$ contain the variables of the connections of the subcircuit to the, in general nonlinear, remainder of the whole circuit. By replacing, in the nonlinear system (1.1), the linear subsystem (1.2) and (1.3) by a reduced-order model of much smaller state-space dimension, the dimension of (1.1) can be reduced significantly before a DAE solver is then applied to such a smaller version of (1.1).

The remainder of this paper is organized as follows. In Section 2, we review some basic facts about time-invariant linear dynamical systems. In Section 3, we discuss reduced-order modeling of linear dynamical systems via Krylov-subspace techniques. In Section 4, we describe the use of Schur interpolation for various reduced-order modeling problems. In Section 5, we discuss Hankel-norm model reduction. Sections 6 and 7 are concerned with reduced-order modeling of second-order and semisecond-order dynamical systems. Finally, in Section 8, we make some concluding remarks.

2. Time-invariant linear dynamical systems

In this section, we review some basic facts about time-invariant linear dynamical systems and introduce reduced-order models defined by Padé or Padé-type approximants. We also discuss stability and passivity of linear dynamical systems.

2.1. State-space description

We consider m-input p-output time-invariant linear dynamical systems given by a *state-space description* of the form

$$E\frac{dx}{dt} = Ax + Bu(t), \tag{2.1}$$

$$y(t) = C^T x(t) + Du(t), \tag{2.2}$$

together with suitable initial conditions. Here, $A, E \in \mathbb{R}^{N \times N}$, $B \in \mathbb{R}^{N \times m}$, $C \in \mathbb{R}^{N \times p}$, and $D \in \mathbb{R}^{p \times m}$ are given matrices, $x(t) \in \mathbb{R}^N$ is the vector of state variables, $u(t) \in \mathbb{R}^m$ is the vector of inputs, $y(t) \in \mathbb{R}^p$ the vector of outputs, N is the state-space dimension, and m and p are the number of inputs and outputs, respectively. Note that systems of the form (1.2) and (1.3) are just a special case of (2.1) and (2.2) with $D = 0$.

The linear system (2.1) and (2.2) is called *regular* if the matrix E in (2.1) is nonsingular, and it is called *singular* or a *descriptor system* if E is singular. Note that, in the regular case, the linear system (2.1) and (2.2) can always be re-written as

$$\frac{dx}{dt} = (E^{-1}A)x + (E^{-1}B)u(t),$$

$$y(t) = C^T x(t) + Du(t),$$

which is just a system (2.1) and (2.2) with $E = I$.

The linear dynamical systems arising in circuit simulation are descriptor systems in general. Therefore, in the following, we allow $E \in \mathbb{R}^{N \times N}$ to be a general, possibly singular, matrix. The only assumption on the matrices $A, E \in \mathbb{R}^{N \times N}$ in (2.1) is that the matrix pencil $A - sE$ is *regular*, i.e., the matrix $A - sE$ is singular for only finitely many values of $s \in \mathbb{C}$.

In the case of singular E, Eq. (2.1) represents a system of DAEs. Solving DAEs is significantly more complex than solving systems of ordinary differential equations (ODEs). Moreover, there are constraints on the possible initial conditions that can be imposed on the solutions of (2.1). For a detailed discussion of DAEs and the structure of their solutions, we refer the reader to CAMPBELL [1980], CAMPBELL [1982], DAI [1989], VERGHESE, LÉVY, and KAILATH [1981]. Here, we only present a brief glimpse of the issues arising in DAEs.

We start by bringing the matrices A and E in (2.1) to a certain normal form. For any regular pencil $A - sE$, there exist nonsingular matrices P and Q such that

$$P(A - sE)Q = \begin{bmatrix} A^{(1)} - sI & 0 \\ 0 & I - sJ \end{bmatrix}, \tag{2.3}$$

where the submatrix J is nilpotent. The matrix pencil on the right-hand side of (2.3) is called the *Weierstrass form* of $A - sE$. Assuming that the matrices A and E in (2.1) are already in Weierstrass form, the system (2.1) can be decoupled as follows:

$$\frac{dx^{(1)}}{dt} = A^{(1)}x^{(1)} + B^{(1)}u(t), \tag{2.4}$$

$$J\frac{dx^{(2)}}{dt} = x^{(2)} + B^{(2)}u(t). \tag{2.5}$$

The first subsystem, (2.4), is just a system of ODEs. Thus for any given initial condition $x^{(1)}(0) = \hat{x}^{(1)}$, there exits a unique solution of (2.4). Moreover, the so-called *free-response* of (2.4), i.e., the solutions $x(t)$ for $t \geqslant 0$ when $u \equiv 0$, consists of combinations of exponential modes at the eigenvalues of the matrix $A^{(1)}$. Note that, in view of (2.3), the eigenvalues of $A^{(1)}$ are just the finite eigenvalues of the pencil $A - sE$. The solutions of the second subsystem, (2.5), however, are of quite different nature. In particular, the free-response of (2.5) consists of $k_i - 1$ independent impulsive motions for each $k_i \times k_i$ Jordan block of the matrix J; see VERGHESE, LÉVY, and KAILATH [1981].

For example, consider the case that the nilpotent matrix J in (2.5) is a single $k \times k$ Jordan block, i.e.,

$$J = \begin{bmatrix} 0 & 1 & 0 & \cdots & 0 \\ 0 & 0 & 1 & \ddots & \vdots \\ \vdots & \ddots & \ddots & \ddots & 0 \\ \vdots & \ddots & \ddots & \ddots & 1 \\ 0 & \cdots & \cdots & 0 & 0 \end{bmatrix} \in \mathbb{R}^{k \times k}.$$

The k components of the free-response $x^{(2)}(t)$ of (2.5) are then given by

$$x_1^{(2)}(t) = -x_2^{(2)}(0-)\delta(t) - x_3^{(2)}(0-)\delta^{(1)}(t) - \cdots - x_k^{(2)}(0-)\delta^{(k-2)}(t),$$

$$x_2^{(2)}(t) = -x_3^{(2)}(0-)\delta(t) - x_4^{(2)}(0-)\delta^{(1)}(t) - \cdots - x_k^{(2)}(0-)\delta^{(k-3)}(t),$$

$$\vdots \quad = \quad \vdots$$

$$x_{k-1}^{(2)}(t) = -x_k^{(2)}(0-)\delta(t),$$

$$x_k^{(2)}(t) = 0.$$

Here, $\delta(t)$ is the delta function and $\delta^{(i)}(t)$ is its ith derivative. Moreover, $x_i^{(2)}(0-)$, $i = 2, 3, \ldots, k$, are the components of the initial conditions that can be imposed on (2.4). Note that there are only $k - 1$ degrees of freedom for the initial condition and that it is not possible to prescribe $x_1^{(2)}(0-)$. In particular, the free-response of (2.5) corresponding to an 1×1 Jordan blocks of J is just the zero solution, and there is no degree of freedom for the selection of an initial value corresponding to that block.

Finally, we remark that, in view of (2.3), the eigenvalues of the matrix pencil $A - sE$ corresponding to the subsystem (2.5) are just the infinite eigenvalues of $A - sE$.

2.2. Reduced-order models and transfer functions

The basic idea of reduced-order modeling is to replace a given system by a system of
the same type, but with smaller state-space dimension. Thus, a *reduced-order model* of
state-space dimension n of a given linear dynamical system (2.1) and (2.2) of dimension
N is a system of the form

$$E_n \frac{dz}{dt} = A_n z + B_n u(t), \tag{2.6}$$

$$y(t) = C_n^T z(t) + D_n u(t), \tag{2.7}$$

where $A_n, E_n \in \mathbb{R}^{n\times n}$, $B_n \in \mathbb{R}^{n\times m}$, $C_n \in \mathbb{R}^{n\times p}$, $D_n \in \mathbb{R}^{p\times m}$, and $n < N$.

The challenge then is to choose the matrices A_n, E_n, B_n, C_n, and D_n in (2.6) and 2.7)
such that the reduced-order model in some sense approximates the original system. One
possible measure of the approximation quality of a reduced-order model is based on the
concept of transfer function.

If we assume zero initial conditions, then by applying the Laplace transform to the
original system (2.1) and (2.2), we obtain the following algebraic equations:

$$sEX(s) = AX(s) + BU(s),$$

$$Y(s) = C^T X(s) + DU(s).$$

Here, the frequency-domain variables $X(s)$, $U(s)$, and $Y(s)$ are the Laplace transforms
of the time-domain variables of $x(t)$, $u(t)$, and $y(t)$, respectively. Note that $s \in \mathbb{C}$. Then,
formally eliminating $X(s)$ in the above equations, we arrive at the frequency-domain
input–output relation $Y(s) = H(s)U(s)$. Here,

$$H(s) := D + C^T(sE - A)^{-1}B, \quad s \in \mathbb{C}, \tag{2.8}$$

is the so-called *transfer function* of the system (2.1) and (2.2). Note that

$$H : \mathbb{C} \mapsto (\mathbb{C}\cup\infty)^{p\times m}, \tag{2.9}$$

is an $p \times m$-matrix-valued rational function.

Similarly, the transfer function, H_n, of the reduced-order model (2.6) and (2.7) is
given by

$$H_n(s) := D_n + C_n^T(sE_n - A_n)^{-1}B_n, \quad s \in \mathbb{C}. \tag{2.10}$$

Note that H_n is a also an $p \times m$-matrix-valued rational function.

2.3. Padé and Padé-type models

The concept of transfer functions allows to define reduced-order models by means of
Padé or Padé-type approximation.

Let $s_0 \in \mathbb{C}$ be any point such that s_0 is not a pole of the transfer function H given
by (2.8). In practice, the point s_0 is chosen such that it is in some sense close to the
frequency range of interest. We remark that the frequency range of interest is usually
a subset of the imaginary axis in the complex s-plane. Since s_0 is not a pole of H, the

function H admits the Taylor expansion

$$H(s) = M_0 + M_1(s - s_0) + M_2(s - s_0)^2 + \cdots + M_j(s - s_0)^j + \cdots \qquad (2.11)$$

about s_0. The coefficients M_j, $j = 0, 1, \ldots$, in (2.11) are called the *moments* of H about the expansion point s_0. Note that the M_j's are $p \times m$ matrices.

A reduced-order model (2.6) and (2.7) of state-space dimension n is called an *nth Padé model* (at the expansion point s_0) of the original system (2.1) and (2.2) if the Taylor expansions about s_0 of the transfer functions H and H_n of the original system and the reduced-order system agree in as many leading terms as possible, i.e.,

$$H(s) = H_n(s) + \mathcal{O}((s - s_0)^{q(n)}), \qquad (2.12)$$

where $q(n)$ is as large as possible. In FELDMANN and FREUND [1995b], FREUND [1995], it was shown that

$$q(n) \geqslant \left\lfloor \frac{n}{m} \right\rfloor + \left\lfloor \frac{n}{p} \right\rfloor,$$

with equality in the "generic" case. The meaning of "generic" will be described more precisely in Section 3.2.

Even though Padé models are defined via the local approximation property (2.12), in practice, they usually are excellent approximations over large frequency ranges. The following single-input single-output example illustrates this statement. The example is a circuit resulting from the so-called PEEC discretization RUEHLI [1974] of an electromagnetic problem. The circuit is an RCL network consisting of 2100 capacitors, 172 inductors, 6990 inductive couplings, and a single resistive source that drives the circuit. Modified nodal analysis is used to set up the circuit equations, resulting in a linear dynamical system of dimension $N = 306$. In turns out that a Padé model of dimension $n = 60$ is sufficient to produces an almost exact transfer function in the relevant frequency range $s = 2\pi \, \mathrm{i}\omega$, $0 \leqslant \omega \leqslant 5 \times 10^9$. The corresponding curves for $|H(s)|$ and $|H_{60}(s)|$ are shown in Fig. 2.1.

It is very tempting to compute Padé models directly via the definition (2.12). More precisely, one would first explicitly generate the $q(n)$ moments $M_0, M_1, \ldots, M_{q(n)-1}$, and then compute H_n and the system matrices in the reduced-order model (2.6) and (2.7) from these moments. However, computing Padé models directly from the moments is extremely ill-conditioned, and consequently, such a procedure is not viable; we refer the reader to FELDMANN and FREUND [1994], FELDMANN and FREUND [1995a] for a detailed discussion and numerical examples.

The preferred way to compute Padé models is to use Krylov-subspace techniques, such as a suitable Lanczos-type process, as we will describe in Section 3. This becomes possible after the transfer function (2.8) is rewritten in terms of a single matrix M, instead of the two matrices A and E. To this end, let

$$A - s_0 E = F_1 F_2, \quad \text{where } F_1, F_2 \in \mathbb{C}^{N \times N}, \qquad (2.13)$$

be any factorization of $A - s_0 E$. For example, the matrices $A - s_0 E$ arising in circuit simulation are large, but sparse, and are such that a sparse LU factorization is feasible.

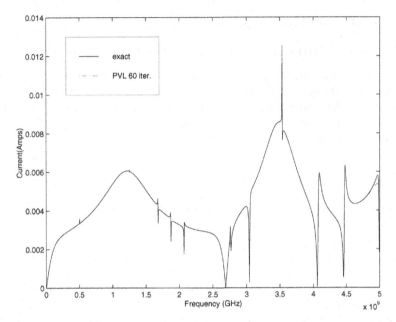

FIG. 2.1. The PEEC transfer function, exact and Padé model of dimension $n = 60$.

In this case, the matrices F_1 and F_2 in (2.13) are the lower and upper triangular factors, possibly with rows and columns permuted due to pivoting, of such a sparse LU factorization of $A - s_0 E$. Using (2.13), the transfer function (2.8) can be rewritten as follows:

$$H(s) = D + C^T(sE - A)^{-1}B \qquad (2.14)$$
$$= D - C^T(A - s_0 E - (s - s_0)E)^{-1}B$$
$$= D - L^T(I - (s - s_0)M)^{-1}R,$$

where

$$M := F_1^{-1}E F_2^{-1}, \quad R := F_1^{-1}B, \quad \text{and} \quad L := F_2^{-T}C. \qquad \cdot \qquad (2.15)$$

Note that (2.14) only involves one $N \times N$ matrix, namely M, instead of the two $N \times N$ matrices A and E in (2.8). This allows to apply Krylov-subspace methods to the single matrix M, with the $N \times m$ matrix R and the $N \times p$ matrix L as blocks of right and left starting vectors.

While Padé models often provide very good approximations in frequency domain, they also have undesirable properties. In particular, in general, Padé models do not preserve stability or passivity of the original system. However, by relaxing the Padé-approximation property (2.12), it is often possibly to obtain stable or passive models. More precisely, we call a reduced-order model (2.6) and (2.7) of state-space dimension n an *nth Padé-type model* (at the expansion point s_0) of the original system (2.1) and (2.2) if the Taylor expansions about s_0 of the transfer functions H and H_n of the original system and the reduced-order system agree in a number of leading terms, i.e.,

$$H(s) = H_n(s) + \mathcal{O}((s - s_0)^{q'}), \tag{2.16}$$

where $1 \leqslant q' < q(n)$.

2.4. Stability

An important property of linear dynamical systems is stability. An actual physical system needs to be stable in order to function properly. If a linear dynamical system (2.1) and (2.2) is used as a description of such a physical system, then clearly, it should also be stable. Moreover, when (2.1) and (2.2) is replaced by a reduced-order model that is then used in a time-domain analysis, the reduced-order model also needs to be stable.

In this subsection, we present a brief discussion of stability of linear descriptor systems. For a more general survey of the various concepts of stability of dynamical systems, we refer the reader to ANDERSON and VONGPANITLERD [1973], WILLEMS [1970].

A descriptor system of the form (2.1) and (2.2) is said to be *stable* if its free-response, i.e., the solutions $x(t)$, $t \geqslant 0$, of

$$E\frac{\mathrm{d}x}{\mathrm{d}t} = Ax, \qquad x(0) = x_0,$$

remain bounded as $t \to \infty$ for any possible initial vector x_0. Recall from the discussion in Section 2.1 that for singular E, there are certain restrictions on the possible initial vectors x_0.

Stability can easily be characterized in terms of the finite eigenvalues of the matrix pencil $A - sE$; see, e.g., MASUBUCHI, KAMITANE, OHARA, and SUDA [1997]. More precisely, we have the following theorem.

THEOREM 2.1. *The descriptor system (2.1) and (2.2) is stable if, and only if, the following two conditions are satisfied*:
  (i) *All finite eigenvalues $\lambda \in \mathbb{C}$ of the matrix pencil $A - sE$ satisfy* $\mathrm{Re}\,\lambda \leqslant 0$;
  (ii) *All finite eigenvalues λ of the matrix pencil $A - sE$ with* $\mathrm{Re}\,\lambda = 0$ *are simple.*

We stress that, in view of Theorem 2.1, the infinite eigenvalues of the matrix pencil $A - sE$ have no effect on stability. The reason is that these infinite eigenvalues result only in impulsive motions, which go to zero as $t \to \infty$.

Recall that the transfer function H of the descriptor system (2.1) and (2.2) is of the form

$$H(s) = D + C^{\mathrm{T}}(sE - A)^{-1}B, \quad \text{where} \tag{2.17}$$

$$A, E \in \mathbb{R}^{N \times N}, \quad B \in \mathbb{R}^{N \times m}, \quad C \in \mathbb{R}^{N \times m}, \quad \text{and} \quad D \in \mathbb{R}^{p \times m}. \tag{2.18}$$

Note that any pole of H is necessarily an eigenvalue of the matrix pencil $A - sE$. Hence, it is tempting to determine stability via the poles of H. However, in general, not all eigenvalues of $A - sE$ are poles of H. For example, consider the following system

$$\frac{dx}{dt} = \begin{bmatrix} 1 & 0 \\ 0 & -1 \end{bmatrix} x + \begin{bmatrix} 0 \\ 1 \end{bmatrix} u(t),$$

$$y(t) = [\,1 \quad 1\,]x(t),$$

which is taken from p. 128 of ANDERSON and VONGPANITLERD [1973]. The pencil associated with this system is

$$A - sI = \begin{bmatrix} 1-s & 0 \\ 0 & -1-s \end{bmatrix}.$$

Its eigenvalues are ± 1, and hence this system is unstable. The transfer function $H(s) = 1/(s+1)$, however, only has the "stable" pole -1. Therefore, checking conditions (i) and (ii) of Theorem 2.1 only for the poles of H is, in general, not enough to guarantee stability. In order to infer stability of the system (2.1) and (2.2) from the poles of its transfer function, one needs an additional condition, which we formulate next.

Let H be a given $m \times p$-matrix-valued rational function. Any representation of H of the form (2.17) with matrices (2.18) is called a *realization* of H. Furthermore, a realization (2.17) of H is said to be *minimal* if the dimension N of the matrices (2.18) is as small as possible. We will also say that the state-space description (2.1) and (2.2) is a minimal realization if its transfer function (2.18) is a minimal realization.

The following theorem is the well-known characterization of minimal realizations in terms of conditions on the matrices (2.18); see, e.g., VERGHESE, LÉVY, and KAILATH [1981]. We also refer the reader to the related results on controllability, observability, and minimal realizations of descriptor systems given in Chapter 2 of DAI [1989].

THEOREM 2.2. *Let H be a $m \times p$-matrix-valued rational function given by a realization (2.17). Then, (2.17) is a minimal realization of H if, and only if, the matrices (2.18) satisfy the following five conditions*:

(i) $\operatorname{rank}[\,A - sE \quad B\,] = N$ *for all $s \in \mathbb{C}$*;
     (Finite controllability)

(ii) $\operatorname{rank}[\,E \quad B\,] = N$;
      (Infinite controllability)

(iii) $\operatorname{rank}[\,A^T - sE^T \quad C\,] = N$ *for all $s \in \mathbb{C}$*;
       (Finite observability)

(iv) $\operatorname{rank}[\,E^T \quad C\,] = N$;
      (Infinite observability)

(v) $A\ker(E) \subseteq \operatorname{Im}(E)$.
     (Absence of nondynamic modes)

For descriptor systems given by a minimal realization, stability can indeed be checked via the poles of its transfer function.

THEOREM 2.3. *Let (2.1) and (2.2) be a minimal realization of a descriptor system, and let H be its transfer function (2.17). Then, the descriptor system (2.1) and (2.2) is stable* *if, and only if, all finite poles s_i of H satisfy* $\operatorname{Re} s_i \leqslant 0$ *and any pole with* $\operatorname{Re} s_i = 0$ *is simple.*

2.5. Passivity

In circuit simulation, reduced-order modeling is often applied to large passive linear subcircuits, such as RCL networks consisting of only resistors, inductors, and capacitors. When reduced-order models of such subcircuits are used within a simulation of the whole circuit, stability of the overall simulation can only be guaranteed if the reduced-order models preserve the passivity of the original subcircuits; see, e.g., CHIRLIAN [1967], ROHRER and NOSRATI [1981]. Therefore, it is important to have techniques to check passivity of a given reduced-order model.

Roughly speaking, a system is *passive* if it does not generate energy. For descriptor systems of the form (2.1) and (2.2), passivity is equivalent to positive realness of the transfer function. Moreover, such systems can only be passive if they have identical numbers of inputs and outputs. Thus, for the remainder of this subsection, we assume that $m = p$. Then, a system described by (2.1) and (2.2) is passive, i.e., it does not generate energy, if, and only if, its transfer function (2.17) is *positive real*; see, e.g., ANDERSON and VONGPANITLERD [1973]. A precise definition of positive realness is as follows.

DEFINITION 2.1. An $m \times m$-matrix-valued function $H : \mathbb{C} \mapsto (\mathbb{C} \cup \infty)^{m \times m}$ is called *positive real* if the following three conditions are satisfied:
  (i) H is analytic in $\mathbb{C}_+ := \{ s \in \mathbb{C} \mid \operatorname{Re} s > 0 \}$;
  (ii) $H(\bar{s}) = \overline{H(s)}$ for all $s \in \mathbb{C}$;
  (iii) $H(s) + (H(s))^H \succeq 0$ for all $s \in \mathbb{C}_+$.

In 2.1 and in the sequel, the notation $M \succeq 0$ means that the matrix M is Hermitian positive semidefinite. Similarly, $M \preceq 0$ means that M is Hermitian negative semidefinite.

For transfer functions H of the form (2.17), condition (ii) of 2.1 is always satisfied since the matrices (2.18) are assumed to be real. Furthermore, condition (i) simply means that H cannot have poles in $\mathbb{C}_+$, and this can be checked easily. For the special case $m = 1$ of scalar-valued functions H, condition (iii) states that the real part of $H(s)$ is nonnegative for all s with nonnegative real part. In order to check this condition, it is sufficient to show that the real part of $H(s)$ is nonnegative for all purely imaginary s. This can be done by means of relatively elementary means. For example, in BAI and FREUND [2000], a procedure based on eigenvalue computations is proposed. For the general matrix-valued case, $m \geqslant 1$, however, checking condition (iii) is much more involved. One possibility is to employ a suitable extension of the classical positive real lemma (see, e.g. ANDERSON [1967], Chapter 5 of ANDERSON and VONGPANITLERD [1973], or Section 13.5 of ZHOU, DOYLE, and GLOVER [1996]) that characterizes positive realness of regular linear systems via the solvability of certain linear matrix inequalities (LMIs). Such a version of the positive real lemma for general descriptor systems is stated in Theorem 2.4 below.

We remark that any matrix-valued rational function H has an expansion about $s = \infty$ of the form

$$H(s) = \sum_{j=-\infty}^{j_0} M_j s^j, \tag{2.19}$$

where $j_0 \geqslant 0$ is an integer. Moreover, the function H has a pole at $s = \infty$ if, and only if, $j_0 \geqslant 1$ and $M_{j_0} \neq 0$ in (2.19).

The positive real lemma for descriptor systems can now be stated as follows.

THEOREM 2.4 (Positive real lemma for descriptor systems (FREUND and JARRE [2004a])). *Let H be a real $m \times m$-matrix-valued rational function of the form* (2.17) *with matrices* (2.18).

(a) (Sufficient condition)

If the LMIs

$$\begin{bmatrix} A^{\mathsf{T}} X + X^{\mathsf{T}} A & X^{\mathsf{T}} B - C \\ B^{\mathsf{T}} X - C^{\mathsf{T}} & -D - D^{\mathsf{T}} \end{bmatrix} \preceq 0 \quad \text{and} \quad E^{\mathsf{T}} X = X^{\mathsf{T}} E \succeq 0 \tag{2.20}$$

have a solution $X \in \mathbb{R}^{N \times N}$, then H is positive real.

(b) (Necessary condition)

Suppose that (2.17) *is a minimal realization of H and that the matrix M_0 in the expansion* (2.19) *satisfies*

$$(D - M_0) + (D - M_0)^{\mathsf{T}} \succeq 0. \tag{2.21}$$

If H is positive real, then there exists a solution $X \in \mathbb{R}^{N \times N}$ of the LMIs (2.20).

The result of Theorem 2.4 allows to check positive realness by solving the semi-definite programming problems of the form (2.20). Note that there are N^2 unknowns in (2.20), namely the entries of the $N \times N$ matrix X. Problems of the form (2.20) can be tackled with interior-point methods; see, e.g., BOYD, EL GHAOUI, FERON, and BALAKRISHNAN [1994], FREUND and JARRE [2004a]. However, the computational complexity of these methods grows quickly with N, and thus, these methods are viable only for rather small values of N.

For the special case $E = I$, the result of Theorem 2.4 is just the classical positive real lemma. In this case, (2.20) reduces to the problem of finding a symmetric positive semidefinite matrix $X \in \mathbb{R}^{N \times N}$ such that

$$\begin{bmatrix} A^{\mathsf{T}} X + X A & X B - C \\ B^{\mathsf{T}} X - C^{\mathsf{T}} & -D - D^{\mathsf{T}} \end{bmatrix} \preceq 0.$$

Moreover, if $E = I$, the condition (2.21) is always satisfied, since in this case $M_0 = 0$ and $D + D^{\mathsf{T}} \succeq 0$.

2.6. Linear RCL subcircuits

In circuit simulation, an important special case of passive circuits is linear subcircuits that consist of only resistors, inductors, and capacitors. Such linear RCL subcircuits

arise in the modeling of a circuit's interconnect and package; see, e.g., FREUND and
FELDMANN [1997], FREUND and FELDMANN [1998], KIM, GOPAL, and PILLAGE [1994],
PILEGGI [1995].

The equations describing linear RCL subcircuits are of the form (2.1) and (2.2) with
$D = 0$ and $m = p$. Furthermore, the equations can be formulated such that the matrices
$A, E \in \mathbb{R}^{N \times N}$ in (2.1) are symmetric and exhibit a block structure; see FREUND and
FELDMANN [1996a], FREUND and FELDMANN [1998]. More precisely, we have

$$A = A^\mathrm{T} = \begin{bmatrix} -A_{11} & A_{12} \\ A_{12}^\mathrm{T} & 0 \end{bmatrix} \quad \text{and} \quad E = E^\mathrm{T} = \begin{bmatrix} E_{11} & 0 \\ 0 & -E_{22} \end{bmatrix}, \tag{2.22}$$

where the submatrices $A_{11}, E_{11} \in \mathbb{R}^{N_1 \times N_1}$ and $E_{22} \in \mathbb{R}^{N_2 \times N_2}$ are symmetric positive
semidefinite, and $N = N_1 + N_2$. Note that, except for the special case $N_2 = 0$, the
matrices A and E are indefinite. The special case $N_2 = 0$ arises for RC subcircuits that
contain only resistors and capacitors, but no inductors.

If the RCL subcircuit is viewed as an m-terminal component with $m = p$ inputs and
outputs, then the matrices B and C in (2.1) and (2.2) are identical and of the form

$$B = C = \begin{bmatrix} B_1 \\ 0 \end{bmatrix} \quad \text{with } B_1 \in \mathbb{R}^{N_1 \times m}. \tag{2.23}$$

In view of (2.22) and (2.23), the transfer function of such an m-terminal RCL subcircuit
is given by

$$H(s) = B^\mathrm{T}(sE - A)^{-1}B, \quad \text{where } A = A^\mathrm{T}, \ E = E^\mathrm{T}. \tag{2.24}$$

We call a transfer function H *symmetric* if it is of the form (2.24) with real matrices A,
E, and B.

We will also use the following nonsymmetric formulation of (2.24). Let J be the
block matrix

$$J = \begin{bmatrix} I_{N_1} & 0 \\ 0 & -I_{N_2} \end{bmatrix}, \tag{2.25}$$

where I_{N_1} and I_{N_2} is the $N_1 \times N_1$ and $N_2 \times N_2$ identity matrix, respectively.

Note that, by (2.23) and (2.25), we have $B = JB$. Using this relation, as well
as (2.22), we can rewrite (2.24) as follows:

$$H(s) = B^\mathrm{T}\left(s\tilde{E} - \tilde{A}\right)^{-1}B, \quad \text{where}$$

$$\tilde{A} = \begin{bmatrix} -A_{11} & A_{12} \\ -A_{12}^\mathrm{T} & 0 \end{bmatrix}, \qquad \tilde{E} = \begin{bmatrix} E_{11} & 0 \\ 0 & E_{22} \end{bmatrix}. \tag{2.26}$$

In this formulation, the matrix $\tilde{A}$ is no longer symmetric, but now

$$\tilde{A} + \tilde{A}^\mathrm{T} \preceq 0 \quad \text{and} \quad \tilde{E} \succeq 0. \tag{2.27}$$

It turns out that the properties are the key to ensure positive realness. Indeed, the following result was established as Theorem 13 in FREUND [2000b].

THEOREM 2.5. *Let* $\tilde{A}$, $\tilde{E} \in \mathbb{R}^{N \times N}$, *and* $B \in \mathbb{R}^{N \times m}$. *Assume that* $\tilde{A}$ *and* $\tilde{E}$ *satisfy* (2.27), *and that the matrix pencil* $\tilde{A} - s\tilde{E}$ *is regular. Then, the* $m \times m$-*matrix-valued function*

$$H(s) = B^{\mathrm{T}}(s\tilde{E} - \tilde{A})^{-1} B$$

is positive real.

3. Krylov-subspace techniques

In this section, we discuss the use of Krylov-subspace methods for the construction of Padé and Padé-type reduced-order models of time-invariant linear dynamical systems. We also point the reader to FREUND [2003] for a more extended survey of Krylov-subspace methods for model reduction.

3.1. Block Krylov subspaces

We consider general descriptor systems of the form (2.1) and (2.2). The key to using Krylov-subspace techniques for reduced-order modeling of such systems is to first replace the matrix pair A and E by a single matrix M. To this end, let $s_0 \in \mathbb{C}$ be any given point such that the matrix $A - s_0 E$ is nonsingular. Then, with M, R, and L denoting the matrices defined in (2.15), the linear system (2.1) and (2.2) can be rewritten in the following form:

$$M\frac{\mathrm{d}x}{\mathrm{d}t} = (I + s_0 M)x + Ru(t), \tag{3.1}$$

$$y(t) = L^{\mathrm{T}}x(t) + Du(t). \tag{3.2}$$

Note that $M \in \mathbb{C}^{N \times N}$, $R \in \mathbb{C}^{N \times m}$, and $L \in \mathbb{C}^{N \times p}$, where N is the state-space dimension of the system, m is the number of inputs, and p is the number of outputs.

The transfer function H of the rewritten system (3.1) and (3.2) is given by (2.14). By expanding (2.14) about s_0, we obtain

$$H(s) = D - \sum_{j=0}^{\infty} L^{\mathrm{T}} M^j R (s - s_0)^j. \tag{3.3}$$

Recall from Section 2.3 that Padé and Padé-type reduced-order models are defined via the leading coefficients of an expansion of H about s_0. In view of (3.3), the jth coefficient of such an expansion can be expressed as follows:

$$-L^{\mathrm{T}} M^j R = -((M^{j-i})^{\mathrm{T}} L)^{\mathrm{T}} (M^i R), \quad i = 0, 1, \ldots, j. \tag{3.4}$$

Notice that the factors on the right-hand side of (3.4) are blocks of the *right* and *left* block Krylov matrices

$$\begin{bmatrix} R & MR & M^2 R & \cdots & M^i R & \cdots \end{bmatrix} \quad \text{and}$$
$$\begin{bmatrix} L & M^{\mathrm{T}} L & (M^{\mathrm{T}})^2 L & \cdots & (M^{\mathrm{T}})^k L & \cdots \end{bmatrix}, \tag{3.5}$$

respectively. As a result, all the information needed to generate Padé and Padé-type reduced-order models is contained in the block Krylov matrices (3.5). However, simply computing the blocks $M^i R$ and $(M^T)^i L$ in (3.5) and then generating the leading coefficients of the expansion (3.3) from these blocks is not a viable numerical procedure. The reason is that, in finite-precision arithmetic, as i increases, the blocks $M^i R$ and $(M^T)^i L$ quickly contain only information about the eigenspaces of the dominant eigenvalue of M. Instead, one needs to employ suitable Krylov-subspace methods that generate numerically better basis vectors for the subspaces associated with the block Krylov matrices (3.5).

Next, we give a formal definition of the subspaces induced by (3.5). Note that each block $M^i R$ consists of m column vectors of length N. By scanning these column vectors of the right block Krylov matrix in (3.5) from left to right and by deleting any column that is linearly dependent on columns to its left, we obtain the *deflated* right block Krylov matrix

$$[\, R_1 \quad M R_2 \quad M^2 R_3 \quad \cdots \quad M^{i_{\max}-1} R_{i_{\max}} \,]. \tag{3.6}$$

This process of detecting and deleting the linearly dependent columns is called *exact deflation*. We remark that the matrix (3.6) is finite, since at most N of the column vectors can be linearly independent. Furthermore, a column $M^i r$ being linearly dependent on columns to its left in (3.5) implies that any column $M^{i'} r$, $i' \geq i$, is linearly dependent on columns to its right. Therefore, in (3.6), for each $i = 1, 2, \ldots, i_{\max}$, the matrix R_i is a submatrix of R_{i-1}, where, for $i = 1$, we set $R_0 = R$.

Let m_i denote the number of columns of R_i. The matrix (3.6) thus has

$$n_{\max}^{(r)} := \leqslant m_1 + m_2 + \cdots + m_{i_{\max}},$$

columns. For each integer n with $1 \leqslant n \leqslant n_{\max}^{(r)}$, we define the *nth right block Krylov subspace* $\mathcal{K}_n(M, R)$ (induced by M and R) as the subspace spanned by the first n columns of the deflated right block Krylov matrix (3.6).

Analogously, by deleting the linearly independent columns of the left block Krylov matrix in (3.5), we obtain a deflated left block Krylov matrix of the form

$$[\, L_1 \quad M^T L_2 \quad (M^T)^2 L_3 \quad \cdots \quad (M^T)^{i_{\max}-1} L_{k_{\max}} \,]. \tag{3.7}$$

Let $n_{\max}^{(l)}$ be the number of columns of the matrix (3.7). Then for each integer n with $1 \leqslant n \leqslant n_{\max}^{(l)}$, we define the *nth left block Krylov subspace* $\mathcal{K}_n(M^T, L)$ (induced by M^T and L) as the subspace spanned by the first n columns of the deflated left block Krylov matrix (3.7).

For a more detailed discussion of block Krylov subspaces and deflation, we refer the reader to ALIAGA, BOLEY, FREUND, and HERNÁNDEZ [2000], FREUND [2000b].

3.2. *Approaches based on Lanczos and Lanczos-type methods*

In this section, we discuss reduced-order modeling approaches that employ Lanczos and Lanczos-type methods for the construction of suitable basis vectors for the right and left block Krylov subspaces $\mathcal{K}_n(M, R)$ and $\mathcal{K}_n(M^T, L)$.

3.2.1. The MPVL algorithm

For the special case $m = p = 1$ of single-input single-output linear dynamical systems, each of the "blocks" R and L only consists of a single vector, say r and l, and $\mathcal{K}_n(M, r)$ and $\mathcal{K}_n(M^T, l)$ are just the standard nth right and left Krylov subspaces induced by single vectors. The classical Lanczos process (LANCZOS [1950]) is a well-known procedure for computing two sets of bi-orthogonal basis vectors for $\mathcal{K}_n(M, r)$ and $\mathcal{K}_n(M^T, l)$. Moreover, these vectors are generated by means of three-term recurrences the coefficients of which define a tridiagonal matrix T_n. It turns out that T_n contains all the information that is needed to set up an nth Padé reduced-order model of a given single-input single-output time-invariant linear dynamical system. The associated computational procedure is called the *Padé via Lanczos* (PVL) algorithm in FELDMANN and FREUND [1994], FELDMANN and FREUND [1995a].

Here, we describe in some detail an extension of the PVL algorithm to the case of general m-input p-output time-invariant linear dynamical systems. The underlying block Krylov subspace method is the *nonsymmetric band Lanczos algorithm* (FREUND [2000a]) for constructing two sets of right and left Lanczos vectors

$$v_1, v_2, \ldots, v_n \quad \text{and} \quad w_1, w_2, \ldots, w_n, \tag{3.8}$$

respectively. These vectors span the nth right and left block Krylov subspaces (induced by M and R, and M^T and L, respectively):

$$\text{span}\{v_1, v_2, \ldots, v_n\} = \mathcal{K}_n(M, R) \quad \text{and}$$
$$\text{span}\{w_1, w_2, \ldots, w_n\} = \mathcal{K}_n(M^T, L). \tag{3.9}$$

Moreover, the vectors (3.8) are constructed to be bi-orthogonal:

$$w_j^T v_k = \begin{cases} 0 & \text{if } j \neq k, \\ \delta_j & \text{if } j = k, \end{cases} \quad \text{for all } j, k = 1, 2, \ldots, n. \tag{3.10}$$

It turns out that the Lanczos vectors (3.8) can be constructed by means of recurrence relations of length at most $m + p + 1$. The recurrence coefficients for the first n right Lanczos vectors define an $n \times n$ matrix $T_n^{(pr)}$ that is "essentially" a band matrix with total bandwidth $m + p + 1$. Similarly, the recurrence coefficients for the first n left Lanczos vectors define an $n \times n$ band matrix $\widetilde{T}_n^{(pr)}$ with total bandwidth $m + p + 1$. For a more detailed discussion of the structure of $T_n^{(pr)}$ and $\widetilde{T}_n^{(pr)}$, we refer the reader to ALIAGA, BOLEY, FREUND, and HERNÁNDEZ [2000], FREUND [2000a].

Algorithm 3.1 below gives a complete description of the numerical procedure that generates the Lanczos vectors (3.8) with properties (3.9) and (3.10). In order to obtain a Padé reduced-order model based on this algorithm, one does not need the Lanczos vectors themselves, but rather the matrix of right recurrence coefficients $T_n^{(pr)}$, the matrices $\rho_n^{(pr)}$ and $\eta_n^{(pr)}$ that contain the recurrence coefficients from processing the starting blocks R and L, respectively, and the diagonal matrix

$$\Delta_n = \text{diag}(\delta_1, \delta_2, \ldots, \delta_n),$$

whose diagonal entries are the δ_j's from (3.10). The following algorithm produces the matrices $T_n^{(pr)}$, $\rho_n^{(pr)}$, $\eta_n^{(pr)}$, and Δ_n as output.

ALGORITHM 3.1 (*Nonsymmetric band Lanczos algorithm*).
INPUT: A matrix $M \in \mathbb{C}^{N \times N}$;
  A block of m right starting vectors $R = [r_1 \quad r_2 \quad \cdots \quad r_m] \in \mathbb{C}^{N \times m}$;
  A block of p left starting vectors $L = [l_1 \quad l_2 \quad \cdots \quad l_p] \in \mathbb{C}^{N \times p}$.
OUTPUT: The $n \times n$ Lanczos matrix $T_n^{(\mathrm{pr})}$, and the matrices $\rho_n^{(\mathrm{pr})}$, $\eta_n^{(\mathrm{pr})}$, and Δ_n.
(0) For $k = 1, 2, \ldots, m$, set $\hat{v}_k = r_k$.
  For $k = 1, 2, \ldots, p$, set $\hat{w}_k = l_k$.
  Set $m_c = m$, $p_c = p$, and $\mathcal{I}_v = \mathcal{I}_w = \emptyset$.
For $n = 1, 2, \ldots$, until convergence or $m_c = 0$ or $p_c = 0$ or $\delta_n = 0$ do:
  (1) (If necessary, deflate $\hat{v}_n$.)
    Compute $\|\hat{v}_n\|_2$.
    Decide if $\hat{v}_n$ should be deflated. If yes, do the following:
    (a) Set $\hat{v}_{n-m_c}^{\mathrm{defl}} = \hat{v}_n$ and store this vector. Set $\mathcal{I}_v = \mathcal{I}_v \cup \{n - m_c\}$.
    (b) Set $m_c = m_c - 1$.
      If $m_c = 0$, set $n = n - 1$ and stop.
    (c) For $k = n, n + 1, \ldots, n + m_c - 1$, set $\hat{v}_k = \hat{v}_{k+1}$.
    (d) Repeat all of Step (1).
  (2) (If necessary, deflate $\hat{w}_n$.)
    Compute $\|\hat{w}_n\|_2$.
    Decide if $\hat{w}_n$ should be deflated. If yes, do the following:
    (a) Set $\hat{w}_{n-p_c}^{\mathrm{defl}} = \hat{w}_n$ and store this vector. Set $\mathcal{I}_w = \mathcal{I}_w \cup \{n - p_c\}$.
    (b) Set $p_c = p_c - 1$.
      If $p_c = 0$, set $n = n - 1$ and stop.
    (c) For $k = n, n + 1, \ldots, n + p_c - 1$, set $\hat{w}_k = \hat{w}_{k+1}$.
    (d) Repeat all of Step (2).
  (3) (Normalize $\hat{v}_n$ and $\hat{w}_n$ to obtain v_n and w_n.)
    Set

$$t_{n,n-m_c} = \|\hat{v}_n\|_2, \qquad \tilde{t}_{n,n-p_c} = \|\hat{w}_n\|_2,$$

$$v_n = \frac{\hat{v}_n}{t_{n,n-m_c}}, \quad \text{and} \quad w_n = \frac{\hat{w}_n}{\tilde{t}_{n,n-p_c}}.$$

  (4) (Compute δ_n and check for possible breakdown.)
    Set $\delta_n = w_n^{\mathrm{T}} v_n$. If $\delta_n = 0$, set $n = n - 1$ and stop.
  (5) (Orthogonalize the right candidate vectors against w_n.)
    For $k = n + 1, n + 2, \ldots, n + m_c - 1$, set

$$t_{n,k-m_c} = \frac{w_n^{\mathrm{T}} \hat{v}_k}{\delta_n} \quad \text{and} \quad \hat{v}_k = \hat{v}_k - v_n t_{n,k-m_c}.$$

  (6) (Orthogonalize the left candidate vectors against v_n.)
    For $k = n + 1, n + 2, \ldots, n + p_c - 1$, set

$$\tilde{t}_{n,k-p_c} = \frac{\hat{w}_k^{\mathrm{T}} v_n}{\delta_n} \quad \text{and} \quad \hat{w}_k = \hat{w}_k - w_n \tilde{t}_{n,k-p_c}.$$

  (7) (Advance the right block Krylov subspace to get $\hat{v}_{n+m_c}$.)

(a) Set $\hat{v}_{n+m_{\mathrm{c}}} = M v_n$.
(b) For $k \in \mathcal{I}_w$ (in ascending order), set

$$\tilde{\sigma} = \left(\hat{w}_k^{\mathrm{defl}}\right)^{\mathrm{T}} v_n, \qquad \tilde{t}_{n,k} = \frac{\tilde{\sigma}}{\delta_n},$$

and, if $k > 0$, set

$$t_{k,n} = \frac{\tilde{\sigma}}{\delta_k} \quad \text{and} \quad \hat{v}_{n+m_{\mathrm{c}}} = \hat{v}_{n+m_{\mathrm{c}}} - v_k t_{k,n}.$$

(c) Set $k_v = \max\{1, n - p_{\mathrm{c}}\}$.
(d) For $k = k_v, k_v + 1, \ldots, n - 1$, set

$$t_{k,n} = \tilde{t}_{n,k} \frac{\delta_n}{\delta_k} \quad \text{and} \quad \hat{v}_{n+m_{\mathrm{c}}} = \hat{v}_{n+m_{\mathrm{c}}} - v_k t_{k,n}.$$

(e) Set

$$t_{n,n} = \frac{w_n^{\mathrm{T}} \hat{v}_{n+m_{\mathrm{c}}}}{\delta_n} \quad \text{and} \quad \hat{v}_{n+m_{\mathrm{c}}} = \hat{v}_{n+m_{\mathrm{c}}} - v_n t_{n,n}.$$

(8) (Advance the left block Krylov subspace to get $\hat{w}_{n+p_{\mathrm{c}}}$.)
(a) Set $\hat{w}_{n+p_{\mathrm{c}}} = M^{\mathrm{T}} w_n$.
(b) For $k \in \mathcal{I}_v$ (in ascending order), set

$$\sigma = w_n^{\mathrm{T}} \hat{v}_k^{\mathrm{defl}}, \qquad t_{n,k} = \frac{\sigma}{\delta_n},$$

and, if $k > 0$, set

$$\tilde{t}_{k,n} = \frac{\sigma}{\delta_k} \quad \text{and} \quad \hat{w}_{n+p_{\mathrm{c}}} = \hat{w}_{n+p_{\mathrm{c}}} - w_k \tilde{t}_{k,n}.$$

(c) Set $k_w = \max\{1, n - m_{\mathrm{c}}\}$.
(d) For $k = k_w, k_w + 1, \ldots, n - 1$, set

$$\tilde{t}_{k,n} = t_{n,k} \frac{\delta_n}{\delta_k} \quad \text{and} \quad \hat{w}_{n+p_{\mathrm{c}}} = \hat{w}_{n+p_{\mathrm{c}}} - w_k \tilde{t}_{k,n}.$$

(e) Set

$$\tilde{t}_{n,n} = t_{n,n} \quad \text{and} \quad \hat{w}_{n+p_{\mathrm{c}}} = \hat{w}_{n+p_{\mathrm{c}}} - w_n \tilde{t}_{n,n}.$$

(9) Set

$$T_n^{(\mathrm{pr})} = [\, t_{i,k} \,]_{i,k=1,2,\ldots,n},$$
$$\rho_n^{(\mathrm{pr})} = [\, t_{i,k-m} \,]_{i=1,2,\ldots,n; k=1,2,\ldots,k_\rho}, \qquad \text{where } k_\rho = m + \min\{0, n - m_{\mathrm{c}}\},$$
$$\eta_n^{(\mathrm{pr})} = [\, \tilde{t}_{i,k-p} \,]_{i=1,2,\ldots,n; k=1,2,\ldots,k_\eta}, \qquad \text{where } k_\eta = p + \min\{0, n - p_{\mathrm{c}}\},$$
$$\Delta_n = \mathrm{diag}(\delta_1, \delta_2, \ldots, \delta_n).$$

(10) Check if n is large enough. If yes, stop.

REMARK 3.1. When applied to single starting vectors, i.e., for the special case $m = p = 1$, Algorithm 3.1 reduces to the classical nonsymmetric Lanczos process (LANCZOS [1950]).

REMARK 3.2. It can be shown that, at step n of Algorithm 3.1, exact deflation of a vector in the right, respectively left, block Krylov matrix (3.5) occurs if, and only if, $\hat{v}_n = 0$, respectively $\hat{w}_n = 0$, in Step (1), respectively Step (2). Therefore, to run Algorithm 3.1 with exact deflation only, one deflates $\hat{v}_n$ if $\|\hat{v}_n\|_2 = 0$ in Step (1), and one deflates $\hat{w}_n$ if $\|\hat{w}_n\|_2 = 0$ in Step (2). In finite-precision arithmetic, however, so-called *inexact deflation* is employed. This means that in Step (1), $\hat{v}_n$ is deflated if $\|\hat{v}_n\|_2 \leqslant \varepsilon$, and in Step (2), $\hat{w}_n$ is deflated if $\|\hat{w}_n\|_2 \leqslant \varepsilon$, where $\varepsilon = \varepsilon(M) > 0$ is a suitably chosen small constant.

REMARK 3.3. The occurrence of $\delta_n = 0$ in Step (4) of Algorithm 3.1 is called a *breakdown*. In finite-precision arithmetic, in Step (4) one should also check for *near-breakdowns*, i.e., if $\delta_n \approx 0$. In general, it cannot be excluded that breakdowns or near-breakdowns occur, although they are very unlikely. Furthermore, by using so-called *look-ahead* techniques, it is possible to remedy the problem of possible breakdowns or near-breakdowns. For the sake of simplicity, we have stated the band Lanczos algorithm without look-ahead only. A look-ahead version of Algorithm 3.1 is described in ALIAGA, BOLEY, FREUND, and HERNÁNDEZ [2000].

The *matrix-Padé via Lanczos* (MPVL) algorithm, which was first introduced in FELD-MANN and FREUND [1995b], FREUND [1995], consists of applying Algorithm 3.1 to the matrices M, R, and L defined in (2.15), and running it for n steps. The matrices $T_n^{(\mathrm{pr})}$, $\rho_n^{(\mathrm{pr})}$, $\eta_n^{(\mathrm{pr})}$, and Δ_n produced by Algorithm 3.1 are then used to set up a reduced-order model of the original linear dynamical system (2.1) and (2.2) as follows:

$$T_n^{(\mathrm{pr})} \frac{\mathrm{d}z}{\mathrm{d}t} = \left(s_0 T_n^{(\mathrm{pr})} - I \right) z + \rho_n^{(\mathrm{pr})} u(t), \tag{3.11}$$

$$y(t) = \left(\eta_n^{(\mathrm{pr})} \right)^{\mathrm{T}} \Delta_n z(t) + D u(t). \tag{3.12}$$

Note that the transfer function of this reduced-order model is given by

$$H_n(s) = D + \left(\eta_n^{(\mathrm{pr})} \right)^{\mathrm{T}} \Delta_n \left(I - (s - s_0) T_n^{(\mathrm{pr})} \right)^{-1} \rho_n^{(\mathrm{pr})}. \tag{3.13}$$

The reduced-order model (3.11) and (3.12) is indeed a matrix-Padé model of the original system.

THEOREM 3.1 (Matrix-Padé model (FELDMANN and FREUND [1995b], FREUND [1995])). *Suppose that* Algorithm 3.1 *is run with exact deflation only and that* $n \geqslant \max\{m, p\}$. *Then, the reduced-order model* (3.11) *and* (3.12) *is a matrix-Padé model of the linear dynamical system* (2.1) *and* (2.2). *More precisely, the Taylor expansions about* s_0 *of the transfer functions,* H, *(2.8) and,* H_n, *(3.13) agree in as many leading coefficients as possible, i.e.,*

$$H(s) = H_n(s) + \mathcal{O}\big((s - s_0)^{q(n)}\big),$$

where $q(n)$ is as large as possible. In particular,

$$q(n) \geqslant \left\lfloor \frac{n}{m} \right\rfloor + \left\lfloor \frac{n}{p} \right\rfloor.$$

A disadvantage of Padé models is that, in general, they do not preserve the stability and possibly passivity of the original linear dynamical system. In part, these problems can be overcome by means of suitable postprocessing techniques, such as the ones described in BAI, FELDMANN, and FREUND [1998], BAI and FREUND [2001a]. However, the reduced-order models obtained by postprocessing of Padé models are necessarily no longer optimal in the sense of Padé approximation. Furthermore, postprocessing techniques are not guaranteed to always result in stable and possibly passive reduced-order models.

For special cases, however, Padé models can be shown to be stable and passive. In particular, this is the case for linear dynamical systems describing RC subcircuits, RL subcircuits, and LC subcircuits; see BAI and FREUND [2001b], FREUND and FELDMANN [1996a], FREUND and FELDMANN [1997], FREUND and FELDMANN [1998].

Next, we describe the SyMPVL algorithm (FREUND and FELDMANN [1996a], FREUND and FELDMANN [1997], FREUND and FELDMANN [1998]), which is a special version of MPVL tailored to linear RCL subcircuits.

3.2.2. The SyMPVL algorithm

Recall from Section 2.6 that linear RCL subcircuits can be described by linear dynamical systems (2.1) and (2.2) with $D = 0$, symmetric matrices A and E of the form (2.22), and matrices $B = C$ of the form (2.23). Furthermore, the transfer function, H, (2.24) is symmetric.

We now assume that the expansion point s_0 for the Padé approximation is chosen to be real and nonnegative, i.e., $s_0 \geqslant 0$. Together with (2.22) it follows that the matrix $A - s_0 E$ is symmetric indefinite, with N_1 nonpositive and N_2 nonnegative eigenvalues. Thus, $A - s_0 E$ admits a factorization of the following form:

$$A - s_0 E = -F_1 J F_1^{\mathrm{T}}, \tag{3.14}$$

where J is the block matrix defined in (2.25). Instead of the general factorization (2.13), we now use (3.14). By (3.14) and (2.15), the matrices M, R, and L, are then of the following form:

$$M = F_1^{-1} E F_1^{-\mathrm{T}} J, \quad R = F_1^{-1} B, \quad \text{and} \quad L = -J F_1^{-1} C.$$

Since $E = E^{\mathrm{T}}$ and $B = C$, it follows that

$$J M = M^{\mathrm{T}} J \quad \text{and} \quad L = -J R.$$

This means that M is J-symmetric and the left starting block L is (up to its sign) the J-multiple of the right starting block R. These two properties imply that all the right and left Lanczos vectors generated by the band Lanczos Algorithm 3.1 are J-multiples of each other:

$$w_j = J v_j \quad \text{for all } j = 1, 2, \dots, n.$$

Consequently, Algorithm 3.1 simplifies in that only the right Lanczos vectors need to be computed. The resulting version of MPVL for computing matrix-Padé models of RCL subcircuits is just the SyMPVL algorithm. The computational costs of SyMPVL are half of that of the general MPVL algorithm.

Let $H_n^{(1)}$ denote the matrix-Padé model generated by SyMPVL after n Lanczos steps. For general RCL subcircuits, however, $H_n^{(1)}$ will not preserve the passivity of the original system.

An additional reduced-order model that is guaranteed to be passive can be obtained as follows, provided that all right Lanczos vectors are stored. Let

$$V_n = [\, v_1 \quad v_2 \quad \cdots \quad v_n \,]$$

denote the matrix that contains the first n right Lanczos vectors as columns. Then, by projecting the matrices in the representation (2.26) of the transfer function H of the original RCL subcircuit onto the columns of V_n, we obtain the following reduced-order transfer function:

$$H_n^{(2)}(s) = \left(V_n^{\mathrm{T}} B \right)^{\mathrm{T}} \left(s V_n^{\mathrm{T}} \tilde{E} V_n - V_n^{\mathrm{T}} \tilde{A} V_n \right)^{-1} V_n^{\mathrm{T}} B. \tag{3.15}$$

The passivity of the original RCL subcircuit, together with Theorem 2.5 implies that the reduced-order model defined by $H_n^{(2)}$ is indeed passive. Furthermore, in FREUND [2000b], it is shown that $H_n^{(2)}$ is a matrix-Padé-type approximation of the original transfer function and that, at the expansion point s_0, $H_n^{(2)}$ matches half as many leading coefficients of H as the matrix-Padé approximant $H_n^{(1)}$.

Next, we illustrate the behavior of SyMPVL with two circuit examples.

3.2.3. A package model

The first example arises is the analysis of a 64-pin package model used for an RF integrated circuit. Only eight of the package pins carry signals, the rest being either unused or carrying supply voltages. The package is characterized as a passive linear dynamical system with $m = p = 16$ inputs and outputs, representing 8 exterior and 8 interior terminals. The package model is described by approximately 4000 circuit elements, resistors, capacitors, inductors, and inductive couplings, resulting in a linear dynamical system with a state-space dimension of about 2000.

In FREUND and FELDMANN [1997], SyMPVL was used to compute a Padé-based reduced-order model of the package, and it was found that a model $H_n^{(1)}$ of order $n = 80$ is sufficient to match the transfer-function components of interest. However, the model $H_n^{(1)}$ has a few poles in the right half of the complex plane, and therefore, it is not passive.

In order to obtain a passive reduced-order model, we ran SyMPVL again on the package example, and this time, also generated the projected reduced-order model $H_n^{(2)}$ given by (3.15). The expansion point $s_0 = 5\pi \times 10^9$ was used. Recall that $H_n^{(2)}$ is only a Padé-type approximant and thus less accurate than the Padé approximant $H_n^{(2)}$. Therefore, one now has to go to order $n = 112$ to obtain a projected reduced-order model $H_n^{(2)}$ that matches the transfer-function components of interest. Figs. 3.1 and 3.2 show

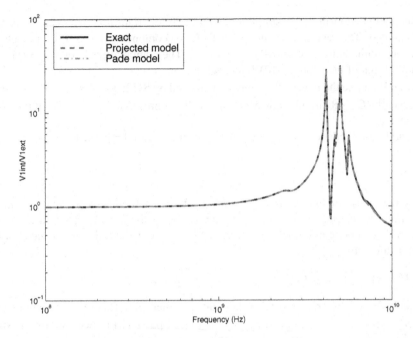

Fɪɢ. 3.1. Package: Pin no. 1 external to Pin no. 1 internal, exact, projected model, and Padé model.

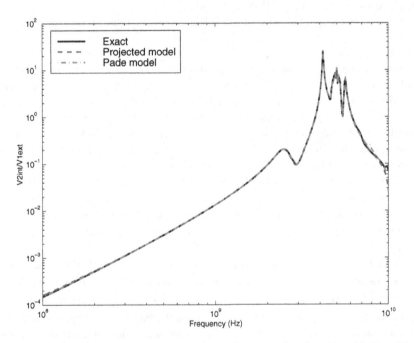

Fɪɢ. 3.2. Package: Pin no. 1 external to Pin no. 2 internal, exact, projected model, and Padé model.

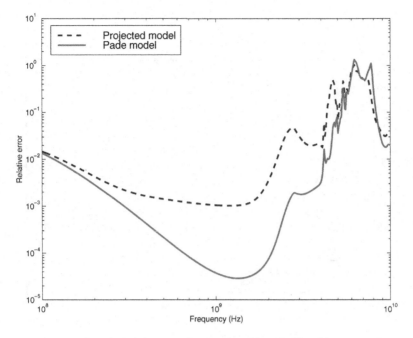

FIG. 3.3. Relative error of projected model and Padé model.

the voltage-to-voltage transfer function between the external terminal of Pin no. 1 and the internal terminals of the same pin and the neighboring Pin no. 2, respectively. The plots show results with the projected model $H_n^{(2)}$ and the Padé model $H_n^{(2)}$, both of order $n = 112$, compared with an exact analysis.

In Fig. 3.3, we compare the relative error of the projected model $H_{112}^{(2)}$ and the Padé model $H_{112}^{(1)}$ of the same size. Clearly, the Padé model is more accurate. However, out of the 112 poles of $H_{112}^{(1)}$, 22 have positive real parts, violating the passivity of the Padé model. On the other hand, the projected model is passive.

3.2.4. An extracted RC circuit

This is an extracted RC circuit with about 4000 elements and $m = 20$ ports. The expansion point $s_0 = 0$ was used. Since the projected model and the Padé model are identical for RC circuits, we only computed the Padé model via SyMPVL.

The point of this example is to illustrate the usefulness of the deflation procedure built into SyMPVL. It turned out that sweeps through the first two Krylov blocks, R and MR, of the block Krylov matrix (3.5) were sufficient to obtain a reduced-order model that matches the transfer function in the frequency range of interest. During the sweep through the second block, 6 almost linearly dependent vectors were discovered and deflated. As a result, the reduced-order model obtained with deflation is only of size $n = 2m - 6 = 34$. When SyMPVL was rerun on this example, with deflation turned off, a reduced-order model of size $n = 40$ was needed to match the transfer function. In Fig. 3.4, we show the $H_{1,11}$ component of the reduced-order model obtained with

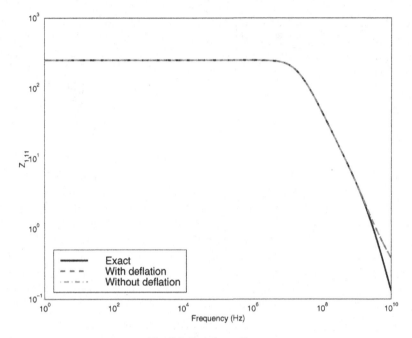

Fig. 3.4. Impedance $H_{1,11}$.

deflation and without deflation, compared to the exact transfer function. Clearly, deflation leads to a significantly smaller reduced-order model that is as accurate as the bigger one generated without deflation.

3.3. Approaches based on the Arnoldi process

The Arnoldi process (ARNOLDI [1951]) is another widely-used Krylov-subspace method. A band version of the Arnoldi process that is suitable for multiple starting vectors can also be used for reduced-order modeling. However, the models generated from the band Arnoldi process are only Padé-type models.

In contrast to the band Lanczos algorithm, the band Arnoldi process only involves one of the starting blocks, namely R, and it only uses matrix–vector products with M. Moreover, the band Arnoldi process only generates one set of vectors, $v_1, v_2, \ldots, v_n$, instead of the two sequences of right and left vectors produced by the band Lanczos algorithm. The Arnoldi vectors span the nth right block Krylov subspace (induced by M and R):

$$\text{span}\{v_1, v_2, \ldots, v_n\} = \mathcal{K}_n(M, R).$$

The Arnoldi vectors are constructed to be orthonormal:

$$V_n^H V_n = I, \quad \text{where } V_n := [\, v_1 \quad v_2 \quad \cdots \quad v_n \,].$$

After n iterations, the Arnoldi process has generated the first n Arnoldi vectors, namely the n columns of the matrix V_n, as well as an $n \times n$ matrix $G_n^{(\text{pr})}$ of recurrence coefficients, and, provided that $n \geqslant m$, an $n \times m$ matrix $\rho_n^{(\mathrm{pr})}$. The matrices $G_n^{(\mathrm{pr})}$ and $\rho_n^{(\mathrm{pr})}$ are projections of the matrices M and R onto the subspace spanned by the columns of V_n, which is just the block Krylov subspace $\mathcal{K}_n(M, R)$. More precisely, we have

$$G_n^{(\mathrm{pr})} = V_n^H M V_n \quad \text{and} \quad \rho_n^{(\mathrm{pr})} = V_n^H R. \tag{3.16}$$

The band Arnoldi process can be stated as follows.

ALGORITHM 3.2 (*Band Arnoldi process*).
INPUT: A matrix $M \in \mathbb{C}^{n \times n}$;
        A block of m right starting vectors $R = [r_1 \quad r_2 \quad \cdots \quad r_m] \in \mathbb{C}^{n \times m}$.
OUTPUT: The $n \times n$ Arnoldi matrix $G_n^{(\mathrm{pr})}$.
        The matrix $V_n = [v_1 \quad v_2 \quad \cdots \quad v_n]$ containing the first n Arnoldi vectors,
        and the matrix $\rho_n^{(\mathrm{pr})}$.
(0) For $k = 1, 2, \ldots, m$, set $\hat{v}_k = r_k$.
    Set $m_c = m$ and $\mathcal{I} = \emptyset$.
For $n = 1, 2, \ldots$, until convergence or $m_c = 0$ do:
(1) (If necessary, deflate $\hat{v}_n$.)
    Compute $\|\hat{v}_n\|_2$.
    Decide if $\hat{v}_n$ should be deflated. If yes, do the following:
    (a) Set $\hat{v}_{n-m_c}^{\mathrm{defl}} = \hat{v}_n$ and store this vector. Set $\mathcal{I} = \mathcal{I} \cup \{n - m_c\}$.
    (b) Set $m_c = m_c - 1$. If $m_c = 0$, set $n = n - 1$ and stop.
    (c) For $k = n, n+1, \ldots, n + m_c - 1$, set $\hat{v}_k = \hat{v}_{k+1}$.
    (d) Repeat all of Step (1).
(2) (Normalize $\hat{v}_n$ to obtain v_n.)
    Set

$$g_{n,n-m_c} = \|\hat{v}_n\|_2 \quad \text{and} \quad v_n = \frac{\hat{v}_n}{g_{n,n-m_c}}.$$

(3) (Orthogonalize the candidate vectors against v_n.)
    For $k = n+1, n+2, \ldots, n + m_c - 1$, set

$$g_{n,k-m_c} = v_n^H \hat{v}_k \quad \text{and} \quad \hat{v}_k = \hat{v}_k - v_n g_{n,k-m_c}.$$

(4) (Advance the block Krylov subspace to get $\hat{v}_{n+m_c}$.)
    (a) Set $\hat{v}_{n+m_c} = M v_n$.
    (b) For $k = 1, 2, \ldots, n$, set

$$g_{k,n} = v_k^H \hat{v}_{n+m_c} \quad \text{and} \quad \hat{v}_{n+m_c} = \hat{v}_{n+m_c} - v_k g_{k,n}.$$

(5) (a) For $k \in \mathcal{I}$, set $g_{n,k} = v_n^H \hat{v}_k^{\mathrm{defl}}$.
    (b) Set

$$G_n^{(\mathrm{pr})} = [g_{i,k}]_{i,k=1,2,\ldots,n},$$
$$\rho_n^{(\mathrm{pr})} = [g_{i,k-m}]_{i=1,2,\ldots,n; \, k=1,2,\ldots,k_\rho},$$
$$\text{where } k_\rho = m + \min\{0, n - m_c\}.$$

(6) Check if n is large enough. If yes, stop.

Note that, in contrast to the band Lanczos algorithm, the band Arnoldi process requires the storage of all previously computed Arnoldi vectors.

Like the band Lanczos algorithm, the band Arnoldi process can also be employed to reduced-order modeling. Let M, R, and L be the matrices defined in (2.15). After running Algorithm 3.2 (applied to M and R) for n steps, we have obtained the matrices $G_n^{(\text{pr})}$ and $\rho_n^{(\text{pr})}$, as well as the matrix V_n of Arnoldi vectors. The transfer function H_n of a reduced-order model H_n can now be defined as follows:

$$H_n(s) = \left(V_n^H L\right)^H \left(I - (s - s_0) V_n^H M V_n\right)^{-1} \left(V_n^H R\right).$$

Using the relations (3.16) for $G_n^{(\text{pr})}$ and $\rho_n^{(\text{pr})}$, the formula for H_n reduces to

$$H_n(s) = \left(V_n^H L\right)^H \left(I - (s - s_0) G_n^{(\text{pr})}\right)^{-1} \rho_n^{(\text{pr})}. \tag{3.17}$$

The matrices $G_n^{(\text{pr})}$ and $\rho_n^{(\text{pr})}$ are directly available from Algorithm 3.2. In addition, one also needs to compute the matrix

$$\eta_n^{(\text{pr})} = V_n^H L.$$

It turns out that the transfer function (3.17) defines a matrix-Padé-type reduced-order model.

THEOREM 3.2 (Matrix-Padé-type model (FREUND [2000b], ODABASIOGLU [1996])). *Suppose that* Algorithm 3.2 *is run with exact deflation only and that* $n \geq m$. *Then, the reduced-order model associated with the reduced-order transfer function* (3.17) *is a matrix-Padé-type model of the linear dynamical system* (2.1) *and* (2.2). *More precisely, the Taylor expansions about* s_0 *of the transfer functions,* H, (2.8) *and,* H_n, (3.17) *agree in at least*

$$q'(n) \geq \left\lfloor \frac{n}{m} \right\rfloor$$

leading coefficients:

$$H(s) = H_n(s) + \mathcal{O}\left((s - s_0)^{q'(n)}\right). \tag{3.18}$$

REMARK 3.4. The number $q'(n)$ is the exact number of terms matched in the expansion (3.18) provided that no exact deflations occur in Algorithm 3.2. In the case of exact deflations, the number of matching terms is somewhat higher, but so is the number of matching terms for the matrix-Padé model of Theorem 3.1; see FREUND [2000b]. In particular, the matrix-Padé model is always more accurate than the matrix-Padé-type model obtained from Algorithm 3.2. On the other hand, the band Arnoldi process is certainly simpler than the band Lanczos process. Furthermore, the true orthogonality of the Arnoldi vectors in general results in better numerical behavior than the bi-orthogonality of the Lanczos vectors.

REMARK 3.5. For the special case of RCL subcircuits, the algorithm PRIMA proposed in ODABASIOGLU [1996], ODABASIOGLU, CELIK, and PILEGGI [1997] can be interpreted as a special case of the Arnoldi reduced-order modeling procedure described here. Furthermore, in FREUND [1999a], FREUND [2000b] it is shown that the reduced-order model produced by PRIMA is mathematically equivalent to the additional passive model produced by SyMPVL. In contrast to PRIMA, however, SyMPVL also produces a true matrix-Padé model, and thus PRIMA does not appear to have any real advantage over or be even competitive with SyMPVL.

REMARK 3.6. An improved variant of PRIMA is the SPRIM reduction algorithm, which was recently proposed by FREUND [2004c]. While PRIMA generates provably passive reduced-order models, it does not preserve other structures, such as reciprocity or the block structure of the circuit matrices, inherent to RCL circuits. This has motivated the development of algorithms such as ENOR (SHEEHAN [1999]) and its variants (CHEN, LUK, and CHEN [2003]) that generate passive and reciprocal reduced-order models, yet still match as many moments as PRIMA. However, the moment-matching property of the PRIMA models is not optimal. SPRIM overcomes these disadvantages of PRIMA. In particular, SPRIM generates provably passive and reciprocal macromodels of multiport RCL circuits, and the SPRIM models match twice as many moments as the corresponding PRIMA models obtained with identical computational work. For a detailed description of SPRIM and its properties, we refer the reader to FREUND [2004c]. Here, we only present one example, which is taken from FREUND [2004c]. The example is a circuit resulting from the so-called PEEC discretization (RUEHLI [1974]) of an electromagnetic problem. The circuit is an RCL network consisting of

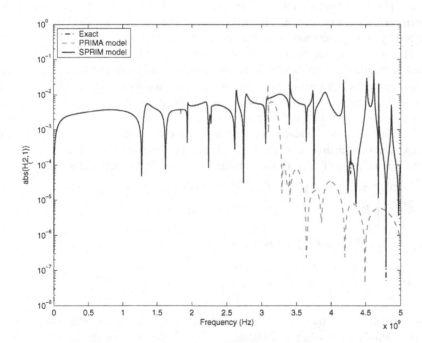

FIG. 3.5. $|H_{2,1}|$ for PEEC circuit.

2100 capacitors, 172 inductors, 6990 inductive couplings, and a single resistive source that drives the circuit. The circuit is formulated as a 2-port. We compare the PRIMA and SPRIM models corresponding to the same dimension n of the underlying block Krylov subspace. The expansion point $s_0 = 2\pi \times 10^9$ was used. In Fig. 3.5, we plot the absolute value of the $(2, 1)$ component, $H_{2,1}$, of the 2×2-matrix-valued transfer function over the frequency range of interest. The dimension $n = 120$ was sufficient for SPRIM to match the exact transfer function. The corresponding PRIMA model of the same dimension, however, has not yet converged to the exact transfer function in large parts of the frequency range of interest. Fig. 3.5 clearly illustrates the better approximation properties of SPRIM due to the matching of twice as many moments as PRIMA.

4. Schur interpolation

4.1. The setting

The modeling of physical effects often produces large, positive definite Hermitian matrices. For example, the modeling of interconnects in an integrated circuit produces in first instance a full elastance matrix G from which a sparse approximating capacitance matrix C has to be derived. Likewise, the behavior of the substrate of an integrated circuit is modeled by a conductivity matrix, and the inductive behavior of the interconnects by an inductance matrix. These matrices are positive definite, because they express either conservation of energy or dissipation. It is a nontrivial problem to find low-complexity approximations to a positive definite matrix, which are positive definite in their own right. For example, if $G = [G_{i,j}]$ is positive definite, then the matrix G_a obtained by putting elements outside a given band equal to zero, i.e., $(G_a)_{i,j} = G_{i,j}$ for $|i - j| < n$ some n, and zero otherwise, will not necessarily be positive definite. If a matrix is diagonally dominant, then putting some off-diagonal elements equal to zero while keeping the Hermitian property would preserve the dominance and hence also the positive definiteness. We shall analyze some of the properties of such schemes soon. An important observation is that properties such as "banded" and "diagonally dominant" are not preserved under inversion: the inverse of a banded matrix is not banded (except when the matrix is block diagonal) and the inverse of a diagonally dominant matrix is not diagonally dominant. Consider for example the matrix (for real a)

$$M_a = \begin{bmatrix} 1 & a & a^2 \\ a & 1 & a \\ a^2 & a & 1 \end{bmatrix}.$$

It is positive definite for $|a| < 1$ with inverse

$$M_a^{-1} = \frac{1}{1 - a^2} \begin{bmatrix} 1 & -a & 0 \\ -a & 1 + a^2 & -a \\ 0 & -a & 1 \end{bmatrix}.$$

If we truncate M_a by putting $(M_a)_{1,3} = (M_a)_{3,1} = 0$, then the resulting matrix will be positive definite only when in addition $a \leqslant 1/\sqrt{2}$. We see that the inverse of M_a is diagonally dominant for $|a| < 1$ while that is only the case for M_a when $a < (\sqrt{5} - 1)/2$. So, why would it be better to truncate a matrix rather than its inverse? A related issue is whether the inverse of a banded matrix has the same computational complexity as the original. Further in this section we shall develop a nice theory that is capable of answering such questions.

Another approach would be to perform the approximation on a Cholesky factor R where $G = R^H R$, R is upper triangular and R^H represents the Hermitian conjugate of R, rather than on the original matrix. Assuming that the off-diagonal elements of R become small the farther they are located from the main diagonal, it makes sense to approximate R by a banded matrix. Also, approximating R by some approximant R_a will produce automatically an approximant $G_a = R_a^H R_a$ that is positive definite. At first sight it would appear that it is not any better to approximate the square root than the original – an ε relative error on the square root of a scalar quantity would roughly produce a 2ε error on the square. The situation with matrices is, however, vastly different, since the condition number of the square root of a (positive definite) matrix, or of its Cholesky factor is just the square root of the original. Still the question arises whether a direct, element-wise approximation of the square root would be a "good" approximation technique, in the sense of either strong norms or complexity? What we need is a theory to gauge both complexity and approximation error. In addition, we would like the approximation procedure to be as simple as possible, for example, it should use a minimal amount of computations in its own right.

We start out this section with the celebrated theory of maximum-entropy interpolation of positive definite matrices. It gives a good stronghold on low-complexity approximation when "low-complexity" is understood as minimizing the number of independent algebraic parameters, e.g., by putting a sufficient number of elements in the matrix or its inverse zero. Immediately the question arises when the sparsity pattern of a positive definite matrix is preserved in its Cholesky factors. This question also has a very neat answer, namely when the matrix entries exhibit a "chordal pattern." In that case, the maximum-entropy interpolant can be found directly, in a minimal number of computations equal to the number of nonzero entries in the matrix, by a matrix interpolation algorithm that is a matrix version of the celebrated Schur interpolation algorithm of complex function analysis. The approximating properties of Schur's algorithm are known and we shall spend a few words explaining them. Finally, we shall show ways of generalizing Schur's algorithm to a more complex situation, namely the so-called "multiple band case."

4.2. Maximum-entropy interpolation of strictly positive definite matrices

Suppose that the following information on an otherwise unknown strictly positive definite (and of course Hermitian) matrix G of size $N \times N$ is given:

- The diagonal elements $G_{k,k}$ for all $k = 1, 2, \ldots, N$;
- Some off-diagonal elements, characterized by a set S: if $(i, j) \in S$ then $G_{i,j}$ is known. Since G is Hermitian, we restrict elements of S to be in the strictly upper triangular zone where $i < j$.

This information is known as "interpolating conditions." The question we ask is: *is it possible to find a positive definite matrix G_a which has the assigned element values on the main diagonal and the set S, and is otherwise in some sense "of minimal complexity?"*

It turns out that this question has a nice definite answer if "complexity" here is understood to mean: "the value of the off-diagonal elements $(G^{-1})_{i,j}$ is zero for (i,j) not in S." A comfortable treatment of the theory leading to this result requires the introduction of the notion of "entropy of a strictly positive matrix H," originating from stochastic system theory and which is given by the (finite) quantity:

$$\mathcal{E}(H) = \log \det H.$$

The following theorem is valid.

THEOREM 4.1. *Suppose that the diagonal elements $G_{k,k}$ and some off-diagonal elements belonging to an off-diagonal set of indices S of a strictly positive definite matrix G are given. Then, there exists a unique strictly positive definite matrix G_a such that G_a interpolates the given entries, i.e., $(G_a)_{i,j} = G_{i,j}$ for $i = j$ and $(i,j) \in S$, and which is such that $(G_a^{-1})_{i,j} = 0$ for (i,j) not in S. This G_a also maximizes the entropy $\mathcal{E}(H) = \log \det H$ over all H that meet the interpolation conditions.*

SKETCH OF PROOF. Suppose that H is a strictly positive definite matrix depending on some parameter ξ. The differential of the entropy with respect to ξ is then given by

$$\frac{\partial}{\partial \xi} \log \det H = \frac{1}{\det H} \frac{\partial \det H}{\partial \xi}.$$

Let us observe that the dependency of $\det H$ on a given entry $H_{i,j}$ can be expressed using the Cramer minor expansion based on the row i:

$$\det H = \sum_{k=1}^{N} H_{i,k} M_{i,k},$$

where $M_{i,k}$ is the minor corresponding to the element at the position (i,k). The minor $M_{i,k}$ does not depend on any element in the ith row of H, in particular it does not depend on $H_{i,j}$ – the determinant is linear in that element. Let now $\xi = G_{i,j}$ for some (i,j) not in S, corresponding to the position of an element that must be determined. Since the $\log \det H$ surface is smooth over the space of parameters to be determined, an extremum will only occur if each possible ξ is chosen so that the variation of the entropy with respect to ξ is zero (or else at the border of feasibility, but that situation cannot lead to a maximum since the border corresponds to matrices whose determinant is zero). The variation for $\xi = G_{i,j}$ on G is now given by:

$$\frac{\partial}{\partial \xi} \log \det G = \frac{1}{\det G} \frac{\partial \det G}{\partial \xi} = \frac{M_{i,j}}{\det G} = (G^{-1})_{j,i}.$$

Hence the top of the entropy surface in the parameter space of the unknown entries of the matrix G_a, i.e., the entries not in S, must correspond to a strictly positive definite extension G_a of G for which $(G_a^{-1})_{i,j} = 0$. The proof now terminates by showing that this top exists and is unique. This must be reasonable in view of the fact that there is a uniform upper bound on the entropy, namely

$$\sum_{k=1}^{N} \log G_{k,k}.$$

This bound can be obtained through recursive evaluation via Cholesky decomposition, and the fact that the interpolating set is convex, if H_1 and H_2 are strictly positive definite and interpolating, so is $kH_1 + (1 - k)H_2$ for $0 \leqslant k \leqslant 1$. $\qquad\Box$

Hence the maximum-entropy extension of entries of a strictly positive definite matrix does exist, and it produces a sparse inverse matrix! This is already a very useful result for model reduction of, for example, capacitive models of IC interconnects, as we shall soon see. However, it is a theoretical result in that the proof of existence does not produce a direct algorithm to compute the result. One may resort to dynamic optimization, and, indeed, that should lead to a solution, but maybe a problematic one, first because it leads to complex computations involving all the elements outside the interpolating set, and second because the entropy surface is most likely very flat, making the optimum hard to find even though there are very good algorithms for convex optimization. Hence it pays to find a way of computing the solution directly on the basis of the known data, if possible. This question is related to the question whether a sparsity pattern in an original, strictly positive definite matrix G is preserved in the Cholesky factor L, where $G = LL^H$, a question which we now address.

4.3. Chordal systems

Assume that we are given a strictly positive definite matrix G whose diagonal elements are known and which is otherwise sparse with upper triangular sparsity pattern $\mathcal{S}$, i.e., $G_{i,j} = 0$ for (i, j) with $i < j$ not belonging to $\mathcal{S}$ (G is of course Hermitian). Connected to $\mathcal{S}$ there is a *sparsity graph* defined as follows:
- Nodes: there are N nodes corresponding to the N rows of the matrix;
- Edges: there is an edge between node i and node j iff $(i, j) \in \mathcal{S}$, assuming $i < j$.

For example, a matrix with fillings

$$\begin{bmatrix} * & * & \cdot & * & \cdot \\ * & * & * & \cdot & * \\ \cdot & * & * & * & \cdot \\ * & \cdot & * & * & * \\ \cdot & * & \cdot & * & * \end{bmatrix} \tag{4.1}$$

has the sparsity graph shown in Fig. 4.1.

We say that a sparsity graph is *chordal* when there is no loop of more than three nodes that has no chord in the graph, a chord being a direct connection between two nodes (with reference to a polygone). The graph shown in Fig. 4.1 is nonchordal, the loop 1-2-3-4-1 has no chords (and there are more such loops). It turns out that the Cholesky

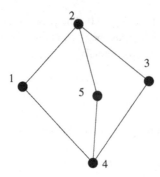

FIG. 4.1. Sparsity graph of the matrix template (4.1).

factorization of a positive definite matrix with chordal sparsity graph will suffer no fill-ins provided it is executed in the right order. To find that order we need another property of chordal graphs.

We shall say that a node of a graph has an *adjacent clique* if the subgraph consisting of that node and the nodes directly connected to it together with the edges connecting these nodes form a clique, i.e., are fully connected. A chordal graph now has the two following properties:

- The graph obtained by deleting one node with the edges connected to it is chordal;
- It has at least one node which has an adjacent clique.

The first property is almost evident, while the second property can be proven recursively on the number of nodes. Hence, a reordering and peeling off of the nodes of a chordal graph is possible whereby each node in turn has an adjacent clique in the remaining graph: start with such a node in the original graph, remove it with its connecting edges and continue recursively. Finding a node with an adjacent clique can be done in less than N^2 steps, hence the complexity of the reordering is certainly polynomial in N.

With this reordering of nodes, performing the Cholesky factorization in the order of peeling will not produce any fill-ins, exactly because of the adjacent clique property at each step. The converse is "generically" true as well, if a Cholesky factorization does not result in fill-ins *generically* (an element might accidentally become zero), then the sparsity graph must be chordal as well. It turns out that the maximum-entropy interpolant of a matrix with chordal sparsity pattern can be computed directly on the given entries, the famous algorithm to do so is the generalized Schur algorithm described in the next subsection. Unfortunately, many problems in modeling or reduced modeling of integrated circuits involve strictly positive definite matrices that do not have chordal sparsity patterns. In particular, multiband patterns are almost essentially nonchordal and hence will need additional, nonexact techniques for reduced modeling. This question is treated in the section on multiband generalization. A special case of a chordal graph is a graph representing a staircase filling, i.e., a filling corresponding to a nonregular band. One would obtain such a graph if in the order of nodes with adjacent cliques, each node in turn belongs to the adjacency set of its predecessor.

4.4. Schur's algorithm in the chordal case

We are now ready to introduce the generalized matrix Schur algorithm, originally presented as an estimation algorithm in DEPRETTERE [1981], and whose matrix properties were analyzed in DEWILDE and DEPRETTERE [1987]. The application of the algorithm to reduced modeling of integrated circuits was given in DEWILDE [1988]. We utilize the algebraic framework of the latter paper, slightly generalizing it to cover chordal sparsity in addition to staircases. Let the original, $N \times N$ strictly positive definite matrix be $G = [G_{i,j}]$ and let D be its main diagonal:

$$D = \text{diag}(G_{1,1}, G_{2,2}, \ldots, G_{N,N}).$$

It is advantageous to work with a normalized version of G, for theoretical purposes if not for numerical ones. Hence, let

$$g = D^{-1/2} G D^{-1/2}.$$

The matrix g will have all its diagonal elements equal to one (the situation could be generalized to the case where all the entries in G are in fact matrices, the block case, but for simplicity of explanation we keep the procedures scalar and shall indicate later on how to handle the block-matrix case). Let us assume, moreover, that the nodes are put in a correct adjacent-clique order, the staircase order will do if available.

4.4.1. A side excursion: the classical Schur parametrization case

Before engaging in the description of the matrix Schur algorithm, let us make a brief side excursion to the original algorithm involved in Schur's parametrization of a contractive, analytic function on the unit disc $\mathbf{D} = \{z: |z| < 1\}$ of the complex plane. Suppose that

$$s(z) = s_0 + s_1 z + s_2 z^2 + \cdots$$

is such a function, represented by its MacLaurin series. The question answered by the Schur parametrization is whether the given MacLaurin series does indeed correspond to a contractive function. To start, either $|s_0| = 1$ and $s(z)$ reduces to a constant of modulus one (by the maximum modulus theorem of complex analysis), or $|s_0| < 1$ and then a new contractive function which is analytic in $\mathbf{D}$ may be derived from $s(z)$ via the recipe:

$$s^{(1)}(z) = \frac{s(z) - s_0}{z(1 - \overline{s_0}s(z))} = s_0^{(1)} + s_1^{(1)} z + \cdots.$$

Notice that the transformation

$$s \mapsto \frac{s - s_0}{1 - \overline{s_0}s}$$

maps the unit disc onto itself. The procedure may be repeated on $s^{(1)}(z)$, yielding a criterion on $s_0^{(0)}$ and a new $s^{(2)}(z)$, and then recursively continued further. Let $\rho_0 = s_0$, $\rho_1 = s_0^{(1)}, \ldots$ be the so-called "Schur parameters" for $s(z)$. In an inverse scattering context where they often appear, the ρ_k's are also called reflection coefficients. The sequence of Schur parameters of a contractive function that is analytic in $\mathbf{D}$ is either finite, in which case the last coefficient is of unit modulus, or infinite, and then all Schur parameters are less than one in modulus. The Schur parameters determine $s(z)$ uniquely, just as the s_k's do, one series can be converted into the other and vice versa. In his famous 1917 paper, SCHUR [1917] demonstrates that $s(z)$ is contractive in the unit disc iff the Schur parametrization satisfies one of these two properties – this is the Schur criterion for contractivity (the proof is in fact pretty straightforward). The transformation that leads from $s^{(k)}(z)$ to $s^{(k+1)}(z)$ is obviously bilinear. It can be linearized if it is put in matrix form. Let us write for that purpose

$$s^{(n)}(z) = \frac{\delta^{(n)}(z)}{\gamma^{(n)}(z)}.$$

Then the following linear recursion produces the same effect as the original Schur parametrization

$$z\left[\gamma^{(n+1)}(z)\ \delta^{(n+1)}(z)\right] = \left[\gamma^{(n)}(z)\ \delta^{(n)}(z)\right]\frac{1}{\sqrt{1-|\rho_n|^2}}\begin{bmatrix} z & -\rho_n \\ -\bar{\rho}_n & 1 \end{bmatrix}$$

when the Schur parameter chosen as

$$\rho_n = \frac{\delta^{(n)}(0)}{\gamma^{(n)}(0)}$$

is less than one in magnitude (the square roots are included for normalization purposes, they may be dispensed with in practical computations). The recursion is started with $[\gamma^{(0)}(z)\ \delta^{(0)}(z)] = [1\ s(z)]$. Aside from a shift represented by z, the Schur recursion involves transformations with a hyperbolic matrix, sometimes called a Halmos transformation and defined as

$$H(\rho) = \frac{1}{\sqrt{1-|\rho|^2}}\begin{bmatrix} 1 & -\rho \\ -\bar{\rho} & 1 \end{bmatrix}.$$

Let us define the signature matrix

$$J = \begin{bmatrix} 1 & \\ & -1 \end{bmatrix}.$$

Then, we compute easily that $H(\rho)J(H(\rho))^H = (H(\rho))^H JH(\rho) = J$, which represents the hyperbolic property.

The original Schur theory works on a contractive function $s(z)$. Alternatively, one could start from what is known as a *positive real function* $\phi(z)$, i.e., a function that is analytic in $\mathbf{D}$ and such that $\text{Re}(\phi(z)) = (\phi(z)+\overline{\phi(z)})/2 \geq 0$ in $\mathbf{D}$. The Cayley transformation relates a contractive function $s(z)$ to a *positive real function* $\phi(z)$ (i.e., a function with positive real part $\text{Re}(\phi(z))$ for all z in the unit disc):

$$s(z) = \frac{\phi(z)-1}{\phi(z)+1}.$$

Schur's parametrization provides a test for positive reality on the sequence defined by the MacLaurin expansion of ϕ, the linearized recursion can now be started with

$$\left[\gamma^{(0)}(z)\ \delta^{(0)}(z)\right] = \tfrac{1}{2}\left[\phi(z) + 1\ \phi(z) - 1\right].$$

After $n + 1$ steps it will yield

$$\tfrac{1}{2}\left[\phi(z) + 1\ \phi(z) - 1\right]\theta_0(z)\theta_2(z)\cdots\theta_n(z) = z^n\left[\gamma^{(n)}(z)\ \delta^{(n)}(z)\right]$$

with each $\theta_i(z)$ representing an elementary Schur step. Let us introduce the para-Hermitian conjugate of a function of z as $f^*(z) = \overline{f(1/\overline{z})}$. In the Schur parametrization theory (see, e.g., DEWILDE, VIEIRA, and KAILATH [1978]), one deduces that the overall Schur matrix $\Theta_n(z) = \theta_0(z)\theta_1(z)\cdots\theta_n(z)$ has the form

$$\Theta_n(z) = \frac{1}{2}\begin{bmatrix} (1 + \phi_n^*(z))T_{rn}^{-*}(z) & (1 - \phi_n(z))T_{fn}^{-1}(z) \\ (1 - \phi_n^*(z))T_{rn}^{-*}(z) & (1 - \phi_n(z))T_{fn}^{-1}(z) \end{bmatrix}$$

in which $\phi_n(z)$ is also PR in $\mathbf{D}$, $T_{rn}(z)$ and $T_{fn}(z)$ are analytic in $\mathbf{D}$ and

$$\frac{\phi_n(z) + \phi_n^*(z)}{2} = T_{rn}(z)T_{rn}^*(z) = T_{fn}^*(z)T_{fn}(z).$$

(Notice that the para-Hermitian conjugate is equal to the Hermitian conjugate only on the unit circle. Outside the unit circle it is its analytic continuation, when definable. Often in the engineering literature, the para-Hermitian conjugate is denoted by a substar, in contrast to the upper star, which is often interpreted as equal to complex conjugation. Here we use upper star, to indicate that the upper-stared quantity corresponds in fact to the analytic continuation of the adjoint in the Fourier domain on the unit circle). One of the central properties of $\phi_n(z)$, resulting from the Schur parametrization, is that it interpolates the original $\phi(z)$ to order n:

$$\phi(z) = \phi_n(z) + z^{n+1}r(z)$$

in which $r(z)$ is analytic in $\mathbf{D}$. Remark also that $\phi_n^{-1}(z)$ is polynomial hence $\phi_n(z)$ is of the "autoregressive type." The theory of maximum entropy interpolation is well developed in complex function theory, and it is satisfied by $\phi_n(z)$ as a maximum entropy interpolant of order n for $\phi(z)$, whereby the entropy measure now must be taken as

$$\int_{-\pi}^{\pi} \log \mathrm{Re}\left(\phi(\xi)\right)\frac{\mathrm{d}\xi}{2\pi}.$$

4.4.2. The matrix case
In the matrix case, the hyperbolic transformation will play a role similar to the complex case. We embed the Halmos transformation in an otherwise unitary matrix and index its position, much as is done in the classical QR algorithm based on Jacobi transformations.

This leads to $2N \times 2N$ hyperbolic matrices $\theta_{i,j}(\rho)$ of the form

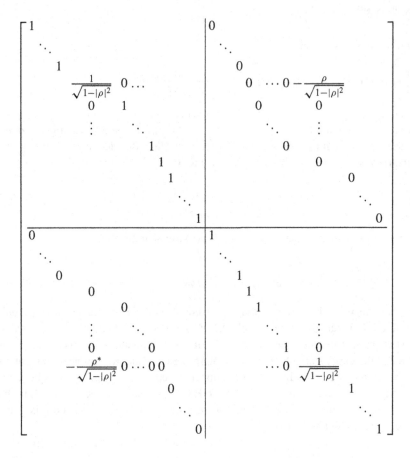

in which the elements of the elementary hyperbolic transformation are on the intersections of rows i and $N + j$ with columns i and $N + j$. Similar as in the complex case we will use it to eliminate entries in an appropriate linearization. We give the analog for Schur's algorithm in the matrix case first before motivating it. This simplifies the discussion, but an alert reader will recognize the similarities with the complex case. First we define the equivalent of the function $\phi(z)$ (as announced earlier we work on a normalized version of G, although this is not strictly necessary):

$$\Phi = \begin{bmatrix} 1 & 2g_{1,2} & 2g_{1,3} & \cdots & 2g_{1,N} \\ 0 & 1 & 2g_{2,3} & \cdots & 2g_{2,N} \\ 0 & 0 & 1 & \cdots & 2g_{3,N} \\ \vdots & \vdots & \vdots & \ddots & \vdots \\ 0 & 0 & 0 & \cdots & 1 \end{bmatrix}.$$

Hence $g = \frac{1}{2}(\Phi + \Phi^H)$, where A^H indicates the Hermitian conjugate of the matrix A. We define as initial data

$$\Gamma_0 = \begin{bmatrix} 1 & g_{1,1} & \cdots & g_{1,N} \\ 0 & 1 & \cdots & g_{2,N} \\ \vdots & \vdots & \ddots & \vdots \\ 0 & 0 & \cdots & 1 \end{bmatrix}, \qquad \Delta_0 = \begin{bmatrix} 0 & g_{1,1} & \cdots & g_{1,N} \\ 0 & 0 & \cdots & g_{2,N} \\ \vdots & \vdots & \ddots & \vdots \\ 0 & 0 & \cdots & 0 \end{bmatrix}.$$

Hence we have

$$[\Gamma_0 \quad \Delta_0] = \tfrac{1}{2}[\Phi + I \quad \Phi - I].$$

Let $S_0 = \Gamma_0^{-1}\Delta_0$. We see that

$$g = \frac{\Phi + \Phi^H}{2} = \tfrac{1}{4}(\Phi + I)(I - S_0 S_0^H)(\Phi^H + I),$$

and hence S_0 is a contractive matrix in the sense that $S_0 S_0^H \preceq I$. We shall say that a couple of $N \times N$ upper triangular matrices $[\Gamma \ \Delta]$ are (*strictly*) *admissible* if Γ is invertible and $\Gamma^{-1}\Delta$ is (strictly) contractive. Define the $2N \times 2N$ signature matrix

$$J = \begin{bmatrix} I_N & \\ & -I_N \end{bmatrix},$$

where I_N is the unit matrix of dimension N. If Θ is a $2N \times 2N$ is a J-unitary matrix, i.e.,

$$\Theta J \Theta^H = \Theta^H J \Theta = J,$$

then any transformation of an admissible $[\Gamma \ \Delta]$ on the right with Θ will yield a new matrix

$$[\Gamma' \quad \Delta'] = [\Gamma \quad \Delta]\Theta,$$

which is (strictly) admissible when the original is (strictly) admissible. A product of J-unitary matrices will itself be J-unitary as well.

The Schur elimination procedure based on the chordal set $\mathcal{S}$ will consist in applying a sequence of elementary Halmos transformations on recursively computed admissible matrices, starting with $[\Gamma_0 \ \Delta_0]$, in the adjacent-clique order on the interpolation data. Each Halmos transformation is intended to eliminate one off-diagonal entry corresponding to a position in the set $\mathcal{S}$. Let the matrices G, g, Γ_0, Δ_0 be ordered in the adjacent-clique order, and suppose that the elements of $\mathcal{S}$ in row i are given by $(i, n_{i,1}), (i, n_{i,2}), \ldots, (i, n_{i,m_i})$ where $i < n_{i,1} < \cdots < n_{i,m_i}$ (the set may even be empty of course). We shall perform the elimination procedure in the strict order $(1, n_{1,1}), (1, n_{1,2}), \ldots, (2, n_{2,1}), \ldots$. Let us number these steps by the integer K. At step K corresponding to, say, the predecessor of $(i, n_{i,k})$, we have available an admissible pair $[\Gamma_K \ \Delta_K]$, which is such that the elements $(\Delta_K)_{i,j}$ have been annihilated for all pairs (i, j)'s in the elimination list preceding $(i, n_{i,k})$. The new step will annihilate $(\Delta_K)_{i,n_{i,k}}$ and use for that purpose an elimination matrix of the Halmos type, namely

$H_{i,n_{i,k}}(\rho_{i,n_{i,k}})$ with

$$\rho_{i,n_{i,k}} = (\Gamma_K)_{i,n_{i,k}}^{-1} (\Delta_K)_{i,n_{i,k}}.$$

At least three remarks are important here:

- The element $(\Delta_{K+1})_{i,n_{i,k}}$ is set equal to zero by the elimination procedure;
- The elements that were put to zero in previous steps remain zero in all the subsequent eliminations because of the adjacent-clique order;
- There are no fill-ins, also due to the adjacent-clique property at each step.

After completion of all the steps, an overall elimination matrix Θ_t results given by

$$\Theta_t = \theta_{1,n_{1,1}} \theta_{1,n_{1,2}} \cdots \theta_{N,n_{N,m_N}}$$

and finally

$$[\,\Gamma_t \quad \Delta_t\,] = [\,\Gamma_0 \quad \Delta_0\,]\,\Theta_t$$

are obtained, in which all the elements belonging to the set S in Δ_t have been annihilated (as well as all the diagonal elements due to the initial normalization). In parallel, the entries in Θ_t are essentially constrained to the diagonal, the set S and its reflection. To make this statement more precise, let

$$\Theta_t = \begin{bmatrix} \Theta_{1,1} & \Theta_{1,2} \\ \Theta_{2,1} & \Theta_{2,2} \end{bmatrix}.$$

Then, the nonzero entries of $\Theta_{1,1}$ are restricted to diagonals and S^*, those of $\Theta_{2,2}$ to diagonals and S while the nonzero entries of $\Theta_{1,2}$ are restricted to S and those of $\Theta_{2,1}$ are restricted to S^*. This follows also from the special structure of S and the order in which the eliminations have been done. We shall call such a J-unitary matrix "S-based." The following theorem from DEWILDE and DEPRETTERE [1987] holds.

THEOREM 4.2. *An S-based J-unitary matrix Θ_t has the form*

$$\frac{1}{2} \begin{bmatrix} (I + \Phi_t^H)L_t^{-H} & (I - \Phi_t)M_t^{-1} \\ (I - \Phi_t^H)L_t^{-H} & (I + \Phi_t)M_t^{-1} \end{bmatrix}$$

in which Φ_t, L_t, and M_t are upper triangular matrices, Φ_t has unit main diagonal, L_t and M_t are invertible, and in addition

$$\frac{\Phi_t + \Phi_t^H}{2} = L_t L_t^H = M_t^H M_t.$$

The Schur procedure executed as detailed above yields the following interpolation result.

THEOREM 4.3 (DEWILDE and DEPRETTERE [1987]). *Let $g_t = \frac{1}{2}(\Phi_t + \Phi_t^H)$ be the result of the Schur elimination procedure based on the chordal set S. Then,*

$$(g - g_t)_{i,j} = 0 \quad for\ (i, j) \in S$$

and, in particular,

$$\Phi - \Phi_t = 2\Delta_t M_t,$$

where Δ_t is defined by $[\Gamma_0 \ \Delta_0]\Theta_t = [\Gamma_t \ \Delta_t]$.

Given the theory developed so far, the two theorems are not too hard to prove. The Schur recursion necessitates a number of elementary Halmos transformations precisely equal to the number of elements in the interpolation set S, and it produces the desired maximum-entropy interpolant, due to the fact that the appropriate entries in the inverse matrix are zero. Notice also that L_t^{-1} and M_t^{-1} have supports on S and the diagonal, while L_t, M_t, and Φ_t are full matrices, which in practical calculations will never be computed – a banded computational scheme exists for vector–matrix multiplication with both L_t and L_t^{-1}, see DEWILDE and DEPRETTERE [1987].

4.5. Generalizations

The preceding theory works only for matrices with a chordal sparsity pattern. Can the theory be extended to more general types of matrices, in particular to matrices with multiple bands, as often occur in 2D or 3D finite element or finite difference problems. We give an indication on how an approximate technique may yield satisfactory results. We refer the reader to the literature (NELIS, DEWILDE, and DEPRETTERE [1989]) for further information. A first remark is that in some, quite common cases, a double banded (or even multibanded) matrix can be chordal. For example, a $2n \times 2n$ matrix with four $n \times n$ blocks with filling pattern as in

$$\begin{bmatrix} * \ * & & & * \ * & & & \\ * \ * \ * & & & * \ * \ * & & \\ & * \ * \ * & & & * \ * \ * & \\ & & * \ * & & & * \ * \\ \hline * \ * & & & * \ * & & & \\ * \ * \ * & & & * \ * \ * & & \\ & * \ * \ * & & & * \ * \ * & \\ & & * \ * & & & * \ * \end{bmatrix}$$

is actually of chordal type and can be solved exactly using Schur matrix interpolation (more general forms can easily be derived using the theory of adjacent cliques described above). This result can be used to factorize more general matrices approximatively. For example, a (positive definite) block matrix of the type

$$\begin{bmatrix} A_{11} & A_{12} & \\ A_{21} & A_{22} & A_{23} \\ & A_{32} & A_{33} \end{bmatrix}$$

in which all the nonzero blocks are only sparsely specified and which is such that the two submatrices

$$A_1 := \begin{bmatrix} A_{11} & A_{12} \\ A_{21} & A_{22} \end{bmatrix}, \qquad A_2 := \begin{bmatrix} A_{22} & A_{23} \\ A_{32} & A_{33} \end{bmatrix}$$

have chordal filling specifications has a sparse approximant for its inverse which can be constructed from sparse approximants of A_1, A_2, and A_{22} as follows. Let A_{ME} indicate the maximum-entropy approximant of a sparsely specified matrix A, then A_{ME}^{-1} has corresponding sparse fillings according to the theory developed above. In addition, let us introduce one more bit of notation: by "$\Box A$" we mean the operation of fitting the matrix A in a larger matrix that extends its range of indices while padding it with zeros. A "good" approximant for the ME inverse of A is then given by

$$A_{ME}^{-1} \approx \Box(A_1)_{ME}^{-1} + \Box(A_2)_{ME}^{-1} - \Box(A_{22})_{ME}^{-1}. \tag{4.2}$$

The significance of this formula is that the inverse of the maximum-entropy interpolant for the matrix A based on the given nonchordal definition pattern is expressed in terms of maximum interpolants of submatrices whose definition pattern is presumably chordal and which can hence be computed by a fast algorithm such as the Schur parametrization given in the previous section. We give a short motivation for this result, a complete theory with proofs is given in NELIS [1989]. The main property used is the fact that for reasonably well-conditioned positive definite matrices with entries specified on a given pattern, the inverse of the ME approximant of a principal submatrix is actually a good approximation of the restriction of the inverse of the ME approximant to the same indices as the submatrix – in matrix notation, let $A(i, j)$ be the principal submatrix obtained by restraining A to the index range $i \cdots j$ then, utilizing the same pattern of specified entries,

$$(A(i, j))_{ME}^{-1} \approx (A_{ME}^{-1})(i, j). \tag{4.3}$$

Notice that the two matrices now have the same sparsity pattern corresponding to the pattern given, but they are not numerically the same. This opens the way for a "calculus of sparse inverse matrices" of the ME type. The formula (4.2) can now be interpreted as defining block-wise approximations on the ME inverse of the original matrix, whereby the middle matrix (corresponding to the "22" block) is repeated trice, each time with a different approximant. There is no guarantee that (4.2) actually defines a positive definite matrix, but since the approximants are assumed close, the approximation should be good when the original matrix is well conditioned, a detailed error analysis can be found in the already cited thesis by NELIS [1989]. The reason why (4.3) holds is the fact that ME approximants actually define strong norm approximants on the Cholesky factors. This seems to have been remarked first in DEWILDE and DYM [1981]. Formula (4.2) generalizes to large matrices with intricate block sparsity patterns and has been used successfully in the modern finite-element modeling program for interconnects of integrated circuits SPACE (see VAN DER MEIJS [1992]).

5. Hankel-norm model reduction

5.1. The setting

In this section we are interested in linear operators – of the type T where T induces a linear map $y = Tu$ – and where T is represented by a "model," more precisely a model that represents the linear computations the computer actually executes, based on its sequential intake of data, use of memory and sequential production of results. Such a model is called a "state-space model" because it is an instance of a classical model for a time-varying dynamical system adapted to the computational context. We start out with a simple but computationally intensive representation of the desired function and we shall proceed to reduce that representation to another one with much lower computational complexity. The model we start with will be a direct derivative of all the known data and will therefore be of much too high complexity, called a "model of high complexity." Our goal will be to reduce that model to one of smallest possible computational complexity, given a specified and acceptable tolerance on the accuracy. Here, T may be a matrix, but it may also be an infinite-dimensional operator, the theory that we shall present is not restricted to finite operators. In our basic framework, T will be a lower triangular operator, it represents a "causal" transfer between the vectors u and y viewed as time series. If it so happens that T does not satisfies this property, e.g., when T is a full matrix, then we would decompose T first either additively or multiplicatively into lower/upper operators: $T = L + U$ or $T = LU$ and then approximate L and dually U separately, but it may also be useful to move the main diagonal up so that the whole matrix becomes lower triangular, see the special case treated later. Our theory does not really become more complicated if we assume that T is in fact a block matrix, i.e., that the entries in T are actually matrices themselves, provided dimensions in rows and columns match. Hence T will look as follows:

$$
T = \begin{bmatrix}
\ddots & & & & \\
\ddots & 0 & & 0 & \\
\ddots & T_{-1,-1} & 0 & & \\
\ddots & T_{0,-1} & \boxed{T_{0,0}} & 0 & \\
\ddots & T_{1,-1} & T_{1,0} & T_{1,1} & \ddots \\
\ddots & & \ddots & \ddots & \ddots
\end{bmatrix}.
$$

Here, the $T_{i,j}$ block has dimension $n_i \times m_j$ and represents a partial map of the vector entry u_j to an additive component of y_i in the output vector in the map:

$$
\begin{bmatrix}
\ddots & & & & \\
\ddots & 0 & & 0 & \\
\ddots & T_{-1,-1} & 0 & & \\
\ddots & T_{0,-1} & \boxed{T_{0,0}} & 0 & \\
\ddots & T_{1,-1} & T_{1,0} & T_{1,1} & \ddots \\
\ddots & & \ddots & \ddots & \ddots
\end{bmatrix}
\begin{bmatrix}
\vdots \\ u_{-2} \\ u_{-1} \\ \boxed{u_0} \\ u_1 \\ u_2 \\ \vdots
\end{bmatrix}
=
\begin{bmatrix}
\vdots \\ y_{-2} \\ y_{-1} \\ \boxed{y_0} \\ y_1 \\ y_2 \\ \vdots
\end{bmatrix}.
$$

We identify the 0th (block-)element in a matrix by putting a square around it, and similarly for the $(0, 0)$th element of a matrix of operator for orientation purposes. The (linear) computation defined by $y = Tu$ as executed by a computer that takes the input data

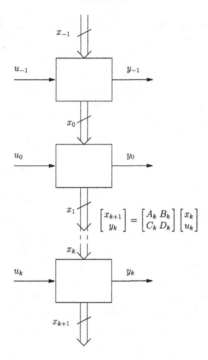

FIG. 5.1. A causal state-space realization of an operator T: the state represents the data available for computation at a given stage.

sequence u and produces the output sequence y can be represented by a "causal model" for T. The transfer from the input vector u to the output vector y can indeed be written in terms of an intermediate sequence $\{x_k\}$ of data which the computer stores in memory, and so-called *realization matrices* representing the computations at the sequence point k, as:

$$x_{k+1} = A_k x_k + B_k u_k,$$
$$y_k = C_k x_k + D_k u_k.$$

This is called a "time-varying state-space representation" of the computation. The dimension δ_k of the vector x_k is called the state dimension at point k, and the dimensions of the realization matrices A_k, B_k, C_k, D_k are respectively $\delta_{k+1} \times \delta_k, \delta_{k+1} \times m_k, n_k \times \delta_k, n_k \times m_k$. A graphical representation of the state representation is shown in Fig. 5.1.

We call A_k the *state transition matrix* at point k, while the other matrices B_k, C_k, and D_k stand for partial local maps input–state, state–output, and input–output, respectively, at point k. The system will be strictly causal when $D_k = 0$. It may happen that some of the vectors and matrices are not present. For example, if a matrix is represented by a state model, then the initial state in the representation (e.g., x_0) will not be present. In that case we say that the dimension of the respective vector is zero, it is represented

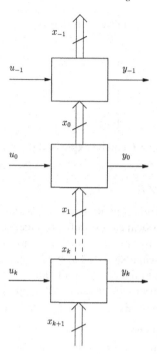

Fig. 5.2. The signal flow in an anticausal system.

by a place holder "·", but there is no numerical value present (not even zero). Similarly, the last state will also not be present when a finite matrix is represented, and disappears accordingly.

An anticausal system – represented by an upper block matrix – may similarly have an anticausal state realization as follows:

$$x'_k = A'_k x'_{k+1} + B'_k u_k,$$
$$y_k = C'_k X'_{k+1} + D'_k u_k.$$

Here, we have chosen to make the realization strictly anticausal by putting $D'_k = 0$. A graphical representation of an anticausal linear system is shown in Fig. 5.2.

It is convenient for notational purposes to assemble the realization matrices in diagonal matrices or operators with appropriate dimensions. So we define:

$$A = \begin{bmatrix} \ddots & & & \\ & A_{-1} & & \\ & & \boxed{A_0} & \\ & & & A_1 \\ & & & & \ddots \end{bmatrix}, \quad B = \begin{bmatrix} \ddots & & & \\ & B_{-1} & & \\ & & \boxed{B_0} & \\ & & & B_1 \\ & & & & \ddots \end{bmatrix},$$

and so forth. Introducing also the causal shift matrix:

$$
Z = \begin{bmatrix} \ddots & & & & \\ & \ddots & 0 & & \\ & & I & \boxed{0} & \\ & & & I & 0 \\ & & & & \ddots & \ddots \end{bmatrix},
$$

we see that the original operator T can be expressed in terms of these diagonal matrices as follows:

$$
T = D + C(I - ZA)^{-1}ZB.
$$

This generalizes the classical representation matrix for stationary discrete time systems, where Z now replaces the classical causal shift z (notice, however, that the unit matrices in Z may have different dimensions and that Z does not commute with matrices in the sense that $ZA \neq AZ$, as the scalar shift "z" would). Some care must be exercised when one interprets these formulas in the case of finite matrices. The block diagonals, A, B, C, D, and Z are all block matrices or operators with appropriate dimensions. So, A will map a state sequence of dimensions $\ldots, \delta_{-1}, \boxed{\delta_0}, \delta_1, \ldots$ to $\ldots, \delta_0, \boxed{\delta_1}, \delta_2, \ldots$, and Z applied to the same state sequence will map to $\ldots, \delta_{-2}, \boxed{\delta_{-1}}, \delta_0, \ldots$. Numerically, Z will be a perfect unit matrix in the finite case, but its block decomposition will make it look like the shift matrix that it is, for a shift on a finite sequence will keep the numerical values of that sequence, but will shift their indices! The inverse in the formula for T can be interpreted in a purely formal sense as meaning $(I - AZ)^{-1} = I + AZ + (AZ)^2 + \cdots$, but the series will of course also converge in the operator sense, if AZ is idempotent (which would be the case with finite matrices) or if $(AZ)^k$ converges to zero quickly enough. To make the notion of convergence more precise, we introduce a norm on the input and output spaces, namely the ℓ_2 or quadratic norm:

$$
\|u\| = \sqrt{\sum_j \|u_j\|^2},
$$

where the $\|u_j\|$ is the usual Euclidean norm on a vector (square root of the sum of magnitudes square of the entries). In this paper we treat operators T that are bounded as maps between input and output spaces endowed with the quadratic norm (this corresponds to the L_∞ norm on the unit circle in the classical case). A sufficient condition for this is that the spectral radius of AZ is strictly less than one, in which case the Neumann series $I + AZ + (AZ)^2 + \cdots$ converges in norm. If that is the case we say that the realization for T is *uniformly exponentially stable* or *ues* (exponential stability of time-varying systems is extensively treated in the time-varying literature). To characterize this case further we define:

$$
\ell_A = \sigma(ZA) = \lim_{n \to \infty} \|(ZA)^n\|^{1/n}
$$

and a system realization will be ues if $\ell_A < 1$. The "Z" can be taken out of the formula for ℓ_A if we define the South-East diagonal shift (with "∗" indicating the adjoint operator):

$$A^{(1)} = ZAZ^*$$

so that

$$\ell_A = \lim_{n \to \infty} \|AA^{(1)}A^{(2)} \cdots A^{(n-1)}\|^{1/n}.$$

The "continuous product" that appears in the formula is useful for other purposes. In particular, if we express the block entries in T in terms of a realization, we obtain, for $i > j$, $T_{i,j} = C_i A_{i-1} \cdots A_{j+1} B_j$, and we see that the entries become (uniformly) exponentially small for large $i - j$ when $\ell_A < 1$. In a later section, we shall see how we can recover a realization from the entries in T, but before doing so we turn to some more definitions and properties in the basic framework.

5.1.1. *Lyapunov transformations*
A state realization for an operator or matrix is not unique, even when it is minimal. In fact, we can permit ourselves a state transformation that introduces at each point k a transformed state x'_k related to the original via $x_k = R_k x'_k$ where the state transformation matrix R_k is nonsingular for each k. In the case of infinite systems we usually require even more, namely R_k and R_k^{-1} should be uniformly bounded over k. Such transformations we call "Lyapunov transformations." They have the nice property that they are preserving the exponential stability of the realization. Under the state transformation, a causal realization transforms as follows:

$$\begin{bmatrix} A_k & B_k \\ C_k & D_k \end{bmatrix} \mapsto \begin{bmatrix} R_{k+1}^{-1} A_k R_k & R_{k+1}^{-1} B_k \\ C_k R_k & D_k \end{bmatrix}, \tag{5.1}$$

or, when expressed in the global diagonal notation:

$$\begin{bmatrix} A & B \\ C & D \end{bmatrix} \mapsto \begin{bmatrix} (R^{(-1)})^{-1} AR & (R^{(-1)})^{-1} B \\ CR & D_k \end{bmatrix}.$$

State transformations are very important not only to achieve canonical representations discussed below, but also to obtain algebraically minimal calculations – see Chapter 14 of DEWILDE and VAN DER VEEN [1998].

5.1.2. *Input/output normal forms*
We say that a realization is in *output normal form* when

$$A^*A + C^*C = I$$

i.e., $A_k^* A_k + C_k^* C_k = I$ for each k. From (5.1) and putting $M_k = R_k^{-*} R_k^{-1}$, we see that a realization can be brought to output normal form if a bounded and invertible solution exists to the recursive set of *Lyapunov–Stein* equations

$$A_k^* M_{k+1} A_k + C_k^* C_k = M_k,$$

or, equivalently, if $A^* M^{(-1)} A + C^* C = M$ has a boundedly invertible diagonal operator M as a solution. The existence of the solution has been much studied in Lyapunov stability theory, we suffice here with some facts. If the original realization is ues (i.e., if $\ell_A < 1$), then the Lyapunov–Stein equation always has a bounded solution M. The solution M can be expressed as the so-called observability Gramian:

$$M = \sum_{k=0}^{\infty} (A^{\{k\}})^* (C^*C)^{(-k)} (A^{\{k\}}),$$

where we have put

$$A^{\{k\}} = A^{(-k+1)} \cdots A^{(-1)} A$$

and the sum converges in norm because of the ues assumption. The state transformation needed to bring the system in output normal is then obtained from $M^{-1} = RR^*$. The problem with its existence is whether M is boundedly invertible. We shall say that the system is *strictly observable* if that is the case. In the sequel we shall normally assume this property to be valid.

5.1.3. Realization theory and canonical spaces

One may wonder when a causal transfer operator T has a finite-dimensional realization at each time point k. It turns out (see DEWILDE and VAN DER VEEN [1998]) that this will be the case iff each kth order operator

$$H_k := \begin{bmatrix} T_{k+1,k} & T_{k+1,k-1} & T_{k+1,k-2} & \cdots \\ T_{k+2,k} & T_{k+2,k-1} & T_{k+2,k-2} & \cdots \\ T_{k+3,k} & T_{k+3,k-1} & T_{k+3,k-2} & \cdots \\ \vdots & \vdots & \vdots & \end{bmatrix}$$

has finite rank δ_k. We call these operators local Hankel matrices, and their rank δ_k actually gives the minimal state dimension needed at point k. The here defined Hankel operators do not have the classical Hankel structure (elements equal along antidiagonals), but they do fit the general functional definition of Hankel operators as exemplified in Fig. 5.3, where the matrices are shown in a graphical way (notice that the columns in the picture are in reverse order, the definition of the H_k fits the classical matrix representation).

Realization theory shows that any collection of minimal factorizations of all H_k will produce a minimal realization. If we express the Hankel operators in terms of a state

$$\begin{bmatrix} \ddots & & \ddots & & & \\ & \ddots & & T_{k,k} & & \\ \hline & & T_{k+1,k-1} & T_{k+1,k} & T_{k+1,k+1} & \\ & & T_{k+2,k-1} & T_{k+2,k} & T_{k+2,k+1} & T_{k+2,k+2} \\ & & & & \ddots & & \ddots \end{bmatrix}$$

FIG. 5.3. Generalized Hankel operators in a matrix or operator.

space representation we have:

$$H_k = \mathcal{O}_k \mathcal{R}_k = \begin{bmatrix} C_k \\ C_{k+1}A_k \\ \vdots \end{bmatrix} [\, B_{k-1} \quad A_{k-1}B_{k-2} \quad \cdots \,],$$

and the "realization theory" is reduced to reading the A_k, B_k, C_k, D_k from the factorization. The columns of $\mathcal{O}_k$ form a basis for the columns of the Hankel matrix H_k while the rows of $\mathcal{R}_k$ form a basis for its rows. We shall obtain a realization in output normal form iff the columns of $\mathcal{O}_k$ have been chosen orthonormal for each k. The realization derived from the factorization is then given by:

$$B_{k-1} = (\mathcal{R}_k)_1, \qquad C_k = (\mathcal{O}_k)_0, \qquad A_k = \mathcal{O}_{k+1}^{\dagger} \mathcal{O}_k^{\downarrow},$$

where $(\mathcal{R}_k)_1$ is the first element of the "reachability" matrix $\mathcal{R}_k$, $(\mathcal{O}_k)_0$ the top element of the "observability" matrix $\mathcal{O}_k$, the "$\dagger$" indicates the Moore–Penrose inverse, and the "downarrow" on $\mathcal{O}_k$ indicates a matrix equal to $\mathcal{O}_k$ except for its first block-element, which has been deleted. The matrix A_k is uniquely defined because of the minimality of the factorization, even when any general inverse is used.

5.1.4. Balanced realization

It is also possible to define a balanced realization, by using a factorization based on a singular value decomposition of the Hankel operator:

$$H_k = U_k \begin{bmatrix} \sqrt{\sigma_1} & & \\ & \ddots & \\ & & \sqrt{\sigma_k} \end{bmatrix} \cdot \begin{bmatrix} \sqrt{\sigma_1} & & \\ & \ddots & \\ & & \sqrt{\sigma_k} \end{bmatrix} V_k.$$

However, balanced realizations and approximations are only of limited use in time-varying theory, they are unable to handle transfer operators of low rank with sparse entries far from the main diagonal adequately (see DEWILDE and VAN DER VEEN [1998]). We give them here for the sake of completeness.

5.1.5. Reachability/observability bases in terms of realizations

It is easy to produce a direct relation between realizations and reachability or controllability bases, in particular we find:

$$\mathbf{F}_0 = C(I - ZA)^{-1} = \begin{bmatrix} \ddots & & & & \\ \ddots & C_{-1} & & & \\ \ddots & C_0 A_{-1} & \boxed{C_0} & & \cdots \\ \ddots & C_1 A_0 A_{-1} & C_1 A_0 & C_1 & \\ \ddots & & \ddots & \ddots & \ddots \end{bmatrix}$$

and dually

$$\mathbf{F} = B^* Z^* (I - A^* Z^*)^{-1}.$$

Each block column of $\mathbf{F}_0$ or $\mathbf{F}$ forms the basis for a local observability or controllability space.

5.2. Hankel-norm model reduction

We are given a lower (block-)operator T (we write: $T \in \mathcal{L}$) that we wish to approximate by a lower operator T_a of minimal complexity and that meets a certain preassigned complexity. First we make the notion "complexity" and "meeting a preassigned norm" more concrete.

5.2.1. Complexity
We identify "complexity" with "local state dimension." Suppose indeed that at stage k the state dimension (the total number of floating-point numbers the system has stored in memory from its past) is δ_k. Then it can be shown that number, together with the dimensions of the local input and output space determines the local computational complexity. It turns out that the number of floating point operations needed at stage k is given by $\frac{1}{2}(m_k + n_k + \delta_k)(m_k + n_k + \delta_{k+1} + 1)$ (see DEWILDE and VAN DER VEEN [1998]), exactly equal to the number of "algebraically free parameters" at that stage.

5.2.2. Norm
What is an adequate approximating norm? In the classical model reduction context an L_∞-type norm is known to be too strong (because the polynomials or rationals are not dense in such a space), while an L_2 norm is usually too weak, because it gives rise to undesirable phenomena like the Gibbs phenomenon. A good compromise, one that also offers quite a bit of flexibility, is provided by the Hankel norm, i.e., the supremum of the norms of the local Hankel operators we defined before. This is the norm we shall be using, hence we define

$$\|T\|_H = \sup_k \|H_k\|.$$

We still need to characterize the approximation accuracy needed. We take as measure for precision a Hermitian, strictly positive diagonal operator Γ – in fact it could be taken as $\Gamma = \varepsilon \cdot I$ for some small epsilon, but we may need the extra freedom of accommodating the precision at each time point.

5.2.3. High-order model
As described earlier, we start out our model reduction by selecting an appropriate representation of the desired computation as a high-complexity or high-order model that can be used computationally. An example of such a high-order model is given by a truncated Taylor-like series of high-enough order so that the truncation error has hardly any impact, but other, more convenient high-order representations may be adequate as well. If

$$T \approx T_0 + Z T_1 + Z^2 T_2 + \cdots + Z^n T_n$$

(with n sufficiently large and where each T_k represents a shifted diagonal of T), then a simple but high-complexity realization for T is given by the generalized companion form (in formal output normal form)

$$
\begin{bmatrix} A & B \\ C & D \end{bmatrix} = \left[\begin{array}{cccc|c} 0 & & & & T_n \\ I & 0 & & & T_{n-1} \\ & \ddots & \ddots & & \vdots \\ & & I & 0 & T_1 \\ \hline 0 & \cdots & 0 & I & T_0 \end{array} \right]. \tag{5.2}
$$

Expression (5.2) should be interpreted as a matrix consisting of block diagonals. At time point k the local realization has the form

$$
\begin{bmatrix} A_k & B_k \\ C_k & D_k \end{bmatrix} = \left[\begin{array}{cccc|c} 0 & & & & T_{n,k} \\ I & 0 & & & T_{n-1,k} \\ & \ddots & \ddots & & \vdots \\ & & I & 0 & T_{1,k} \\ \hline 0 & \cdots & 0 & I & T_{0,k} \end{array} \right].
$$

Also the shift matrix Z must be interpreted in a block fashion and now has the form

$$
\begin{bmatrix} Z & & & \\ & Z & & \\ & & \ddots & \\ & & & Z \end{bmatrix}
$$

conformal with the block-diagonal decomposition of A. Given the higher model for T and the precision Γ, the model reduction problem becomes:

Find a causal operator T_a of minimal state complexity such that

$$
\|(T - T_a)\Gamma^{-1}\|_H \leqslant 1,
$$

i.e., T_a approximates T up to a precision given by Γ. It is customary to take the higher model T so that it is strictly causal, i.e., $T_0 = 0$ and to require the same of the low-order approximation. We follow that habit since it does not impair generality and simplifies some properties. Before embarking on the solution and its properties, we introduce the main ingredients needed.

5.2.4. Ingredient #1: Nehari reduction

The Nehari theorem adapted to our context is as follows.

THEOREM 5.1. *For any bounded, strictly causal operator T,*

$$
\|T\|_H = \min_{T'' \in \mathcal{U}} \|T + T''\|,
$$

where the norm in the second member is the operator norm and T'' is a bounded, anticausal operator.

A proof of the Nehari theorem in the general context of nest algebras (to which our setup conforms) goes back to the work of ARVESON [1975]. For a proof restricted to our specific context, see DEWILDE and VAN DER VEEN [1998]. Application of the Nehari theorem reduces the problem to: *Find a (general) bounded operator T' so that its causal part $T_a = \mathbf{P}T'$ is of minimal complexity and*

$$\|(T - T')\Gamma^{-1}\| \leqslant 1.$$

5.2.5. Ingredient #2: external factorization

We are given $T \in \mathcal{L}$. An "external factorization" consists of finding $\Delta \in \mathcal{L}$ and $U \in \mathcal{L}$ unitary such that $T = U \Delta^*$ (a more general type relaxes the requirement on U, see further). This type of factorization is reminiscent of the coprime factorization of classical system theory, where U is an all-pass function that collects the "poles" of T and Δ^* is obtained as $U^*T - U^*$ pushes the poles of T to anticausality. It is easy to perform an external factorization on the state-space representation of T, especially when it is given in output normal form. So suppose that the realizations are given as (we use the $\approx$ sign to represent realizations).

$$\mathbf{T}_k \approx \begin{bmatrix} A_k & B_k \\ C_k & D_k \end{bmatrix}$$

in which, for all k,

$$A_k^* A_k + C_k^* C_k = I.$$

Then, the kth realization matrix for U is found by completing the first block column to form unitary matrices:

$$\begin{bmatrix} A_k & B_{Uk} \\ C_k & D_{Uk} \end{bmatrix}$$

thereby producing B_{Uk} and D_{Uk} as completing matrices. The "remainder" Δ_k is then given by

$$\Delta_k \approx \begin{bmatrix} A_k & B_{Uk} \\ B_k^* A_k + D_k^* C_k & \\ & B_k^* B_{Uk} + D_k^* D_{Uk} \end{bmatrix}.$$

Algorithmically, a simplified "Householder-type" algorithm will provide the missing data. Numerical analysts would write, somewhat equivocally

$$\begin{bmatrix} B_{Uk} \\ D_{Uk} \end{bmatrix} = \begin{bmatrix} A_k \\ C_k \end{bmatrix}^{\perp}.$$

5.2.6. Ingredient #3: J-unitary operators

In interpolation and approximation theory, J-unitary operators of various types play a central, if not crucial role. Causal J-unitary operators map input spaces of the type $\ell_2^{\mathcal{M}_1} \times \ell_2^{\mathcal{M}_2}$ to output spaces of the type $\ell_2^{\mathcal{N}_1} \times \ell_2^{\mathcal{N}_2}$, hence they are of the block type:

$$\Theta = \begin{bmatrix} \Theta_{11} & \Theta_{12} \\ \Theta_{21} & \Theta_{22} \end{bmatrix}.$$

These spaces are endowed with a nondefinite metric. We denote

$$J_{\mathcal{M}} = \begin{bmatrix} I_{\mathcal{M}_1} & \\ & -I_{\mathcal{M}_2} \end{bmatrix}, \qquad J_{\mathcal{N}} = \begin{bmatrix} I_{\mathcal{N}_1} & \\ & -I_{\mathcal{N}_2} \end{bmatrix}.$$

The J-unitary operators that we shall use will all be bounded and causal. The J-unitarity means

$$\Theta J_{\mathcal{N}} \Theta^* = J_{\mathcal{M}}, \qquad \Theta^* J_{\mathcal{M}} \Theta = J_{\mathcal{N}}.$$

It has important consequences for the block entries of Θ:

- Θ_{22} is boundedly invertible and $\|\Theta_{22}^{-1}\| \ll 1$;
- $\|\Theta_{22}^{-1}\Theta_{21}\| \ll 1$.

The operator Θ_{22}^{-1} turns out to be of great importance in model-reduction theory. It is most likely of mixed type (causal/anticausal). We return later to its state-space analysis.

5.2.7. Method of solution
With the ingredients previously detailed, the actual method to generate the solution appears very straightforward. It consists of two steps:

Step 1: Perform a coprime external factorization:

$$T = U\Delta^* \tag{5.3}$$

with $\Delta \in \mathcal{L}$ and $U \in \mathcal{L}$.

Step 2: Perform an external factorization of the type:

$$\Theta \begin{bmatrix} U^* \\ -\Gamma^{-1}T^* \end{bmatrix} = \begin{bmatrix} A' \\ -B' \end{bmatrix}. \tag{5.4}$$

Here, Θ is a block lower-triangular J-unitary operator of dimensions conforming to $\begin{bmatrix} U^* \\ -\Gamma^{-1}T^* \end{bmatrix}$, $A' \in \mathcal{L}$, and $B' \in \mathcal{L}$. The solution of the interpolation problem is now given by

$$T' = B'^* \Theta_{22}^{-*} \Gamma,$$
$$T_a = \text{strictly lower part of } T'. \tag{5.5}$$

Before embarking on computational issues, we show first that this recipe indeed produces a T' and a T_a that satisfies the norm and the minimality conditions. The norm condition is easy to treat directly. As to the study of complexity, it will be based on the state-space properties of the operator Θ appearing in the special J-unitary external factorization that have to be studied first.

5.2.8. The norm condition
From the second block row in (5.5), we obtain

$$\Theta_{21} U^* - \Gamma^{-1}\Theta_{22}T^* = -B',$$

and since Θ_{22} is invertible, it follows immediately by reordering of terms that

$$(T - T')\Gamma^{-1} = [\Theta_{22}^{-1}\Theta_{21}U^*]^*,$$

where we have put $T' = B'^* \Theta_{22}^{-*} \Gamma$. Hence,

$$\|(T - T')\Gamma^{-1}\| < 1$$

since $\|U^*\| = 1$ and $\|\Theta_{22}^{-1}\Theta_{21}\| < 1$.

5.2.9. The construction of the special J-external factorization

We are looking for a minimal Θ that meets the factorization condition expressed in (5.4). As is the case of the regular external factorization, it will be based on the completion of appropriate reachability operators. A realization for $[U \ -T\Gamma^{-1}]$ is given by

$$\begin{bmatrix} A & B_U & -B\Gamma^{-1} \\ C & D_U & 0 \end{bmatrix}$$

whose reachability part is given by $[A \ B_U \ -B\Gamma^{-1}]$, based on the realization

$$\begin{bmatrix} A & B_U \\ C & D_U \end{bmatrix}$$

for U (notice that in case one starts out with a companion form as detailed above, this part of the procedure is actually trivial, we simply have $U = Z^n$). From the realization theory we can deduce next that a bounded, causal ues J-unitary operator has the property that it possesses a realization which is J-unitary for some, still to be determined state signature

$$J_B = \begin{bmatrix} I_{B_+} & \\ & -I_{B_-} \end{bmatrix}. \tag{5.6}$$

Hence, an appropriate state transformation should be able to produce the desired J_B and J-unitarity based on such a signature matrix on the state. As is the case for the regular external factorization, a somewhat special reachability Gramian will play a central role in finding this transformation. Indeed, let $\{R_k\}$ be the set of state-transformation matrices needed. Then the reachability matrices transform as

$$[\, R_{k+1}^{-1} A_k R_k \quad R_{k+1}^{-1}(B_U)_k \quad -R_{k+1}^{-1} B\Gamma^{-1}\,],$$

and we wish each of these matrices to be part of a J-unitary matrix, i.e., they have each to be J-isometric for an adequate local signature matrix. Suppose that we already have the signature matrices $(J_B)_k$, and let $\Lambda_k = R_k (J_B)_k R_k^*$, then the J-unitarity of the Gramian can be expressed as follows:

$$A_k \Lambda_k A_k^* + (B_U)_k (B_U)_k^* - B_k \Gamma_k^{-2} B_k^* = \Lambda_{k+1}. \tag{5.7}$$

A solution for Λ will exist if this Lyapunov–Stein equation has a definite solution that is also boundedly invertible. Note that because of the ues condition on A, the equation has a unique bounded solution; the question is whether the solution is also boundedly invertible. The existence of the solution can be studied directly in terms of the original data by eliminating B_U, since

$$AA^* + B_U B_U^* = I.$$

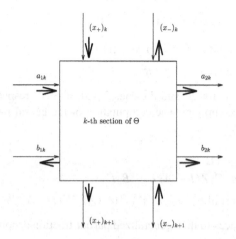

FIG. 5.4. The dataflow in a Theta section is shown with normal arrows, the "energy flow" indicating the sign of the quadratic norms is indicated with fat arrows.

Setting $M = I - \Lambda$ the equation turns into

$$M_{k+1} = A_k M_k A_k^* + B_k \Gamma_k^{-2} B_k^*.$$

Here, M is the reachability Gramian of $T\Gamma^{-1}$, and we find that a solution to the J-unitary embedding problem exists iff $(I - M)^{-1}$ exists and is bounded, i.e., iff the eigenvalues of M_k are bounded away from 1, uniformly over k. In the case the solution is not definite, a "borderline" solution may exist, and thus the case becomes singular. Although that singular case is beyond the present treatment, we shall devote some words to it in the discussion at the end of this section. Let us now assume that a strictly definite solution does exist and analyze it further. Let the inertia of Λ_k be given by

$$\Lambda_k = R_k \begin{bmatrix} (I_{\mathcal{B}_+})_k & \\ & -(I_{\mathcal{B}_-})_k \end{bmatrix} R_k^*.$$

After application of the state transformation $R_{k+1}^{-1} \cdots R_k$, the dataflow for Θ looks as in Fig. 5.4.

Associated with the various signature matrices, we can also imagine an "energy flow" representing the conservation of quadratic norm or energy which follows from the J-unitarity imposed on Θ. The energy flow corresponding to the signature is shown by fat arrows in Fig. 5.4.

5.2.10. Complexity analysis
We have as proposed solution

$$T_a = \text{strictly causal part of } B'^* \Theta_{22}^{-*} \Gamma.$$

In this expression, B'^* is anticausal while Θ_{22}^{-*} is of mixed causality. We first establish that the complexity of T_a is essentially determined by the (strictly) causal part of Θ_{22}^{-*}.

Next we shall analyze the complexity of the latter. Let

$$B' = d + cZ(I - aZ)^{-1}b, \quad \text{causal part of}$$
$$\Theta_{22}^{-*} = D_2 + C_2Z(I - A_2Z)^{-1}B_2,$$

be minimal realizations for B' and the causal part of Θ_{22}^{-*}, respectively (for the existence of the latter, see further). The computation of the causal part for the product is straightforward:

$$\text{causal part of } B'^* \Theta_{22}^{-*} \Gamma$$
$$= d^* D_2 \Gamma + d^* C_2 Z(I - A_2Z)^{-1}B_2\Gamma$$
$$\quad + \text{causal part of } b^*(I - Z^*a^*)^{-1}(c^*C_2)^{(-1)}(I - A_2Z)^{-1}B_2\Gamma.$$

The computation reduces to the "generalized partial-fraction decomposition" of the last part. This is handled in the following generic lemma.

LEMMA 5.1. *Let a and A_2 be transition operators with $\ell_a \leqslant 1$, $\ell_{A_2} \leqslant 1$ and at least one less than one, then*

$$(I - Z^*a^*)^{-1}(c^*C_2)^{(-1)}(A - A_2Z)^{-1}$$
$$= (I - Z^*a^*)^{-1}Z^*a^*m + m + mA_2Z(I - A_2Z)^{-1},$$

where m is the unique bounded solution of the Lyapunov–Stein equation

$$m^{(1)} = c^*C_2 + a^*mA_2.$$

PROOF. The proof of the lemma is by direct computation, after chasing the denominators and identifying the entries.   □

Applying the lemma to the product that defines T_a, we obtain

$$T_a = \left(d^* D_2 \Gamma + b^* m B_2 \Gamma\right) + \left(d^* C_2 + b^* m\right)Z(I - A_2Z)^{-1}B_2\Gamma.$$

We see that T_a inherits the complexity of Θ_{22}^{-*}, at least essentially (further cancellations are theoretically possible but not very likely). In fact, they have the same reachability space based on $\{A_2, B_2\Gamma\}$. The complexity analysis hence proceeds with the analysis of the complexity of Θ_{22}^{-*}. This can be done in a particularly elegant way by studying the strict-past/future decomposition of the operator Θ. We decompose an arbitrary signal (say a belonging to some ℓ_2-space) in its strict-past part and its future part ($a_k = a_{pk} + a_{fk}$). Let the corresponding operators be denoted by $\mathbf{P}_k$ for the projection on the future and $\mathbf{P}'_k = I - \mathbf{P}_k$ for the projection on the strict past, then the splitting of the operator Θ happens as shown in Fig. 5.5, where we also have indicated the sign decomposition of the state discussed earlier. The arrows in Fig. 5.5 indicate flow of energy in the sense that each block satisfies the energy balance with respect to incoming and outgoing energetic contributions (isometric or J-isometric depending on whether a signal is considered an input or an output in the formulation at hand). The causal part of Θ_{22}^{-*} will of course

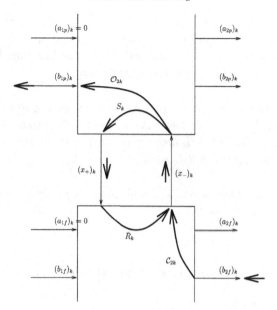

FIG. 5.5. The figure shows the signal flow for $(\Theta_{22}^{-1})_k$. The energy flow of Fig. 5.4 applies, here the relevant signal propagation is indicated with fat arrows.

correspond to the anticausal part of Θ_{22}^{-1}. Writing out

$$\Theta_{22}^{-1} = B_2^* Z^* (I - A_2^* Z^*)^{-1} C_2^* + D_2^* + \text{causal part,}$$

we see that the relevant state dimension is given by the state dimension needed for the operator represented by the first term that produces the map b_{2f} to b_{1p} with $a_1 = 0$ and $b_{2p} = 0$ since Θ_{22}^{-1} maps b_2 to b_1 under the assumption $a_1 = 0$ and the portion b_{1f} in b_1 is to be neglected by the restriction to the lower part of the result (with a slight abuse of notation we can handle all time points k in the same global formula – see DEWILDE and VAN DER VEEN [1998] for details). With reference to the situation in Fig. 5.5, let us define two new diagonal operators $S : x_- \mapsto x_+$ (in the past) and $R : x_+ \mapsto x_-$ (in the future). It is not hard to see (and a more detailed analysis would show) that both these operators are causal and strictly contractive. With b_{2f} as only nonzero input in this configuration, and with energy conservation in vigor, we see that both b_{1p} and x_+ are solely dependent on x_-. In fact, we have

$$x_- = (I - RS)^{-1} C_2 b_{2f},$$
$$b_{1p} = O_2 x_-,$$
$$x_+ = S x_-,$$

where C_2 and O_2 are appropriate reachability and observability maps derived from the anticausal part of Θ_{22}^{-1} (and which we do not detail any further here). The map from b_{2f} to b_{1p} then factors as

$$b_{2p} = O_2 \cdot (I - RS)^{-1} R_2 b_{2f},$$

and its state complexity is determined by the dimension of the "anticausal" state x_-. Hence, T_a has the same complexity as the strict lower part of Θ_{22}^{-*}, which is locally equal to the dimension δ_{k-} of x_-. This dimension is now easy to gauge from the original construction of Θ and is given by the following theorem.

THEOREM 5.2. *Assuming that there exists an ε so that all singular values of all H_k, Hankel matrices of $T\Gamma^{-1}$, are at least ε distant from 1, the dimension δ_{k-} is given by the number of singular values of H_k larger than one. This is also the minimal dimension of any strictly causal approximant T_a satisfying $\|(T - T_a)\Gamma^{-1}\| < 1$.*

PROOF. Recall that the dimension of x_{k-} is given by the number of eigenvalues of M_k larger than one, where M_k satisfies

$$M_{k+1} = A_k M_k A_k^* + B_k \Gamma_k^{-2} B_k^*.$$

Since we started out with a system in output normal form, and $H_k = \mathcal{O}_k \mathcal{R}_k$, we have

$$H_k^* H_k = \mathcal{R}_k^* \mathcal{O}_k^* \mathcal{O}_k \mathcal{R}_k = \mathcal{R}_k^* \mathcal{R}_k = M_k,$$

where $\mathcal{O}_k^* \mathcal{O}_k = I$ since we assumed the system in output normal form, and the singular values of H_k equal the eigenvalues of M_k. This proves the first statement. As for the second assertion, its proof is much more complex, and based on the fact that all approximants which meet the norm condition can be generated by loading Θ in a contractive and causal operator S_L, more precisely, all T' have the form

$$T' = T + U S^* \Gamma.$$

Here,

$$S = (S_L \Theta_{21} + \Theta_{22})^{-1}(S_L \Theta_{11} + \Theta_{12}),$$

and U is as defined earlier. It turns out that the complexity of its lower part is then at least equal to the complexity of the lower part of Θ_{22}^{-*}. For a complete treatment, see Chapter 10 of DEWILDE and VAN DER VEEN [1998], in particular Theorem 10.18.   □

These are the basic results on Hankel-norm approximation of a lower operator. Many more properties can be derived on this new and interesting method, in particular, state-space representations for the approximants are relatively easy to derive, for details we refer to the literature cited.

5.3. The recursive Schur algorithm for Hankel-norm approximation

A low-complexity Hankel-norm approximation to a strictly upper but otherwise general matrix can be derived from an elementary Schur-type elimination algorithm using both orthogonal and hyperbolic elementary matrices. It is a direct application of the previous theory to finite matrices and was first presented in DEWILDE and VAN DER VEEN [1998]. Here, we present the result without proof.

Suppose that the original matrix to be approximated is given by

$$
T = \begin{bmatrix} \boxed{0} & & & \\ t_{21} & 0 & & \\ \vdots & & \ddots & \\ t_{n1} & t_{n2} & \cdots & 0 \end{bmatrix},
$$

then a trivial external factorization for $T = U \Delta^*$ is given by

$$
U = \begin{bmatrix} \boxed{1\,0\,0\,0} \\ 0\,1\,0\,0 \\ 0\,0\,1\,0 \\ 0\,0\,0\,1 \end{bmatrix}, \qquad
\Delta = \begin{bmatrix} \boxed{0\; t_{21}^* \; \cdots \; t_{n1}^*} \\ 0 \cdots t_{n2}^* \\ \ddots \vdots \\ 0 \end{bmatrix}.
$$

According to the theory in the previous sections, the Θ matrix necessary for the Hankel-norm approximation must now have the following three properties:

1. It must be J-unitary for appropriate signature matrices;
2. It must be block lower;
3. It must make the product

$$
\Theta \begin{bmatrix} U^* \\ -T^* \end{bmatrix}
$$

lower. (Point 1 may seem cryptic but will be partly justified in the sequel.)

The right-hand side signature of Θ is certainly given by $J_2 = I_n \oplus -I_n$, in accordance with the right factor, the left-hand side signature will follow from the construction and will differ case by case. It is possible at this point to determine the local arrow dimensions of Θ but not yet the signs of the state and output arrows. To illustrate the point, let us assume that the entries in T are scalar. Because of the structure of U^* and T^*, the first block in a realization for Θ will have n positive inputs (from U^*) and one negative input (from $-T^*$), and it will have $n-1$ states going to the next stage. This means that this first stage must have two outputs (the signs of the outgoing states and outputs are yet to be determined – see Fig. 5.5 for extra information).

The matrix to be block lowered using elementary operations is given by:

$$
\begin{bmatrix} U^* \\ -T^* \end{bmatrix} =
\begin{matrix} + \\ + \\ \vdots \\ + \\ - \\ \vdots \\ - \end{matrix}
\begin{bmatrix} \boxed{1} & 0 & & \\ 0 & 1 & & \\ \vdots & \vdots & \ddots & \\ 0 & 0 & \cdots & 1 \\ \boxed{-t_{11}^*} & -t_{21}^* & \cdots & -t_{n1}^* \\ 0 & -t_{22}^* & \cdots & -t_{n2}^* \\ \vdots & \vdots & \ddots & \vdots \\ 0 & 0 & \cdots & -t_{nn}^* \end{bmatrix}.
$$

The elimination procedure now starts with the elimination of $-t_{n1}^*$ in the first row of the second block, using the last row in the first block. We indicate this state of affairs with the pair of indices $\langle n, 1 \rangle$. Since the sign of the last row of the first block is positive, and that of the first row of the second block negative, a hyperbolic rotation must be used, which can be of two forms, depending on the magnitude of $-t_{n1}^*$. One possibility is

$$\frac{1}{\sqrt{1 - |\rho_{n1}|^2}} \begin{bmatrix} 1 & \overline{\rho_{n1}} \\ \rho_{n1} & 1 \end{bmatrix},$$

in which case $\rho_{n1} = t_{n1}$ has to be smaller than one in magnitude and the target signature is $\langle +, - \rangle$, while the other possibility, when $|t_{n1}| > 1$, is

$$\frac{1}{\sqrt{1 - |\rho_{n1}|^2}} \begin{bmatrix} \rho_{n1} & 1 \\ 1 & \overline{\rho_{n1}} \end{bmatrix},$$

in which case $\rho_{n1} = 1/t_{n1}$, again of magnitude smaller than one. The case where $|t_{n1}| = 1$ is not allowable in the present state of the theory (for an extension, see DEWILDE [1995]), the respective coefficient in Γ then has to be adapted (the condition on the singular values of the Hankel operator is not satisfied). It may happen that in the course of the elimination procedure, a signature of the type $\langle +, + \rangle$ or $\langle -, - \rangle$ is encountered. In that case a regular (unitary) Jacobi rotation will do, and if $\langle -, + \rangle$ as initial signature is found, then the mirror case of the case detailed above holds. The type of rotations used in the scheme will determine the actual flow of energy between the stages of the realization for Θ. The resulting complexity can also be deduced directly from the signature resulting at the output. For example, if the output sequence is $\langle +, - \rangle$, $\langle +, - \rangle$, ..., then all state transitions have positive signs and Θ_{22} is causally invertible. The low-complexity approximant then reduces to a diagonal matrix. At the opposite side, and taking for example the 4×4 case, the output sequence $\langle +, + \rangle$, $\langle +, + \rangle$, $\langle -, - \rangle$, $\langle -, - \rangle$ will result in a state sequence given by $\langle +, +, - \rangle$, $\langle -, - \rangle$, $\langle - \rangle$, resulting in an "approximant" of maximal complexity. The principle involved is that at each state there must be an equal number of incoming and outgoing arrows on the one hand, and an equal number of incoming and outgoing energy arrows as well. A connection will bear a "+" sign if the two arrows point in the same direction and a "−" sign in the opposite case. From the resulting diagram, a realization for Θ_{22}^{-*} can be derived, and from there a realization for the approximant T_a, we refer to the literature cited for details. Although the algorithm does provide for an optimal solution, the computational details are still somewhat extensive.

6. Second-order linear dynamical systems

Second-order models arise naturally in the study of many types of physical systems, such as electrical and mechanical systems; see, e.g., BAI [2002] and the references given there. A *time-invariant multiinput multioutput second-order system* is described by equations of the form

$$M \frac{d^2 q}{dt^2} + D \frac{dq}{dt} + Kq = Pu(t), \tag{6.1}$$

$$y(t) = L^{\mathrm{T}} q(t), \tag{6.2}$$

together with initial conditions $q(0) = q_0$ and $\frac{dq}{dt}(0) = \dot{q}_0$. Here, $q(t) \in \mathbb{R}^N$ is the vector of state variables, $u(t) \in \mathbb{R}^m$ is the input force vector, and $y(t) \in \mathbb{R}^p$ is the output measurement vector. Moreover, $M, D, K \in \mathbb{R}^{N \times N}$ are system matrices, such as mass, damping, and stiffness matrices in structural dynamics, $P \in \mathbb{R}^{N \times m}$ is the input distribution matrix, and $L \in \mathbb{R}^{N \times p}$ is the output measurement matrix. Finally, N is the state-space dimension, and m and p are the number of inputs and outputs, respectively. In most practical cases, m and p are much smaller than N.

The second-order system (6.1) and (6.2) can be reformulated as an equivalent linear first-order system in many different ways. We will use the following equivalent linear system:

$$E\frac{dx}{dt} = Ax + Bu(t), \tag{6.3}$$

$$y(t) = C^{\mathrm{T}} x(t), \tag{6.4}$$

where

$$x = \begin{bmatrix} q \\ \frac{dq}{dt} \end{bmatrix}, \quad A = \begin{bmatrix} -K & 0 \\ 0 & W \end{bmatrix}, \quad E = \begin{bmatrix} D & M \\ W & 0 \end{bmatrix},$$

$$B = \begin{bmatrix} P \\ 0 \end{bmatrix}, \quad C = \begin{bmatrix} L \\ 0 \end{bmatrix}.$$

Here, $W \in \mathbb{R}^{N \times N}$ can be any nonsingular matrix. A common choice is the identity matrix, $W = I$. If the matrices M, D, and K are all symmetric and M is nonsingular, as it is often the case in structural dynamics, we can choose $W = M$. The resulting matrices A and E in the linearized system (6.3) are then symmetric, and thus preserve the symmetry of the original second-order system.

Assume that, for simplicity, we have zero initial conditions, i.e., $q(0) = q_0$, $\frac{dq}{dt}(0) = 0$, and $u(0) = 0$ in (6.1) and (6.2). Then, by taking the Laplace transform of (6.1) and (6.2), we obtain the following system:

$$s^2 M Q(s) + D Q(s) + K Q(s) = P U(s),$$
$$Y(s) = L^{\mathrm{T}} Q(s).$$

Eliminating $Q(s)$ results in the frequency-domain input–output relation $Y(s) = H(s)U(s)$, where

$$H(s) := L^{\mathrm{T}} (s^2 M + s D + K)^{-1} P$$

is the transfer function. In view of the equivalent linearized system (6.3) and (6.4), the transfer function can also be written as

$$H(s) = C^{\mathrm{T}} (s E - A)^{-1} B.$$

If the matrix K in (6.1) is nonsingular, then $s_0 = 0$ is guaranteed not to be a pole of H. In this case, H can be expanded about $s_0 = 0$ as follows:

$$H(s) = M_0 + M_1 s + M_2 s^2 + \cdots,$$

where the matrices M_j are the so-called *low-frequency moments*. In terms of the matrices of the linearized system (6.3) and (6.4), the moments are given by

$$M_j = -C^T(A^{-1}E)^j A^{-1}B, \quad j = 0, 1, 2, \dots.$$

6.1. Frequency-response analysis methods

In this subsection, we describe the use of eigensystem analysis to tackle the second-order system (6.1) and (6.2) directly.

We assume that the input force vector $u(t)$ of (6.1) is time-harmonic:

$$u(t) = \tilde{u}(\omega)\,e^{i\omega t},$$

where ω is the frequency of the system. Correspondingly, we assume that the state variables of the second-order system can be represented as follows:

$$q(t) = \tilde{q}(\omega)\,e^{i\omega t}.$$

The problem of solving the system of second-order differential equations (6.1) then reduces to solving the parameterized linear system of equations

$$(-\omega^2 M + i\omega D + K)\tilde{q}(\omega) = P\tilde{u}(\omega) \tag{6.5}$$

for $\tilde{q}(\omega)$. This approach is called the *direct frequency-response analysis method*. For a given frequency ω_0, one can use a linear system solver, either direct or iterative, to obtain the desired vector $\tilde{q}(\omega_0)$.

Alternatively, we can try to reduce the cost of solving the large-scale parameterized linear system of Eq. (6.5) by first applying an eigensystem analysis. This approach is called the *modal frequency-response analysis* in structural dynamics. The basic idea is to first transfer the coordinates $\tilde{q}(\omega)$ of the state vector $q(t)$ to new coordinates $p(\omega)$ as follows:

$$q(t) \cong W_k p(\omega)\,e^{i\omega t}.$$

Here, W_k consists of k selected modal shapes to retain the modes whose resonant frequencies lie within the range of forcing frequencies. More precisely, W_k consists of k selected eigenvectors of the underlying quadratic eigenvalue problem $(\lambda^2 M + \lambda D + K)w = 0$. Eq. (6.5) is then approximated by

$$(-\omega^2 M W_k + i\omega D W_k + K W_k)p(\omega) = P\tilde{u}(\omega).$$

Multiplying this equation from the left by W_k^T, we obtain a $k \times k$ parameterized linear system of equations for $p(\omega)$:

$$(-\omega^2(W_k^T M W_k) + i\omega(W_k^T D W_k) + (W_k^T K W_k))p(\omega) = W_k^T P(\omega).$$

Typically, $k \ll n$. The main question now is how to obtain the desired modal shapes W_k. One possibility is to simply extract W_k from the matrix pair (M, K) by ignoring the contribution of the damping term. This is called the *modal superposition method* in structural dynamics. This approach is applicable under the assumption that the damping term is of a certain form. For example, this is the case for so-called Rayleigh damping $D = \alpha M + \beta K$, where α and β are scalars (see CLOUGH and PENZIEN [1975]). In general, however, one may need to solve the full quadratic eigenvalue problem $(\lambda^2 M + \lambda D + K)w = 0$ in order to obtain the desired modal shapes W_k. Some of these techniques have been reviewed in the recent survey paper by TISSEUR and MEERBERGEN [2001] on the quadratic eigenvalue problem.

6.2. Reduced-order modeling based on linearization

An obvious approach to constructing reduced-order models of the second-order system (6.1) and (6.2) is to apply any of the model-reduction techniques for linear systems to the linearized system (6.3) and (6.4). In particular, we can employ the Krylov-subspace techniques discussed in Section 3.

The resulting approach can be summarized as follows:
(1) Linearize the second-order system (6.1) and (6.2) by properly defining the $2N \times 2N$ matrices A and E of the equivalent linear system (6.3) and (6.4). Select an expansion point s_0 "close" to the frequency range of interest and such that the matrix $A - s_0 E$ is nonsingular.
(2) Apply a suitable Krylov process, such as the nonsymmetric band Lanczos algorithm described in Section 3.2, to the matrix $M := (A - s_0 E)^{-1} E$ and the blocks of right and left starting vectors $R := (A - s_0 E)^{-1} B$ and $L := C$ to obtain bi-orthogonal Lanczos basis matrices V_n and W_n for the nth right and left block-Krylov subspaces $\mathcal{K}_n(M, R)$ and $\mathcal{K}_n(M^T, L)$.
(3) Approximate the state vector $x(t)$ by $V_n z(t)$ where $z(t)$ is determined by the following linear reduced-order model of the linear system (6.3) and (6.4):

$$E_n \frac{dz}{dt} = A_n z + B_n u(t), \qquad y(t) = C_n^T z(t).$$

Here, $E_n = T_n$, $A_n = \Delta_n + s_0 T_n$, $B_n = \rho_n^{(pr)}$, $C_n = \eta_n^{(pr)}$, and $T_n, \Delta_n, \rho_n^{(pr)}, \eta_n^{(pr)}$ are the matrices generated by the nonsymmetric band Lanczos algorithm.

In Fig. 6.1, we show the results of this approach applied to the linear-drive multimode resonator structure described in CLARK, ZHOU, and Pister [1998]. The solid lines are the Bode plots of the frequency response of the original second-order system, which is of dimension $N = 63$. The dashed line in the left, respectively right, plot is the Bode plot of the frequency response of the reduced-order model of dimension $n = 8$, respectively $n = 12$. The relative error between the transfer functions of the original system and the reduced-order model of dimension $n = 12$ is less than 10^{-4} over the frequency range shown in Fig. 6.1.

There are a couple of advantages of the linearization approach. First, one can directly employ existing reduced-order modeling techniques developed for linear systems. Second, one can also exploit the structures of the linearized system matrices A and E in a Krylov process to reduce the computational cost. However, the linearization approach also has disadvantages. In particular, it ignores the physical meaning of the original system matrices, and more importantly, the reduced-order models are no longer in a second-order form. For engineering design and control of structural systems, it is often

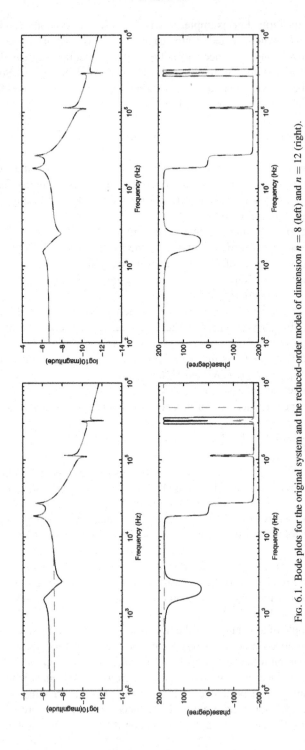

FIG. 6.1. Bode plots for the original system and the reduced-order model of dimension $n = 8$ (left) and $n = 12$ (right).

desirable to have reduced-order models that preserve the second-order form; see, e.g., Su and Craig [1991].

While the straightforward linearization approach has the above disadvantages, it is possible to exploit the inherent structure of the Krylov subspaces associated with the linearized system to construct reduced-order models of second-order, or even higher-order, systems that preserve the higher-order structure. Such structure-preserving linearization approaches are described in Freund [2004a], Freund [2004b].

6.3. *Reduced-order modeling based on second-order systems*

In this section, we discuss a Krylov-subspace technique that produces a reduced-order model of second-order form. This approach is based on the work of Su and Craig [1991].

The key observation is the following. In view of the linearization (6.3) and (6.4) of the second-order system (6.1) and (6.2), the desired Krylov subspace for reduced-order modeling is

$$\text{span}\{\widetilde{B}, (A^{-1}E)\widetilde{B}, (A^{-1}E)^2\widetilde{B}, \ldots, (A^{-1}E)^{n-1}\widetilde{B}\}.$$

Here, $\widetilde{B} := -A^{-1}[B\ C]$. Moreover, we have assumed that the matrix A in (6.3) is nonsingular. Let us set

$$R_j = \begin{bmatrix} R_j^d \\ R_j^v \end{bmatrix} := (-A^{-1}E)^j\, \widetilde{B},$$

where R_j^d is the vector of length N corresponding to the displacement portion of the vector R_j, and R_j^v is the vector of length N corresponding to the velocity portion of the vector R_j, see Su and Craig [1991]. Then, in view of the structure of the matrices A and E, we have

$$\begin{bmatrix} R_j^d \\ R_j^v \end{bmatrix} = (-A^{-1}E)\begin{bmatrix} R_{j-1}^d \\ R_{j-1}^v \end{bmatrix} = \begin{bmatrix} K^{-1}DR_{j-1}^d + K^{-1}MR_{j-1}^d \\ -R_{j-1}^d \end{bmatrix}.$$

Note that the jth velocity-portion vector R_j^v is the same (up to its sign) as the $(j-1)$st displacement-portion vector R_{j-1}^d. In other words, the second portion R_j^v of R_j is the "one-step" delay of the first portion R_{j-1}^d of R_j. This suggests that one may simply choose

$$\text{span}\{R_0^d, R_1^d, R_2^d, \ldots, R_{n-1}^d\} \tag{6.6}$$

as the projection subspace used for reduced-order modeling.

In practice, for numerical stability, one may opt to employ the Arnoldi process to generate an orthonormal basis Q_n of the subspace (6.6). The resulting procedure can be summarized as follows.

ALGORITHM 6.1 (*Algorithm by Su and Craig Jr.*).
   (0) (Initialization)

Set $R_0^d = K^{-1}[P \ L]$, $R_0^v = 0$, $U_0 S_0 V_0^{\mathrm{T}} = (R_0^d)^{\mathrm{T}} K R_0^d$ (by computing an SVD),

$$Q_1^d = R_0^d U_0 S_0^{-1/2}, \text{ and } Q_1^v = 0.$$

(1) (Arnoldi loop)

For $j = 1, 2, \ldots, n-1$ do:

  Set $R_j^d = K^{-1}(D Q_{j-1}^d + M Q_{j-1}^v)$ and $R_j^v = -Q_{j-1}^d$.

(2) (Orthogonalization)

For $i = 1, 2, \ldots, j$ do:

  Set $T_i = (Q_i^d)^{\mathrm{T}} K R_j^d$, $R_j^d = R_j^d - Q_i^d T_i$, and $R_j^v = R_j^v - Q_i^v T_i$.

(3) (Normalization)

Set $U_0 S_0 V_0^{\mathrm{T}} = (R_j^d)^{\mathrm{T}} K R_j^d$ (by computing an SVD),

$$Q_{j+1}^d = R_j^d U_0 S_0^{-1/2}, \text{ and } Q_{j+1}^v = R_j^v U_0 S_0^{-1/2}.$$

An approximation of the state vector $q(t)$ can then be obtained by constraining $q(t)$ to the subspace spanned by the columns of Q_n, i.e., $q(t) \approx Q_n z(t)$. Moreover, the reduced-order state vector $z(t)$ is defined as the solution of the following second-order system:

$$M_n \frac{\mathrm{d}^2 q}{\mathrm{d}t^2} + D_n \frac{\mathrm{d}q}{\mathrm{d}t} + K_n q = P_n u(t), \tag{6.7}$$

$$y(t) = L_n^{\mathrm{T}} q(t), \tag{6.8}$$

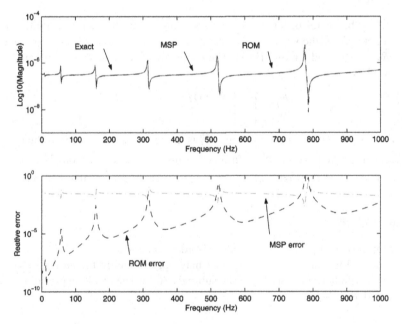

FIG. 6.2. Frequency-response analysis (top plot) and relative errors (bottom plot) of a finite-element model of a shaft.

where $M_n := Q_n^{\mathrm{T}} M Q_n$, $D_n := Q_n^{\mathrm{T}} D Q_n$, $K_n := Q_n^{\mathrm{T}} K Q_n$, $P_n := Q_n^{\mathrm{T}} P$, and $L_n := Q_n^{\mathrm{T}} L$. Note that (6.7) and (6.8) is a reduced-order model in second-order form of the original second-order system (6.1) and (6.2).

In SU and CRAIG [1991], a number of advantages of this approach are described. Here, we present some numerical results of a frequency-response analysis of a second-order system of order $N = 400$, which arises from a finite-element model of a shaft on bearing support with a damper. In the top of Fig. 6.2, we plot the magnitudes of the transfer function H computed exactly, approximated by the model-superposition (MSP) method, and approximated by the Krylov-subspace technique (ROM). For the MSP method, we used the 80 modal shapes W_{80} from the matrix pencil (M, K). The reduced-order model (6.7) and (6.8) is also of dimension $n = 80$. The bottom plot of Fig. 6.2 shows the relative errors between the exact transfer function and its approximations based on the MSP method (dash-dotted line) and the ROM method (dashed line). The plots indicate that no accuracy has been lost by the Krylov subspace-based method.

7. Semisecond-order dynamical systems

In some applications, in particular in the simulation of MEMS devices (SENTURIA, ALURU and WHITE [1997]), the underlying mathematical models are second-order systems with nonlinear excitation forces of the following type:

$$
M \frac{\mathrm{d}^2 q}{\mathrm{d}t^2} + D \frac{\mathrm{d}q}{\mathrm{d}t} + K q = P u \left(q, \frac{\mathrm{d}q}{\mathrm{d}t}, t \right),
$$

$$
y(t) = L^{\mathrm{T}} q(t).
$$

(7.1)

Here, the system matrices M, D, K, P, and L have the same interpretation as in the standard second-order system (6.1) and (6.2). However, excitation force u is now a nonlinear function of q, and possibly $\frac{\mathrm{d}q}{\mathrm{d}t}$.

Systems of the form (7.1) and (7.1) are called *semisecond-order* time-invariant multiinput multioutput linear dynamical systems. Such systems are used as the underlying mathematical models in SUGAR [2001], which is a system-level simulation package for MEMS devices. For example, Fig. 7.1 shows a simple electrostatic gap-closing actuator, which is used as a demo in SUGAR. In this case, the excitation force u includes the electrostatic potential between the plates and is proportional to $(v(t)/\mathrm{gap}(q))^2$, where $v(t)$ is the voltage between electrodes and $\mathrm{gap}(q)$ is a scalar function of q for the distance between the two place electrodes. For mode details about the model used for the electrostatic gap-closing actuator, see BAI, BINDEL, CLARK, DEMMEL, PISTER, and ZHOU [2000].

Instead of treating the semisecond-order system (7.1) and (7.1) as a general nonlinear system, we can exploit the structure of the system and apply the idea of "nonlinear dynamics using linear modes." This approach is suggested in ANANTHASURESH, GUPTA, and SENTURIA [1996], where a nondamped system, i.e., $D = 0$ is considered and the eigenmodes of M and K are used to extract a reduced-order model. In BAI, BINDEL, CLARK, DEMMEL, PISTER, and ZHOU [2000], we described a Krylov-subspace based reduced-order modeling technique for systems (7.1) and (7.1). The idea is to first ignore

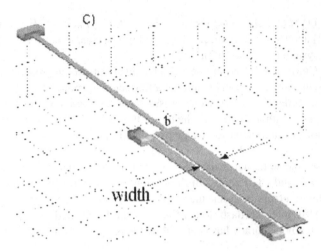

FIG. 7.1. Electrostatic gap-closing actuator.

the nonlinearity in the force term u, and treat the system as a second-order system. Using the approach discussed in Section 6.2, a projection space V_n is constructed, which may be regarded as the *linear Krylov modes*. The vector q is then expanded in terms of the constructed subspace, namely $q(t) \approx V_n z(t)$, and we obtain the following reduced-order model in terms of the vector $z(t)$:

$$E_n \frac{dz}{dt} = A_n x + B_n u\,(V_n z(t), t)\,,$$

$$y(t) = C_n^T z(t).$$

Here, the definitions of E_n, A_n, B_n, and C_n are the same as in Section 6.2. Note that the excitation force term $u(q, t)$ of the full-order system is replaced by $u(V_n z(t), t)$ in the reduced-order model. When the reduce-order model is solved by a numerical method, it is necessary that $u(V_n z_j, t)$ can be evaluated for the given z_j, which may be regarded as the approximation of $z(t)$ at time step $t = t_j$.

In Fig. 7.2, we illustrate this approach for the transient analysis of the electrostatic gap-closing actuator shown in Fig. 7.1. The first plot shows the output $y(t)$ of the original system and the output $\tilde{y}(t)$ of the reduced-order system of dimension $n = 6$. The original systems has dimension $N = 30$. The second plot shows the accuracy of the reduced-order model of dimension $n = 6$ in terms of the relative error $\|y(t) - \tilde{y}(t)\| / \|y(t)\|$.

We remark that, as indicated in GABBAY, MEHNER, and SENTURIA [2000], the use of linear (eigen or Krylov) modes may not adequately capture all the features of nonlinear behavior. It is the subject of current research to further understand the approach sketched in this section and its limitations.

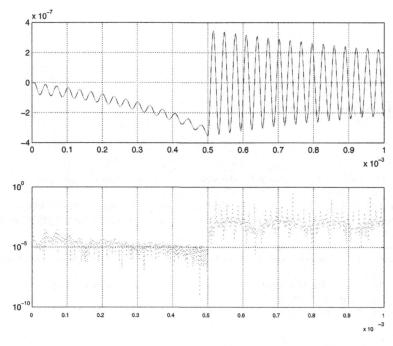

FIG. 7.2. Transient responses of the gap-closing actuator.

8. Concluding remarks

We presented a survey of the most common techniques for reduced-order modeling of large-scale linear dynamical systems. By and large, the area of linear reduced-order modeling is fairly well explored, and we have a number of efficient techniques at our disposal. Still, some open problems remain. One such problem is the construction of reduced-order models that preserve stability or passivity and at the same time, have optimal approximation properties. In particular in circuit simulation, reduced-order modeling is used to substitute large linear subsystems within the simulation of even larger, in general nonlinear systems. It would be important to better understand the effects of these substitutions on the overall nonlinear simulation.

Finally, the systems arising in the simulation of electronic circuits are nonlinear in general, and it would be highly desirable to apply nonlinear reduced-order modeling techniques directly to these nonlinear systems. However, the area of nonlinear reduced-order modeling is in its infancy compared to the state-of-the-art of linear reduced-order modeling. We expect that further progress in model reduction will mainly occur in the area of nonlinear reduced-order modeling.

References

ALIAGA, J.I., BOLEY, D.L., FREUND, R.W., HERNÁNDEZ, V. (2000). A Lanczos-type method for multiple starting vectors. *Math. Comp.* **69**, 1577–1601.

ANANTHASURESH, G.K., GUPTA, R.K., SENTURIA, S.D. (1996). An approach to macromodeling of MEMS for nonlinear dynamic simulation. In: *Microelectromechanical Systems (MEMS)*. In: ASME Dynamics Systems & Control (DSC) Ser. **59**, pp. 401–407.

ANDERSON, B.D.O. (1967). A system theory criterion for positive real matrices. *SIAM J. Control* **5**, 171–182.

ANDERSON, B.D.O., VONGPANITLERD, S. (1973). *Network Analysis and Synthesis* (Prentice-Hall, Englewood Cliffs, NJ).

ARNOLDI, W.E. (1951). The principle of minimized iterations in the solution of the matrix eigenvalue problem. *Quart. Appl. Math.* **9**, 17–29.

ARVESON, W. (1975). Interpolation problems in nest algebras. *J. Funct. Anal.* **20**, 208–233.

BAI, Z. (2002). Krylov subspace techniques for reduced-order modeling of large-scale dynamical systems. *Appl. Numer. Math.* **43** (1–2), 9–44.

BAI, Z., BINDEL, D., CLARK, J., DEMMEL, J., PISTER, K.S.J., ZHOU, N. (2000). New numerical techniques and tools in SUGAR for 3D MEMS simulation. In: *Technical Proc. 4th Internat. Conf. on Modeling and Simulation of Microsystems*, pp. 31–34.

BAI, Z., FELDMANN, P., FREUND, R.W. (1998). How to make theoretically passive reduced-order models passive in practice. In: *Proc. IEEE 1998 Custom Integrated Circuits Conference* (IEEE, Piscataway, NJ), pp. 207–210.

BAI, Z., FREUND, R.W. (2000). Eigenvalue-based characterization and test for positive realness of scalar transfer functions. *IEEE Trans. Automat. Control* **45** (12), 2396–2402.

BAI, Z., FREUND, R.W. (2001a). A partial Padé-via-Lanczos method for reduced-order modeling. *Linear Algebra Appl.* **332**, 139–164.

BAI, Z., FREUND, R.W. (2001b). A symmetric band Lanczos process based on coupled recurrences and some applications. *SIAM J. Sci. Comput.* **23** (2), 542–562.

BOYD, S., EL GHAOUI, L., FERON, E., BALAKRISHNAN, V. (1994). *Linear Matrix Inequalities in System and Control Theory* (SIAM Publications, Philadelphia, PA).

CAMPBELL, S.L. (1980). *Singular Systems of Differential Equations* (Pitman, London, UK).

CAMPBELL, S.L. (1982). *Singular Systems of Differential Equations II* (Pitman, London, UK).

CHEN, T.-H., LUK, C., CHEN, C.C.-P. (2003). In: *Technical Digest of the 2003 IEEE/ACM Int. Conf. on Computer-Aided Design* (IEEE Computer Society Press, Los Alamitos, CA), pp. 786–792.

CHIRLIAN, P.M. (1967). *Integrated and Active Network Analysis and Synthesis* (Prentice-Hall, Englewood Cliffs, NJ).

CLARK, J.V., ZHOU, N., PISTER, K.S.J. (1998). MEMS simulation using SUGAR v0.5. In: *Proc. Solid-State Sensors and Actuators Workshop* (Hilton Head Island, SC), pp. 191–196.

CLOUGH, R.W., PENZIEN, J. (1975). *Dynamics of Structures* (McGraw-Hill, New York).

DAI, L. (1989). *Singular Control Systems*, Lecture Notes in Control and Information Sciences **118** (Springer-Verlag, Berlin, Germany).

DEPRETTERE, E. (1981). Mixed-form time-variant lattice recursions. In: *Outils et Modèles Mathématiques pour l'Automatique, l'Analyse de Systèmes et le Traitement du Signal* (CNRS, Paris), pp. 545–562.

DEWILDE, P. (1988). New algebraic methods for modeling large-scale integrated circuits. *Circuit Theory Appl.* **16**, 473–503.

DEWILDE, P. (1995). J -unitary matrices for algebraic approximation and interpolation – the singular case. In: Moonen, M., Moor, B.D. (eds.), *SVD and Signal Processing, III, Algorithms, Architectures and Applications* (Elsevier, Amsterdam), pp. 209–223.

DEWILDE, P., DEPRETTERE, E.F. (1987). Approximate inversion of positive matrices with applications to modelling. In: Curtain, R.F. (ed.), *Modelling, Robustness and Sensitivity Reduction in Control Systems.* In: NATO ASI Series **F34** (Springer-Verlag, Berlin), pp. 211–238.

DEWILDE, P., DYM, H. (1981). Schur recursions, error formulas, and convergence of rational estimators for stationary stochastic sequences. *IEEE Trans. Inf. Theory* **27** (4), 446–461.

DEWILDE, P., VAN DER VEEN, A.-J. (1998). *Time-Varying Systems and Computations* (Kluwer, Dordrecht).

DEWILDE, P., VIEIRA, A., KAILATH, T. (1978). On a generalized Szegö-Levinson realization algorithm for optimal linear predictors based on a network synthesis approach. *IEEE Trans. Circuits Syst.* **25** (9), 663–675.

FELDMANN, P., FREUND, R.W. (1994). Efficient linear circuit analysis by Padé approximation via the Lanczos process. In: *Proceedings of EURO-DAC'94 with EURO-VHDL'94* (IEEE Computer Society Press, Los Alamitos, CA), pp. 170–175.

FELDMANN, P., FREUND, R.W. (1995a). Efficient linear circuit analysis by Padé approximation via the Lanczos process. *IEEE Trans. Computer-Aided Design* **14**, 639–649.

FELDMANN, P., FREUND, R.W. (1995b). Reduced-order modeling of large linear subcircuits via a block Lanczos algorithm. In: *Proc. 32nd ACM/IEEE Design Automation Conference* (ACM, New York, NY), pp. 474–479.

FREUND, R.W. (1995). Computation of matrix Padé approximations of transfer functions via a Lanczos-type process. In: Chui, C., Schumaker, L. (eds.), *Approximation and Interpolation.* In: Approximation Theory VIII **1** (World Scientific Publishing Co Inc., Singapore), pp. 215–222.

FREUND, R.W. (1999a). Passive reduced-order models for interconnect simulation and their computation via Krylov-subspace algorithms. In: *Proc. 36th ACM/IEEE Design Automation Conference* (ACM, New York, NY), pp. 195–200.

FREUND, R.W. (1999b). Reduced-order modeling techniques based on Krylov subspaces and their use in circuit simulation. In: Datta, B.N. (ed.), Applied and Computational Control, Signals, and Circuits **1** (Birkhäuser, Boston), pp. 435–498.

FREUND, R.W. (2000a). Band Lanczos method (Section 7.10). In: Bai, Z., Demmel, J., Dongarra, J., Ruhe, A., van der Vorst, H. (eds.), *Templates for the Solution of Algebraic Eigenvalue Problems: A Practical Guide* (SIAM Publications, Philadelphia, PA), pp. 205–216. Also available online from http://cm.bell-labs.com/cs/doc/99.

FREUND, R.W. (2000b). Krylov-subspace methods for reduced-order modeling in circuit simulation. *J. Comput. Appl. Math.* **123** (1–2), 395–421.

FREUND, R.W. (2003). Model reduction methods based on Krylov subspaces. *Acta Numer.* **12**, 267–319.

FREUND, R.W. (2004a). Krylov subspaces associated with higher-order linear dynamical systems. Technical report, Department of Mathematics, University of California, Davis, CA, submitted for publication.

FREUND, R.W. (2004b). Padé-type model reduction of second-order and higher-order linear dynamical systems. Technical report, Department of Mathematics, University of California, Davis, CA, USA, submitted for publication.

FREUND, R.W. (2004c). SPRIM: structure-preserving reduced-order interconnect macromodeling. In: *Tech. Dig. 2004 IEEE/ACM International Conference on Computer-Aided Design* (IEEE Computer Society Press, Los Alamitos, CA), pp. 80–87.

FREUND, R.W., FELDMANN, P. (1996a). Reduced-order modeling of large passive linear circuits by means of the SyPVL algorithm. In: *Tech. Dig. 1996 IEEE/ACM International Conference on Computer-Aided Design* (IEEE Computer Society Press, Los Alamitos, CA), pp. 280–287.

FREUND, R.W., FELDMANN, P. (1996b). Small-signal circuit analysis and sensitivity computations with the PVL algorithm. In: IEEE Trans. Circuits and Systems – II: Analog and Digital Signal Processing **43**, pp. 577–585.

FREUND, R.W., FELDMANN, P. (1997). The SyMPVL algorithm and its applications to interconnect simulation. In: *Proc. 1997 Internat. Conf. on Simulation of Semiconductor Processes and Devices* (IEEE, Piscataway, NJ), pp. 113–116.

FREUND, R.W., FELDMANN, P. (1998). Reduced-order modeling of large linear passive multiterminal circuits using matrix-Padé approximation. In: *Proc. Design, Automation and Test in Europe Conference 1998* (IEEE Computer Society Press, Los Alamitos, CA), pp. 530–537.

FREUND, R.W., JARRE, F. (2004a). An extension of the positive real lemma to descriptor systems. *Optim. Methods Softw.* **18** (1), 69–87.

FREUND R.W., JARRE, F. (2004b). Numerical computation of nearby positive real systems in the descriptor case. Technical Report, Department of Mathematics, University of California, Davis, California, USA, in preparation.

GABBAY, L.D., MEHNER, J.E., SENTURIA, S.D. (2000). Computer-aided generation of nonlinear reduced-order dynamic macromodels – I: Non-stress-stiffened case. *J. Microelectromech. Syst.* **9** (2), 262–269.

KIM, S.-Y., GOPAL, N., PILLAGE, L.T. (1994). Time-domain macromodels for VLSI interconnect analysis. *IEEE Trans. Computer-Aided Design* **13**, 1257–1270.

LANCZOS, C. (1950). An iteration method for the solution of the eigenvalue problem of linear differential and integral operators. *J. Res. Natl. Bur. Stand.* **45**, 255–282.

MASUBUCHI, I., KAMITANE, Y., OHARA, A., SUDA, N. (1997). H_∞ control for descriptor systems; a matrix inequalities approach. *Automatica J. IFAC* **33** (4), 669–673.

NELIS, H. (1989). Sparse approximations of inverse matrices. Ph.D. thesis, Delft Univ. Techn., The Netherlands.

NELIS, H., DEWILDE, P., DEPRETTERE, E. (1989). Inversion of partially specified positive definite matrices by inverse scattering. In: The Gohberg Anniversary Collection. Operator Theory: Advances and Applications **40** (Birkhäuser Verlag, Basel), pp. 325–357.

ODABASIOGLU, A. (1996). Provably passive RLC circuit reduction. M.S. thesis, Department of Electrical and Computer Engineering, Carnegie Mellon University.

ODABASIOGLU, A., CELIK, M., PILEGGI, L.T. (1997). PRIMA: passive reduced-order interconnect macro-modeling algorithm. In: *Tech. Dig. 1997 IEEE/ACM International Conference on Computer-Aided Design* (IEEE Computer Society Press, Los Alamitos, CA), pp. 58–65.

PILEGGI, L.T. (1995). Coping with RC(L) interconnect design headaches. In: *Tech. Dig. 1995 IEEE/ACM International Conference on Computer-Aided Design* (IEEE Computer Society Press, Los Alamitos, CA), pp. 246–253.

ROHRER, R.A., NOSRATI, H. (1981). Passivity considerations in stability studies of numerical integration algorithms. *IEEE Trans. Circuits Syst.* **28**, 857–866.

RUEHLI, A.E. (1974). Equivalent circuit models for three-dimensional multiconductor systems. *IEEE Trans. Microwave Theory Tech.* **22**, 216–221.

SCHUR, I. (1917). Über Potenzreihen, die im Innern des Einheitskreises beschränkt sind, I. *J. Reine Angew. Math.* 147, 205–232. English transl. In: Operator Theory: Advances and Applications **18** (Birkhäuser Verlag, Basel), 1986, pp. 31–59.

SENTURIA, S.D., ALURU, N., WHITE, J. (1997). Simulating the behavior of MEMS devices: Computational methods and needs. *IEEE Comput. Sci. Eng. Mag.* **4**, 30–43.

SHEEHAN, B.N. (1999). ENOR: model order reduction of RLC circuits using nodal equations for efficient factorization. In: *Proc. 36th ACM/IEEE Design Automation Conference* (ACM, New York, NY), pp. 17–21.

SU, T.-J., CRAIG JR., R.R. (1991). Model reduction and control of flexible structures using Krylov vectors. *J. Guidance Control Dynam.* **14**, 260–267.

SUGAR (2001). A MEMS simulation program. Available at 2001; http://www-bsac.eecs.berkeley.edu/cadtools/sugar.

TISSEUR, F., MEERBERGEN, K. (2001). The quadratic eigenvalue problem. *SIAM Rev.* **43** (2), 235–286.

VAN DER MEIJS, N. (1992). Accurate and efficient layout extraction. Ph.D. thesis, Delft Univ. Techn., The Netherlands.

VERGHESE, G.C., LÉVY, B.C., KAILATH, T. (1981). A generalized state-space for singular systems. *IEEE Trans. Automat. Control* **26** (4), 811–831.

VLACH, J., SINGHAL, K. (1994). *Computer Methods for Circuit Analysis and Design*, second ed. (Van Nostrand Reinhold, New York, NY).

WILLEMS, J.L. (1970). *Stability Theory of Dynamical Systems* (John Wiley & Sons, Inc., New York, NY).

ZHOU, K., DOYLE, J.C., GLOVER, K. (1996). *Robust and Optimal Control* (Prentice-Hall, Upper Saddle River, NJ).

Subject Index

Printed in the United States
By Bookmasters